Dieter Stoye, Werner Freitag
(Editors)

Paints, Coatings and Solvents

WILEY-VCH

Other Titles of Interest:

Industrial Inorganic Pigments
Edited by Gunther Buxbaum
Second, Completely Revised Edition 1998
ISBN: 3-527-28878-3

W. Herbst, K. Hunger
Industrial Organic Pigments
Second, Completely Revised Edition 1997
ISBN: 3-527-28836-8

Automotive Paints and Coatings
Edited by Gordon Fettis
First Edition 1995
ISBN: 3-527-28637-3

Hans G. Völz
Industrial Color Testing
First Edition 1995
ISBN: 3-527-28643-8

Heinrich Zollinger
Color Chemistry
Second, Revised Edition 1991
ISBN: 3-527-28352-8

Dieter Stoye, Werner Freitag (Editors)

Paints, Coatings and Solvents

Second, Completely Revised Edition

WILEY-VCH

Weinheim · New York · Chichester · Brisbane · Singapore · Toronto

Dr. Dieter Stoye
Am Gecksbach 42
D-46286 Dorsten
Federal Republic of Germany

Dr. Werner Freitag
Creanova Spezialchemie GmbH
Gebäude 1328/16
D-45764 Marl
Federal Republic of Germany

This book was carefully produced. Nevertheless, authors, editors and publisher do not warrant the information contained therein to be free of errors. Readers are advised to keep in mind that statements, data, illustrations, procedural details or other items may inadvertently be inaccurate.

First Edition 1993
Second, Completely Revised Edition, 1998

Library of Congress Card No.:

British Library Cataloguing-in-Publication Data:

Die Deutsche Bibliothek - CIP-Einheitsaufnahme

Paints, coatings and solvents / Dieter Stoye ; Werner Freitag (ed.). - 2., completely rev. ed.
- Weinheim ; New York ; Basel ; Cambridge ; Tokyo : Wiley-VCH, 1998
  ISBN 3-527-28863-5

© WILEY-VCH Verlag GmbH, D-69469 Weinheim
(Federal Republic of Germany), 1998

Printed on acid-free and chlorine-free paper.

All rights reserved (including those of translation into other languages). No part of this book may be reproduced in any form – by photoprinting, microfilm, or any other means – nor transmitted or translated into a machine language without written permission from the publishers. Registered names, trademarks, etc. used in this book, even when not specifically marked as such, are not to be considered unprotected by law.

Composition, Printing and Bookbinding: Graphischer Betrieb, Konrad Triltsch, D-97070 Würzburg

Printed in the Federal Republic of Germany.

Preface to the Second Edition

The work at hand offers a wealth of information about coating materials and coating processes in a form that is clearly laid out. The swift pace of developments in the past few years has made a revised edition seem appropriate. The organization and structure of the work have been maintained, but changes and additions to content have been made where necessary. In particular, attention has been paid to updating economic data and information on standards, laws, and regulations. Commercially available products and their producers have also been subject to clearly recognizable changes, and these changes have been in part caused by the growing tendency of companies to merge and concentrate on their core businesses.

Among products and processes, the trend to environmentally friendly alternatives has also increased, even though the share of solvent-containing coating materials still dominates the market. Therefore, the article on solvents will remain indispensable for some time to come. The second edition will serve to confirm the book in its role as a standard reference for anyone working with coatings.

Marl, April 1998   WERNER FREITAG

Preface to the 1st edition

Paints and coatings are used to protect substrates against mechanical, chemical, and atmospheric influences. At the same time, they serve to decorate and color buildings, industrial plants, and utensils.

Coatings are of high economic importance because they provide protection against corrosive and atmospheric attack. It is therefore understandable that in industrialized countries such as the European Community, the United States, and Japan the annual consumption per capita is high and is continuing to rise.

There are numerous paint systems, production and process technologies due to the many demands made on quality, processibility, and economical importance. These have been fully discussed in this book, which presents the articles "Paints and Coatings" and "Solvents" as published in the 5th Edition of Ullmann's Encyclopedia of Industrial Chemistry.

Comprehensive information on all paint systems and binders, pigments, fillers, and additives has been given in individual chapters. Modern, low-emission paints such as high-solids paints, water-borne paints, powder paints, and radiation-curing systems are also discussed in detail.

There are special sections which deal with different production and processing technologies. Recommendations for each target application of a coating system are provided. Finally, special treatment of state-of-the-art paint testing, analysis, environmental protection, recycling, and toxicology is offered.

Although the paint industry has made great efforts to substitute volatile and organic solvents for environmental reasons, the majority of paints today still contain these solvents since they are useful processing agents. A knowledge of their physical data, their toxicological and environmental properties as well as the interaction between solvent and binder forms the basis for practice-oriented paint development. The inclusion of the chapter on Solvents is an ideal addition to this presentation of coating systems.

The special value of this book is that it provides a concise, up-to-date overview of all the properties of paints and coatings, their production and processing technologies, and applications for a wide readership. The book is generously illustrated with numerous figures that aid further understanding, and the extensive literature references serve to deepen one's knowledge of the topics described.

The publisher has successfully gathered together authors of international renown. Undoubtedly, the book will become a standard work for all producers of raw materials, paints and coatings, for users of paints and coatings, as well as for institutes and public authorities.

August 1993   Dieter Stoye

Authors

Uwe Biethan, Hüls AG, Marl, Federal Republic of Germany (Chap. 1)

Werner Funke, Institut für Technische Chemie, Stuttgart, Federal Republic of Germany (Section 2.1)

Lutz Hoppe, Wolff Walsrode AG, Walsrode, Federal Republic of Germany (Section 2.2.1)

Jürgen Hasselkus, Krahn Chemie GmbH, Hamburg, Federal Republic of Germany (Section 2.2.2)

Larry G. Curtis, Eastman Chemical Products, Kingsport, Tennessee 37662, United States (Section 2.2.2)

Klaus Hoehne, Bayer AG, Leverkusen, Federal Republic of Germany (Section 2.3)

Hans-Joachim Zech, Hüls AG, Marl, Federal Republic of Germany (Sections 2.4.1 and 2.4.2)

Peter Heiling, Wacker-Chemie GmbH, Burghausen, Federal Republic of Germany (Sections 2.4.1–2.4.7, apart from Section 2.4.3.3)

Masaaki Yamabe, Asahi Glass Co. Ltd., Hazawa-cho, Kanagawa-ku, Yokohama, Japan (Section 2.4.3.3)

Klaus Dören, Hüls AG, Marl, Federal Republic of Germany (Section 2.4.8)

Hans Schupp, BASF AG, Ludwigshafen, Federal Republic of Germany (Section 2.5)

Rolf Küchenmeister, Bayer AG, Leverkusen, Federal Republic of Germany (Section 2.6)

Martin Schmitthenner, Hüls AG, Witten, Federal Republic of Germany (Section 2.7)

Wolfgang Kremer, Bayer AG, Krefeld-Uerdingen, Federal Republic of Germany (Section 2.8)

Wolfhart Wieczorrek, Bayer AG, Leverkusen, Federal Republic of Germany (Section 2.9)

Hans Gempeler, Wolfgang Schneider, Ciba-Geigy AG, Basel, Switzerland (Section 2.10)

James W. White, Anthony G. Short, Dow Corning Ltd., Barry, South Glamorgan, CF6 7YL, United Kingdom (Section 2.11)

Werner J. Blank, Leonard J. Calbo, King Industries, Norwalk, Connecticut 06852, United States (Section 2.12)

DIETER PLATH, Hoechst AG, Wiesbaden, Federal Republic of Germany (Section 2.13)

FRIEDRICH WAGNER †, Sika-Chemie, Stuttgart, Federal Republic of Germany (Section 2.14)

WERNER HALLER, Henkel KGaA, Düsseldorf, Federal Republic of Germany (Section 2.15.1)

KARL-MARTIN RÖDDER †, Hüls AG, Troisdorf, Federal Republic of Germany (Section 2.15.2)

HANS-JOACHIM STREITBERGER, BASF Corporation, Southfield, Michigan 48086, United States (Sections 3.1 and 3.8)

EDMUND URBANO, Vianova Kunstharz AG, Graz, Austria (Section 3.2)

RICHARD LAIBLE, Akzo Coatings GmbH, Stuttgart, Federal Republic of Germany (Section 3.3)

BERND D. MEYER, Akzo Powder Coatings GmbH, Reutlingen, Federal Republic of Germany (Sections 3.4 and 8.3.5)

ENGIN BAGDA, Deutsche Amphibolin-Werke, Ober-Ramstadt, Federal Republic of Germany (Section 3.5)

FREDERICK A. WAITE, ICI Paints, Slough, Berkshire SL2 5DS, United Kingdom (Section 3.6)

MICHEL PHILIPS, Radcure Specialties, Drogenbos, Belgium (Section 3.7)

KLAUS KÖHLER, Bayer AG, Krefeld-Uerdingen, Federal Republic of Germany (Section 4.1)

PETER SIMMENDINGER, Ciba-Geigy AG, Basel, Switzerland (Section 4.2)

WOLFGANG ROELLE, Bassermann & Co., Mannheim, Federal Republic of Germany (Section 4.3)

WILFRIED SCHOLZ, WOLFGANG KORTMANN, BYK-Chemie GmbH, Wesel, Federal Republic of Germany (Chap. 5, apart from Section 5.7)

ANDREAS VALET, MARIO SLONGO, Ciba-Geigy AG, Basel, Switzerland (Section 5.7)

THOMAS MOLZ, Henkel KGaA, Düsseldorf, Federal Republic of Germany (Chap. 6)

RAINER HILLER, DIETMAR MÖLLER, BASF Lacke + Farben AG, Münster-Hiltrup, Federal Republic of Germany (Chap. 7)

KLAUS WERNER THOMER, Adam Opel AG, Rüsselsheim, Federal Republic of Germany (Chap. 8, apart from Section 8.3.5)

KLAUS VOGEL, Herberts GmbH, Wuppertal, Federal Republic of Germany (Chap. 9)

ULRICH SCHERNAU, BERNHARD HÜSER, BASF Coatings AG, Münster, Federal Republic of Germany (Chap. 10)

ALFRED BRANDT, ICI Lacke Farben GmbH, Hilden, Federal Republic of Germany (Sections 11.1–11.3, 11.5–11.8)

ALEX MILNE, Courtaulds Coatings Ltd., Felling, Gateshead NE10 0JY, United Kingdom (Section 11.4)

HELMUT WEYERS, ICI Lacke Farben GmbH, Hilden, Federal Republic of Germany (Section 11.9)

WOLFGANG PLEHN, Umweltbundesamt, Berlin, Federal Republic of Germany (Chap. 12)

HANNS-ADOLF LENTZE, CEPE, Brussels, Belgium (Chap. 13)

MARTINA ORTELT, Creanova Spezialchemie GmbH, Marl, Federal Republic of Germany (Revision of Chapter 14)

Contents

1.	**Introduction**	1
1.1.	**Fundamental Concepts**	1
1.2.	**Historical Development**	2
1.3.	**Composition of Paints**	3
1.3.1.	Binders and Resins	3
1.3.2.	Plasticizers	4
1.3.3.	Pigments and Extenders	4
1.3.4.	Paint Additives	5
1.3.5.	Solvents	7
1.4.	**Paint Application**	7
1.5.	**Drying and Film Formation**	8
1.6.	**Multicoat Systems**	9
1.7.	**Economic Aspects**	10
1.8.	**Future Outlook**	10
2.	**Types of Paints and Coatings (Binders)**	11
2.1.	**Oil-Based Coatings**	11
2.2.	**Cellulose-Based Coatings**	12
2.2.1.	Nitrocellulose Lacquers	12
2.2.1.1.	Raw Materials	13
2.2.1.2.	Application and Uses	15
2.2.2.	Organic Cellulose Ester Coatings	16
2.2.2.1.	Cellulose Acetate Butyrate	16
2.2.2.2.	Cellulose Acetate Propionate	19
2.3.	**Chlorinated Rubber Coatings**	19
2.3.1.	Starting Products	19
2.3.2.	Chlorinated Rubber Paints	20
2.3.3.	Chlorinated Rubber Combination Paints	22
2.4.	**Vinyl Coatings**	23
2.4.1.	General Properties	23
2.4.2.	Coatings Based on Polyolefins and Polyolefin Derivatives	24
2.4.3.	Poly(Vinyl Halides) and Vinyl Halide Copolymers	25
2.4.3.1.	Poly(Vinyl Chloride) and Vinyl Chloride Copolymers	25
2.4.3.2.	Vinylidene Chloride Copolymers	27
2.4.3.3.	Fluoropolymer Coatings	27

2.4.4.	Poly(Vinyl Esters)	31
2.4.4.1.	Solid Resins	31
2.4.4.2.	Dispersions	32
2.4.5.	Poly(Vinyl Alcohol)	33
2.4.6.	Poly(Vinyl Acetals)	34
2.4.7.	Poly(Vinyl Ethers)	35
2.4.8.	Polystyrene and Styrene Copolymers	35
2.5.	**Acrylic Coatings**	37
2.6.	**Alkyd Coatings**	41
2.6.1.	Alkyd Resin Binders and Uses	42
2.6.2.	Additional Raw Materials	47
2.6.3.	Production	49
2.6.4.	Environmental and Health Protection Measures	50
2.7.	**Saturated Polyester Coatings**	50
2.7.1.	Properties	51
2.7.2.	Production of Polyester Resins and Coatings	53
2.7.3.	Cross-Linking of Polyester Resins	54
2.7.4.	Uses	55
2.8.	**Unsaturated Polyester Coatings**	57
2.8.1.	Unsaturated Polyester Binders	57
2.8.2.	Other Raw Materials	58
2.8.3.	Formulation, Application, Use, Properties	60
2.8.4.	Storage, Transport, Toxicology	63
2.9.	**Polyurethane Coatings**	63
2.9.1.	Raw Materials	64
2.9.2.	Polyurethane Systems	65
2.9.2.1.	One-Pack Systems	66
2.9.2.2.	Two-Pack Systems	67
2.9.3.	Properties and Uses	68
2.10.	**Epoxy Coatings**	69
2.10.1.	Epoxy Resin Types	69
2.10.2.	Curing Agents	70
2.10.3.	Chemically Modified Epoxy Resins	73
2.10.4.	Uses	73
2.10.4.1.	Curing at Ambient Temperature	73
2.10.4.2.	Curing at Elevated Temperature	75
2.10.4.3.	Radiation Curing	77
2.10.5.	Toxicology	77
2.11.	**Silicone Coatings**	78
2.12.	**Urea, Benzoguanamine, and Melamine Resins for Coatings**	80
2.13.	**Phenolic Resins for Coatings**	86
2.13.1.	Resols	87
2.13.2.	Novolacs	89
2.13.3.	Modified Phenolic Resins	90

2.14.	**Asphalt, Bitumen, and Pitch Coatings**	91
2.14.1.	Asphalt and Asphalt Combination Coatings	91
2.14.2.	Bitumen Coatings	92
2.14.3.	Bitumen Combination Coatings	93
2.14.4.	Pitch Coatings	94
2.15.	**Silicate Coatings**	94
2.15.1.	Water Glass Coatings	94
2.15.2.	Alkyl Silicates	96
3.	**Paint Systems**	101
3.1.	**Solventborne Paints**	101
3.1.1.	General Information	101
3.1.2.	Properties and Raw Materials	102
3.1.3.	Environmental Protection and Application Technology	104
3.2.	**Solvent-Free and Low-Solvent (High-Solids) Paints**	105
3.2.1.	Principles	105
3.2.2.	Production and Uses	107
3.3.	**Waterborne Paints**	109
3.3.1.	Properties	109
3.3.2.	Production and Application	113
3.3.3.	Uses and Environmental Aspects	114
3.4.	**Coating Powders**	115
3.4.1.	Introduction and Economic Importance	115
3.4.2.	Production	117
3.4.3.	Properties	117
3.4.4.	Testing	121
3.4.5.	Storage and Transportation	122
3.4.6.	Environmental Aspects and Safety	122
3.4.7.	Uses	123
3.5.	**Waterborne Dispersion Paints (Emulsion Paints)**	125
3.6.	**Nonaqueous Dispersion Paints**	129
3.7.	**Radiation-Curing Systems**	135
3.7.1.	Introduction	135
3.7.2.	Radiation-Curable Systems Based on Acrylates	136
3.7.3.	Equipment	137
3.7.4.	Fields of Application	138
3.8.	**Electrodeposition Paints**	139
4.	**Pigments and Extenders**	143
4.1.	**Inorganic Pigments**	143
4.2.	**Organic Pigments**	148
4.3.	**Extenders**	150
4.3.1.	Introduction	150

4.3.2.	Properties	152
4.3.3.	Modification of Extenders	157
5.	**Paint Additives**	**159**
5.1.	**Defoamers**	**160**
5.2.	**Wetting and Dispersing Additives**	**161**
5.3.	**Surface Additives**	**163**
5.4.	**Driers and Catalysts**	**165**
5.5.	**Preservatives**	**165**
5.6.	**Rheology Additives**	**166**
5.7.	**Light Stabilizers**	**167**
5.8.	**Corrosion Inhibitors**	**170**
5.9.	**Use and Testing of Additives**	**171**
6.	**Paint Removal**	**173**
6.1.	**Paint Removal from Metals**	**173**
6.1.1.	Chemical Paint Removal	173
6.1.2.	Thermal Paint Removal	174
6.1.3.	Mechanical and Low-Temperature Paint Removal	175
6.2.	**Paint Removal from Wood and Mineral Substrates**	**175**
7.	**Production Technology**	**177**
7.1.	**Principles**	**177**
7.2.	**Paint-Making Processes**	**182**
7.2.1.	Varnishes	182
7.2.2.	Paints	183
7.2.3.	Coating Powders	184
7.3.	**Apparatus**	**185**
7.3.1.	Mixers	185
7.3.2.	Dissolvers	186
7.3.3.	Kneaders and Kneader Mixers	186
7.3.4.	Media Mills	188
7.3.5.	Roller Mills	191
7.3.6.	Filter Systems	192
8.	**Paint Application**	**195**
8.1.	**Types of Substrate**	**195**
8.2.	**Pretreatment of Substrate Surfaces**	**195**
8.2.1.	Pretreatment of Metallic Substrates	196
8.2.1.1.	Cleaning	196
8.2.1.2.	Degreasing	198
8.2.1.3.	Formation of Conversion Layers	198
8.2.2.	Pretreatment of Plastics	201
8.2.3.	Pretreatment of Wood	202

8.3.	**Application Methods**	203
8.3.1.	Spraying (Atomization)	203
8.3.2.	Electrostatic Atomization	205
8.3.3.	Dipping	207
8.3.4.	Miscellaneous Wet Paint Coating Methods	210
8.3.5.	Powder Coating	214
8.3.6.	Coating of Plastics and Wood	216
8.4.	**Paint Curing Methods**	216
9.	**Properties and Testing**	219
9.1.	**Properties of Coating Materials**	219
9.2.	**Properties of Coatings**	222
9.2.1.	Films for Testing	222
9.2.2.	Optical Properties	226
9.2.3.	Mechanical Properties	229
9.2.4.	Chemical Properties	231
9.2.5.	Weathering Tests	232
10.	**Analysis**	235
10.1.	**Analysis of Coating Materials**	235
10.1.1.	Separation of the Coating Material into Individual Components	235
10.1.2.	Analysis of Binders	236
10.1.3.	Analysis of Pigments and Extenders	239
10.1.4.	Analysis of Solvents	240
10.1.5.	Analysis of Additives	240
10.2.	**Analysis of Coatings**	241
11.	**Uses**	243
11.1.	**Coating Systems for Corrosion Protection of Large Steel Constructions (Heavy-Duty Coatings)**	243
11.2.	**Automotive Paints**	245
11.2.1.	Car Body Paints	245
11.2.2.	Other Automotive Coatings	248
11.3.	**Paints Used for Commercial Transport Vehicles**	249
11.3.1.	Railroad Rolling Stock	249
11.3.2.	Freight Containers	251
11.3.3.	Road Transport Vehicles	251
11.3.4.	Aircraft Coatings	252
11.4.	**Marine Coatings**	252
11.4.1	Substrate, Surface Preparation, and Priming	253
11.4.2.	Ship Paint Systems	255
11.4.3.	Fouling and Antifouling	257
11.5.	**Coil Coating**	258
11.6.	**Coatings for Domestic Appliances**	259

11.7.	Coatings for Packaging (Can Coatings)	260
11.8.	Furniture Coatings	261
11.9.	Coatings for Buildings	262
11.9.1.	Exterior-Use Coatings	263
11.9.2.	Interior-Use Coatings	265
12.	**Environmental Protection and Toxicology**	267
12.1.	Clean Air Measures	267
12.2.	Wastewater	270
12.3.	Solid Residues and Waste	271
12.4.	Toxicology	272
13.	**Economic Aspects**	275
14.	**Solvents**	277
14.1.	Definitions	277
14.2.	Physicochemical Principles	278
14.2.1.	Theory of Solutions	278
14.2.2.	Dipole Moment, Polarity, and Polarizability	285
14.2.3.	Hydrogen Bond Parameters	286
14.2.4.	Solvation	287
14.2.5.	Solvents, Latent Solvents, and Non-Solvents	287
14.2.6.	Dilution Ratio and Dilutability	288
14.2.7.	Influence of Molecular Mass on Solubility	290
14.2.8.	Dissolution and Solution Properties	291
14.3.	Physical and Chemical Properties	293
14.3.1.	Evaporation and Vaporization	293
14.3.2.	Hygroscopicity	296
14.3.3.	Density and Refractive Index	297
14.3.4.	Viscosity and Surface Tension	298
14.3.5.	Vapor Density	300
14.3.6.	Thermal and Electrical Data	300
14.3.7.	Flash Point, Ignition Temperature, and Ignition Limits	301
14.3.8.	Heats of Combustion and Calorific Values	303
14.3.9.	Chemical Properties	304
14.4.	Toxicology and Occupational Health	305
14.4.1.	Toxicology	305
14.4.2.	Occupational Health	309
14.5.	Environmental and Legal Aspects	311
14.5.1.	Environmental Protection	311
14.5.2.	Laws Concerning Dangerous Substances	314
14.5.3.	Fire Hazard	315
14.5.4.	Waste	316
14.6.	Purification and Analysis	318

14.7.	**Uses**	318
14.7.1.	Solvents in Paints	318
14.7.2.	Solvents in Paint Removers	322
14.7.3.	Solvents in Printing Inks	322
14.7.4.	Extraction	323
14.7.5.	Extractive Distillation	323
14.7.6.	Chromatography	323
14.7.7.	Solvents for Chemical Reactions	324
14.7.8.	Solvents for Recrystallization	324
14.7.9.	Solvents in Film Production	325
14.7.10.	Solvents for Synthetic Fibers	325
14.7.11.	Solvents for Rubber, Plastics, and Resin Solutions	325
14.7.12.	Solvents for Degreasing	325
14.7.13.	Solvents for Dry Cleaning	326
14.7.14.	Solvents in Aerosol Cans and Dispensers	326
14.8.	**Economic Aspects**	326
14.9.	**Solvent Groups**	327
14.9.1.	Aliphatic Hydrocarbons	327
14.9.2.	Cycloaliphatic Hydrocarbons	350
14.9.3.	Terpene Hydrocarbons and Terpenoids	350
14.9.4.	Aromatic Hydrocarbons	351
14.9.5.	Chlorinated Hydrocarbons	352
14.9.6.	Alcohols	353
14.9.7.	Ketones	358
14.9.8.	Esters	362
14.9.9.	Ethers	366
14.9.10.	Glycol Ethers	368
14.9.11.	Miscellaneous Solvents	372
15.	**References**	375
	Index	401

1. Introduction

1.1. Fundamental Concepts

Paints or coatings are liquid, paste, or powder products which are applied to surfaces by various methods and equipment in layers of given thickness. These form adherent films on the surface of the substrate.

Film formation can occur physically or chemically. Physical film formation from liquid coatings is known as drying, whereas for powder coatings, it is melting process. Drying is always associated with evaporation of organic solvents or water. Physical film formation is only possible if the coating components remaining on the substrate are solid and nontacky. Chemical film formation is necessary if the coating components are liquid, tacky, or pasty; conversion to a solid nontacky film takes place by chemical reaction between the components. The reactive components can be constituents of the coating, and the reaction can be initiated by energy (heat or radiation) after application of the coating. However, it is also possible to add a reaction partner while applying the coating (multipack paints). A special case of chemical film formation is the oxidation of coating component(s) by atmospheric oxygen (air drying). Physical and chemical film formation are often combined, e.g., in solvent-containing stoving paints, where the first stage is solvent evaporation, after which the film is cured by stoving. The properties of a paint are determined by its qualitative and quantitative composition, suitable choice of which enables the viscosity, electrical conductivity, and drying behavior to be matched to the application conditions. Also, the properties of the coating film (luster, elasticity, scratch resistance, hardness, adhesion, and surface structure) are determined by the paint properties. However, the condition of the substrate surface (cleanliness and freedom from dust and grease) is also important.

Coatings must fulfill many requirements. They protect the substrate against corrosion, weathering, and mechanical damage; have a decorative function (automotive coatings, household appliances, furniture); provide information (traffic signs, information signs, advertising); or have other specific properties.

"Coating" is a general term denoting a material that is applied to a surface. "Paint" indicates a pigmented material, while "varnish" refers to a clear lacquer (ISO 4618/1; DIN 55945).

1.2. Historical Development

The earliest evidence of well-preserved prehistoric paintings, dating from the 16th millenium B.C. can be found in caves in Southern France (Font-de-Gaume, Niaux, Lascaux), Spain (Altamira), and South Africa. The colors used were pure oil paints prepared from animal fat mixed with mineral pigments such as ocher, manganese ore (manganese dioxide), iron oxide, and chalk. The oldest rock paintings from North Africa (Sahara, Tassili n'Ajjer) data from between the 5th and the 7th millennium B.C. Many examples of paintings from Babylon, Egypt, Greece, and Italy dating from the 1st and 2nd millenium B.C. are also known.

The first painted objects come from China. Furniture and utensils were covered with a layer of paint in an artistic design. The oldest tradition work dates from around 200 B.C. The lacquer used was the milky juice from the bark of the lacquer tree (*Rhus vernicifera*). This was colored black or red with minerals, and later also with gold dust or gold leaf.

The oldest recipe for a lacquer, from linseed oil and the natural resin sandarac, dates from 1100 A.D. and was due to the monk ROGERUS VON HELMERSHAUSEN. Natural products such as vegetable oils and wood resins remained the most important raw materials for paint production, into the early 1900s. Only the introduction of faster production equipment such as belt conveyors made the development of new paints necessary. Initially, the rapid-drying binder used was nitrocellulose, which after World War I could be manufactured on a large scale in existing guncotton plants. Phenolic resins were the first synthetic binders (ca. 1920), followed by the alkyd resins (1930). The large number of synthetic binders and resins now available are tailored for each application method and area of use. These paint raw materials are based on petrochemical primary products. Vegetable and animal oils and resins are now seldom used in their natural form, but only after chemical modification. The tendency to use such "renewable" raw materials is increasing. Consumer demand has led to a marked renaissance of natural products ("biopaints").

The use of organic solvents in paint technology was linked to the development of modern rapid-drying binders. Whereas the liquid components previously used in coatings were vegetable oils or water and possibly ethanol, it now became necessary to use solvent mixtures to give accelerated drying and optimized paint-application properties. Production of a wide range of solvents began worldwide in the chemical industry in the 1920s.

Methods of applying paints also underwent major changes in the 1900s. Whereas up to this time coatings were applied manually with a brush, even in industry, this technique is today only used in the handicraft and DIY areas. Modern mechanized and automated application methods are used today for industrial-scale application because of greater efficiency, low material losses, qualitatively better results, and lower labor costs. They include high-pressure spraying using compressed air or electrostatic charging, modern automatic and environmentally friendly dipping and electrophoretic processes, and application by rollers.

Problems of environmental pollution also followed from the introduction of solvents. These were recognized by the late 1960s and became the subject of development work. Waterborne coatings, low-solvent coatings, solvent-free powder coatings, and new radiation-curing coating systems with reactive solvents that are bound chemically during the hardening process were developed. These environmentally friendly coating systems have gained a considerable market share. However, in some areas solvent-containing coatings are difficult to replace without affecting quality. For this reason, solvent-recycling and solvent-combustion plants have been developed to recover or incinerate the solvents in the waste air.

1.3. Composition of Paints

Paints are made of numerous components, depending on the method of application, the desired properties, the substrate to be coated, and ecological and economic constraints. Paint components can be classified as volatile or nonvolatile.

Volatile paint components include organic solvents, water, and coalescing agents. Nonvolatile components include binders, resins, plasticizers, paint additives, dyes, pigments, and extenders. In some types of binder, chemical hardening can lead to condensation products such as water, alcohols, and aldehydes or their acetals, which are released into the atmosphere, thus being regarded as volatile components.

All components fulfill special functions in the liquid paint and in the solid coating film. Solvents, binders, and pigments account for most of the material, the proportion of additives being small. Low concentrations of additives produce marked effects such as improved flow behavior, better wetting of the substrate of pigment, and catalytic acceleration of hardening.

Solvents and pigments need not always be present in a coating formulation. Solvent-free paints and pigment-free varnishes are also available.

The most important component of a paint formulation is the binder. Binders essentially determine the application method, drying and hardening behavior, adhesion to the substrate, mechanical properties, chemical resistance, and resistance to weathering.

1.3.1. Binders and Resins

Binders are macromolecular products with a molecular mass between 500 and ca. 30 000. The higher molecular mass products include cellulose nitrate and polyacrylate and vinyl chloride copolymers, which are suitable for physical film formation. The low molecular mass products include alkyd resins, phenolic resins, polyisocyanates, and epoxy resins. To produce acceptable films, these binders must be chemically hardened after application to the substrate to produce high molecular mass cross-linked macromolecules.

Increasing relative molecular mass of the binder in the polymer film improves properties such as elasticity, hardness, and impact deformation, but also leads to higher solution viscosity of the binder. While the usefulness of a coating is enhanced by good mechanical film properties, low viscosity combined with low solvent content are also desirable for ease of application and for environmental reasons. Therefore, a compromise is necessary.

The low molecular mass binders have low solution viscosity and allow low-emission paints with high solids contents or even solvent-free paints to be produced. Here, the binder consists of a mixture of several reactive components, and film formation takes place by chemical drying after application of the paint. If chemical hardening occurs even at room temperature, the binder components must be mixed together shortly before or even during application (two- and multicomponent systems).

Today, most binders are synthetic resins such as alkyd or epoxy resins.

The natural resin most commonly used as a binder today is rosin, which is often tailored by chemical modification to suit specific applications. Also, many synthetic hard resins mainly based on cyclohexanone, acetophenone, or aldehydes, are used in the paints industry. Hard resin binders increase the solids content, accelerate drying, and improve surface hardness, luster, and adhesion.

Most synthetic binders are softer and more flexible thant hard resins. Consequently, they impart good elasticity, impact resistance, and improved adhesion, even to critical undercoats, as well as offering adequate resistance to weathering and chemicals. These binders are produced with a property profile tailored to suit particular application methods and to comply with a range of technical requirements, including environmental protection, low toxicity, and suitability for recycling and disposal.

1.3.2. Plasticizers

Plasticizers are organic liquids of high viscosity and low volatility. The esters of dicarboxylic acids (e.g., dioctyl phthalate) are well-known examples. Plasticizers lower the softening and film-forming temperatures of the binders. They also improve flow, flexibility, and adhesion properties. Chemically, plasticizers are largely inert and do not react with the binder components. Most binders used today are inherently flexible and can be regarded as "internally plasticized" resins. For this reason, use of plasticizers has declined.

1.3.3. Pigments and Extenders

Pigments and extenders in coatings are responsible for their color and covering power, and in some cases give the coating film improved anticorrosion properties.

Pigments and extenders are finely ground crystalline solids that are dispersed in the paint. They are divided into inorganic, organic, organometallic, and metallic pigments. By far the most commonly used pigment is titanium dioxide. As a rule, mixtures of pigments are used for technical and economic reasons. The hiding power and tinting strength of a paint depend on the particle size of the pigment. The usual size range aimed at is 0.1–2.0 µm, which means that the pigment has a high surface area that must be wetted as effectively as possible by the binder components to give the coating film good stability, weathering resistance, and luster. This is achieved by bringing the pigment and binder into intimate contact under the influence of high shear forces. The high hiding power of some pigments enables them to be partially replaced by the cheaper extenders such as barium sulfate, calcium carbonate, or kaolin. Extenders have a particle size distribution similar to that of the pigments and are incorporated into the coating in the same way. The concentration of pigment in coating films is expressed by the pigment volume concentration (PVC). This is the ratio of the volume of pigments and extenders to the total volume of the nonvolatile components. Each coating system has a critical pigment volume concentration (CPVC) at which the binder just fills the free space between the close-packed pigment particles. At higher pigment concentrations, the pigment particles in the coating film are no longer fully wetted by the binder, leading to a marked deterioration in coating film properties such as luster, stability, strength, and anticorrison properties.

1.3.4. Paint Additives

Paint additives are auxiliary products that are added to coatings, usually in small amounts, to improve particular technical properties of the paints or coating films. Paint additives are named in accordance with their mode of action.

Leveling agents promote formation of a smooth, uniform surface on drying of the paint. Suitable materials include certain high-boiling solvents such as butyl ethers of ethylene glycol, propylene glycol and diglycols, as well as cyclohexanone and alkylated cyclohexanones, and in some cases aromatic and aliphatic hydrocarbons. Low molecular mass resins (e.g., some polyacrylates and silicones) are also used. Solid leveling agents, such as special low molecular mass resins, are also useful for improving the surface properties of films produced from powder coatings. Flow agents act by reducing the paint viscosity during drying. The effectiveness of a particular flow agent depends on the type of binder and the drying or hardening temperature.

Film-formation promoters, which are closely related to flow agents, reduce the film-forming temperature for film formation from dispersions, leading to a surface that is as pore-free and uniform as possible. Certain high-boiling glycol ethers and glycol ether esters are used, often in combination with hydrocarbons.

Wetting Agents, Dispersants, and Antisetting Agents. Wetting agents from one of the largest groups of coating additives. These are surfactants which aid wetting of the pigments by the binders and prevent flocculation of the pigment particles. This leads to the formation of a uniform, haze-free color and a uniformly high luster of the coating film. This group also includes the dispersants, which give good pigment wetting and hence optimum dispersion of the pigments in the paint, thereby preventing sedimentation particularly of high-density pigments. As well as good wetting properties, some pseudoplasticity is also necessary. Antisetting agents have similar characteristics to dispersants.

Antifoaming agents are used to prevent foaming during paint manufacture and application and to promote release of air from the coating film during drying. Various products are used, including fatty acid esters, metallic soaps, mineral oils, waxes, silicon oils, and siloxanes, sometimes combined with emulsifiers and hydrophobic silicas.

Catalysts are added to paints to accelerate drying and hardening. They include drying agents (driers, siccatives), which, in the case of the air-drying binders (including some alkyd resins or unsaturated oils), accelerate decomposition of the peroxides and hydroperoxides that form during the drying process, thereby enabling radical polymerization of the binders to take place. The driers used are mainly metallic soaps such as cobalt naphthenate; manganese, calcium, zinc, and barium salts; and zirconium compounds.

Various products are used to catalyze the cross-linking of binder systems at room temperature. For acid-catalyzed systems such as polyester – melamine resin systems, free acids, their ammonium salts, or labile esters are suitable, while for base-catalyzed systems such as polyester – isocyanate, tertiary amines or dibutyltin dilaurate are used. The amount of catalyst used must be such that the pot life is not impaired.

Antifloating and antiflooding agents prevent horizontal and vertical segregation of pigments with different densities and surface properties. This prevents differences in the color and luster of the surface of the film, which can lead to a blotchy appearance.

Antiskinning agents are added to air-drying paints to prevent surface skin formation caused by contact with atmospheric oxygen. In the film, they produce uniform drying and prevent shrinkage (wrinkling). Chemically, these materials are antioxidants such as oximes, which evaporate with the solvents during the drying process.

Matting agents are used to produce coatings with a matt, semi-matt, or silk finish. They include natural mineral products such as talc or diatomites and synthetic materials such as pyrogenic silicas or polyolefin waxes. Matting can also be obtained by special formulations that exploit the incompatibility between binder components and their cross-linked structures.

Neutralizing agents are used in waterborne paints to neutralize binders and stabilize the product. Ammonia and various alkylated aminoalcohols are used, depending on the type of binder and method of application. On hardening, the amines mainly evaporate along with the water.

Thickening agents control the rheological properties of paints of various types. They include inorganic (mainly silicates), organometallic (titanium and zirconium chelates), naturally occurring organic (mainly cellulose ethers) and synthetic organic products (polyacrylates, polyvinylpyrrolidone, polyurethanes).

Preservatives (biocides, fungicides) prevent the attack of paint systems, principally water-based, by microorganisms.

Corrosion inhibitors are used to prevent the formation of corrosion products when waterborne paints are applied to metallic substrates (flash rust). They include oxidizing salts such as chromates, metaborates, nitrites, and nitrates; organic amines or sulfur-containing products; and organic salts (benzoates, naphthenates, octoates).

1.3.5. Solvents

Solvents are compounds that are normally liquid at room temperature. Those most commonly used in coatings technology are aromatic and aliphatic hydrocarbons, esters of acetic acid, glycol ethers, alcohols, and some ketones. Solvents dissolve solid and highly viscous binder components. They enable incompatibility between paint components to be overcome, improve pigment wetting and dispersion, and control storage stability and viscosity of the coating. They promote the release of included air from the liquid coating film, control the drying behavior of the coating, and optimize flow properties and luster. Organic solvents are used in most liquid coatings systems, including, waterborne coatings, in which they perform important fluctions.

After paint application, the solvents should evaporate as quickly as possible, leaving the film. If no special precautions are taken, the solvents enter the atmosphere as pollutants. To protect operating personnel from the toxic effects of evaporating solvents, safety measures such as ventilation and air exhaust are necessary. To protect the environment, incineration and sometimes solvent-recovery plant is installed to prevent solvents entering the atmosphere. Other measures for the protection of the workplace and the environment from solvent vapors include the development and use of new low-solvent or solvent-free coatings, e.g., high-solids paints, waterborne coatings, and powder coatings.

1.4. Paint Application

Paint application can be performed manually, for example with brushes or rollers, or by mechanical methods such as spraying, atomization by rotating disks or cones, dipping, pouring, rotating drums and tumbling equipment, and automated application by rollers. Powder coatings are applied by electrostatic spraying or by dipping components into the powders. Multicomponent coatings are applied with multicomponent spraying equipment.

1.5. Drying and Film Formation

As the paint dries on the substrate, a firmly bonded film is formed. The properties of this film are determined both by the substrate and its pretreatment (cleaning, degreasing) and by the composition of the coating and the application method used.

Drying of the paint on the substrate takes place physically (1–3) or chemically (4):

1) Evaporation of the organic solvents from solvent-containing paints
2) Evaporation of water from waterborne paints
3) Cooling of the polymer melts (powder coatings)
4) Reaction of low molecular mass products with other low or medium molecular mass binder components (polymerization or cross-linking) to form macromolecules

Physical Drying. Physical drying takes place mainly for paints with high molecular mass polymer binders such as cellulose nitrate, cellulose esters, chlorinated rubber, vinyl resins, polyacrylates, styrene copolymers, thermoplastic polyesters and polyamide and polyolefin copolymers. These materials give good flexibility and stability because of their high molecular mass. Their glass transition temperature should be above room temperature to ensure adequate hardness and scratch resistance. With these polymers, film formation can also take place from solutions or dispersions in organic solvents or water, from which the solvent or water evaporates, leaving behind the chemically unchanged polymer film.

Film formation can be accelerated by drying at elevated temperatures (forced drying). Physically drying solvent-containing paints have a low solids content because the molecular mass of the binder is relatively high. Higher solids contents are obtained by dispersing the binder in water (dispersions, emulsions) or in organic solvents (nonaqueous dispersion or NAD systems). Films formed from physically drying paints, especially those formed from solutions, are sensitive to solvents (dissolution or swelling). The physically drying coatings also include many powder coatings that contain thermoplastic binders. Film formation takes place by heating the powder that has been applied to the substrate above its melting point. This ensures that a sealed film of polymer is formed.

Plastisols and organosols are a special case of physically drying coatings systems in which the binders consist of finely dispersed poly(vinyl chloride) or thermoplastic poly(meth)acrylates suspended in plasticizers. Organosols also contain some solvent. On drying at elevated temperatures, the polymer particles are swollen by the plasticizer, a process known as gelation.

Chemical Drying. Chemically drying paints contain binder components that react together on drying to form cross-linked macromolecules. These binder components have a relatively low molecular mass, so that their solutions can have a high solids content and a low viscosity. In some cases, solvent-free liquid paints are possible. Chemical drying can occur by polymerization, polyaddition, or polycondensation.

When *polymerization* is used as the hardening principle, reactive components combine to form the binder, e.g., unsaturated polyesters with styrene or acrylate monomers. Here, one component often behaves as a reactive solvent for the other, and low-emission coating systems are the result. Cross-linking can be carried out at room temperature (cold curing) or by radiation curing.

In drying by *polyaddition*, low molecular mass reactive polymers such as alkyd resins, saturated polyesters, or polyacrylates react with polyisocyanates or epoxy resins to form cross-linked macromolecules. Because this reaction can take place at room temperature, the binder components must be mixed shortly before application. The period of time during which a coating of this type remains usable after mixing of the components is known as the pot life. These are known as two-pack coatings, differing from the one-pack systems, which can be stored for months or even years.

Chemically blocking one of the polyaddition binder components (e.g., the polyisocyanate) gives a coating system stable at room temperature. Heat is required to deblock the component and enable cross-linking to occur. Stoving paints of this type are used in industry and in powder coatings.

Polycondensation drying requires the addition of catalysts or the use of higher temperatures. Acid-catalyzed coatings are well-known cold-curing paint systems used in the furniture industry, while heat-curing and stoving paints are used as industrial and automotive coatings. The binding agents used are functional alkyd resins, saturated polyesters, or polyacrylates in combination with urea resins, melamine resins, or phenolic resins. On cross-linking, water, low molecular mass alcohols, aldehydes, acetals, and other volatile compounds are released.

In practice drying of coatings and paints does not take place by one method alone. With solvent-containing and waterborne heat-curing coatings, physical drying by solvent evaporation always precedes chemical drying. Depending on the composition of the binder system, physical and chemical drying can take place simultaneously, and the various mechanisms of chemical drying can proceed concurrently or consecutively, depending on the nature of the binder. A knowledge of binder composition is important in order to assess the drying of a coating and able to accelerate it by heat, radiation, and addition of catalysts.

1.6. Multicoat Systems

Because dried coating films are not always pore-free, optimal protection of the substrate is not always ensured by one coat. A single coat can seldom fulfill all requirements such as good adhesion, corrosion protection, elasticity, hardness, decorative effect, coloration, and resistance to weathering and chemicals. Coatings with different compositions and functions are therefore often applied in succession. For example, primers provide good adhesion to the substrate and maximum corrosion protection, whereas color stability, gloss, and resistance to weathering are better provided by a top coat which is specially designed for this purpose but may not have particularly good corrosion resistance.

Intermediate coatings between the top coat and the primer are also applied if the highest quality is required, e.g., in the automobile industry. These have the task of providing adhesion between the primer and the top coat, and they also smooth out irregularities on the substrate, thereby indirectly helping to ensure good flow of the top coat and a high gloss with no defects.

1.7. Economic Aspects

In the industrialized countries of Europa and in North America, annual paint consumption of coatings per capita is > 20 kg, and high growth rates can be expected in the less industrialized countries of Eastern Europe, Asia, and South East Asia. The volume of coatings produced in Western Europa in 1994 was ca. 5.4×10^6 t, and in the United States, 6.2×10^6 t. The annual growth rates of solvent-containing conventional coatings are estimated to be $1-2\%$, and, of environmentally friendly coatings (high solids, water-based, powder coatings, etc.), ca. 5%.

1.8. Future Outlook

In the past, the development of coatings was mainly based on technical, quality, and economic considerations. These factors are just as important today from a business point of view and will continue to be so in the future. However, other considerations are now very much in the foreground, i.e., environmental protection, toxicology, environmentally friendly disposal of paint residues and coated articles at the end of their life cycle, the recycling of coated articles, and the conservation of raw materials and energy.

Thus, numerous low-emission paints have been developed, including high-solids paints, waterborne paints, aqueous dispersions for industrial use, powder coatings, and radiation-curing coatings. At the forefront in adopting these environmentally friendly products is heavy industry, in particular the automobile and household appliance industries. Medium-sized and smaller businesses will profit from this experience, adapting it for their own needs.

To conserve raw materials based on mineral oil, renewable raw materials derived from natural oils and resins will be investigated and assessed for potential use in paints.

Nevertheless, the principal development goal is still to secure further improvements in the quality of paint systems and to prolong the durability of coating films. The longer the renewal of a coated surface can be delayed, the less the environment is polluted, and the smaller are the amounts of waste produced and of raw materials and energy consumed.

The continuous increase in automation and electronic control of paint production and application are equally relevant, enabling products to be manufactured that are consistently of the highest quality.

2. Types of Paints and Coatings (Binders)

In this chapter paints and coatings will be discussed according to their binders.

2.1. Oil-Based Coatings [2.1]

Composition. Oil-based paints (oil paints) are among the oldest organic coating materials; in China, they have been known for more than 2000 years. Oil paints consist of natural drying oils (e.g., linseed oil, China wood or tung oil, and soybean oil) which undergo autoxidative polymerization in the presence of catalytic driers and atmospheric oxygen. Further constituents may include hard resins (e.g., alkylphenolic resins) that generally react with the drying oils at elevated temperature (230–280 °C) to form oleoresinous binders. On account of the air sensitivity of the oils, heating mainly takes place under an inert gas atmosphere.

Auxiliaries may be added to oil paints to improve their wetting and flow properties. The desired handling consistency is generally adjusted with aliphatic hydrocarbon solvents such as mineral spirits and in certain cases with toluene or xylenes.

With clear varnishes 5–10 wt % of solvent is sufficient, with paints 10–20 wt % is sufficient. There are very few restrictions in the choice of pigment; basic pigments (e.g., zinc oxide) can be used.

Conventional dispersion equipment (e.g., ball, roller, or sand mills) are suitable for producing oil paints.

Oil paints are relatively environmentally friendly as long as harzardous solvents and toxic pigments (e.g., red lead or zinc chromate) are not used. The oils used in such paints have a low viscosity. They are therefore particularly suitable for priming coats on manually derusted steel surfaces since they wet and penetrate the residual layers of rust well, resulting in thorough coverage. Oil paints are easily applied by conventional methods (e.g., brushing, roller coating, spraying, and dipping).

During film formation (curing), atmospheric oxygen reacts with the oil to form hydroperoxides which decompose into radicals and then initiate polymerization of the binder. Driers (metallic soaps such as cobalt, lead, and manganese naphthenates or octoates) catalyze formation and decomposition of the hydroperoxides and thereby accelerate film formation. A combination of several driers is normally used to control the curing reaction at the surface and in the interior of the coating.

The thickness of an oil-paint coating is restricted on account of the atmospheric oxygen required for curing. With thick layers (25–30 µm on vertical surfaces and

40–50 µm on horizontal surfaces), the oxygen penetrates too slowly and the lower region of the paint layer remains soft. Since the shrinkage of the coating differs in various layer regions during oxidative drying, wrinkles may form if the layer is too thick. The drying time is highly temperature dependent and may increase substantially in the absence of light. At room temperature, oil paint films dry in ca. 12–24 h depending on the amount of drier added, whereas several weeks are required in the vicinity of the freezing point of water.

During drying the films take up ca. 10–20 wt% of oxygen (relative to the pure oil). The smell detected during drying is partly due to decomposition products of the binder that are formed during autoxidative polymerization.

Coatings derived from oil paints are tough but not excessively hard, and exhibit limited weather resistance. They lose their high gloss relatively quickly (ca. two years) and yellow much more than other binders, both in the light and dark as well as at elevated temperature. The coatings are readily hydrolyzed and are therefore unsuitable, at least as a topcoat, in applications involving exposure to strong chemical influences.

On account of these disadvantages and the relatively long drying time, oil paints have almost completely lost their former importance over the last 30 years in favor of oxidatively drying alkyd resins. Being "naturally-based paints", renewed interest has, however, recently been shown in oil paints owing to ecological reasons.

Chemically modified, oxidatively drying oils (e.g., polyurethane oils, Section 2.9.2.1) are being increasingly used as binders for high-solids coatings (e.g., for wood protection). Binders for oil paints include low molecular mass 1,4-*cis*-polybutadienes, known as polyoils and produced by Hüls. Relatively short drying times (8–12 h) are achieved as a result of the polybutadiene component. The polyoils are heated with oxidatively drying oils or modified with maleic anhydride; the maleic anhydride units being converted into imide groups with amines. Such oil paints are suitable as priming coats for corrosion protection of manually derusted steel surfaces because they stabilize residual rust layers.

2.2. Cellulose-Based Coatings

2.2.1. Nitrocellulose Lacquers

Nitrocellulose (cellulose nitrate) lacquers are a mixture of binders (nitrocellulose and resins), plasticizers, and (optionally) pigments dissolved/dispersed in organic solvents. The nonvolatile components are:

1) Nitrocellulose
2) Resin
3) Plasticizer
4) Pigment (extender, dye)

The volatile components are

1) Active solvents
2) Latent solvents
3) Nonsolvents (diluents such as benzene, toluene, or xylene)

Physical evaporation of the solvents results in formation of the desired solid film on the substrate surface. Films may also be obtained from aqueous, low-solvent, or solvent-free nitrocellulose emulsions or dispersions [2.2].

2.2.1.1. Raw Materials

The compositions of cellulose nitrate lacquers are summarized in Table 2.1.

Nitrocellulose is an outstanding film-forming substance which displays rapid solvent evaporation (short drying time). It is compatible with most coating raw materials.

Nitrocellulose is characterized by its nitrogen content and solubility. The nitrogen contents are:

Ester-soluble nitrocellulose   11.8–12.2 wt%
Alcohol-soluble nitrocellulose 10.9 –11.3 wt%

High-viscosity, medium-viscosity, and low-viscosity formulations of each type are available [2.3]. Important producers of nitrocellulose used in lacquers include Hercules (USA), ICI (UK), BNC (France), Wolff Walsrode (FRG), and NQB (Brazil).

Films formed from high-viscosity nitrocellulose have good flexibility combined with a high crack resistance. They are therefore employed where high mechanical stress is to be expected (e.g., in leather coatings, putty, adhesives). Only lacquers with low solids contents can be obtained from high-viscosity nitrocellulose.

Low-viscosity nitrocellulose is used to prepare high-solids lacquers. Since low-viscosity nitrocellulose produces hard to brittle coating films, plasticizers and plastifying resins must be added to the lacquer formulation. They are used in putty, dipping paints, and printing inks.

The medium-viscosity nitrocelluloses have the broadest application range, a major field being furniture lacquers. They are also employed in paper and metal coatings as well as in reaction lacquers (e.g., acid-catalyzed lacquers and polyurethane paints).

Ester-soluble nitrocelluloses are mainly used in the lacquers described above. Alcohol-soluble nitrocellulose (which is also soluble in esters and ketones) is used for odorless lacquers, particularly for printing inks and sealing waxes.

According to international agreement, industrial nitrocelluloses have a maximum nitrogen content of 12.6 wt% and are stabilized (phlegmatized) for commercial use. Wetted nitrocellulose cotton (with water, ethanol, 2-propanol, or butanol) contains 65 (or 70) wt% nitrocellulose and 35 (or 30) wt% wetting agent.

Nitrocellulose is also available in the form of chips containing ≤ 82 wt% nitrocellulose and ≥ 18 wt% plasticizers (e.g., dibutyl phthalate); pigments may also be incorporated if desired.

Table 2.1. Formulation of nitrocellulose and lacquers

Lacquer type	Ingredients	Quantitative (weight) ratio
Primer sealer	DBP, blown castor oil, maleate resin	NC:PL:resin = 1:0.3:0.5–1
Film lacquer	Dicyclohexyl phthalate, dammar resin (dewaxed), maleate resin, paraffin	NC:PL:resin = 1:0.5:0.5 + 8% paraffin (*mp* 57–62 °C)
Gel dipping lacquer	DOP, amino resin/alkyd resin (1:1)	NC:PL:resin = 1:0.2:2 alcohol-soluble NC:solvent 2-propanol:toluene = 60:40 to 30:70
Incandescent lamp lacquer	tricresyl phosphate, alkyd resin	NC:PL:resin = 1:0.5:0.6 colored with ceres dyes
Primer	DBP, unrefined castor oil	NC:PL = 1:0.4; ratio of DBP to castor oil (plasticizer) = 1:1
Wood lacquer	DBP/unrefined castor oil (1:1), alkyd resin/maleate resin (2:1)	NC:PL:resin = 1:0.5:3
Leather lacquer	DBP, blown castor oil	NC:PL = 1:0.9 high-viscosity NC (e.g., E 950, E 840, Wolff Walsrode)
Light-metal lacquer	DOP, vinyl chloride copolymer	NC:PL:resin = 1:0.5:1
Open-pore wood varnish	DBP/blown castor oil (0.5:1), melamine resin, dammar or ketone resin	NC:PL:resin = 1:0.5:1
Furniture lacquer	DBP, polyacrylate resin, resin ester, alkyd resin	NC:PL:resin = 1:0.25–0.5:3–4
Nail polish	plasticizer	NC:PL = 1:1.2
Paper lacquer	DBP/castor oil (1:2)	NC:PL = 1:0.8
Polishing lacquer	DBP/DOP (1:1.7), peanut oil alkyd resin	NC:PL:resin = 1:0.3:0.7
Polish	DBP, shellac	NC:PL:resin = 1:0.2:1
Sanding primer	DBP/castor oil (2:1)	NC:PL:resin = 1:0.3:1 + ca. 0.1 zinc stearate
Buffing lacquer	DBP/DOP (1:2), peanut oil alkyd resin	NC:PL:resin = 1:0.3:0.5
Zapon lacquer	plasticizer	NC:PL = 1:0.1–0.3 high-viscosity NC (e.g., E 1160, Wolff Walsrode)

DBP = dibutyl phthalate; DOP = dioctyl phthalate; NC = nitrocellulose; PL = plasticizer.

Wetted nitrocellulose must not be allowed to dry out because of the risk of explosion.

Plasticizers. Plasticizers that are compatible with nitrocellulose and resins are used in coatings for the following purposes:

1) To improve adhesive strength and gloss

2) To improve mechanical properties such as elongation, pliability, buckling strength, crease resistance, and deep-drawing ability
3) To increase resistance to light, heat, cold, and sudden temperature changes (cold-check test)

Plasticizers may be solvents or nonsolvents for nitrocellulose. The type used depends on the application. Nitrocellulose is for example soluble in dibutyl phthalate, dioctyl phthalate, dicyclohexyl phthalate, tricresyl phosphate, and triphenyl phosphate. Plasticizers in which nitrocellulose is insoluble include crude and blown vegetable oils, stearates, and oleates.

Resins. A large number of synthetic coating resins (e.g., alkyd, ketone, urea, maleate, and acrylic resins) are available for formulating nitrocellulose combination lacquers. Selection criteria include price, color, influence on solvent release, gloss, hardness, sandability, yellowing, and durability of the final coating.

Nitrocellulose (generally in the form of chips) is used in polyurethane coatings to improve drying behavior, to increase body, and to obtain good flow.

Solvents. The solvent mixture has a large influence on the quality of the coated film. The solvent that evaporates last should be a solvent for all raw materials in the lacquer formulation. The most important active (true) solvents are acetate esters (e.g., ethyl, butyl, or propyl acetate) and ketones (e.g., acetone, methyl ethyl ketone, methyl isobutyl ketone).

Latent solvents, which become effective only in the presence of active or true solvents, include alcohols (e.g., methanol, ethanol, and propanol). Like the nonsolvents these are used to reduce costs. The lower alcohols (e.g., methanol or ethanol) are, of course, true solvents for alcohol-soluble nitrocellulose.

2.2.1.2. Application and Uses

The preparation of nitrocellulose lacquers is simple and involves dissolution and mixing procedures. The viscosity should be compatible with the equipment used.

Nitrocellulose lacquers can be sprayed efficiently with compressed air or by an "airless" technique. Electrostatic spraying is employed to reduce the overspray and for good coverage (e.g., when coating chairs). Flat articles, thin sheets (foils), or paper can be coated inexpensively on casting machines. High-viscosity lacquers are frequently applied by roller coating. Smaller objects are often coated by the dipping method. The pushing-through process is used for coating pencils.

An important use of nitrocellulose lacquers is in printing inks employed in flexographic, gravure, or silk-screen printing.

The most important areas of use of nitrocellulose lacquers are for coating wood, metal (e.g., automotive repair), paper, foil (cellophane, aluminum), leather, and textiles and in nail polish.

Aqueous nitrocellulose lacquers that contain small amounts of solvent are used in the form of emulsions or dispersions to coat leather [2.4] and decorative foils. Solvent-free dispersions are cured by UV radiation after evaporation of the water and are used to coat furniture, profiled boards, and paper.

2.2.2. Organic Cellulose Ester Coatings [2.5]–[2.7]

Cellulose acetate [*9004-35-7*], the simplest organic cellulose ester, offers excellent properties in coating films (e.g., flame resistance, high melting point, toughness, and clarity). These esters have limited solubility and compatibility with other resins; this is, however, necessary for widespread use.

Cellulose butyrate contains the bulkier butyryl group; these esters are more compatible and soluble than acetates, but are too soft for most coating applications. Cellulose esterified with blends of alkyl groups can provide many intermediate properties needed in coatings. Selection of the appropriate cellulose acetate butyrate [*9004-36-8*] (CAB) and cellulose acetate propionate [*9004-39-1*] (CAP) content must be based on specific application requirements.

Production of organic cellulose esters starts by mixing the appropriate organic acids and anhydrides, sulfuric acid catalyst, and purified cellulose. Esterification proceeds rapidly until all three anhydroglucose hydroxyls are esterified with acyl groups. Anhydride mixtures produce mixed esters (e.g., CAB and CAP). Fully acylated cellulose is of limited value in the coatings and plastics industries. Some free hydroxyl groups along the cellulose chain are necessary to provide solubility, flexibility, compatibility, and toughness. Since termination of the esterification reaction is not feasible, the fully acylated triester is slowly hydrolyzed to give the desired hydroxyl content.

Following esterification and hydrolysis, the product undergoes additional manufacturing steps that include filtration, precipitation, washing, and drying. The final product is usually a dry, free-flowing powder.

2.2.2.1. Cellulose Acetate Butyrate

Tennessee Eastman is presently the world's only manufacturer of CAB and CAP. Table 2.2 lists the properties of the commercially available CAB and CAP products.

Properties. The large size and low polarity of the butyryl groups separate the cellulose chains and lowers the attraction between them. As butyryl content increases, properties are affected as follows:

1) Solubility increases
2) Tolerance for diluents increases (tolerance signifies the ability to withstand inclusion of a nonsolvent in the solvent system before haze or precipitation occurs)
3) Water tolerance decreases
4) Compatibility increases
5) Flexibility increases
6) Moisture resistance increases
7) Grease resistance decreases
8) Tensile strength decreases
9) Hardness decreases
10) Melting range decreases

Table 2.2. Properties of cellulose acetate butyrates (CAB) and cellulose acetate propionates (CAP) (Tennessee Eastman)

Cellulose ester	Viscosity*, Pa·s	Acetyl** content, wt%	Butyryl content, wt%	Propionyl content, wt%	Hydroxyl content, wt%	Melting range, °C	M_r	T_g, °C	Density, g/cm³
CAB-171-15S	5.7	29.5	17		1.1	230–240	65 000	161	1.26
CAB-321-0.1	0.038	17.5	32.5		1.3	165–175	12 000	127	1.20
CAB-381-0.1	0.038	13.5	38		1.3	155–165	20 000	123	1.20
CAB-381-0.5	0.19	13.5	38		1.3	155–165	30 000	130	1.20
CAB-381-2	0.76	13.5	38		1.3	171–184	40 000	133	1.20
CAB-381-2BP	0.836	14.5	35.5		1.8	175–185	40 000	130	1.20
CAB-381-20	7.6	13.5	37		1.8	195–205	70 000	141	1.20
CAB-381-20BP	6.08	15.5	35.5		0.8	185–195	70 000	128	1.20
CAB-500-5	1.9	4.0	51		1.0	165–175	57 000	96	1.18
CAB-531-1	0.722	3.0	50		1.7	135–150	40 000	115	1.17
CAB-551-0.01	0.0038	2.0	53		1.5	127–142	16 000	85	1.16
CAB-551-0.2	0.076	2.0	52		1.8	130–140	30 000	101	1.16
CAB-553-0.4	0.114	2.0	46		4.8	150–160	20 000	136	1.20
CAP-482-0.5	0.152	2.5		45	2.6	188–210	25 000	142	1.22
CAP-482-20	7.6	2.5		46	1.8	188–210	75 000	147	1.22
CAP-504-0.2	0.076	0.6		42.5	5.0	188–210	15 000	159	1.26

*ASTM D 817 (Formula A) and D 1343. **ASTM D 817.

The *hydroxyl content* of CAB is perhaps the most important chemical variable on the cellulose chain. It affects properties as follows:

1) *Solubility*. At hydroxyl levels of <1%, solubility is limited. Solubility increases as the hydroxyl content increases. At hydroxyl levels of ca. 5%, CAB is soluble in lower molecular mass alcohols.
2) *Moisture Resistance*. The greater the hydroxyl content, the more hydrophilic are films formed from it.
3) *Toughness*. Fully substituted esters are not as tough and flexible as those with a low hydroxyl content.
4) *Reactivity*. The degree of reactivity increases as hydroxyl content increases. When cross-linked with other resins, the cross-link density of resultant films increases correspondingly.

The higher viscosity form of a particular ester type is associated with a higher molecular mass and longer chain length. Increasing molecular mass of a particular ester (and coatings formulated with it) slightly lowers its solubility and compatibility but does not affect hardness or density. Melting ranges and toughness increase with molecular mass.

Uses. Protective and decorative *coatings for metals* can be formulated as converting or curing systems or as air-drying lacquer systems. Cellulose acetate butyrate is included in many such coatings as a modifying resin to impart specific properties. It can also be used as the primary film-forming resin. Cellulose acetate butyrate is usually included in coatings for metal to accelerate solvent release from the film. This significantly reduces the dry-to-touch time and consequently reduces dirt pickup.

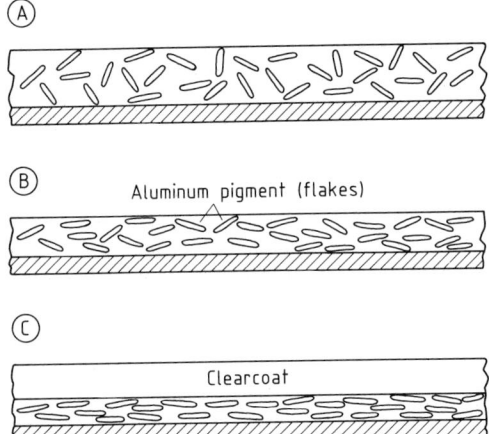

Figure 2.1. The function of CAB in automotive clear-on-base finishes
A) Viscosity of CAB permits application of heavy basecoat; B) Solvent evaporates, film shrinks, and flake orientation begins; C) The CAB prevents solvents in the clear topcoat from redissolving the basecoat

The cellulose resin may be added at levels of 1–5 wt% to improve film leveling and reduce cratering.

Clear-on-base (basecoat–clearcoat) automotive coatings are stoving enamels that are used worldwide. Proper application of the basecoat is critical for obtaining the desired appearance of the coating. Normally, aluminum flake is used for pigmentation and must be oriented parallel to the substrate. This can be achieved by inducing maximum shrinkage of the basecoat during solvent evaporation. Cellulose acetate butyrate (20–30 wt% of total resin solids) greatly assists film shrinkage and metallic flake orientation by increasing coating viscosity following atomization during painting operations (Fig. 2.1). The viscosity increase permits the application of a relatively thick, wet coating without sagging and running. Subsequent solvent evaporation results in film shrinkage causing the aluminum to assume a position relatively parallel to the substrate. The CAB prevents redissolution of the basecoat when the clear topcoat is subsequently applied.

Cellulose acetate butyrate is used in a wide variety of *coatings for wood* because they provide many desirable properties (fast solvent release, flowout, and leveling; excellent spray characteristics, nonyellowing, and cold crack resistance).

The surface of molded plastic parts is often coated to obtain properties that the plastic does not have (e.g., mar resistance, solvent resistance, reduced dirt pickup in barrier coatings). Mar resistance is the ability to withstand scratching and scuffing caused by sliding a rough object or cracking upon impact of a hard object. Cellulose acetate butyrate is used in *coatings for plastic* because of its toughness, low color, color stability, good abrasion resistance, and generally good bonding characteristics.

The development of urethane elastomers has allowed the formulation of tough, abrasion-resistant coatings for many flexible substrates including textiles. One widely used application of *CAB-modified urethane elastomeric systems* is for coating lightweight outdoor backpacking and camping equipment (e.g., portable tents).

Cellulose acetate butyrate is beneficial in *radiation-curing systems*. CAB 551-0.01 with its high butyryl content and low viscosity is soluble in many vinyl monomers used in this area. Levels as low as 1 – 5 wt% provide good flowout and leveling of the coating which often tends to form craters and pull back at edges.

Another very important application is the *dispersion of pigments* that are difficult to disperse (e.g., carbon black, transparent iron oxides, phthalocyanine blue and green, and perylene red). The use of CAB and two-roll milling is the most efficient method of dispersion.

Application. Cellulose acetate butyrate lacquers are usually applied by spraying (air atomization, airless, spinning disk). Application by brush or dip is possible but less commonly used.

2.2.2.2. Cellulose Acetate Propionate

Cellulose acetate propionates (CAP) have the same characteristics as CAB, including high solubility and compatibility with other resins. They also have a very low odor; this is important in printing applications and in reprographic processes. Commercially available products and their typical properties are listed in Table 2.2.

Cellulose acetate propionate is used mainly in printing inks where a low odor is required (e.g., in food packaging). It is also used for coating leather clothing and for printing gift wrapping paper.

2.3. Chlorinated Rubber Coatings

2.3.1. Starting Products [2.8]–[2.10]

To manufacture chlorinated rubber (CR) natural or synthetic rubber such as polyethylene, polypropylene or polyisoprene is degraded to low molecular mass compounds by mastication or addition of radical formers and dissolved in carbon tetrachloride (CTC). Chlorine contents are typically 64–68 wt%. Chlorine gas is introduced into this solution and reacts with the raw material to form CR. The solution is then introduced into boiling water. The CR is precipitated, and the solvent vaporizes. The CR is separated from water, rinsed, dried and ground to form a white powder which is the saleable product. After removal of the water, chlorine, hydrochloric acid and other impurities the solvent is reused.

Commercial Products. Chlorinated rubber is only produced by a few manufacturers. Trade names include Aquaprene (Asahi Denka), Chlortex (Caffaro), Pergut (Bayer) [2.8], Superchlon (Nippon Papers). These products are available in various viscosity grades, whose ranges largely coincide for the aforementioned commercial products (table 2.3).

Table 2.3. Viscosity grades of Pergut, an example of commerical chlorinated rubber product

Designation	Viscosity*, mPa s	Mean molecular mass**
Pergut S 5	3.5–6.5	60.000
Pergut S 10	9.0–13.0	124.000
Pergut S 20	16.0–24.0	160.000
Pergut S 40	33.0–51.0	213.000
Pergut S 90	74.0–110.0	302.000
Pergut S 130	120.0–150.0	327.000
Pergut S 170	130.0–200.0	359.000

* measured in a 18.5% solution in toluene at 23 °C in a Höppler viscometer (DIN 53 015)
** measured by a combination of gel permeation chromatography and viscometry.

As CTC attacks the ozone layer, CTC-emission from modern plants are almost zero. The CTC-contents in chlorinated rubber from these plants is as low as 10 ppm (Bayer). CR from old or low standard plants has a CTC content of up to 10%. This product and products produced with this must be labelled downstream according to the relevant regulations in the different countries.

Recently, an aqueous process has been developed to produce CR. Unfortunately, the CTC generated in this process leads to a CTC-content in CR of 100–500 ppm.

Properties. The high degree of chlorination substantially alters the properties of the starting polymers. A hard, granular, white powder with the following properties is obtained: high resistance to oxidizing agents (e.g. ozone or peroxide), water, inorganic salts, acids, alkalis and gases; good solubility in almost all conventional solvents except water, aliphatic hydrocarbons, and alcohols; good compatibility with a wide range of paint resins and plasticizers; low flammability; fungistatic and bacteriostatic behavior; pigmentability with almost all inorganic pigments and extenders, as well as many organic pigments.

Disadvantages of the pure CR resulting from the high chlorine content include low temperature resistance (60 °C wet, 90 °C dry) on account of elimination of hydrochloric acid. Chlorinated rubber also tends to undergo yellowing where exposed to atmospheric influences.

2.3.2. Chlorinated Rubber Paints

Chlorinated rubber and related chlorinated polymers form coating films by physical drying. Plasticizers or resins have to be added since otherwise brittle films are formed.

Composition. The *binder* consists of ca. 65% chlorinated rubber (usually low-viscosity grades) and ca. 35% plasticizer. Chlorinated paraffins are delivered by ICI (Cereclor) and Clariant, Muttens (CH). Special nonhydrolyzable plasticizers may be added if necessary, e.g., bisphenoxyethylformal (Desavin, Bayer) or resin-modified phenyl alkylsulfonates (Leromoll, Bayer). This composition ensures the "nonhydrolyzability" of the binder (resistance to water, acid, and alkali), which is not the case if hydrolyzable phthalate or adipate plasticizers are used. Nonhydrolyzable resins (e.g., coumarone–indene resins or other hydrocarbon resins) are often added as "extenders".

Red lead has proved outstandingly suitable as a *pigment* for priming coats on steel, and is fully effective in chlorinated rubber coatings. For reasons of environmental protection and occupational health, the use of toxic lead compounds is diminishing. Zinc phosphate is used instead, although it does not have the same corrosion protection effect. Conventional metal pigments (e.g., lead dust, aluminum bronze, and zinc dust) produce diffusionproof coatings with good mechanical properties. In the case of aluminum bronze and zinc dust, stabilization of the paint is required to prevent gelatinization. Iron oxide, chromium oxide, and titanium dioxide pigments, commonly used in the paint industry, are suitable for finishing and topcoats. Zinc oxide, white lead, and lithopone are, however, unsuitable.

All inert minerals are suitable as *extenders*. Carbonate-containing extenders may only be used if no stringent requirements have to be satisfied as regards resistance to water and chemicals.

The choice of *solvent* is practically unlimited. Xylene or other alkylbenzenes are generally recommended. Mixtures of esters and mineral spirit can be used to avoid compulsory warning labels.

Hydrogenated and modified castor oil is used as an *additive* to adjust the viscosity and facilitate application with a brush or spray gun (compressed air or airless); layer thicknesses of ≥ 100 μm are thereby achieved [2.11]–[2.13].

Production. Chlorinated rubber paints are produced by conventional means. The plasticizer, resins, and in some cases a proportion of the chlorinated rubber are first dissolved in the solvent. The high-boiling solvent contained in the formulation is preferred for this step. The hydrogenated castor oil is then added and the resultant mixture is dispersed in a dissolver. In order to obtain optimum "digestion", the instructions of the castor oil supplier should be strictly observed; the temperature should not be allowed to exceed ca. 60 °C. Dispersion is followed by the formation of a paste with the pigments and extenders, and grinding. Conventional apparatus including dissolvers is suitable as grinding equipment; grinding with steel balls should be avoided since the iron dust that is formed can cause the final paint to gelatinize after prolonged storage. The ground material is then combined with the separately prepared chlorinated rubber solution.

Application. Chlorinated rubber paints can be applied with all conventional coating equipment. The suppliers' (manufacturers') instructions must, however, be observed since the coating material (chlorinated rubber paint) is specifically formulated for the recommended application equipment.

Uses. On account of their high water resistance, chlorinated rubber paints are used for underwater coatings on steel and concrete (e.g., water storage vessels, swimming pools, sewage systems, harbor installations, and docks). The chemical resistance is exploited in vessels, tanks, and constructional parts used in mines, chemical plants, etc., in which aqueous solutions of inorganic chemicals are handled. Coatings for concrete require chlorinated rubber as a binder due to the alkalinity of the concrete surface.

The main area of use of chlorinated rubber paints is for underwater coatings on ships (see also Section 11.4). Favorable properties for this application are high water resistance, rapid drying (which is independent of the external temperature in the shipyard), good mutual adhesion of the individual layers, and the fact that old coats of paint can easily be renewed.

2.3.3. Chlorinated Rubber Combination Paints

Composition. Chlorinated rubber combination paints contain a second resin as the property-determining binder. The chlorinated rubber is added to an alkyd resin, acrylic resin, or bituminous substances to improve properties such as drying rate, water resistance, or chemical resistance. This application only accounts for a small proportion of the total chlorinated rubber consumption.

The proportion of chlorinated rubber in the binder varies from 10 to 50 wt% depending on the intended application; plasticizers and/or alkyd resins and/or acrylic resins account for the remainder.

In combinations with bituminous substances the proportion of chlorinated rubber ranges from 1:10 to 10:1. The ratio depends on whether the goal is to improve the bitumen-based coating without any substantial increase in cost, or to reduce the cost of the chlorinated rubber coating. Adhesion is improved but with the disadvantage of darker shades caused by the black bitumen.

Production corresponds to that of pure chlorinated rubber paints (see Section 2.3.2).

Chlorinated Rubber–Alkyd Resin Combinations. In these combinations chlorinated rubber accounts for 25–50% of the binder. Chlorinated rubber is used to increase the drying rate and/or improve the chemical resistance against inorganic chemicals like acidic or basic compounds. These paints also exploit the benefits of the alkyd resin, e.g., good brushability and nonsolubilization. They are used for corrosion protection in industrial plants or marine environments to protect steel, galvanized steel, and aluminum; air-drying or forced-dried industrial paints (e.g., for agricultural machinery); and road marking paints.

Chlorinated Rubber–Acrylic Resin Combinations. Physically drying acrylic resins are used for these combinations. These combinations have the same drying rates as normal chlorinated rubber paints (see Section 2.3.2). They have improved flow properties (particularly when applied by pouring techniques), improved weather resistance (chalking and yellowing), and favorable mechanical properties (adhesion and extensibility). Applications include topcoats for ship superstructures and priming coats on galvanized surfaces.

Combinations with Bituminous Substances. Chlorinated rubber can be combined with bitumen and tars but compatibility has to be checked. Addition of chlorinated rubber reduces thermoplasticity, accelerates drying, and prevents cracking of the final coating in adverse weather conditions, without, however, adversely affecting the good adhesion, water resistance, and chemical resistance of the bituminous substance.

Bituminous coatings reinforced with chlorinated rubber are used in silos, tanning pits, drinking water containers, and on ships' hulls (on the underwater part). Bituminous substances for coatings are supplied as special products that are free from carcinogenic constituents.

2.4. Vinyl Coatings

This section deals with paints based on vinyl resins (including vinyl copolymers) which are synthesized by polymerization of monomers containing terminal $CH_2=CH$ groups. Polyolefins, poly(vinyl halides) and vinyl halide copolymers, poly(vinyl esters), poly(vinyl alcohol), poly(vinyl acetals), poly(vinyl ethers), and polystyrene are discussed. Polyacrylates (acrylic resins) are treated in Section 2.5.

2.4.1. General Properties

Paints and coating materials based on vinyl resins are generally physically drying. Only in a few cases vinyl resins can be chemically cross-linked with other reactants via incorporated reactive groups. The properties of the paints are therefore primarily determined by the chemical and physical nature of the vinyl resin. Despite the large number of available vinyl resins this class of binders has some common features.

All vinyl resins have a linear carbon chain with lateral substituents and exhibit a range of molecular masses. Increasing molecular mass is accompanied by improved mechanical properties, a decrease in solubility, and an increase in the viscosity of their solutions. Vinyl resins of high molecular mass can therefore only be used in the form of dispersions or powders for paint applications. Solvent-containing paints require vinyl resins of considerably lower molecular mass than plastics, since only then a sufficient binder content can be achieved in the viscosity range required for paint application.

The properties of vinyl resins, paints, and coatings are chiefly determined by the nature and number of substituents. The substituents influence the crystallization behavior and thus the properties of interest in paint technology such as the softening range, mechanical properties (film flexibility, cold embrittlement tendency, film hardness), the film-forming temperature in dispersions, solubility, and compatibility with other binders. Chemical behavior also depends on the substituents: ester groups can be hydrolyzed, free carboxyl groups improve adhesion to metals, and hydroxyl groups permit cross-linking with reactants such as isocyanates. Pigment wetting, pigment loading, water swelling capacity, diffusion of water vapor and other gases, solvent retention, and many other phenomena are also largely determined by the substituents.

Copolymerization and the associated introduction of further substituents allows individual properties to be modified. For example, internal plasticization can be achieved by copolymerizing "rigid" monomers with "soft" monomers.

Vinyl paints are produced by conventional techniques (see Chap. 7) and can be applied by all conventional methods (e.g., spraying, brushing, roller coating, and dipping). They are dried at room temperature; heating can be used to shorten the drying process.

2.4.2. Coatings Based on Polyolefins and Polyolefin Derivatives

Polyethylene can only dissolve in hydrocarbons above its melting point. On account of its low solubility it is used in coatings solely in the form of powders and dispersions.

On account of their paraffinic nature, polyethylene coatings are highly resistant to chemicals. They are, however, attacked by strong oxidizing agents. Polyethylene coatings are elastically tough, flexible (even at low temperature), resistant to hot water, and nontoxic.

Polyethylenes of lower molecular mass are added as slip and matting agents to paints and printing inks and can also produce dirt-repellent and abrasion-resistant effects. The wax dispersions are produced by hot dissolution and precipitation in aromatic hydrocarbons. This is not necessary with commercial microcrystalline grades. Aqueous dispersions of polyethylene are important as polishing agents.

Polyisobutenes do not crystallize. Polyisobutenes of low molecular mass are flexible resins, and those of high molecular mass are elastomers. Their use in the paint sector is restricted on account of their aliphatic nature and limited compatibility with other binders. They are employed in combination with hydrocarbons such as paraffins, rubber, and bitumen. They are used as plasticizing components.

Ethylene Copolymers. *Ethylene–vinyl acetate copolymers* differ according to their vinyl acetate content and molecular mass. With increasing vinyl acetate content, compatibility with paraffin waxes decreases, but that with other binders increases. Low molecular mass types containing 25–40% vinyl acetate are readily or sufficiently soluble in solvents. With a 40% vinyl acetate content, they can be combined with polar resins and nitrocellulose. Terpolymers with free carboxyl groups exhibit improved adhesion. Ethylene–vinyl acetate copolymers are primarily added to waxes to improve their properties, but are also used to increase flexibility and adhesion in paints, printing inks and adhesives, and for hot melt coatings. Ethylene–vinyl acetate copolymers have low water vapor and gas permeabilities (barrier effect).

Powder coatings are also formed from ethylene–vinyl acetate copolymers (see Section 3.4). *Copolymers of ethylene with maleic acid (anhydride)* of low molecular mass are water-soluble, form salts, and undergo cross-linking reactions.

Chlorinated Polyethylene and Polypropylene. Totally chlorinated polyethylene and polypropylene have a chlorine content of 64 – 68 %. Their properties largely correspond to those of chlorinated rubber (see Section 2.3). Chlorinated polypropylenes can be used for chemical-resistant and weather-resistant coatings. These binders are important in adhesion priming coats and heat-sealing lacquers for polypropylene foils.

Commercial products include chlorinated polyolefins CP (Eastman) and Hardlen (Toyo Kasei Kogyo).

Chlorosulfonated polyethylene is obtained from polyethylene by simultaneous treatment with chlorine and chlorosulfonic acid. The binder is cross-linked via its sulfonyl group with metal oxides, preferably lead oxide, in the presence of organic acids and sulfur-containing organic accelerators. The paints may be formulated as air- and oven-drying, one- and two-pack systems with a pot life of one to two weeks. Cross-linked films of chlorosulfonated polyethylene have an extremely high chemical resistance, particularly against oxidizing agents (it is used in internal coatings of chromic acid baths).

2.4.3. Poly(Vinyl Halides) and Vinyl Halide Copolymers

2.4.3.1. Poly(Vinyl Chloride) and Vinyl Chloride Copolymers

Poly(vinyl chloride) [*9002-86-2*] (PVC) is sparingly soluble in the solvents used in the paint industry, and so is rarely used as a paint binder. It is used to a significant extent in the form of paste-forming PVC powders in plastisols and organosols. Plastisols are PVC dispersions in plasticizers that also contain stabilizers, extenders, pigments, and processing agents. Organosols additionally contain solvent-soluble binders. Since plastisols and organosols adhere poorly to metals, adhesion promoters are necessary; organosols can also be combined with adhesive resins when applied to metals.

Film formation takes place at gelation temperatures from 160 to 200 °C. The properties of the coating depend on the type and amount of PVC, plasticizer, and extender. A foaming effect can be achieved during gelation by adding blowing agents. Poly(vinyl chloride) is also used in powder coatings, which are applied by fluidized-bed coating and electrostatic spraying (corrosion protection for metal furniture, wire, aluminum front elements for buildings, and road-marking posts). PVC coatings have favorable mechanical properties, a high abrasion resistance, and high chemical resistance.

Commercial products include Ekavyl, Lucovyl (Atochem); Geon (Goodrich); Solvic (Solvay); Vestolit (Hüls); and Vinnolit (Vinnolit).

Chlorinated Poly(Vinyl Chloride). Post-chlorinated PVC combines the advantageous properties of PVC, e.g., good chemical and weather resistance with good solubility in most conventional solvents. Its importance has, however, continually decreased in the paints sector.

Vinyl Chloride Copolymers. Vinyl chloride copolymers can be used in a wide variety of paint technology applications. The solubility of vinyl chloride copolymers is considerably higher than that of the PVC homopolymer. Important examples are copolymers without additional functional groups formed with vinyl acetate, dibutyl maleate, or isobutyl vinyl ether; terpolymers with carboxyl groups formed with dibutyl maleate or vinyl acetate and a dicarboxylic acid; and copolymers and terpolymers with hydroxyl groups formed with hydroxyacrylates or with vinyl acetate and vinyl alcohol.

Commercially available vinyl chloride copolymers differ in composition and molecular mass. They are physically drying binders that undergo film formation by solvent evaporation. (The types with hydroxyl groups can, however, also be used as combination binders in reactive systems. They can be cross-linked, for example, with melamine resins or polyisocyanate resins.) The films are tough, abrasion resistant, thermoplastic, of low flammability, colorless, odorless, tasteless, and physiologically harmless. Flexibility and abrasion resistance improve with increasing molecular mass. Vinyl chloride copolymers exhibit a good water resistance and outstanding resistance to alkalis, dilute mineral acids, salt solutions, oils, fats, greases, alcohols, and gasoline. Paints can readily be formulated with this group of binders and can be applied by conventional methods.

Preferred solvents include ketones, esters, and chlorinated hydrocarbons. Aromatic hydrocarbons have a swelling effect on most vinyl chloride copolymers, but are widely used as diluents. Normally alcohols and aliphatic hydrocarbons do not dissolve vinyl chloride copolymers.

Both monomeric and polymeric plasticizers are suitable for plasticization. Practically all monomeric plasticizers for PVC can be used. Suitable polymeric plasticizers include polyadipates, chlorinated paraffins, carbamide resins, and epoxides. Vinyl chloride copolymers are compatible with most conventional pigments and extenders. Despite their high intrinsic stability, paints based on vinyl chloride copolymers have to be stabilized against dehydrochlorination in the presence of heat and/or UV radiation for some applications. Epoxy compounds are often sufficient for thermal stabilization.

The composition of vinyl chloride copolymers without functional groups influences their solubility behavior and compatibility with other paint binders. For example, copolymers with isobutyl vinyl ether or maleate esters dissolve in aromatic hydrocarbons, whereas copolymers with vinyl acetate merely swell in these solvents. Paint films formed from vinyl chloride copolymers without functional groups are heat sealable on account of their thermoplastic character. Since the films adhere poorly to nonabsorbing substrates such as metals, they are suitable as binders for strippable coatings. On account of their good chemical resistance, vinyl chloride copolymers are also extremely suitable as binders for exterior-use paints, traffic paints, and paper and foil lacquers; their lack of taste and odor means that they can be used as pasteurization-resistant coatings for can interiors.

Vinyl chloride terpolymers containing carboxyl groups adhere extremely well to metals. Due to their special properties (outstanding adhesion to aluminum, good chemical resistance, heat-sealable from ca. 140 °C—the sealing temperatures can be lowered by adding plasticizers) these copolymers are ideal binders for heat-sealable finishes of aluminum foils used in the packaging sector.

Vinyl chloride copolymers containing hydroxyl groups can be used in combination with other binders in reactive systems. The non-cross-linked paint films are thermoplastic and therefore heat-sealable. Cross-linking (e.g., with melamine resins or polyisocyanate resins) lowers the thermoplasticity and improves adhesion and resistance. The hydroxyl groups are also responsible for very good adhesion to organic substrates; these copolymers are therefore used as intermediate layers in marine coatings. In two-component polyurethane lacquers, hydroxyl-containing vinyl chloride copolymers can be used alone or in combination with polyols. In the latter case, lacquer viscosity can be adjusted and initial physical drying can be accelerated. As a result of their outstanding pigment wetting and stabilization properties, their high pigment loading, and compatibility with polyester and polyurethane resins, hydroxyl-containing vinyl chloride copolymers are used as binders for magnetic storage media.

Commercial products include Hostaflex (Vianova); Laroflex, Lutofan (BASF); S-Lec (Sekisui); Ucar, Vinylite (Union Carbide); Vilit (Hüls); and Vinnol (Wacker).

2.4.3.2. Vinylidene Chloride Copolymers

On account of its low thermal stability, poly(vinylidene chloride) is seldom used in paints. Vinylidene chloride copolymers with vinyl chloride, acrylonitrile, or acrylates are mainly employed. These heat-sealable copolymers are efficient gas barriers and have an outstanding resistance to chemicals. They are marketed as solid resins and dispersions. Vinylidene chloride copolymers are mainly used for coating food-packaging foils. They are also important in paint coatings where good chemical resistance is required.

Commercial products include Diofan (BASF), Haloflex (ICI), Ixan (Solvay), and Saran (Dow Chemical).

2.4.3.3. Fluoropolymer Coatings

Organic fluoropolymers have been used in many fields because they have special properties that no other polymers can provide.

Coating is one of the important uses of fluoropolymers, since it enables them to exhibit their characteristics on the surface of a substrate. Some of the conventional fluoropolymers such as polytetrafluoroethylene [*9002-84-0*] (PTFE), tetrafluoroethylene–hexafluoropropylene copolymer [*25067-11-2*] (FEP), and ethylene–tetrafluoroethylene copolymer [*25038-71-5*] (ETFE) have been used as antistick or anticorrosive coatings. Only poly(vinylidene fluoride) [*9002-58-1*] (PVDF) has so far been used in paints. The major difficulties in employing thermoplastic fluoropolymers in paints and coatings result from their poor solubility in organic solvents and also from the necessity to bake them at a rather high temperature (> 200 °C). In recent years, however, novel fluoropolymers with curable characteristics have been developed (one of them was successfully commercialized), mainly as highly weather-resistant paints [2.14], [2.15].

Coatings with Thermoplastic Fluoropolymers. *Poly(vinylidene fluoride)*, PVDF, is the only conventional thermoplastic fluoropolymer that is used as a commercial product for weather-resistant paints. This crystalline polymer is composed of -CH_2CF_2- repeating units; it is soluble in highly polar solvents such as dimethylformamide or dimethylacetamide. Poly(vinylidene fluoride) is usually blended with 20–30 wt% of an acrylic resin such as poly(methyl methacrylate) to improve melt flow behavior at the baking temperature and substrate adhesion. The blended polymer is dispersed in a latent solvent (e.g., isophorone, propylene carbonate, dimethyl phthalate). The dispersion is applied to a substrate and baked at ca. 300 °C for ca. 40–70 s. The weather resistance of the paints exceeds 20 years [2.16]–[2.18].

Commercial products include Kynar (Elf Atochem), Hylar (Ausimont), and Soles (Solvay).

Copolymers of vinylidene fluoride with tetrafluoroethylene or hexafluoropropylene have recently been developed as an air-drying paint mainly for repair coating of the PVDF finish [2.19]–[2.21]. Poly(vinylidene fluoride) is now widely used for coil coating of galvanized iron and aluminum sheets, and as a maintenance-free coating for walls of skyscraper buildings and roofing of industrial constructions.

Polytetrafluoroethylene. An aqueous dispersion of PTFE is produced by emulsion polymerization of tetrafluoroethylene followed by thermal concentration of the latex up to ca. 60 wt%. Since the polymer has a very high melt viscosity and does not adhere to many substrates, the properties of the finish are greatly influenced by pretreatment of the substrate. On physically or chemically pretreated iron or aluminum, the PTFE coating is rather vulnerable to scratching (soft coat). If, however, the treated substrate is first coated with a ceramic-powder primer to form a coarse, hard surface, the subsequently applied PTFE finish becomes tough and scratch-resistant (hard coat). These coatings are used in hot cooking ware (e.g., frying pans) with the advantage of being nonstick and easily lubricated [2.22], [2.23]. Recently, mixtures of PTFE dispersions and heat-resistant hydrocarbon polymers (e.g., polyimide, polyether sulfone, or polyphenylene sulfide) have been developed to improve the poor adhesion of fluoropolymer to a substrate and applied as a primer or one-coat enamel [2.24].

Tetrafluoroethylene Copolymers. Tetrafluoroethylene–hexafluoropropylene copolymer (FEP) and tetrafluoroethylene–perfluoroalkoxyethylene (PFA) are used as dispersion coatings in the same way as PTFE, taking advantage of their low melt viscosity and low viscosity at baking temperature.

Powder coating is another popular technique in fluoropolymer coating. Tetrafluoroethylene ethylene copolymer (ETFE) has a highly alternating sequence with a low melting point (280 °C) and low melt viscosity. Tetrafluoroethylene–ethylene copolymer powder has a better melt processability than PTFE; electrostatic coating gives thick finishes without pinholes that have excellent anticorrosive and antistick characteristics. Tetrafluoroethylene–perfluoroalkoxyethylene is also applied as a powder coating, exploiting its low melt viscosity in the same way as ETFE.

Coatings with Amorphous Cyclic Perfluoropolymers. Recently, novel cyclic perfluoropolymers have been reported [2.24a].

$$\underset{\text{CYTOP (Asahi glass)}}{-(CF_2-CF-CF-CF_2)_n-\atop \underset{CF_2}{O\diagdown\diagup CF_2}} \qquad \underset{\text{Teflon AF (du Pont)}}{-(CF_2-CF_2)_x-(CF-CF)_y-\atop \underset{CF_3\;\;CF_3}{O\;\;\;\;\;\;O\atop \diagdown C\diagup}} \qquad [2.24a]$$

These polymers have some unique properties (e.g., solubility in specific perfluoro solvents, high tranparency, low refractive index, low dielectric constant and low water absorption) due to amorphous morphology attributed to cyclic structure. They can be coated by various methods and form uniform, pinholeless finishes.

Excellent electrical and optical properties of these polymers enable to apply to protecting coat for electric devices and anti-reflectin coat for display device.

Commercial Products include CYTOP (Asahi Glass) and Teflon AF (duPont).

Coatings with Curable Fluoropolymers. In order to facilitate the application of fluoropolymers, extensive studies have been performed on curable fluoropolymers. A block copolymer containing 65% vinylidene fluoride, 25% tetrafluoroethylene, and 10% vinyl ester forms a highly weatherable, strongly adhering, solventborne coating on metals or cellulosic materials after photoinitiated cross-linking (UV curing) [2.25], [2.26].

Hydroxyl-containing fluoropolymers made of fluoroolefin and hydroxyalkyl vinyl ether, hexamethoxymethylmelamine, and silica form highly cross-linkable liquid mixtures and can be applied as an excellent scratch-resistant coating on plastic objects [2.27], [2.28].

Although these fluoropolymers are soluble in organic solvents and can be applied at a rather low temperature ($< 200\,°C$), characteristic properties of the coating seem to be too specialized to be widely used in the coating area.

A fluoroolefin–vinyl ether terpolymer (Lumiflon) has recently been developed [2.14]. This polymer is an amorphous, alternating copolymer of a fluoroolefin with several vinyl monomers (Fig. 2.2). The alternating sequence is responsible for the high performance of the resultant finish. The combination of vinyl monomers provides the polymer with various properties necessary for a coating material.

The outstanding characteristics of this polymer as a coating material are its excellent weatherability and its ease of handling and processing [2.23], [2.29]–[2.31]. The hydroxyalkyl groups in the polymer react with polyisocyanates at room temperature and with melamine resin or blocked isocyanates at higher temperature. Lumiflon can therefore be formulated for both on-site coatings that are cured at ambient temperature and for thermoset coatings in the factory. Practical application to plastics, buildings, various architectural structures, bridges, and automotives is now proceeding and the market is expanding annually. Recently, two kinds of JIS (Japanese Industrial Standard) for fluoropolymer coatings were established in Japan. One is for architectural coatings [JIS K 5658] and the other for heavy duty coatings [JIS K 5659].

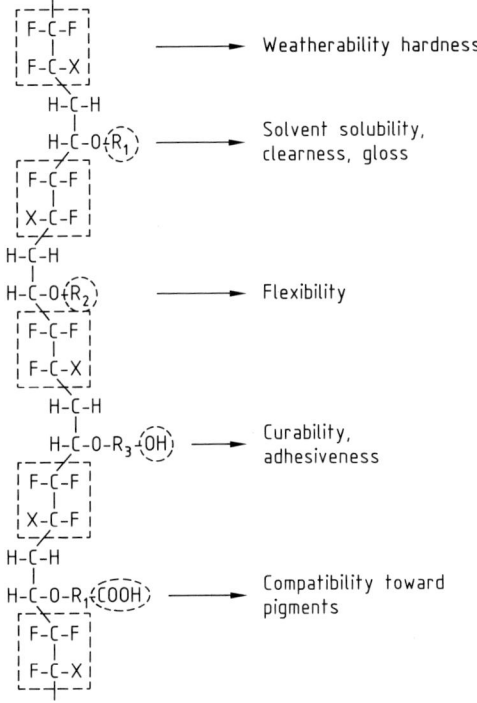

Figure 2.2. Structure of Lumiflon polymer

Several fluoroolefin–vinyl terpolymers have been developed based on the Lumiflon polymer [2.32]–[2.35]:

$$\mathrm{+(CF_2CF-CH_2CH)_{\mathit{n}}(CF_2CF-CH_2CH)_{\mathit{m}}}\atop{\mathrm{XOR^1XOR^2}}$$
[2.14]
[2.29]

$$\mathrm{+(CF_2CF-CH_2CH)_{\mathit{n}}(CF_2CF-CH_2CH)_{\mathit{m}}}\atop{\mathrm{XOR^1XOCR^2}\atop\|\atop\mathrm{O}}$$
[2.32]

$$\mathrm{+(CF_2CF-CH_2CH)_{\mathit{n}}(CF_2CF-CH_2CH)_{\mathit{m}}}\atop{\mathrm{XOR^1XCH_2OR^2}}$$
[2.33]

$$\mathrm{+(CF_2CF-CH_2CH)_{\mathit{n}}(CF_2CF-CH_2CH)_{\mathit{m}}}\atop{\mathrm{XOR^1XSi-OR^2}\atop\mathrm{OR^3}}$$
[2.34]

$$\mathrm{+(CF_2CF-CH_2CH)_{\mathit{n}}(CF_2CF-CH_2CH)_{\mathit{m}}}\atop{\mathrm{XOR^1XOCH_2CF_2CF_2H}}$$
[2.35]

$$\mathrm{(CF_2-CF-CH_2-CH)_{\mathit{n}}(CF_2-CF-\overset{CH_3}{\overset{|}{C}H}-CH)_{\mathit{m}}}\atop{\mathrm{XOCR^1XCO_2R^2}\atop\|\atop\mathrm{O}}$$
[2.35a]

Recently waterborne coatings based on fluoroolefin–vinyl ether terpolymers have been reported, in which a macromonomer with a hydrophilic side chain was copolymerized with fluoroolefin [2.36], [2.36a].

Commercial products include Lumiflon (Asahi Glass), Cefralcoat (Central Glass), Fluonate (Dainippon Ink and Chemicals) and Zaflon (Toagosei Chemical Industry).

A unique fluoroepoxy compound is liquid at ambient temperature and can be cured by incorporation of suitable agents (e.g., amino silicone compounds) [2.37]. It forms finishes with low friction, oil and water repellency, and antifouling properties [2.38], [2.39].

2.4.4. Poly(Vinyl Esters)

Poly(vinyl esters) used in paints and adhesives are available as homopolymers and copolymers in the form of solid resins, solutions, and dispersions.

2.4.4.1. Solid Resins

Poly(vinyl acetate) and vinyl acetate copolymers with crotonic acid, vinyl laurate, and dibutyl maleate are important solid resins; some are available in solution.

Poly(Vinyl Acetate). Commercial grades of poly(vinyl acetate) [9003-20-7] differ in molecular mass and therefore in their viscosity when dissolved. Poly(vinyl acetate) is a physically drying binder that forms transparent, lightfast films with good hardness, gloss, and adhesive strength. It dissolves in lower alcohols, glycols, esters, ketones, and toluene. Thanks to their elastic properties, paints and adhesives based on poly(vinyl acetate) generally require little plasticizer. Plasticizers not only increase elasticity but also lower the glass transition temperature, which adversely affects blocking stability and can lead to sticky surfaces.

The neutral behavior of poly(vinyl acetate) allows the use of all conventional pigments. Poly(vinyl acetate) is highly compatible with ester-soluble nitrocellulose and improves the adhesion and lightfastness of the latter. Poly(vinyl acetate) can also be readily combined with phenolic resins, ketone resins, and colophony resins.

On account of their excellent lightfastness, high gloss, and physiological harmlessness, poly(vinyl acetates) are used as binders in nitrocellulose combination paints for paper, labels, cardboard, wood, leather, and certain plastics. Low molecular mass grades are used in impregnation coatings that are resistant to oil, grease, and moisture, and in priming coats on cardboard or masonry. On account of the thermoplastic properties, heat-sealable lacquers can also be formulated. Poly(vinyl acetate) is an extremely important raw material in the adhesives industry.

Commercial products include Gelva (Monsanto), Mowilith (Clariant), Rhodopas (Rhône–Poulenc), Vinac (Air Products), Vinavil (Montedison), and Vinnapas (Wacker).

Vinyl Acetate Copolymers. Copolymerization of vinyl acetate with other monomers allows specific improvement of certain properties. Copolymers generally exhibit a broader compatibility than the homopolymer. For example, softer, permanently flexible polymers with a lower water uptake and higher alkali resistance are obtained by polymerizing vinyl acetate with vinyl laurate. On account of their thermoplastic properties these copolymers are used in heat-sealable finishes on paper, cardboard, and aluminum foil. In cellulose nitrate lacquers they increase adhesion, lightfastness, and the body fullness of the paint film. They can also be used for priming coats and for stabilizing porous or absorbent substrates.

Copolymers of vinyl acetate and dibutyl maleate are used in adhesives and as binders in deep sealers. Copolymers of vinyl acetate and crotonic acid dissolve in aqueous alkalis with salt formation. The carboxyl groups confer better metal adhesion.

These copolymers are also used as a raw material for wash-off adhesives, textile finishing agents, and marking inks with extremely good adhesion on a wide variety of substrates.

Commercial products include Mowilith (Clariant) and Vinnapas (Wacker).

2.4.4.2. Dispersions

Poly(vinyl ester) dispersions are quantitatively more important than solid resins. Homopolymer and copolymer dispersions are used for binders in emulsion (dispersion) paints, plastic-bonded plasters, and water-thinnable adhesives. Poly(vinyl acetate) dispersions are less important than vinyl acetate copolymer dispersions. The most important comonomers of vinyl acetate are vinyl laurate, dibutyl maleate, Versatic Acid esters (VeoVa, Shell), ethylene, vinyl chloride, and butyl acrylate. Poly(vinyl propionate) and copolymers of vinyl propionate with butyl acrylate, styrene, or vinyl chloride are also marketed and used as dispersions.

The properties of the films and coatings obtained from dispersions depend primarily on the polymer composition, stabilization system, and particle size. The type of polymer determines film-forming properties, resistance to hydrolysis, and to some extent the water resistance, flammability, and mechanical properties such as flexibility, elongation at break, and tensile strength. The stabilization system [protective colloids such as poly(vinyl alcohol) and cellulose derivatives or emulsifiers] influences behavior under mechanical stress, pigment and extender compatibility, pigment loading, water resistance, and rheology.

Film formation in polymer dispersions occurs as a result of the agglomeration and fusion of the polymer particles (diameter 100–1000 nm) after evaporation of the water. The minimum temperature at which the particles fuse to form a film (minimum film-forming temperature) is related to the glass transition temperature. It can be lowered to facilitate paint application by adding plasticizers or solvents.

Poly(vinyl acetate) dispersions form lightfast, dry, hard, brittle films. Plasticizers therefore have to be used (external plasticization), which are, however, volatile and lead to embrittlement of the films after a relatively short time. Internally plasticized dispersions of copolymers of vinyl acetate with vinyl laurate, butyl maleate, Versatic Acid esters, or ethylene form permanently flexible, nonaging films that are not, however, always sufficiently resistant to hydrolysis. Terpolymer (vinyl acetate – ethylene – vinyl chloride) dispersions form films that are more resistant to hydrolysis than homopolymer and copolymer dispersions. The films also have a higher mechanical strength and lower flammability. The glass transition temperature of the terpolymer can be varied within wide limits and properties can be matched to requirements by using a suitable choice of comonomers. The same is true of vinyl propionate copolymer dispersions.

Poly(vinyl ester) dispersions are important binders for indoor (conventional, solvent-free) and outdoor paints, special coatings, and textured finishes. Special types are used for wood paints and for coating paper and cardboard. Poly(vinyl ester) dispersions are also important in the adhesives and textile finishing industries.

Commercial products include Airflex (Air Products), Dilexo (Condea), Elotex (Ebnöther), Emultex (Revertex), Ertimul (ERT), Mowilith (Clariant), (BASF), Ravemul (ANIC), Rhodopas (Rhône – Poulenc), Ubatol (Cray Valley Kunstharze), Ucar (Union Carbide), Vinamul (Vinyl Products), Viking (Kirkless Chemicals), Vinnapas (Wacker), and Walpol (Reichhold).

2.4.5. Poly(Vinyl Alcohol)

Poly(vinyl alcohol) [*9002-89-5*] is obtained by hydrolysis of poly(vinyl acetate). Commercial grades of poly(vinyl alcohol) differ in the degree of polymerization (molecular mass) and degree of hydrolysis [residual poly(vinyl acetate) content].

The water solubility of these polymers declines with decreasing hydrolysis and increasing molecular mass. Poly(vinyl alcohols) containing up to 20 wt% vinyl acetate are insoluble in organic solvents. Poly(vinyl alcohol) is resistant to oils, fats, greases, and waxes. Films obtained from aqueous solutions are clear and colorless, have a high crack resistance, exhibit good lightfastness, and are impermeable to water vapor.

Water-soluble organic compounds with highly polar groups and a high boiling point may be used as plasticizers. Poly(vinyl alcohol) has a good pigment binding capacity and is compatible with the pigments and extenders conventionally used in the paint industry. Resistance to water can be improved by reacting the hydroxyl groups with aldehydes or by cross-linking with urea resins or melamine resins in the presence of acid catalysts.

Poly(vinyl alcohol) is used in the coating sector (e.g., as a thickening agent for aqueous systems), in adhesives, for finishing paper and cardboard, for coating paper, as a binder for strippable coatings, and as a protective colloid for dispersions.

Commercial products include Airvol (Air Products), Elvanol (Du Pont), Ertivinol (ERT), Gohsenol (Nippon Gohsei), Mowiol (Clariant), Polyviol (Wacker), Poval (Kuraray, Denka, Shinet-Su), and Rhodoviol (Rhône – Poulenc).

2.4.6. Poly(Vinyl Acetals)

Poly(vinyl acetals) are produced by reacting poly(vinyl alcohol) with aldehydes. Since acetalation does not proceed quantitatively and a proportion of the acetyl groups remains after hydrolysis of poly(vinyl acetate) to poly(vinyl alcohol), poly(vinyl acetals) may be regarded as terpolymers of vinyl alcohol, vinyl acetal, and vinyl acetate.

Poly(vinyl formals) and poly(vinyl butyrals) are of importance in the coating industry. Commercial products differ in the degree of polymerization and acetalation, but especially in the residual poly(vinyl alcohol) content.

Poly(Vinyl Formals). Poly(vinyl formals) [*63148-64-1*] are physically drying binders. The films exhibit high resistance to chemicals and favorable mechanical properties. Relatively powerful organic solvents are required for dissolution (e.g., chlorinated hydrocarbons, cyclic ethers, and mixtures of alcohols and aromatic hydrocarbons). The free hydroxyl groups of the polymers confer good pigment wetting and adhesion to various substrates. Poly(vinyl formals) can be cross-linked via the hydroxyl groups; this improves chemical resistance.

Poly(vinyl formals) are compatible with a range of plasticizers (that are used, for example, to improve the low-temperature flexibility) and binders, in particular polyisocyanate, phenolic, epoxy, and melamine resins.

The most important use of poly(vinyl formals) is for coating wire. They are also used for coating magnetic recording media and as adhesives for metal–metal composites used in construction.

Commercial products include Polyvinylformal (Siva) and Vinylec F (Chisso).

Poly(Vinyl Butyrals). Poly(vinyl butyrals) [*63148-65-2*] are physically drying binders. On account of their free hydroxyl groups they can also be used in reactive systems. Non-cross-linked films are thermoplastic and therefore heat-sealable. As well as having a high heat resistance and lightfastness, poly(vinyl butyrals) also exhibit good resistance to fats, greases, oils, bitumen, and gasoline. Outstanding properties are their good adhesion to metals, glass, absorbent substrates, and plastics foils, as well as their excellent pigment wetting properties. Alcohols are preferred as solvents; esters and ketones have a somewhat lower solvent power.

Poly(vinyl butyrals) form very flexible films. Plasticizers can be used to improve cold flexibility and reduce solution viscosity. Poly(vinyl butyrals) are also compatible with a wide range of resins (e.g., epoxy, urea, melamine, phenolic, polyisocyanate, nitrocellulose, polyethyleneimines, and ketone resins). The resistance of poly(vinyl butyral) to chemicals can be improved by cross-linking. As an additive poly(vinyl butyral) reduces the brittleness of highly cross-linked coatings and improves the adhesion, particularly to metals. Polymer compatibility, the solubility, and plasticizer compatibility depend on molecular mass and polarity (degree of acetalation). Poly(vinyl butyrals) are compatible with the pigments and extenders normally used in the paint industry. The free hydroxyl groups confer outstanding pigment wetting.

Poly(vinyl butyrals) are an important class of binders. One of their main uses is in priming coats for corrosion protection. Poly(vinyl butyrals) are employed together with corrosion protection pigments and phosphoric acid as adhesion priming coats (wash primers); combination with a suitable finishing coat confers protection against corrosion and subsurface rusting in metals. Reactive primers or reinforced primers are obtained by combining poly(vinyl butyrals) with phenolic resins that contain corrosion protection pigments and phosphoric acid. On account of their very high solubility in solvents such as ethanol, their good flow properties, and excellent pigment wetting, poly(vinyl butyrals) are used for formulating flexographic and gravure printing inks for food packaging. High molecular mass grades are used to produce glass composites (laminated safety glass) on account of their lightfastness and good adhesive power. Poly(vinyl butyrals) are also important as binders for primer sealers (e.g., to prevent the migration of bitumen) and as a temporary binder in ceramics production.

Commercial products include Butvar (Monsanto), Denka Butyral (Denki Kagaku), Mowital (Clariant), Pioloform B (Wacker), and S-Lec-B (Sekisui).

2.4.7. Poly(Vinyl Ethers)

Poly(vinyl ethers) formed from methyl ethyl ether or isobutyl ether are used as soft plasticizing resins. Solubility and compatibility depend on the alkyl group. In the paint sector, poly(vinyl ethers) are used mainly as plasticizing and in some cases as adhesion-improving resins for chlorine-containing binders, styrene polymers, nitrocellulose, and brittle resins. An example of a commercial product is Lutonal (BASF).

2.4.8. Polystyrene and Styrene Copolymers

Polystyrene Dispersions. On account of their glass transition temperature (T_g) of ca. 100 °C, polystyrene dispersions do not form films at room temperature. These rigid polymers can only be applied with means of heat drying (e.g., to stiffen fabrics and nonwovens). Film formation is not required in agents used to protect floor coverings and paper coatings (plastic pigments); in this case polystyrene is therefore applied in the form of a dispersion at room temperature.

Styrene Copolymer Dispersions. The T_g and hardness of polystyrene can be adjusted over a wide temperature range by copolymerization of styrene with soft monomers such as butadiene and acrylate esters. Styrene–butadiene (SB) dispersions are quantitatively the most important. With a styrene–butadiene weight ratio of 85:15 the T_g is ca. 80 °C, at a ratio of 45:55 the T_g is ca. −25 °C. On account of the cross-linking capability of butadiene, SB copolymers are not thermoplastics, but elastomers. Elasticity can be modified by controlling the molecular mass and degree of cross-linking.

Styrene–butadiene dispersions are generally stabilized with anionic, or anionic and nonionic emulsifiers. Carboxylation (incorporation of a small proportion of unsaturated carboxylic acids) of SB dispersions increases their stability and improves adhesion to various substrates. Almost all SB dispersions used in the coating sector are carboxylated.

Styrene–butadiene dispersions may undergo oxidative post-cross-linking via the double bond in the butadiene unit. Uncontrolled oxidation leads to embrittlement and, finally, to breakdown of the binder (chalking). The dispersions are thus unable to satisfy stringent requirements regarding color stability and UV resistance.

In conventional exterior-use paints SB dispersions have largely been replaced by styrene–acrylate dispersions, and their use is now restricted to special applications (corrosion protection primers, wood primers, mortar modification) where low film permeability to gases, water vapor, etc., and complete resistance of the polymer to hydrolysis are necessary. In order to achieve a uniform surface and thus improve printability, paper and card are coated with paper-coating colors. Carboxylated SB dispersions are used in these paints as binders.

The many uses of SB copolymer dispersions in the textile sector include tufting, textile, and carpet coatings; needle felt and card web reinforcement; textile finishes; and impregnation.

In addition to butadiene, other monomers can be used for internal plasticization of polystyrene. In the paper sector, styrene–butadiene–acrylate terpolymers have advantages in special applications.

Soluble Polystyrene. Styrene homopolymers (generally in admixture with other binders) are used in special applications such as zinc dust primers and bronze lacquers, in paints for mineral substrates and paper, as well as in adhesives. Polystyrene and poly(α-methylstyrene) have good water resistance, low permeability, resistance to chemicals and light, and also dry quickly. Suitable solvents include aliphatic and aromatic hydrocarbons, in some cases containing ketones or esters.

Soluble Styrene Copolymers. Copolymers internally plasticized by acrylate esters, butadiene, maleate esters, or acrylonitrile are of greater importance than the homopolymers. The proportion, nature, and molecular mass of the comonomer determine the softening point, solubility, flexibility, drying rate, and resistance to water, UV light, and chemicals. Incorporation of butadiene decreases lightfastness, while acrylic acid confers solubility in alcohol. The range of potential uses extends from soft resins for adhesives and paper, to rigid resins for paints and hot-melt adhesives. Styrene– and vinyltoluene–acrylate polymers are used particularly in exterior-use paints, paints for concrete and road-marking, metal lacquers, and printing inks. Styrene–allyl alcohol copolymers may also be used as cross-linkable polyol components.

Commercial Products. *Soluble systems* include Amoco Resins (Amoco); Hercoflex, Kristaflex, Piccolastic, Piccotex (Hercules); Pliolite, Pliowag (Goodyear); RJ-Products (Montesano); Supraval (BASF).

Dispersions. The main suppliers of SB dispersions are: BASF, Bayer, Doverstrand, Dow Chemical, Enichem, Hüls, Rhône–Poulenc, Synthomer. Commercial coating products include Butonal, Styrofan (BASF); Dow-Latices (Dow Chemical); Lipaton (Hüls); Rhodopas (Rhône–Poulenc); Synthomer-Latices (Synthomer).

2.5. Acrylic Coatings

General Properties. Paints containing acrylic resins as binders have been known ince the 1930s. They are now one of the largest product classes in the paint and coatings sector.

Polyacrylates as binders consist of copolymers of acrylate and methacrylate esters. Other unsaturated monomers (e.g., styrene and vinyltoluene) may also be incorporated, but usually to a lesser extent. Copolymers formed exclusively from acrylates and/or methacrylates are termed straight acrylics. The comonomers differ as regards the alcohol residues of the ester group, which also allow incorporation of additional functional groups. Choice of suitable monomers allows wide variation of the physical and chemical properties of the resulting polymer. Hydrophilicity, hydrophobicity, acid–base properties as well as T_g can be adjusted; resins containing hydroxyl, amine, epoxy, or isocyanate groups can also be produced.

The resin products may be solids, solutions in organic solvents or water, emulsions, or dispersions.

Acrylate resins have several advantages over other paint binders [2.40]–[2.44]:

1) Polyacrylates are only slightly attacked by chemicals, and confer a high degree of resistance to paints
2) Polyacrylates are colorless, transparent, and do not yellow, even after prolonged thermal stress
3) Polyacrylates do not absorb above 300 nm and are therefore not degraded by UV radiation (as long as they do not contain styrene or similar aromatic compounds)
4) Polyacrylates do not have unstable double bonds
5) Polyacrylates have outstanding gloss and gloss retention
6) Acrylates, and especially methacrylates, are stable to hydrolysis

The following properties of the coating can be ascribed to individual monomers [2.41]. Methyl methacrylate promotes weather resistance, lightfastness, hardness, gloss, and gloss retention. Styrene increases hardness and resistance to water, chemicals, and salt spray, but reduces lightfastness and gloss retention. Alkyl acrylates and alkyl methacrylates impart flexibility and hydrophobicity, while acrylic acid and methacrylic acid improve adhesion to metals.

The increasing importance of environmental considerations places new requirements on paint resins and has broadened the range of paint systems. Paints are now required that have a low solvent content (medium-solids, high-solids coatings) or are solvent-free (powder coatings), that can be adjusted by dilution with water (waterborne paints), and that are thermoplastic or capable of undergoing cross-linking. All of these properties must be obtained via the polymer structure of the binders. Important parameters are described below.

The *glass transition temperature* (T_g) affects adhesion, flaking, and peeling from the substrate, crack formation, and resistance to impact shock [2.44]. In acrylates adjustment of T_g is achieved relatively easily e.g. via the ratio of methyl methacrylate

(T_g of the homopolymer $+105\,°C$) to *n*-butyl acrylate (T_g of the homopolymer $-54\,°C$) [2.40], [2.41], [2.45]. The T_g also influences properties of dispersions [2.45] and the viscosity of solutions [2.46]–[2.48]. A high T_g value is associated with a faster drying rate [2.49]. In the low molecular mass range (< ca. 6000), which is of interest particularly for high-solids paints, the T_g depends on the molecular mass. Subsequent cross-linking leads to an increase of T_g which is dependent on the cross-linking density.

The *styrene content* in the binder reduces resistance to yellowing and weathering [2.49], but improves resistance to chemicals, hydrophobic properties, adhesion, and pigment wetting. Styrene is therefore largely avoided in topcoat paints for exterior use and in clearcoats.

The development of low-solvent (high-solids) paints requires resins with a very low *viscosity*. The principal viscosity-determining parameters for such binders are the molecular mass and molecular mass distribution [2.45]–[2.47], [2.50]–[2.53]. Oligomers with a molecular mass of ca. 1000–3000 are required for high-solids paints [2.48], [2.54]. An acrylate binder with a molecular mass of 100 000 can be processed to form a paint with 12.5% solids content at the application viscosity; a molecular mass of ca. 6000 results in a paint with 50% solids content [2.46]. A narrow molecular mass distribution is beneficial in achieving low viscosity [2.45], [2.50]. However, the mechanical properties of a paint are favored by a high molecular mass [2.44]. Low molecular mass binders that cross-link after application are therefore used exclusively for high-solids paints. When it is applied, the paint contains low-viscosity oligomers; a highly polymeric system is formed after cross-linking and curing [2.48], [2.50]. Further possibilities of reducing the viscosity include specific interactions between the binder molecules [2.55] and choice of a low-viscosity solvent [2.47] that does not interact significantly with the resin [2.48]. The melt viscosity is an especially important criterion in powder coatings; acrylic resins have disadvantages in this respect when compared with polyesters [2.45].

Incorporation of functional groups in the polymer skeleton is necessary for the *production of dispersions*. Most water-dispersable systems are polymers with free carboxyl groups. Water thinnability is achieved by neutralizing the acid groups with aqueous alkali or amines. Binders may also contain basic nitrogen-containing groups; dispersion can then occur after neutralization (e.g., with acetic or lactic acid). Since the viscosity of dispersions is very low irrespective of the molecular mass, polymers of very high molecular mass are generally used; dispersions are therefore ideal for physically drying coatings. Crosslinking can be achieved by incorporating functional groups.

Solvent emission from paints can be reduced without lowering the molecular mass by using *nonaqueous dispersions* (NAD) (see Section 3.6). Acrylates have been described as binders for NADs, but apart from a low viscosity they offer few advantages over conventional coatings [2.50] and moreover must compete with high-solids paints and powder coatings [2.56].

Cross-Linking of Acrylate Resins. In contrast to thermoplastic polymers, cross-linked polymers are insoluble and are also harder and more resistant to chemicals; these properties are extremely important for high-grade coatings. Cross-linking reactions became important in the 1950s with the introduction of acrylic resins in the automotive sector.

A further impetus was given by increasingly stringent environmental legislation; lower solvent contents and the replacement of conventional solventborne paints by medium-solids and high-solids paints mean that the molecular mass of the binders had to be lowered to a range where the required paint properties (e.g., film formation, hardness, and flexibility) no longer exist. These properties must be obtained by increasing the molecular mass by cross-linking after application. Chemical reaction after application also provides advantages for high molecular mass dispersions; the glass transition temperature and film hardness are increased.

A widely used method for cross-linking paint films consists of reacting hydroxyl-containing acrylates with melamine–formaldehyde resins or urea–formaldehyde resins [2.40], [2.53], [2.57]. The hydroxyl-containing acrylates are prepared using comonomers such as hydroxyethyl (meth)acrylate or butanediol monoacrylate. Amino resins are to some extent self-cross-linking, but also form cross-links with acrylates via the hydroxyl groups [2.52]. Cross-linking can be effected by curing at ca. 130 °C or by acid catalysis. The paints exhibit outstanding gloss and weather resistance [2.58].

A second important cross-linking method is the combination of hydroxyl-containing acrylates with polyisocyanates as hardener. This mixture forms cross-links at room temperature and must therefore be produced and stored as a two-component (two-pack) system [2.40]. Reaction between aromatic isocyanates and the hydroxyl-containing acrylates proceeds extremely rapidly; as aliphatic isocyanates react slowly, the reaction has to be catalyzed (e.g., with dibutyl tin dilaurate, amines, or acids) [2.59]. The properties of such polyurethane paints (see also Section 2.9) are superior to those of most other coating materials, and they are used in an increasing range of applications. One-pack polyurethane paints are also available and are based on hydroxyl-containing acrylates; blocked isocyanates are used as hardeners. Such systems generally require relatively high stoving temperature (> 150 °C).

A third group of cross-linking reactions involves acrylic resins containing free carboxylic acid groups. Polyepoxides are mainly used as cross-linking agents in paints dissolved in organic solvents or powder coatings. This combination is superior to other combinations (e.g., isocyanate cross-linking or melamine resins) as regards hardness and resistance to detergents, lyes, and solvents. Extremely high stoving temperatures (≥ 200 °C) are required; the temperature can be reduced to ca. 120–150 °C if tetrabutylammonium iodide or tertiary amines are used as catalysts [2.57]–[2.60]. Use of catalysts is, however, offset by a low storage stability (only a few weeks).

If less stringent requirements with regard to chemical resistance, abrasion, and hardness are placed on the cross-linking density, carboxyl-containing acrylates can also be cross-linked by salt formation (e.g., with diamines [2.44] or metal complexes [2.61]). This procedure is widely employed, particularly for aqueous dispersions. Cross-linking with bisoxazolines has also been reported [2.45]. Epoxy groups can be incorporated into the binder via glycidyl (meth)acrylate and cross-linked with dicarboxylic acids [2.40].

Aqueous acrylic dispersions are being increasingly used for wood coatings or corrosion protection coatings. These paints are often not baked; instead, mechanical properties are improved by cross-linking at room temperature. Aziridines or bishydrazides are generally used as the cross-linking agent and are mixed with the dispersion after the production process.

Many other cross-linking processes exist that are not widely adopted or have only recently been developed. Cross-linking of epoxy-containing acrylates with amino resins and reactions with polysulfonazides have been reported [2.57]. Reactions of unsaturated acrylic groups with ketimines and reaction of acetoacetate-containing polymers with ketimines should also be mentioned [2.52], [2.61 a].

An alternative way of curing paints is to produce self-cross-linking acrylic resins that react internally at elevated temperature without the addition of an external cross-linking agent. Such systems have advantages as regards resistance to chemicals, hardness, and elasticity, but are less variable in their formulation and can present problems as regards storage stability. Furthermore, in order to achieve a sufficiently high degree of cross-linking a minimum molecular mass is required that is higher than that of resins which do not self-cross-link; high-solids paints accordingly cannot be produced with such systems.

Application. Acrylic paints are used in many different areas and applied by all commonly used methods. Recent developments (low-solvent paints and aqueous dispersions) require special formulation.

Acrylic paints with a low-solvent content (medium- and high-solids systems) present some difficulties in application. Only a very small amount of solvent evaporates from the wet film, the increase in viscosity during drying is slight, and the paint remains liquid with a low viscosity for a long time. If an excessively thick coating is applied, this can easily lead to sagging and evaporation is also hindered; control of the layer thickness therefore becomes difficult. During stoving, viscosity initially falls; viscosity and thus paint stability only increase after initiation of the cross-linking reaction and the associated increase in molecular mass. The rheology of high-solids paints must therefore be finely controlled with appropriate additives during application and stoving. These additives act as thixotropic agents and ensure good leveling during application without adversely affecting the stability of thick layers [2.47], [2.52]. Sagging can also be avoided by applying the paint at elevated temperature (e.g., 50–60 °C). The low molecular mass of high-solids paints also affects the stoving process. The cure window (i.e., the temperature and time ranges used for stoving) is therefore smaller the lower the molecular mass of the binder [2.53]. Paints produced from carboxyl-containing acrylic resins and diepoxides as cross-linking agents exhibit better rheological behavior during application and stoving, presumably on account of the reversible formation of intermolecular hydrogen bonds [2.60].

Problems are also associated with *waterborne acrylic paints*. Water has an evaporation enthalpy of 2.26 kJ/g and thus requires substantially more energy for drying than organic solvents (evaporation enthalpies ca. 500 J/g). Pitting and blistering often occur at stoving temperatures above 100 °C. In physically drying paints the drying behavior depends strongly on the atmospheric humidity and the application zone often has to be air-conditioned. The stability of acrylate dispersions depends on a delicate equilibrium between factors such as pH, emulsifier content, and sedimentation stability [2.61 a]. The formulation of a paint dispersion (emulsion paint) therefore requires additional components (e.g., amines as solubility promoters, dispersants, defoamers, thickening agents) and is much more complicated than that of a conventional solvent-based paint [2.46]. Aqueous dispersions generally contain a few per cent of high-boiling solvents which act as temporary plasticizers and lower the minimum film-forming temperature, thus allowing film formation to occur. A film of sufficient hardness is obtained only after complete evaporation of the solvent, which may take up to several days. The film may, however, be somewhat hydrophilic due to the presence of carboxyl groups and emulsifier residues; this reduces its water resistance, gloss, and gloss retention.

Uses. The largest application sector for acrylate binders are emulsion paints for ceilings, walls, and building fronts. These emulsion paints are generally physically drying and only contain a small amount of binder; the main constituents are pigments and extenders. Acrylate dispersions for paints have a good water vapor permeability and good water resistance.

Acrylic resins have been used in the automotive sector since 1957. They are now important binders in automotive finishes and topcoats, and have replaced alkyd resins in nearly all cases. Advantages for automotive finishes and coatings are high transparency, weather resistance, gloss retention, and yellowing stability [2.40]. Automotive finishes are always cross-linked; melamine resins are generally used as hardeners [2.44], polyisocyanates are being increasingly used for clearcoats [2.52]. Automotive fillers and metallic basecoats may be formulated with acrylate dispersions to reduce solvent emission. Solvent-containing systems are, however, still indispensable in the topcoat sector.

In the industrial coatings sector, acrylates are used to satisfy special requirements. Self-cross-linking acrylates are used for domestic appliances on account of their good resistance to detergents and lyes as well as their temperature and yellowing stability. Metal substrates are stoved with combinations of acrylate and melamine or urea resins; recently, stoving dispersions have also become available. Wood surfaces subjected to a high degree of wear (e.g., in kitchens) are being increasingly coated with polyurethane paints based on acrylic resins and isocyanates.

2.6. Alkyd Coatings

The binders used in the production of alkyd paints, namely alkyd resins, are oil-modified or fatty-acid-modified condensation polymers of polybasic acids and polyhydric alcohols. Alkyd resins account for ca. 45% of the total world production of paint raw materials, excluding plastics latexes and polyvinyl dispersions.

Production of alkyd resins began in 1930. They owe their importance to their universal applicability. In the architectural coatings sector, alkyd resins replaced stand oils on account of their faster drying and curing rates as well as their better film hardness and gloss retention.

Alkyd paints are used for protection and decoration in virtually all sectors including coatings in the steel, sheet, and metal processing sector, house and decorative paints, do-it-yourself paints, wood varnishes, wood protection agents, and automotive finishes. Alkyd resins are extremely important for corrosion protection coatings.

2.6.1. Alkyd Resin Binders and Uses

DIN 53 183 defines an alkyd resin as follows: "Alkyd resins are synthetic polyester resins produced by esterifying polyhydric alcohols with polybasic carboxylic acids. At least one of the alcohols must be trihydric or higher. Alkyd resins are always modified with natural fatty acids or oils and/or synthetic fatty acids. In order to obtain particular application technology properties, alkyd resins may be additionally modified with compounds such as resin acids, benzoic acid, styrene, vinyltoluene, isocyanates, acrylic, epoxy, or silicone compounds."

Worldwide alkyd resin production (including captive-use amounts) is estimated to be $2-3 \times 10^6$ t. Detailed information on commercially available alkyd resins can be found in [2.62].

The properties of alkyd resins can be widely modified by selecting appropriate vegetable oils and fats/fatty acids, varying their content in the resin, and by using special synthetic acids and a variety of modifying agents.

Classification. Normal alkyd resins containing oil/fatty acid, dicarboxylic acid (mainly phthalic acid or phthalic anhydride), and polyhydric alcohol are classified as short oil ($\leq 40\%$ oil content), medium oil (41–60% oil content), and long oil (61–70% oil content) resins. Very long oil alkyd resins with 71–85% oil content are mainly synthesized using isophthalic acid. The term oil content denotes the triglyceride (fatty acid) content relative to the solvent-free resin. Alkyd resins are subdivided into drying and nondrying resins according to the nature of the oils/fatty acids used.

Long oil drying alkyd resins are used as binders for architectural and do-it-yourself paints, and are mainly modified with soybean oil. In the absence of direct light, they do not yellow as much as linseed oil alkyd resins and are less expensive than safflower oil types. Long oil linseed oil alkyd resins are principally used in corrosion protection coatings and colored architectural coatings.

Medium oil alkyd resins (oil content ca. 50%) are based on drying oils or fatty acid mixtures, and are used as binders for air-drying and forced-drying machinery coatings and industrial coatings. They are also used in car repair finishes (refinishes), for original equipment manufacturing (OEM) finishes, and refinishes of trucks and large-capacity vehicles.

Short oil alkyd resins are used in combination with amino resins for industrial stoving finishes (e.g., for metal furniture, radiators, bicycles, garage doors, and small steel items). Dehydrated castor oil types mixed with alkyd resins based on synthetic acids are combined with melamine resin for OEM automotive finishes. Short oil resins are also combined with urea resins to produce acid-curing paints for wood and mixed with nitrocellulose for inexpensive, easily applied furniture lacquers and automotive refinishing systems.

Combination of Alkyd Resins with Other Binders. Alkyd resins can be combined in many different ways with other paint raw materials. In one method alkyd resins are mixed with preformulated products; this is normally carried out in the paint factory. Alternatively, the reactivity of the initial products can be employed to modify the alkyd resin during production.

Alkyd resins can be mixed with nitrocellulose, chlorinated rubber, PVC-copolymers, amino resins, and colophony-modified phenolic and maleic resins. Chlorinated rubber and PVC-copolymers are combined with alkyd resins with the addition of plasticizers for corrosion protection coatings. These combinations have a higher resistance to chemicals than pure alkyd resin coatings. Combinations with amino resins are used in oven-drying industrial coatings and automotive finishes. The gloss, body, adhesion, and the resistance to preserving wax are determined by the alkyd resin. Amino resins allow rapid drying at elevated temperature and improve the mechanical resistance of the coatings.

Chemical reactions with alkyd resins can take place via their hydroxyl or carboxyl groups as well as via the double bonds of the unsaturated fatty acids. Isocyanates, epoxy resins, or colophony, for example, may be reacted with the hydroxyl groups. The carboxyl groups can be reacted with polyamidoamines (reaction products formed from dimerized linoleic acid and ethylenediamine) to form thixotropic resins, or can react with hydroxy-functional silicone precondensates. The double bonds of the unsaturated fatty acids permit copolymerization with vinyl compounds [e.g., styrene or (meth)acrylic acid derivatives].

Styrene-Modified Alkyd Resins. Styrene was first copolymerized with drying oils in the early 1940s. These binders had a substantially improved drying behavior and resistance to water and chemicals compared with untreated oils. The first patents for the production of styrenated alkyd resins were granted in the United Kingdom in 1942 [2.63]. Mixtures of styrene and α-methylstyrene are also used for a more reliable reaction; vinyltoluene yields copolymers with improved thinnability in mineral spirit and pigment absorption.

Advantages of styrenated alkyd resins compared with nonstyrenated resins are faster drying, better resistance to water and chemicals, and a lower yellowing tendency of their films. Disadvantages include the poor drying and low solvent resistance of the paint films, which frequently lead to "lifting" during recoating. A defect of styrene–alkyd resin topcoats is their high surface sensitivity, especially to scratching; they are therefore mainly used in the primer sector where the danger of lifting can be avoided by appropriate pigmentation.

Modification with Other Vinyl Compounds. Low molecular mass acrylate and methacrylate esters are used to modify alkyd resins containing oil/fatty acid [2.64], [2.65]. Methacrylates give products that are more interesting from the paint technology aspect. Acrylated alkyd resins have a good pigment absorption capacity and yield paints with fast drying properties, a high permanent elasticity, and very good adhesion (e.g., to aluminum, brass, and some plastics). Special advantages include better weather resistance and gloss retention; through drying and surface (scratch) resistance are superior to those of styrenated alkyd resins.

Silicone-Modified Alkyd Resins. Hydroxy-functional silicone precondensates can be used to produce air-drying alkyd resins. Methoxy-functional silicone precondensates are used for oven-drying silicone-modified alkyd resins [2.66]. Air-drying coatings have outstanding chalking resistance and gloss retention, and thus substantially improved weather resistance. They are therefore often used on inaccessible objects that can only be repainted at long intervals. The U.S. Navy Specification TTE 049 prescribes silicone alkyd resins for marine coatings. In the oven-drying coatings sector, the improved gloss retention and the substantially higher thermostability/yellowing resistance are important.

Thixotropic Alkyd Resins. Thixotropic flow is always associated with structural viscosity. Thixotropic paints can be obtained by various methods. Use of a thixotropic alkyd resin is the best way of ensuring reproducible flow properties. The alkyd resins become thixotropic after reaction with 10–20% polyamidoamines at elevated temperature (ca. 200 °C) [2.67], [2.68].

Thixotropic resins are mainly used for decorative paints and primers, and in corrosion protection coatings. Thixotropic alkyd resins prevent the formation of precipitates, improve brushability, and reduce the sagging of paints on vertical surfaces and from edges and corners. Thixotropic coatings (including those based on thixotropic urethane alkyd resins) are frequently used in corrosion protection because the required layer thickness and good edge covering can be achieved with a small number of layers.

Thixotropic air-drying and oven-drying alkyd resins with low oil contents are used as binders for one-coat industrial paints with a high dry-film thickness of 50–75 µm. In special-effect paints (e.g., hammer finishes, soft-feel coatings, textured finishes) they facilitate fixation of the desired effect and prevent smudging.

Urethane Alkyd Resins. Attempts to modify drying oils with isocyanates date from the early 1940s [2.69] and led to the development of urethane alkyd resins. Urethane alkyd resins are generally produced by reacting long oil alkyd resins containing excess hydroxyl groups with diisocyanates. Since conversion is quantitative, the properties of the long oil alkyd resins are largely retained in the final products. For example, urethane alkyd resins are readily soluble in mineral spirit, are compatible with many other paint raw materials, and have a good storage stability. Their advantages in paint technology are rapid drying, high hardness, very good film elasticity, and above-average abrasion resistance. Paint films are highly waterproof and show improved resistance to chemicals, even in the alkaline range.

Toluene diisocyanate and isophorone diisocyanate are mainly used to modify alkyd resins. Isophorone diisocyanate yields products with a lower yellowing tendency, higher pigment absorption, and considerably improved chalking resistance under weathering conditions; however, its use is limited by its considerably higher price.

Urethane alkyd resins are used as varnish binders for the interior and exterior coating of wood. Further applications include parquet floor seals, marine paints, and varnishes for the joinery and do-it-yourself sectors. They are used in pigmented paints to produce high-gloss and matt decorative coatings with extremely good mechanical resistance as well as resistance to cleaning agents and many disinfectants. They allow sanding to be carried out at an early stage in rapid-drying primers. They open up interesting possibilities in the corrosion protection sector because their drying is less sensitive to temperature and atmospheric humidity than that of standard resins.

Alkyd Resins with Highly Branched Carboxylic Acids. Highly branched carboxylic acids are produced industrially from olefins, carbon monoxide, and water [2.70]. They are marketed by Shell under the name Versatic Acids. Their esters have an outstanding resistance to hydrolysis. Highly branched fatty acids are, however, difficult to esterify; their sodium salts are therefore reacted with epichlorohydrin to form glycidyl esters which can then react as the epoxides, for example, with carboxylic acids [2.71]. These products are marketed by Shell under the name Cardura resins. The glycidyl esters are esterified with phthalic anhydride or adipic acid and glycerol. Cardura resins are nondrying alkyd resins modified with Versatic Acid. Their fatty acid content is generally ca. 40%. They are soluble in aromatic solvents, ketones, and esters, and are compatible with many amino resins, short oil alkyd resins, oven-drying acrylic resins, epoxy resins, and nitrocellulose. Their main areas of use are stoving finishes, combining extremely good hardness with high elasticity and resistance to chemicals. The coatings are resistant to overstoving and have a good gloss retention and chalking resistance under weathering conditions. Their very good adhesion (even on untreated metal) is an important advantage. They are used in enamels and coatings for domestic appliances, washing machines, and refrigerators, as well as for OEM automotive finishes.

Waterborne Alkyd Resins [2.72]. Alkyd resins can be converted into a waterborne form in two ways:

1) Alkyd resins with high acid numbers (> 50) are neutralized with amines (normal air-drying resins have an acid number < 10, oven-drying resins $20-35$); solubility in water is due to salt formation
2) Alkyd resins are emulsified in water after addition of emulsifying agents and stabilizers or after chemical modification with special monomers e.g. polyglycols.

Neutralization with amines is mainly employed with medium and short oil resins that are modified with drying and semidrying oils/fatty acids and supplied as $70-80\%$ solutions in butyl glycol, or mixed with 1-methoxy-2-propanol or N-methylpyrrolidone. Neutralization is generally performed with triethylamine to obtain air-drying alkyd resins; dimethylethanolamine is used for oven-drying alkyd resins [2.73]. The amines used for neutralization represent an ecotoxicological problem, and attempts are therefore being made to find more environmentally compatible agents. In contrast to alkyd resin emulsions, waterborne alkyd resins permit the production of high-gloss paints. Sensitivity of the resins to hydrolysis during storage of the paints may cause problems and generally results in a decrease in viscosity. With air-drying systems the drying ability can also be adversely affected. These disadvantages are largely overcome by incorporating acids such as trimellitic acid which is less susceptible to hydrolysis for steric reasons; paints based on this system are therefore more stable. Coating materials based on amine-neutralized alkyd resins are used for forced-drying and oven-drying industrial paints (e.g., in combination with waterborne amino resins), for wood protection, and for corrosion protection.

Alkyd resin emulsions or "self-emulsifying" alkyd resins can be produced relatively easily (resin content 40–60%). However, optimization of their properties requires a great deal of experience and careful selection of additives [2.74]. In order to obtain satisfactory water resistance and drying behavior, the content of emulsifiers should be kept as low as possible. The size of the emulsified particles is in the range 5–15 µm. A small particle size ensures good storage stability of the coating materials and improved gloss of the dried films; it also improves miscibility with polymer dispersions. The more homogeneous the system, the better the weather resistance of the resultant coating films. Alkyd resin emulsions are mainly used in architectural paints, corrosion protection coatings, wood protection agents, and house paints. They are particularly suitable for combination with polymeric, especially acrylate, dispersions. The use of alkyd resin emulsions in industrial coatings is still insignificant, an exception being furniture coatings and varnishes.

Alkyd Resins for High-Solids Paints. See also Section 3.2. Alkyd resins have been developed for producing high-solids paints to reduce solvent emission. One-pack, oven-drying high-solids paints have solids contents of 65–70 wt% in the ready-for-use formulation. Air-drying high-solids paints have solids contents of 85–90 wt%. A high solids content can be obtained in three ways:

1) Use of solvents with stronger solvent power, so that resin solutions with lower viscosities are obtained with less solvent. This method is used for stoving paints based on conventional alkyd–melamine resin combinations; it can also be applied with certain restrictions to air-drying systems.
2) Use of reactive diluents that are incorporated into the film by condensation when the stoving paints harden.
3) Use of low-viscosity, reactive alkyd resins that have a lower solvent requirement than conventional products.

With high-solids stoving paints, sagging generally has to be prevented when coating vertical surfaces. Rheology-controlling additives can provide a remedy, but generally lead to a reduction in the solids content.

The solids contents of air-drying paints (i.e., architectural and do-it-yourself paints) can be adjusted relatively easily to 75–80%. However, for a solids content of 90%, very low viscosity binders with a low degree of condensation are necessary and often do not satisfy requirements concerning drying, hardness, and resistance of the coatings. Developments have been aimed at overcoming the deterioration in drying and mechanical properties by incorporating 10–20% of an allyl ether [2.75]–[2.77].

Commercial products include Alftalat (Vianova), Alkydal (Bayer), Beckosol (Reichhold Chemie), Dynotal (Dyno Industries), Jägalyd (Jäger) Rhenalyd (RWE–DEA), Setal (Synthese), Synthalat (Synthopol Chemie), Uralac (DSM Resins), and Vialkyd (Vianova), Worleekyd (Worlee).

Testing. See also Chapters 9 and 10. DIN 53 183 (Alkyd Resin Testing) describes the most important tests for alkyd resins in the Federal Republic of Germany. A comparable U.S. standard is ASTM D 2689-88 (Standard Practices for Testing Alkyd Resins). ISO 6744 (Binders for Paints and Varnishes – Alkyd Resins – General Methods of Tests) defines and classifies the term alkyd resins and contains a summary of international standards for testing alkyd resins. ASTM D 4712-87 (Standard Guide for Testing Industrial Water-Reducible Coatings) may also be mentioned; it includes standards covering the testing of wet paints (properties, storage, application, film formation) and paint films.

2.6.2. Additional Raw Materials (See also Chapter 5)

Solvents. Alkyd resins are generally supplied as concentrated ($\geq 60\%$) solutions (except in the case of aqueous dispersions and emulsions). Further solvent must be added to bring the paints into a ready-for-use state. The choice of solvent is determined by the solubility of the alkyd resin and ecological compatibility.

Mineral spirit is the most commonly used solvent for *long oil alkyd resins*. Additions of other solvents during paint production modify specific properties. The brushability and wet edge time can be improved by addition of higher boiling mineral spirit fractions (*bp* 180–210 °C), aromatics-free petroleum fractions, or isoparaffins [2.78]. (The wet edge time denotes the period that the paint can be brushed side by side without leaving overlap marks.) Additions of small amounts of glycol ethers (2.5–5% of total solvent content) also improve the leveling and application of the paints. Lower alcohols and glycol ethers reduce and stabilize the viscosity. Additions of aromatic hydrocarbons of the solvent naphtha or xylene type improve adhesion of corrosion protection coatings to existing coats. Polar solvents should not generally be used in thixotropic paints because they destroy the thixotropy.

Mineral spirit is the most important solvent for *medium oil alkyd resins* in industrial air-drying and forced-drying paints. On account of the high viscosity of these resins, mineral spirit must be combined with solvents of stronger solvent power or with viscosity-lowering solvents (e.g., xylene, solvent naphtha, lower alcohols, and glycol ethers). These solvents promote leveling and stabilize the paints during storage.

Short oil alkyd resins require special solvent mixtures for each area of use. Principal solvents include aromatic hydrocarbons (e.g., xylene), solvent naphtha fractions, and glycol ethers. Lower alcohols, even in small amounts, have a powerful viscosity-lowering effect. The choice of solvent depends on the solvent power for the resins and the application method. The last solvent to evaporate during drying must be a true solvent for the binders because otherwise the paint film becomes opaque or inhomogeneous.

For their largest use sector (oven-drying industrial coatings) the short oil alkyd resins are combined with amino resins (melamine, urea, or benzoguanamine resins). The paints are cured at 80–250 °C. The amino resins are generally soluble in alcohols and glycol ethers; aromatic hydrocarbons are used as diluents for amino resins. Effective solvent mixtures often consist of aromatic hydrocarbons, glycol ethers, and alcohols in a volume ratio of 7:1:2 in the ready-for-use paint. The glycol ethers and alcohols improve the leveling and surface of the coat as well as the storage stability of the paints. With spraying and roller paints, relatively high-boiling solvents should also be used to achieve good leveling and avoid boiling phenomena (bubble formation). In dipping paints, small additions of low-boiling solvents can prevent sagging.

Short oil alkyd resins can also be combined with nitrocellulose. They are readily soluble in typical nitrocellulose solvents (esters and ketones). Special short oil alkyd resins that are soluble in lower alcohols are often processed into furniture lacquers and acid-curing paints, and are also used in paper (spray) coatings and in flexoprinting inks.

The choice of solvents for *waterborne alkyd resins* is restricted. Resins neutralized with amines are normally supplied in glycol ethers or glycol ether–alcohol mixtures [2.79]. After neutralization they are generally diluted with water. Small amounts of propylene glycol are added to alkyd resin emulsions to prolong the wet edge time. Coalescing agents are used in dispersions with high latex contents. These specially formulated high-boiling solvents lower the film-forming temperature.

Driers are added to drying oils and drying alkyd resins to accelerate oxidative drying. Driers include oil-soluble and solvent-soluble metallic soaps. The naphthenate and octoate salts of cobalt, lead, and manganese with the auxiliary metals calcium, zinc, and cerium are generally used. Calcium and zinc salts are added to the comminuted material during paint production to exploit their action as wetting agents.

For environmental reasons modern alkyd paints are dried without lead driers and with minimum amounts of cobalt driers [2.80]. Manganese driers are only used in colored paints and corrosion protection coatings because they affect the shade of white and pastel coatings. Special nondiscoloring driers are used to obtain particularly pale white paints or honey-yellow varnishes.

Antiskinning agents suppress skin formation on air-drying alkyd paints in the can, particularly after the can has been opened. Antiskinning agents may be phenol derivatives (e.g., 4-*tert*-butylphenol) or oximes (e.g., methyl ethyl ketone oxime or butanaloxime). Oximes are used almost exclusively, at a concentration of 0.5–1.5% relative to the solvent-free alkyd resin. They are colorless, do not affect paint color, and have a negligible effect on drying time.

Antisettling Agents. High-density pigments and extenders tend to settle, especially in low-viscosity paints. They form deposits that are difficult to dislodge. Antisettling agents are used to overcome this problem. They may be minerals (e.g., pyrogenic silica), organically treated minerals (e.g., montmorillonite treated with stearic acid), or organic compounds (e.g., hydrogenated castor oil). Excessive amounts of antisettling agents frequently cause thickening and have an adverse effect on rheological behavior. They should therefore be used with care.

Wetting Agents. A stable mixture of binder solution, pigments, and extenders is essential for the production of storable pigmented paints and primers. Pigments and extenders are usually covered by a layer of water molecules that makes wetting difficult or even impossible. Carbon black and some organic pigments can only be wetted with difficulty by weakly polar or nonpolar binder solutions due to their polar structure.

Wetting agents are surfactants and reduce the surface tension between the binder solution and the pigment particles. Wetting agents used in paint technology include calcium, zinc octoate, or naphthenate, fatty alcohol sulfonates, castor oil acid derivatives, and conversion products, poly(glycol ethers), and low-viscosity silicone oils.

Wetting agents are also often used as antifloating agents. Floating of pigments may occur in combinations of pigments of widely differing densities, particle sizes, differing polarities, and with unsatisfactory wetting.

2.6.3. Production (See also Chapter 7)

In liquid coating materials, pigments and extenders have to be uniformly distributed and fully wetted in the binder solution or dispersion. The pigment agglomerates are therefore comminuted (grinding, dispersion) into primary particles and simultaneously incorporated into the binder solution. Grinding is usually carried out with a dissolver and agitator ball mill, but in some cases only with a dissolver. Grinding can be carried out in roller mills after predispersion of the pigments and extenders with kneaders or stirrers. Three-roll mills are used to produce highly viscous materials (e.g., fillers or printing inks).

Agitator ball mills (i.e., sand or bead mills) permit continuous operation and shorter dispersion times. Particle sizes of < 10 µm can be achieved with relatively fast throughputs. These mills are therefore also used for high-quality paints (e.g., for automotive finishes or car refinishing).

Production with a dissolver and a sand mill proceeds as follows:

1) Dilution of the required amount of binder solution
2) Addition of wetting and antisettling agents required for incorporation of pigments and extenders
3) Addition of the pigments and extenders in a relatively slowly running dissolver
4) Dispersion of the pigments and extenders with the sand mill
5) Monitoring of particle fineness
6) Addition of the remaining auxiliaries (e.g., driers and antiskinning agents)
7) Adjustment of the viscosity
8) Sieving
9) Testing, approval, and packaging

2.6.4. Environmental and Health Protection Measures

The solvents in the as-supplied alkyd resin solutions and in the paint composition represent a potential hazard due to their (eco)toxicological effects and flammability. Manufacturers and suppliers of resin solutions and paints must indicate the potential dangers involved in the handling, transportation, and storage of solvents by labeling drums and containers according to national and international regulations.

Health risks depend on the nature of the solvents and their concentration in the paint. Risks arise from solvent vapors during paint application. Working areas must be adequately ventilated to prevent vapor accumulation and respirators must be worn during spraying to prevent inhalation of paint aerosols.

Paint residues require special handling and disposal (special tips). Paints, solvents, and resins must under no circumstances be allowed to pollute the ground or effluent. Legal provisions concerning emission limits must be observed when applying paints.

Solvents differ as regards their miscibility with water and flash point and thus their potential fire hazards (flammability, combustibility). Explosive solvent vapor–air mixtures may be formed when applying paints. Spark formation (e.g., by static charging) must be avoided and hot surfaces or naked flames must be removed.

Combustible materials contaminated with air-drying and oven-drying alkyd paints, and spray dusts, may tend to exhibit spontaneous ignition due to autoxidation. Storage under exclusion of air and regular cleaning of the spray cabins improves safety.

2.7. Saturated Polyester Coatings

Polyester resins are condensation products of di- or polyfunctional monomers containing hydroxyl groups and carboxyl groups. The development of saturated polyesters began in 1901 with "Glyptal resins", formed from glycerol and phthalic anhydride (Smith, United States). Soluble resins obtained with fatty acids were first employed in 1925. Alkyd resins formed from unsaturated fatty acids can be cured by atmospheric oxidation (see Section 2.6). Other unsaturated polyester resins are discussed in Section 2.8.

The subsequent development of saturated polyester resins for paints and other uses (e.g., adhesives, foams, fibers) was largely dependent on the introduction of new raw materials. A broad range of raw materials are now available, examples follow:

Polycarboxylic acids
    Terephthalic acid (dimethyl ester)   hard
    Isophthalic acid   hard
    Phthalic anhydride   hard
    Hexahydrophthalic anhydride   hard
    Trimellitic anhydride   hard

Adipic acid	soft
Sebacic acid	soft
Dodecanoic acid	soft
Dimerized fatty acid	soft
Polyalcohols	
Ethylene glycol	hard
1,2-Propanediol	hard
Neopentyl glycol	hard
1,4-Cyclohexanedimethanol	hard
Trimethylolpropane	hard
Diethylene glycol	soft
1,4-Butanediol	soft
1,6-Hexanediol	soft
Glycerol	soft

Paint polyesters are generally copolyesters composed of several polycarboxylic acids and polyalcohols. The glass transition temperature of copolyester resins can be varied within wide limits by choosing appropriate raw materials. Different polymer units (e.g., polyethers, polysiloxanes) can be incorporated by condensation into the molecular chain. Instead of terminal hydroxyl or carboxyl groups, the molecules may also contain other functional groups (e.g., acrylic or epoxide groups) that can be cross-linked with suitable reactants during paint hardening (curing) to form thermosetting paints. Saturated polyester resins for paints are available in various formulations [2.81]: low to high molecular mass, linear or branched, as solid or liquid resins, soft/elastic or rigid/brittle, and amorphous or crystalline. This versatility, together with their good resistance to light, moisture, heat, oxygen, and many other substances, accounts for the importance of saturated polyester resins as paint binders.

2.7.1. Properties

Molecular Mass. High molecular mass copolyesters (M_r 10 000 – 30 000) are generally linear. They are formed from terephthalic and isophthalic acids, aliphatic dicarboxylic acids, and various diols. Good solubility in conventional paint solvents is achieved by suitable formulation. In some cases (heat-sealable foil lacquers, printing inks, etc.), high molecular mass copolyesters are used as physically drying binders. Optimum paint film properties are, however, only obtained in combination with cross-linking resins. Special, crystalline, high molecular mass copolyesters are ground and used in thermoplastic powder coatings.

Linear, low molecular mass polyester resins are available with molecular mass up to 7000; branched grades have a molecular mass of up to 5000. These resins are not suitable for physically drying paints; they should be regarded as prepolymers for reaction systems with cross-linking resins. Classes of polyester prepolymers, their preferred cross-linking agents, and uses are summarized in Table 2.4.

Table 2.4. Classification of saturated polyester resins for paints

Structure	Class	M_r (weight average)	Cross-linking agent	Uses
HO~~~OH	linear, high molecular mass	10 000 – 30 000	melamine resins, benzoguanamine resins	coil/can coating, flexible packaging, printing inks
HO~~OH	linear, low molecular mass	1000 – 7000	melamine resins, blocked polyisocyanate resins	coil/can coating, automotive/industrial paints
HO~⋏~OH (OH)	branched, low molecular mass, hydroxy-functional	1000 – 5000	melamine resins, free/blocked polyisocyanate resins	automotive/industrial paints, polyurethane/powder coatings
HOOC~⋏~COOH (COOH)	branched, low molecular mass, carboxy-functional	1000 – 5000	triglycidylisocyanurate, epoxy resins, melamine resins	powder coatings, waterborne paints
(acrylate diester)	low molecular mass, acrylate group content	1000 – 5000	UV and electron-beam curing	paper/plastic coatings, printing inks

Terminal Groups. The polyester–amine system, consisting of hydroxy-functional polyester resins and, for example, melamine resins, is extremely important for the production of heat-curable industrial stoving enamels [2.82].

Polyurethane systems based on hydroxypolyester resins and free polyisocyanate resins (two-pack polyurethane paints, Section 2.9) are used particularly for finishes of road vehicles, railroad vehicles, aircraft, and plastics. Blocked polyisocyanate resins are available for formulating one-pack stoving enamels which are particularly suitable for coil and can coatings [2.83].

The reaction system consisting of "acid" (i.e., carboxy-functional) polyester resins and epoxy resins is important for powder coatings [2.84].

Acid polyester resins can be neutralized (e.g., with amines), diluted with water, and processed into aqueous stoving enamels [2.85].

Acrylate-modified polyester resins have proved suitable as binders for radiation-curable coatings [2.86].

Glass Transition Temperature. The T_g of saturated polyester resins can be varied by a suitable choice of aromatic or aliphatic raw materials. The T_g of nonplasticized aromatic copolyesters is ca. 70 °C, and that of copolyesters formed from cycloaliphatic glycols exceeds 100 °C. Aliphatic polyesters with long methylene chains between the ester groups have T_g values as low as −100 °C (limiting value, polyethylene −125 °C).

Solubility, Crystallinity, and Compatibility. Polyester solubility is largely determined by the nature and proportions of the constituent monomers. Polyesters with a regular structure are crystalline. Examples of highly crystalline polyesters are poly(ethylene terephthalate) and poly(butylene terephthalate). Although medium to highly crystalline copolyesters are insoluble in paint solvents, they can be processed into adhesives and paints in the hot-melt process [2.87]. Slightly crystalline copolyesters are soluble, for example, in ketones and are mainly used for laminate adhesives.

A low molecular mass and a low T_g generally favor the compatibility of polyester resins with other binders (e.g., acrylic resins, epoxy resins, precondensed amino resins, cellulose esters). Not all polyester resins are mutually compatible; for example, polyesters based on phthalic acid are not always compatible with other polyester resins.

2.7.2. Production of Polyester Resins and Paints

The polycondensation of polycarboxylic acids and polyalcohols proceeds via esterification with the elimination of water [2.81]. Transesterification reactions should be taken into account in copolyester resins derived from more than two monomers. Esterification and transesterification can be accelerated with metal-based catalysts.

Low molecular mass linear and branched polyester resins are produced in a one-stage process at 125–240 °C. The volatile condensation products are removed in vacuo (melt condensation process) or by passing a stream of inert gas through the resin melt (gas stream condensation process). Polycondensation in solution with azeotropic removal of water by solvent distillation (azeotropic process) is of lesser importance. High molecular mass copolyesters are produced in two stages as is used for poly(ethylene terephthalate). A precondensate is first obtained by transesterification of dimethyl terephthalate with an excess of diols. In the second stage, the molecular mass of the precondensate is adjusted to the desired value by polycondensation in special reactors with the maximum possible elimination of water and excess diols in vacuo at ca. 250 °C.

Saturated paint polyester resins are soluble in esters, ketones, and in some cases aromatic hydrocarbons, but are insoluble in aliphatic hydrocarbons and alcohols. Solvent-containing paints based on high molecular mass resins have solids contents of 35–50 wt% at application viscosity. Paint solids contents for low molecular mass polyesters are 50–70 wt%.

Polyester resins for waterborne paints mostly contain anionic groups (e.g., carboxyl groups) and become water soluble after amine neutralization. Waterborne paints based on polyesters also contain cosolvents, preferably glycol ethers. Powder coatings based on low molecular mass resins are solvent-free, like radiation-curable polyester paints.

Conventional grinding methods are used to incorporate pigments into solvent-borne and waterborne polyester resins as well as into radiation-curable polyester paints containing reactive diluents. Pigmented powders are produced by coextrusion of polyester granulate with pigments, followed by grinding.

2.7.3. Cross-Linking of Polyester Resins

Amino Resins. Formaldehyde-modified amino resins (i.e., melamine, benzoguanamine, and urea resins) are the most important resins for the heat curing of hydroxyfunctional polyester resins. These resins are readily available with a low molecular mass (very good polyester compatibility, but less reactive) or in precondensed form (limited compatibility, but very reactive). In order to prevent premature reaction in the wet paint, the amino resins are blocked by etherification with, for example, methanol or butanol. Sulfonic acids (e.g., *p*-toluenesulfonic acid, dodecylbenzenesulfonic acid) have proved suitable for accelerating the deblocking of amino resins during heat curing. These acids must also be used in blocked form (ammonium salts, thermolabile adducts).

The polyester–amino resin ratio is generally in the range 90:10 to 70:30. High molecular mass polyester resins can be cross-linked with hexamethoxymethylmelamine resins (HMMM) above ca. 190 °C. Amino resins undergo selfcondensation above this temperature [2.82]. Low molecular mass linear polyesters react above 160 °C, and branched types above ca. 100 °C. The latter generally contain acid groups (acid number 5–25 mg KOH/g) in addition to the hydroxyl groups to enhance reactivity with amino resins. Stoving enamels based on high molecular mass polyester resins and HMMM or benzoguanamine resins have excellent flexibility and surface hardness as well as outstanding adhesion to metal substrates [2.88], [2.89]. Low molecular mass linear polyesters combined with HMMM resins have lower flexibility, but have the advantages of higher reactivity, better pigmentability, and higher paint solids content [2.90], [2.91]. Low molecular mass branched resins form particularly reactive stoving enamels, but the resulting films are less flexible and do not adhere as well to metal substrates.

Polyisocyanate Resins. Hydroxy-functional polyester resins play a significant role in two-pack polyurethane paints. Combinations with nonblocked polyisocyanate resins are discussed in Section 2.9.

Storage-stable one-pack polyester stoving enamels can be formulated with polyisocyanate resins that are thermoreversibly blocked. Tin catalysts (e.g., dibutyl tin dilaurate) are particularly suitable for accelerating hardening with hydroxypolyester resins.

Although linear polyester resins do not readily undergo heat curing, they form extremely flexible paint films with excellent adhesion to metal substrates. Polyisocyanate resins are more resistant to hydrolysis than amino resins. Polyester–polyurethane paints therefore exhibit improved resistance to moisture and weathering but are more expensive.

Epoxy Hardeners. The heat curing of carboxy-functional polyester resins with epoxy resins is mainly used in powder coatings. Catalysts (e.g., benzyltrimethylammonium chloride or 2-methylimidazole) are used to accelerate the reaction during heat curing. Triglycidyl isocyanurate plays a special role as cross-linking agent. Powder coatings with outstanding weather resistance and very good mechanical properties are obtained [2.84]. Hybrid powder coatings based on epoxy resins of the bisphenol A type should also be mentioned.

Radiation Curing. The cross-linking of acrylate-modified polyester resins with UV radiation or an electron beam is an energy-saving alternative to the heat-curing systems. Reactive diluents (e.g., polyfunctional acrylates) are required to adjust the viscosity for application; although they increase reactivity, they reduce the flexibility and substrate adhesion of the paint film.

2.7.4. Uses

Coil Coating. On account of their outstanding hardness–elasticity balance, good adhesion to metals, good application properties, and favorable cost, paints based on saturated polyester resins represent the most important system for coil coating.

On account of their insensitivity to solar radiation, some polyester resins are particularly suitable for weather-resistant topcoats (e.g., facade claddings, blades of shutters and venetian blinds, vehicle claddings, metal signs). Melamine-formulated paints are economical and have a well-balanced property profile. Polyurethane-formulated paints have an even better processability and weather resistance (colorfastness, chalking resistance) [2.92].

Specially adapted polyester resins are available for topcoats for interior architectural use (e.g., underfloor ceilings, light fittings), for appliance claddings (e.g., domestic appliances, hifi equipment), or for automobile fittings. Hard, high molecular mass copolyesters are suitable for primers containing corrosion protection pigments on account of their low oxygen permeability and very good adhesion to metals [2.93].

Can Coating. Polyester stoving enamels are widely used for coating can exteriors (metal decorating) but also for protecting can interiors. Polyester resins are available that satisfy the FDA regulations for internal coatings of cans used for food packaging [2.94]. Polyester coatings do not affect the taste of the food.

High molecular mass polyester resins, cross-linked with benzoguanamine and melamine resins, are particularly suitable for highly drawable, sterilization-resistant decorative stamping enamels [2.88], [2.89]. Low molecular mass polyester resins also exhibit good processability and sterilization resistance in polyurethane formulations.

Automotive Finishes. Important industrial automotive topcoats are produced from alkyd resins. Two-coat metallic finishes consist of an acrylic clearcoat and a metallic basecoat. Solventborne basecoats are generally composed of cellulose acetobutyrate and a low molecular mass, branched polyester resin which exhibits high reactivity with precondensed melamine cross-linking resins.

Surfacers are mainly produced from low molecular mass polyester resins cross-linked with amino resins. Blocked polyisocyanate resins may also be used for high stone chip resistance. Water-soluble polyester resins are increasingly used for water-thinnable stoving surfacers. Spray and dip paints based on polyester resins are used to coat automobile accessories (e.g., windscreen wipers, axle parts).

Industrial Paints. Stoving enamels used to spray or dip-coat machinery, domestic appliances, vehicles, and office furniture are termed industrial paints. On account of their good adhesion and resistance to hydrolysis, polyester–melamine stoving enamels provide very good protection of metal surfaces against undercoat corrosion creep. Polyester resin grades are available for special mechanical requirements (high impact resistance, hardness, abrasion resistance) or for exterior use (weather resistance).

Polyester paints with high solids contents (65–75 wt% at application viscosity) can be produced from very low molecular mass resins [2.95]. Low-pollution paints can also be produced from water-soluble polyester resins [2.85]. On account of their good water solubility, HMMM resins are particularly suitable for cross-linking these resins. Organic cosolvents (mostly glycol ethers) must be added to waterborne polyester paints to control their viscosity and applicability (leveling, substrate wetting).

Two-Pack Polyurethane Paints. Polyester–polyurethane paints are described in detail in Section 2.9. Polyester polyols for polyurethane systems are generally adjusted to be "softer" than those used for stoving enamels. Their hydroxyl number and degree of branching are adapted to the recommended polyisocyanate cross-linking agent.

Polyester resins combined with aliphatic polyisocyanate resins have outstanding weather resistance, good substrate adhesion, and high flexibility even under very large temperature changes (aircraft finishes).

Polyester–polyurethane paints are particularly suitable for coating plastics (maximum drying temperature generally 80 °C) on account of their very high flexibility and good adhesion.

Powder Coatings. Polyester resins are available for two different reactive powder coating systems:

1) Hydroxy-functional resins for combination with blocked polyisocyanate resins (based on isophorone diisocyanate) [2.96]
2) Carboxy-functional (acid) resins for combination with triglycidyl isocyanurate or epoxy resins

The resulting powder coatings have good weather resistance as well as excellent impact strength and adhesion to metals (even under humid conditions) and are therefore suitable for many uses (e.g., in exterior and interior architectural applications, for coating machinery, domestic appliances, steel furniture, garden tools).

Radiation-Curable Polyester Coatings. Radiation-curable polyester paints (see also Section 3.7) are particularly suitable for coating flat surfaces (strips, boards, sheets) and for temperature-sensitive substrates (paper, plastics). Acrylate-modified polyester resins are used for UV printing inks, UV varnishes for paper and plastics, as well as for pigmentable, electron beam-curable gloss and matt paints [2.86]. Radiation-curable polyester paints are also suitable for metal coatings with good elasticity.

Special Uses. High molecular mass copolyester resins are used in the manufacture of flexible packaging. Terephthalate resins are particularly suitable as adhesion promotors for printing inks, lacquers, and adhesives on poly(ethylene terephthalate) films. Some polyester printing inks adhere directly to these sheets. Lacquers that can be heat sealed at relatively low temperature can be produced from high molecular mass, soft copolyester resins. Special linear copolyester resins are used for magnetic tape coatings [2.97].

2.8. Unsaturated Polyester Coatings

Unsaturated polyester coatings are used on, for example, furniture, vehicles and mineral substrates. They are formulated using unsaturated polyester (UP) resins, stabilisers, accelerators, hardeners and, possibly, pigments, extenders, barrier agents, promoters, deaeration agents, flow promoters, thixotropic agents and photoinitiators. The coatings primarily contain monomers. In addition to styrene, the most frequently used monomer, it has become increasingly common in recent years to use acrylates as the copolymerisable monomers, especially in coatings for UV curing. However, monomer-free UP resins have also gained some significance – either 100% or dissolved in, for example, butyl acetate or dispersed in water [2.98].

Coatings containing styrene cure in virtually any thickness as the styrene which initially acts as a solvent polymerises with the double bonds of the UP resin and is incorporated into the paint film. As only a small proportion of the styrene evaporates, this virtually "solvent-free" coating yields films with extremely good body [2.99].

2.8.1. Unsaturated Polyester Binders

UP resins are soluble linear polycondensation products made from polyvalent – usually unsaturated – acids (e.g. maleic or fumaric acids) and bivalent alcohols (e.g. ethylene glycol and/or 1,2-propylene glycol). For special applications, it is common to substitute some of the α, β-unsaturated dicarboxylic acids with phthalic acid and/or adipic acid. Seminal work on unsaturated polycondensation products made from maleic acid, maleic acid anhydride and glycols and on their copolymerisation with styrene is listed in [2.100].

The wide range of UP resins on the market covers products which require the addition of paraffin wax through air-drying, UV-curing, amine-accelerated or flexibilising products to styrene-free and/or water-dispersible resins.

World production of UP resins in 1995 – including those for captive use – is estimated at around 200,000 tonnes. Of this amount, 60,000 tonnes are produced in western Europe where roughly two-thirds are used in furniture coatings and the remaining one-third in fillers.

Commercial Products. UP resins are supplied by the manufacturers under the following tradenames [2.101]: Alpolit (Hoechst), Crystic (Scott Bader), Distroton (Alusuisse), Estratil (Rio Rodano), Gohselac (Nippon Gohsei), Ludopal (BASF), Poloral, Verton (Galstaff), Polylite (Reichhold), Unidic (Dainippon Ink), Roskydal (Bayer), Silmar (Silmar), Synolite (DSM), Vestopal (Hüls), Viapal (Vianova).

The testing of UP resins is governed by, for example, DIN 53 184.

Paraffin-type UP Resins. In the case of the UP resins described above, atmospheric oxygen inhibits polymerisation by chain rupture, with the result that the paint film remains tacky. The addition of small amounts of paraffin wax [2.102] eliminates the inhibiting effect of atmospheric oxygen. Initially, the paraffin wax dissolved in the UP coating loses its solubility as polymerisation progresses. It accumulates on the film surface and forms a layer which prevents contact with oxygen. Coatings containing paraffin wax can be force-dried under heat once they have gelled at room temperature. A further advantage of using paraffin wax is that the wax layer on the surface of the paint film reduces the evaporation of styrene to less than 5% [2.103]. This is of particular significance given the low threshold values which now apply for styrene.

UP Resins for Fillers. Tack-free films are also obtained if phthalic acid anhydride is substituted with tetrahydrophthalic acid anhydride [2.104]. The addition of endomethyl tetrahydrophthalic acid anhydride (a cycloaddition product of maleic acid and cyclopentadiene) also has a positive effect on the air-drying properties [2.105]. Both product classes have become popular above all in the manufacture of binders for highly extended fillers for vehicle repair applications.

Air-drying UP Resins. Air-drying UP resins, also known as gloss polyesters, are produced by incorporating β, γ-unsaturated ethers such as the diallyl ether of glycerol or trimethylol propane and the di- and triallyl ethers of pentaerythrite, in the polymer network. These cure to yield a tack-free film.

Investigations of the drying mechanism of this significant class of resins have been undertaken [2.106] (more at "Polyesters").

2.8.2. Other Raw Materials

The formulation of coatings must be matched to the required property profile in terms of the resin/hardener system as well as the additives which influence the rheology, flow, deaeration, mechanical and chemical properties, resistance to yellowing and film colour.

Hardeners. The polymerisation reaction is initiated by peroxide radicals. Typical examples are cyclohexanone peroxide and methyl ethyl ketone peroxide (hydroperoxides), benzoyl peroxide, perbenzoates and peroctoates (acyl peroxides). It is standard to use around 4% (calculated on the resin and styrene content) of a desensitised supply form (e.g. 50% peroxide in flexibiliser). The manufacturer's instructions are to be observed in handling peroxides.

Accelerators. To allow the degradation of the peroxides at room temperature, the activation energy must be reduced. This is done by adding accelerators. Hydroperoxides are cleaved by heavy metal salts and acyl peroxides by tertiary aromatic amines. Up to 2% of the latter are added by the manufacturers to resins for fillers. These are known as amine-accelerated UP resins. It is common to add 0.02–0.05% cobalt (calculated as cobalt metal) in the form of cobalt naphthenate or cobalt octoate dissolved in aromatics (not white spirit) to the systems hardened with hydroperoxides. The accelerator should only be added shortly before application for reasons of storage stability and drift.

Promoters. Substances such as acetyl acetone, ethyl acetoacetate, amides of acetoacetic acid [2.107], acetyl cyclopentanone [2.108] or tertiary aromatic amines have an accelerating effect on the curing reaction initiated by the hydroperoxide/cobalt octoate. Ethyl acetoacetate is most frequently used. The amount added is 1–3%, calculated on the resin supply form.

Photoinitiators. Derivatives of benzoin and benzil are added in amounts of 1–3% as photoinitiators in UV-curing systems. These have differing effectiveness in the UP systems [2.99]. Special initiators are available for pigmented systems. The UV light splits them into radicals which in turn initiate polymerisation. The UV radiation is generated using superactinic fluorescent lamps and/or high-pressure mercury vapour lamps [2.109].

Special fluorescent lamps or gallium-doped high-pressure lamps are available for pigmented systems.

Stabilisers. If stabilisation is necessary, e.g. to prolong the pot life or increase the storage stability, 0.01–0.03% (solid) hydroquinone or tertiary butyl catechol can be added as a 1 to 5% solution in a suitable solvent. It should be borne in mind that the use of such substances has an effect on the drying time of the coating.

Barrier agents. To prevent inhibition by air, it is common practice to add special paraffin waxes which are suitable for UP resins [2.110]. Paraffin waxes with a high melting point are used at higher application temperatures (max. 40 °C) and products with a low melting point at low ambient temperatures (min. 15 °C). The melting points are between 40 and 70 °C. The usual addition is around 0.1% (solid, calculated on the coating), preferably as a 10% solution in toluene. When coatings and fillers containing paraffin wax cool, there is the risk of crystallisation. The paraffin wax solution is usually added with the cobalt accelerator (as a solution) just before application.

Deaeration Agents and Flow Promoters. Because of the many potential formulations and applications, the type, amount and time of addition must be tailored to the individual system. The storage of coatings containing special silicones may result in cratering.

Solvents and Other Binder Constituents. Coatings based on UP resins generally contain no solvents with the exception of the solvents used in the cobalt, paraffin wax or peroxide solutions. In special cases where inert or co-reacting binders are to be used, low-boiling solvents are required. The type and quantity must be selected according to the application and curing conditions. In thin-coat varnishes containing nitrocellulose or cellulose acetobutyrate, the solvent content can be as high as 75% and special attention should be given to compatibility. In the case of high-build coatings, the solvent content should not exceed 4%. To improve the chemical resistance and the adhesion to the substrate, small amounts of polyisocyanate can be added to UP systems. The changes to the pot life must be borne in mind.

Pigments and extenders. The only pigments and extenders which can be used are those which have virtually no effect on the curing process. The influence on storage stability, shade and surface structure must also be considered. Extenders have the primary task of reducing the shrinkage which occurs during curing to between 7 and 10%. This also improves the adhesion. Talc, barytes, chalk and dolomite are normally used as extenders. It should be ensured that the formulated fillers and coatings have adequate storage stability.

So that coatings based on UP resins can be applied on vertical surfaces, they are made thixotropic. This is done either by using thixotropic resins or by combining standard resins with, for example, highly disperse silicic acid, bentonite and/or hydrated caster oil derivatives. To produce matt coatings, waxes (polyethylene and polypropylene waxes) and/or silicic acid are added in amounts up to 10%.

2.8.3. Formulation, Application, Use, Properties

Fillers and Putties for Wood Materials. Fillers are highly extended (resin/extender ratio, 1:2.5) compounds of high viscosity which are applied to cover any unevenness in a substrate such as chipboard. Putties contain less extender (1:1.5) and are normally used as an intermediate coat between the filler and the top coat. Fillers and putties are formulated using dissolvers, butterfly mixers or attrition mills.

Whereas fillers are applied using a roller machine, putties are usually applied by reverse roller, spray gun or curtain coater. Depending on the substrate, the application rate is between 50 and 200 g/m². Fillers are mainly formulated with paraffin-type UP resins or air-drying types which are UV-cured. For fillers cured with hydroperoxide/cobalt, air-drying UP resins are normally used in combination with flexibilising resins.

UV curing is carried out at a conveyor speed of approximately 3 m/min. The conveyor speed is usually calculated on the basis of a high-pressure mercury vapour lamp with an output of 80 W/cm. Conventional curing of spray fillers in a circulating air oven takes 10 min at 80 °C. These systems provide virtually ideal substrates for a number of top coats as, unlike other systems, they have good hold-out.

Roller, Curtain and Spray Coatings for Laminated Panels and Films (Thin-coat). Laminated panels which are to be given an open-pored coating are first primed and sanded.

2.8. Unsaturated Polyester Coatings 61

Priming process:

a) With UV-curing roller primers: based on an air-drying UP resin applied at a rate of 20–40 g/m^2; curing at a conveyor speed of 3 m/min.
b) With solvent-borne UV-curing spray primers: like a) with a solids content of < 30 % applied at a rate of 50–70 g/m^2; flash-off for 4–5 min at 50 °C followed by UV curing at a conveyor speed of 3 m/min.

Because their solvent content is as high as 75 %, varnishes containing cobalt/hydroperoxide have a pot life of several hours. The reduction in the solids content is achieved primarily through the addition of highly viscous nitrocellulose. The storage stability must be tested if styrene resins are used.

Given flow times of 20–40 s (DIN 4 cup), an application rate of 20–100 g/m^2 is used. The drying time at room temperature is approx. six hours. After flash-off of the solvent (4–5 min at 50 °C), the coating can be cured in approx. 10 min at 80 °C in a circulating air oven, in 1–2 min under IR radiation or at a rate of, for example, 3 m/min under UV radiation. Coatings applied in this way have outstanding properties such as good scratch resistance, adhesion and resistance to household chemicals.

In the case of film coating (application using doctor blades), aqueous and/or solvent-borne UP coatings containing no monomers and dried by conventional means are becoming increasingly significant. The high-gloss or matt coatings have a long pot life (several hours or days) and cure in approx. 30 s at 150 °C.

Aqueous UV-curing UP spray coatings can be recycled [2.111]. They can be recovered by ultrafiltration, wet-on-wet or scraping techniques.

Roller, curtain and spray coatings for wood and wood materials (high-build). When working with conventional curing processes, UP resins requiring paraffin wax are normally used for high-build coatings, either clear or pigmented, with or without matting agent. In many cases, gloss polyesters are used. After the films have been stored overnight or longer, they are sanded and polished. This yields films with high gloss, hold-out and body. For this reason, these systems are favoured as base coats for top coats based on polyurethanes or other binders.

Because of their rapid curing, UV-curing UP resins are being used increasingly for clear high-build coatings, either gloss or matt. Pigmented UV-curing coatings can be applied using either the mono- or double-cure process [2.112]. When using the double-cure process, the pigmented coating is first pre-gelled using conventional cobalt/peroxide accelerators. The film is then cured under UV lamps. In the mono-cure process, the paint film is cured directly (after flash-off) under special UV lamps or specially formulated coatings are used. An overview of the applications of UP and UA resins in wood and furniture coatings is given in [2.112].

Depending on the application process, the viscosity of the coatings varies from 20–100 s (DIN 4 cup). UP coatings which cure conventionally can be applied at a rate of 500 g/m^2 and UV-curing systems at a rate of 250 g/m^2.

The following application processes are employed industrially.

a) Roller coating. Up to around 150 g/m^2 of coating are applied using a reverse roller. As styrene evaporates on the roller, UP systems containing reactive thinners are being used increasingly.

b) Curtain coating. When applying UP formulations on a curtain coater, the main problem of the short pot life is overcome by separating the cobalt and hydroperoxide. For example, in the active primer process which is normally used [2.113], nitrocellulose coatings containing peroxide are applied first. These may also contain styrene-free UP resins. The primer is applied either by roller or curtain coating. Once the solvent has evaporated, the UP coating containing cobalt is applied. Other less frequently used processes are the double-head (1:1), sandwich and transfer processes [2.114]. As UV-curing coatings (one-pack systems) do not have a pot life, they are ideally suited for application by curtain coating.

c) Spraying. UV-curing coatings without a pot life can be applied using conventional spray equipment. Two-component units are used for the application of coatings containing cobalt/ hydroperoxide which have a short pot life.

Curing. When applied at a rate of 450 g/m², coatings containing paraffin wax gel after approx. 20 min at room temperature. They are wipe-resistant after approx. 1 hour and can be stacked after approx. 3 hours (see also 2.8.1.2). In contrast, air-drying UP coatings can be force-dried after a brief flash-off at room temperature and without pre-gelling. UV-curing coatings are cured at a rate of 1.25–3 m/min per 80 W lamp after 30–90 s (curtain coatings) or 5 min (spray coatings) flash-off at room temperature.

Knifing, Spot and Spray Fillers for Metal and Marble (here also coatings). Knifing and spot fillers are especially suited for filling large areas of unevenness in just a few operations in quick succession. Knifing resins (hard to flexible) are mixed to a paste-like consistency with the extenders described. Benzoyl peroxide and tertiary aromatic amine are used as the hardener system (the latter is usually a component of the resin supply form).

Before application, the necessary amount (2–3%) of benzoyl peroxide paste is mixed in well and the formulation applied with a knife. The pot life is usually around 5 min. The curing time is approx. 15 min at room temperature. The filler can then be dry-sanded. The shrinkage of the cured filler is 1–2%. This means that the system adheres on metal. To ensure the complete absence of pores, a spray filler and a non-sanding sealer must be applied over the knifing filler. The spray filler (based on an air-drying UP resin and cured with hydroperoxide and cobalt octoate) serves to fill the marks left by sanding. The polyurethane-based non-sanding sealer prevents blistering after application of the top coat, especially in warm and humid climates. Knifing and spray fillers have been used for many years in vehicle repair and machine finishing (on sanded or sandblasted iron).

The cured fillers are characterised by their very good adhesion (even if the surface is extremely uneven), good through-curing in both thick and thin coats and rapid curing (even at temperatures down to 0 °C). The system described above (knifing/ spot filler, spray filler and non-sanding sealer) can be overcoated with any type of coating.

One application for special UP resins is in stone putties, marble fillers and coatings. The consistency of the formulations ranges from paste-like to pourable. They are cured with benzoyl peroxide and tertiary aromatic amine and can be sanded and polished after 15–20 min. It is important that these hard reactive formulations discolour minimally on curing and have good mechanical properties. The top coats (applied by spray or curtain coating) are based on air-drying UP resins. Coatings containing paraffin wax do not adhere on marble. The coatings are cured with cobalt/hydroperoxide to yield gloss films.

2.8.4. Storage, Transport, Toxicology

UP resins do not have unlimited storage stability. When protected from light and heat, the manufacturers generally guarantee storage stability of 6 months (23 °C). UP resins are supplied in hobbocks, drums, containers and road tankers made from stainless steel, aluminium, tinplate or sheet iron - sometimes coated inside. Depending on the styrene content, the flash point of UP resins is 28–35 °C. The necessary precautions must be taken to prevent electrostatic pick-up.

The following regulations apply for the transport of UP resins: GGVSee/IMDG code: 3.3; UN No.: 1866; MFAG: 310; EmS: 3 05; GGVE/GGVS: Class 3 No. 31 C; RID/ADR: Class 3 No. 31 C; ADNR: Class 3 No. 31 C, Cat.; ICAO/IATA-DGR: 3, 1866 III; exception can be applied for viscous substances in accordance with note re. margin no. 301 E (GGVE/RID) / 2301 E (GGVS/ADR, ADNR).

When formulating, handling and applying coatings based on UP resins dissolved in styrene, particular attention should be paid to the toxicology of styrene. Styrene is harmful when inhaled. It irritates the skin and the eyes. The widely varying threshold values in individual countries must be observed. To prevent inhalation of the coating aerosols during spray application, face masks must be worn.

2.9. Polyurethane Coatings [2.115]–[2.118]

The term polyurethane paints (coatings) originally referred to paint systems that utilized the high reactivity of isocyanates groups with compounds containing acidic hydrogen atoms (e.g., hydroxyl groups) for chemical hardening (curing). However, this term now includes a large variety of binders. The amount of polyurethane raw materials processed into coatings is steadily increasing, and was estimated to be more than 500 000 t worldwide.

Polyurethane paint films all have a polymeric structure with urethane, urea, biuret, or allophanate coupling groups. Coupling can occur during paint hardening (curing) as the result of polyaddition of relatively low molecular mass starting products. Alternatively the paints may already contain high molecular mass polymers synthesized by the coupling of appropriate monomers. High molecular mass adducts with excess isocyanate groups or adducts in which curing occurs via oxidation of conjugated double bonds are also common.

Other variants of technical interest include blocked reactive groups, that can be activated by heat or atmospheric moisture. Microencapsulation of polyisocyanates is also becoming a deactivation method.

Depending on their chemical composition, polyurethane paints are formulated as two-component (two-pack) or one-component (one-pack) mixtures. They are applied from the liquid phase, which may be solvent-containing, aqueous, or solvent-free, or from the solid phase as a powder coating. They may be cured under a wide range of conditions: drying above 0 °C to stoving at ca. 200 °C.

2.9.1. Raw Materials

Polyisocyanates. Numerous diisocyanates may be considered as a basis for paint polyisocyanates. However, only toluene diisocyanate (TDI), hexamethylene diisocyanate (HDI), isophorone diisocyanate (IPDI), methylenediphenyl diisocyanate (MDI), and 1,1-methylenebis(4-isocyanato)cyclohexane (HMDI) are of commercial importance. High molecular mass polyisocyanates or prepolymers derived from such products are the basis of most polyurethane paint formulations. They contain only very small amounts (<0.5%) of monomers that are volatile at room temperature and thus permit safe application from the point of view of industrial hygiene.

Polyisocyanates differ as regards their chemical structure, reactivity, functionality, and isocyanate content. They are the principal curing constituent of two-pack polyurethane paints. Curing agent solutions have an isocyanate content of 5–16 wt%. Solvent-free, liquid curing agents may contain up to 30 wt% isocyanate.

Aliphatic and cycloaliphatic polyisocyanates can be used to produce lightfast, weather-resistant coatings for extremely severe conditions. Aromatic polyisocyanates are of limited suitability for exterior applications and react more quickly than aliphatic polyisocyanates. The different reactivities of aromatic and aliphatic polyisocyanates influence their use because they affect drying and curing.

Polyisocyanate types with free, excess isocyanate groups include adducts of diisocyanates with polyols, isocyanurates formed by trimerization of diisocyanates, and high molecular mass products containing biuret or allophanate groups. Monomeric MDI and its mixtures with oligomers have a low vapor pressure and can be safely handled at room temperature.

Blocked polyisocyanates are obtained by chemical addition of compounds that contain acidic or potentially acidic hydrogen atoms (e.g., alcohols). These isocyanates are not reactive at room temperature since they do not contain free isocyanate groups. Such polyisocyanates can, however, be regenerated by heating to eliminate the blocking agent.

Commercial products include Coronate (Nippoly), Desmodur (Bayer), Vestanat (Hüls), Takenate (Takeda), and Tolonate (Rhône–Poulenc).

Polyols. In conventional two-pack polyurethane paints, high molecular mass compounds containing hydroxyl groups form the "base" that cross-links and hardens with polyisocyanates. Polyester polyols, hydroxyl-containing acrylic copolymers, and polyether polyols are particularly important. They may be linear or branched and their hydroxyl group content ranges from 0.5 to 12%.

The structure and content of reactive groups determine the property profile of the polyurethane formed with the polyisocyanate. The higher the hydroxyl group content and the degree of branching, the better the hardness and solvent and chemical resistance of the paint films.

Commercial products include Desmophen (Bayer), Lumitol (BASF), Macrynal (Hoechst), and Phtalon (Galstaff).

2.9.2. Polyurethane Systems

Polyurethane systems for paints and coatings are summarized in Table 2.5.

Table 2.5. Polyurethane systems for paints and coatings

System designation	Hardening principle	Application form	Film-forming temperature
One-Pack Systems			
1	Oxidation with atmospheric oxygen	solvent-containing	up to 80 °C
2	Reaction of NCO groups with atmospheric moisture	solvent-containing	ambient temperature
3	After unblocking reaction of NCO groups with OH or NH_2 groups	solvent-containing, solvent-free, or powder	110–200 °C
4	After activation of microencapsulated polyisocyanate with OH or NH_2 groups	solvent-free	100–160 °C
5	Physical evaporation	solvent-containing	up to 80 °C
6	Physical evaporation (if necessary additional cross-linking with melamine resin)	aqueous (waterborne)	ambient temperature (with melamine resin cross-linking at 140 °C)
7	Reaction of oxazolidine with atmospheric moisture and further reaction with NCO groups	solvent-containing	ambient temperature
Two-Pack Systems			
8	Reaction of NCO groups with OH groups	solvent-containing, solvent-free, or aqueous	up to 130 °C
9	Reaction of ketimines with atmospheric moisture and further reaction with NCO groups	solvent-containing	up to 130 °C

2.9.2.1. One-Pack Systems

Curing with Atmospheric Oxygen. Type 1 in Table 2.5. "Urethane oils" are produced by reacting diisocyanates (e.g., TDI or IPDI) with polyol-modified drying or semidrying oils. "Urethane alkyds" correspond in structure to alkyd resins, in which a proportion of the dicarboxylic acid is replaced by a diisocyanate. Urethane oils and alkyds do not contain free isocyanate groups. Drying and film formation occur oxidatively, as with alkyd resins (see Section 2.6). Compared to the latter, however, they generally dry more quickly and the resultant films have better mechanical properties and a higher solvent resistance.

Curing with Atmospheric Moisture. Type 2 in Table 2.5. High molecular mass polyaddition products of polyols with excess diisocyanate contain reactive isocyanate groups. They are used to formulate one-pack polyurethane paints that cross-link with the formation of urea groups under the influence of atmospheric moisture to produce paint films with excellent resistance to chemical and mechanical attack. Solvent-free products and dissolved products with isocyanate contents of 5–15% (based on the solid resin) are commercially available.

Since products that contain free isocyanate groups react with water, moisture may affect storage stability when producing pigmented coatings. This has to be prevented, and use of water-binding additives has proved effective (e.g., special monoisocyanates such as Additive TI (Bayer)).

Blocked Polyisocyanate Systems. Type 3 in Table 2.5. The isocyanate groups of polyisocyanates can be "blocked" with agents such as phenol, butanone oxime, ε-caprolactam, or diethyl malonate that are easily eliminated or that rearrange under the action of heat. Mixtures of blocked polyisocyanates with polyols or polyamines are stable, and thus permit the production of storable one-pack stoving finishes. Stoving urethane resins containing both reaction components are commercially available. The stoving temperatures vary from 120 to 220 °C depending on the blocking agent, but can be lowered in certain cases by 10–20 °C using catalysts (e.g., dibutyl tin dilaurate).

The resulting films have a very high mechanical resistance. They are particularly important for stone impact protection in automobile finishes, for coating high-quality industrial goods, for electrical insulation, as well as for coil coating applications. Solvent-free products are used in liquid form as thick-layer systems or in solid form for powder coatings.

Microencapsulated Polyisocyanate Systems. Type 4 in Table 2.5. Some powdered diisocyanates or polyisocyanates can be deactivated when dispersed in a liquid solvent-free medium and protected by a suitable coating. Diamines or polyamines can, for example, be added to produce a stable polyurea layer that requires ca. 1% of the isocyanate groups. Since microencapsulated polyisocyanates do not cross-link with reactive groups under normal conditions, they form stable one-pack systems at room temperature. At higher temperature (> 80 °C) the stabilizing coating melts and dissolves, allowing curing to occur. Such systems (e.g., those based on dimerized TDI) are used as underbody coatings in automobile construction.

Physically Drying Systems. Type 5 in Table 2.5. These systems are based on linear or slightly branched polyurethanes without reactive groups that are soluble in organic solvents; MDI and IPDI are usually used as starting isocyanates. On account of their high molecular mass these polyurethane resins generally yield highly viscous solutions with a relatively low solids content. The rapidly drying paints are extremely flexible and elastic and have a high resistance to mild solvents. Important uses include the coating of flexible substrates such as leather imitates and magnetic tapes.

Aqueous Systems. Type 6 in Table 2.5. Aqueous dispersions of polyurethanes that are slightly supported by ionic groups also undergo physical film formation. The binder consists of polymer chains that are coupled via urethane and urea groups, and contain basic or acid groups. Neutralization by salt formation provides the necessary hydrophilicity if the "self-emulsifying" properties resulting from incorporation of hydrophilic polyether radicals are insufficient.

Anionic dispersions based on IPDI, HMDI, and HDI are particularly important and are largely used for industrial applications (e.g., as glass fiber sizing or as a finish for leather and leather imitates). Chemical cross-linking is achieved by the use of water-soluble melamine resins. Stoving temperatures of ca. 140 °C are necessary.

Systems with Polyisocyanates and Blocked Reactants. Type 7 in Table 2.5. Instead of blocking the isocyanate groups, they can be combined with latent reactants (e.g., oxazolidines) to obtain storage-stable one-pack systems. After application, curing is achieved not by heat, but by atmospheric moisture which induces ring opening and the formation of reactive hydroxyl and amino groups.

2.9.2.2. Two-Pack Systems

Polyisocyanate and Polyhydroxyl Systems. Type 8 in Table 2.5. This is by far the most important system and comprises conventional polyurethane paint. The principal binder constituents (polyisocyanates and polyhydroxyl compounds) must be kept separate before application. In additon to polyols, many other hydroxyl-containing components can be used for the formulation (e.g., alkyd resins, epoxy resins, castor oil, and cellulose nitrate). (See Section 2.9.1). The mixing ratio of base (hydroxyl component) to curing agent (polyisocyanate) is generally between 1:1 and 10:1. The two components are mixed homogeneously immediately before application. Industrially, mixing can be performed automatically in two-component spray equipment.

Two-pack polyurethane paints can be applied by all conventional methods apart from dipping. Curing takes place at ambient temperature, but can be accelerated if necessary by heat. Stoving is not necessary (energy saving).

The reaction between the isocyanate and the hydroxyl groups results in 100% cross-linking if the reactants are converted in stoichiometric proportions. Optimum paint films are generally formed when NCO:OH equals 1. However, it may be advantageous to deviate from this ratio (under- or over-cross-linking). With aliphatic polyisocyanates, a catalyst (accelerator) is generally necessary. Small amounts (0.05 – 0.5 %) of tertiary amines or metal-containing compounds are normally added to the binder. Amine accelerators can also be added to the air stream used for spraying. Alternatively, the paint films may be cured in an amine-containing atmosphere.

Esters, ketones, and ether esters of polyurethane grade (i.e., absence of reactive constituents and a maximum water content of 0.05%) are required for solvent-containing paints. Aromatic hydrocarbons are suitable as diluents. High-solids two-pack paints containing 30–40 wt% of volatile components are becoming more popular on account of their low solvent emission.

Solvent-free two-pack systems require liquid reaction components and are normally based on MDI and its homologues. In contrast to solvent-containing paints, pigmented solvent-free two-pack systems require the use of water-binding agents (e.g., zeolites) to obtain thick, bubble-free coatings.

Aqueous two-pack polyurethane paints are obtained by combining, for example, hydroxyl-containing polyacrylate dispersions with water-emulsifiable, low-reactive aliphatic polyisocyanates.

Systems with Polyisocyanates and Blocked Reactants. Type 9 in Table 2.5. In special cases, compounds with blocked amine groups (e.g., ketimines) can be used as reaction partners for polyisocyanates. The chemical curing of these two-pack systems is initiated by humidity which causes regeneration of the reactive amine groups.

2.9.3. Properties and Uses

Polyurethane coatings have three main advantages: high mechanical resistance, outstanding chemical resistance, and (in the case of aliphatic polyisocyanates) excellent lightfastness and weather resistance. The properties of the paint systems which harden at ambient temperature are unsurpassed. They can also be combined and formulated in a large number of ways. Specific properties can be optimally adjusted even for extreme conditions.

On account of their broad range of properties, polyurethane paints are used industrially and on a trade scale in virtually all sectors. Practical applications range from paper coating to the protection of equipment in industrial plants. Major areas of use include transportation (large motor vehicles, rail vehicles, aircraft, automobile finishes and refinishes), the building sector (parquet floor coatings, outdoor coatings, floor coatings, sealing membranes), the industrial paints sector (equipment, machinery, furniture), as well as steel structural engineering and hydraulic steel structures. Polyurethane paints are particularly important in the surface treatment of plastics.

2.10. Epoxy Coatings

Epoxy resins are generally not used alone but require a reaction partner in order to be cured (hardened). A large number of reaction partners may be used for curing at elevated or at room temperature. The cured films have high adhesion, flexibility, hardness, abrasion resistance, resistance to chemicals, and corrosion protection. In 1995 ca. 700 000 t of epoxy resins were used worldwide by the coating industry (including civil engineering applications).

2.10.1. Epoxy Resin Types

Bisphenol A Resins. Most epoxy resins are condensation products of bisphenol A and epichlorohydrin. Depending on the ratio of bisphenol A to epichlorohydrin, resins of varying chain length are obtained, which differ in molecular mass, melting point, viscosity, solubility, and content of epoxy and hydroxyl groups. The low molecular mass types (M_r ca. 360–500) are liquid and the medium molecular mass types (M_r ca. 500–7000) are solid at room temperature. Low and medium molecular mass resins require reaction partners to give useful coatings. High molecular mass resins give useful, physically drying binders, mainly primers. These resins contain very few epoxy groups but many hydroxyl groups. The hydroxyl groups can be reacted at room temperature with polyisocyanates or at elevated temperature (180–300 °C) with amino resins and phenolic resins [2.119]–[2.121].

Bisphenol F Resins. Bisphenol F is produced by condensing phenol with formaldehyde in an acid medium. Bisphenol F is a mixture of isomeric and oligomeric products (novolacs). Resins produced by reaction of bisphenol F with epichlorohydrin have lower viscosities and a somewhat higher functionality (more epoxy groups) than the corresponding bisphenol A resins. Like the liquid bisphenol A resins, they tend to crystallize. Mixing of bisphenol A and F resins prevents or reduces crystallization. On account of their slightly higher epoxy content, coatings based on bisphenol F resins have a somewhat higher solvent resistance than similarly formulated bisphenol A resins. Coatings based on bisphenol F tend to yellow more than those based on bisphenol A.

Epoxy Novolacs. Epoxy novolacs are made by condensing formaldehyde with phenolic substances in an acid medium, followed by epoxidation. Only the epoxy phenol novolac and cresol novolac resins have gained importance in the market. Although bisphenol A novolac resins are also available, they have not succeeded in making a commercial breakthrough. All of these epoxy novolacs have a substantially higher functionality than bisphenol A resins, ranging from ca. 2.2 (phenol novolacs, e.g., D.E.N. 431, Dow Chemical) to ca. 5.8 (cresol novolacs, e.g., Araldite ECN 1299, Ciba-Geigy). The high functionality of these resins allows formulation of coating systems with a high solvent resistance (high network density). The aromatic structure of the resins is responsible for the high glass transition temperature and good resistance to aqueous and acid solutions of the coatings. The chemical structure and high functionality lead to coatings with limited flexibility and a somewhat lower adhesion to metallic substrates than bisphenol A epoxy resins.

Aliphatic Epoxy Compounds. Epoxidized alcohols are used as reactive diluents for epoxy resins. The reaction products of diols with epichlorohydrin (diglycidyl ethers of butanediol, polypropylene glycol, etc.) are also used as reactive plasticizers. Polyglycidyl ethers (e.g., of sorbitol and pentaerythritol) are commercially available. They have a higher viscosity than the aliphatic mono- and diglycidyl ethers. They show good solvent resistance when cured with cycloaliphatic amines. Such resins are also cured with carboxy-functional polyesters, polyacrylates, or polyanhydrides to give coatings with good weather resistance.

Aliphatic epoxy resins generally have higher color stability and reactivity than aromatic epoxy resins, but their resistance to aqueous acid solutions is much lower.

Cycloaliphatic Epoxy Resins. Cycloaliphatic epoxy resins are produced by oxidizing olefins with peracids. They are of only minor importance in surface coatings. They are cured with anhydrides, carboxy-functional substances, or Lewis acids at 150–200 °C. Ultraviolet curing with triarylsulfonium salts and ferrocenes has become technically important. Coatings based on cycloaliphatic epoxy resins have a high gloss and good weather resistance. They adhere to metallic substrates better than the acrylate esters normally used for UV curing but are considerably more expensive.

Glycidyl Esters. Glycidyl ester resins were originally developed for electrical applications. Glycidyl esters of phthalic acid, hexahydro phthalic acid, terephthalic acid or trimellitic acid (e.g. Araldite PY 284, PT 910) cured with carboxy functional polyesters or polyacrylates at elevated temperatures give coating with both excellent colour stability and outdoor resistance.

Heterocyclic Epoxy Compounds. The technically most important heterocyclic epoxy resin is triglycidyl isocyanurate, obtained by reacting cyanuric acid with epichlorohydrin (Araldite PT 810, Ciba-Geigy; Tepic, Nissan). This trifunctional epoxy resin is combined with carboxy-terminated polyesters to give weather-resistant powder coatings [2.122]–[2.124].

2.10.2. Curing Agents

The most important combinations of epoxy resins and curing agents are summarized in Table 2.6. Depending on their molecular mass, bisphenol A epoxy resins can be cured by polyaddition via their epoxy or hydroxyl groups. Polyamines, polythiols, and polyisocyanates are suitable for room temperature cure. Polyanhydrides, polyphenols, acids, and carboxy-functional polyesters are suitable for hot cure.

Table 2.6. The most important epoxy resin–curing agent combinations and their uses

Curing agent	Curing mechanism	Resin type*	M_r	Typical use
Amines and polyamidoamines	via epoxy group / curing at room temperature	liquid	400	two-component systems for heavy-duty corrosion protection and floorings building blocks for chemical modification
Polyamidoamines and amine adducts		solid, type 1 solid, type 2	1 000	industrial maintenance and marine coatings
Latent amines Polyesters		solid, type 3		powder coatings
Phenolic hardeners	hot curing	solid, type 4	2 000	powder coatings
Amino resins		solid, type 4	2 000	epoxy esters for industrial finishes and can coating
Phenolic resins		solid, type 6		can coatings and finishes
Amino resins		solid, type 7		
Polyanhydrides		solid, type 8	4 000	coil coatings
Phenolic resins	via hydroxyl group	solid, type 9		can coatings and finishes
Amino resins				two-pack polyurethane paints
Isocyanates			5 000	
Isocyanates		high molecular mass epoxy resins		primers (cold curing)
Phenolic resins			10 000	can and coil coatings
Amino resins				

* Types 1–9 indicate increasing molecular mass of the respective solid resins.

Epoxy resins can also be cured by polycondensation with amino resins or phenolic resins. Epoxy resins can be polymerized with catalysts such as tertiary amines, boron trifluoride complexes, ferrocenes, and triarylsulfonium salts.

Polyamines. Aliphatic polyamines cure epoxy resins via their epoxy groups at ambient temperature. Excess amine is generally prereacted with the epoxy resin to form a polyamine adduct. Amines react with monomeric or dimeric fatty acids to form polyamidoamines. Polyamine-adduct-cured epoxy resins have a high chemical resistance. Polyamidoamine-cured epoxy resins exhibit good adhesion and flexibility.

Cycloaliphatic polyamines are less reactive than aliphatic amines, and an accelerator must therefore be used for curing at room temperature (e.g., salicylic acid). Aromatic polyamines are even less reactive than cycloaliphatic amines. Therefore, accelerators must be used to cure aromatic amines with epoxy resins at room temperature [2.125].

Thiols. Thiols have an extremely unpleasant smell, and are only of industrial importance for adhesives and joint groutings. They give a high, permanent flexibility to epoxy resins.

Isocyanates. Whereas polyamines and thiols cure epoxy compounds via their epoxy groups, isocyanates cross-link high molecular mass epoxy resins via their hydroxyl groups to form polyurethanes. The reaction takes place at ambient temperature. These combinations cure more rapidly and at lower temperature than epoxy resins cured with polyamines.

Anhydrides. Polyanhydrides and not monoanhydrides must be used to cure epoxy surface coatings. They are used in powder form for powder coatings and in solution for can coatings; both forms are hot curing. The films have a good acid resistance, and do not impart an undesirable taste to foods.

Acids and Carboxy-Functional Polyesters. Curing of epoxy resins with acids and carboxy-functional polyesters requires heat. Industrially, these systems are most important in the formulation of powder coatings: more than 70% of powder coatings are based on epoxy resins and carboxy-functional polyesters (see also Sections 3.4.2 and 3.4.3).

Polyphenols. Polyphenols react with epoxy resins on heating, but require an accelerator (e.g., tertiary amines, imidazoles). On account of their low color stability, powder coatings based on this combination are not used for decorative purposes, but exclusively for functional purposes where high thermal, mechanical, and chemical resistance is required (e.g., pipe coatings).

Amino Resins. Urea, melamine, and benzoguanamine resins react with high molecular mass epoxy resins on heating. Their hydroxymethyl groups react with one another and with the hydroxyl groups of the epoxy resins to form ether bonds. Such systems are used for coating domestic appliances as well as for packaging.

Phenolic Resins. Phenolic resins react with high molecular mass epoxy resins on heating. Such systems are used as "gold lacquers" to line food containers. The similarly synthesized bisphenol A resols produce colorless coatings with a higher chemical resistance, less odor during cure, and less alteration of the taste of food in contact with the coating. Flexibility is, however, lower than with standard phenolic resins.

Catalytically Curing Compounds. Although tertiary amines can polymerize epoxides even at room temperature, the degree of polymerization is too low to obtain useful coatings. Boron trifluoride complexes polymerize epoxides on heating. Such combinations are sometimes used in powder coatings if high solvent resistance is required.

Ultraviolet curing of epoxy resins has become important. Cycloaliphatic epoxy resins are combined with substances that induce polymerization under the influence of UV radiation or an electron beam. Industrially important products include triarylsulfonium salts (UVE 1014, 1016, General Electric) and arene–ferrocenium compounds (CG 24-061, Ciba-Geigy). To ensure good film flexibility and chemical resistance, polyols should be added. The films should be postcured for 30–120 s at ca. 100 °C [2.126].

2.10.3. Chemically Modified Epoxy Resins

Epoxy Resin Esters. Air- or oven-drying esters can be produced by esterification of epoxy resins (M_r 1000–2000) with fatty or oleoresinous acids (fatty acids of linseed oil, soybean oil, tall oil, coconut oil, or dehydrated castor oil).

Plasticized epoxy resins can be produced by reacting low molecular mass epoxy resins with dimeric fatty acids. These resins (e.g., Epikote 872, Shell) generally exhibit better substrate wetting than the unmodified epoxy resins. This reaction is, however, associated with the formation of ester bonds, and the resins therefore have a poor resistance to alkaline solutions. Instead of fatty acids, carboxy-functional polybutadiene-acrylonitrile elastomers can also be used to flexibilize epoxy resins.

Esters of Epoxy Resins and Acrylic Acid. If low molecular mass liquid to semisolid epoxy resins are heated with acrylic acid at ca. 120 °C, epoxy acrylates are produced. These resins can be dissolved in low molecular mass acrylate esters and used as raw materials for UV-curable finishes. They are mainly used for paper and cardboard (record sleeves), but also for metals.

Esters of Epoxy Resin and Methacrylic Acid. Epoxy resins can also be reacted with methacrylic acid on heating. The resulting methacrylate ester resins are dissolved in polymerizable solvents and catalyzed with peroxides (e.g., Derakane, Dow Chemical). These catalyzed resins are applied as thick-layer laminates in combination with glass fibers, fabrics, or mats. The laminates are generally cured at ca. 60 °C and have extremely good chemical resistance; they are therefore used in the chemical industry for the internal lining of reactors and storage tanks [2.127]–[2.130].

2.10.4. Uses

2.10.4.1. Curing at Ambient Temperature

Solvent-Free Coatings. Solvent-free applicable binder systems which cure at ambient temperature are not only be used for coating applications. By the addition of quarz sand to low viscosity binder systems floorings can be formulated. Depending on the choosen filler/binder mixing ratio these systems are applied either as self-levelling floorings or as mortar screeds. Adhesives and joint fillers for concrete applications are formulated in a similar way. For all these uses only the low molecular mass liquid bisphenol A and bisphenol F epoxy resins (M 340–400) or their mixtures are suitable. While solvent-free coatings are applied in film thicknesses between 300 to 1000 microns, are floorings applied in 2 to 10 mm layers, depending on the quarz sand content and the application method used. Reactive diluents (cresyl and isooctyl glycidyl ethers, diglycidyl ethers of 1,4-butanediol or 1,6-hexanediol) are often added in amounts of 10–20% to these resins. They react with the curing agent and lower the viscosity of the epoxy resin (1000–3000 mPa · s at 25 °C) to improve processing properties. Polyamines and modified polyamines are used as curing agents. Aliphatic polyamines, (EDA, DETA, TETA, Dow; Vestamin TMD, Hüls; Lavomin A 327, BASF) are industrially very important because they are highly reactive and can cure epoxy resins at ambient temperature. Epoxy resin coatings that are cross-linked with aliphatic amines have the highest solvent resistance. Cycloaliphatic amines (Laromin C 252, C 260, BASF; Vestamin IPD, Hüls) have to be modified and accelerated to permit curing at room temperature. These amines give a very attractive appearance to the cured film and are mainly used for decorative wall and floor finishes. Aromatic amines (e.g., diaminodiphenylmethane) are solid, but can be used for cold curing by adduct formation and acceleration. Films based on such curing agents (Ancamin LT, Anchor; HY 830, HY 850, Ciba-Geigy) have a glasslike appearance and extremely good resistance to acid.

Solvent-free epoxy systems must be applied within an extremely short time (30 – 60 min), which presents problems, particularly in hot countries. Such systems are also frequently applied with heatable two-component spraying equipment. Ketimines give a longer pot life. They are produced by condensing ketones and polyamines (Epikure H 3, Shell). When applied as a thin layer, they react with atmospheric moisture and are converted back into the ketone and polyamine. The ketone evaporates from the film and the amine cross-links the epoxy resin. Ketimines can only be used if a sufficiently high ambient temperature ($>20\,°C$) and high atmospheric moisture ($>75\%$ R. H.) can be guaranteed.

Solvent-Containing Coatings. Semisolid to solid bisphenol A resins (M_r 500 – 1500) are used to formulate solvent-containing paints. They are sometimes mixed with epoxy novolacs to increase the network density and thus the solvent resistance of the coating. Aromatic hydrocarbons, often mixed with alcohols, are generally used as solvents. The epoxy resins are cross-linked with modified aliphatic polyamines, in the form of in situ amine adducts or as isolated amine adducts dissolved in xylene and butanol. Polyamine adduct – epoxy resin combinations have an extremely good hardness and good resistance to aqueous alkaline solutions and organic solvents.

In practice, polyamidoamines are the most important curing agents for these types of epoxy resins. Polyamidoamines and polyamidoamine adducts with epoxy resins confer maximum flexibility and adhesion to epoxy coatings. Such systems are frequently used as primers on metallic and mineral substrates. As topcoats, epoxy paints are only used for interior applications since they tend to yellow, become matt, and exhibit chalking under the influence of light and weather. The cured films exhibit good resistance to aqueous, alkaline, and neutral solutions; their resistance to organic solvents is limited. In order to reduce the use of organic solvents in paints low VOC (high solids) coatings are becoming more and more importance in both Europa and USA. The solids content of these ambient cured coatings can be increased by the partial replacement of the solid epoxy resins by liquid bisphenol A – or bisphenol F epoxy resins. Some raw material supplier offer modified bisphenol A resins (e.g. Epikote 874 × 90, Shell; Araldite XB 290, Ciba) and curing agents (Witco, various modified polyamidoamine based hardeners), which allow the formulation of coatings with a low VOC content (below 300 g/lt).

High molecular mass epoxy resins (M_r 3000 – 20 000) may be regarded as polyols, and are cross-linked with isocyanates to form extremely chemically resistant coatings

(Desmodur N, L, Bayer; Vestanat IPDI, Hüls). Epoxy resin–isocyanate combinations are also used to formulate rapidly curing shop primers.

Solvent-Containing, Air-Drying Epoxy Esters. In order to produce epoxy resin esters that can be cured at room temperature, solid epoxy resins (M_r 1500–2000; e.g., Araldite GT 6084, Ciba-Geigy; Epikote 1004, Shell) are reacted with air-drying fatty acids (e.g., linseed oil or soybean oil fatty acid). The area of application of these resins is very similar to that of air-drying alkyd resins, but they have a higher chemical resistance and improved flexibility. Normal driers (e.g., cobalt octoate) are used to accelerate drying.

Waterborne Coatings. Low molecular mass liquid bisphenol A and bisphenol F epoxy resins (M 340–400) and their mixtures are made water emulsifiable by either adding nonionic emulsifiers or epoxy group containing emulsifiers which take part in the curing reaction (Araldite PY 340-2, PY 3960, Ciba; Beckopox EP 147 w, Vianova). Water-soluble polyamidoamines or modified polyimidazolines (e.g. Casamide 360, 362, Anchor former AKZO, Hardener HZ 340, Ciba) are used as curing agents. Such two component systems are environmentally friendly and are mainly used for coating concrete and other mineral substrates as well as pretreated plastics. Up till now more than 80% of the used water-borne binder systems are applied onto these substrates, this because of the fact that water-borne epoxy systems adhere well on wet concrete surfaces and do not attack plastics like solvent-containing coatings do. Aqueous emulsions based on solid resins (M ca. 1000) containing ca. 10% organic solvents are commercially available (Beckopox EP 384, Vianova; Araldite PZ 3961, Ciba). These emulsions are less reactive than those based on liquid bisphenol A epoxy resins. Therefore they can be combined with water-soluble polyamine adducts (e.g. Beckopox Hardener EH 623, EH 613, Vianova; Hardener HZ 3980, HZ 3981, HZ 3982; Ciba; Hardener XE 36, Witco) so that even white coatings with good chemical resistance properties can be formulated. Due to the higher film forming temperature – compared to the liquid resins based systems – such binder systems should not be used for civil engineering applications. They are mainly developed for the surface coating protection of metallic substrates. Coatings with improved flexibility and adhesion on steel can be obtained by the use of modified bisphenol A epoxy resin emulsions (e.g. Beckopox EP 385, Vianova; Araldite PZ 3962, Ciba) which allow the formulation of highly filled primers. Their corrosion protection properties are similar to primers formulated on solvent-containing epoxy resins.

2.10.4.2. Curing at Elevated Temperature

Powder Coatings. Powder coatings mainly contain solid bisphenol A resins with M_r between 1000 and 2000 (Araldite GT 6063, 6064, 7004, Ciba-Geigy; DER 662, 663, 664, Dow Chemical; Epikote 1055, Shell). Carboxy-functional polyesters are the most important curing agents (Grilesta, Emser Werke; Grylcoat, UCB; Uralac, DSM). Modified and accelerated dicyanodiamides and polyphenols, and in some cases polyanhydrides may also be used. For a detailed description of powder coatings, see Section 3.4.

Powder coatings based on aromatic epoxy resins have high chemical and mechanical resistance, but are not resistant to weathering. To obtain weather-resistant powders, triglycidyl isocyanurate (Araldite PT 810, Ciba-Geigy; Tepic, Nissan) must be used as the epoxy component with carboxy-functional polyesters.

Due to the fact that animal studies have shown that triglycidyl, isocyanurate is not only a severe eye irritant and a mild skin nasal irritant but also that TGIC can cause genetic damage giving rise to concern over potential reproductive or carcinogenic effects therefore alternative less toxic epoxy resins have been introduced. Most similar to the well known film properties of the TGIC/carboxy-functional polyester powder coatings are those based on glycidyl esters (e.g. Araldite PT 910, Ciba) which are cured with carboxy-functional polyesters too. Both systems dominate the architectural and exterior coatings sector since they combine good weather stability with good flexibility and excellent adhesion to steel and aluminium (2.213, 2.123). Latest development by DSM combines epoxydized linseed – oil with carboxy-functional polyesters, so that binder systems are available which do not need to be labelled (Patent DE 4340974 C2).

Solvent-Containing Paints. Butylated urea–formaldehyde resins, butylated or methylated melamine–formaldehyde resins, and butylated benzoguanamine–formaldehyde resins react with the hydroxyl groups of epoxy resins (M_r 3000–10000) at elevated temperature (160–200 °C) under elimination of alcohols. Ether bonds are formed that are largely responsible for the good chemical resistance of such paints. The coatings have outstanding adhesion, impact resistance, good flexibility, and high surface hardness. Epoxy resin–urea resin combinations are mainly used as primers or one-coat systems in the packaging sector, whereas combinations with melamine and benzoguanamine resins are mostly used as topcoats [2.131].

Phenol–formaldehyde resins (in some cases butylated) and to a small extent bisphenol A–formaldehyde resins have become important as curing agents for epoxy resins in can coatings. Curing occurs primarily via the hydroxyl group of the epoxy resin and the etherified methylol groups of the phenolic resin. The epoxy groups also react with the phenolic hydroxyl groups of the phenolic resin. The coatings are cured at ca. 200 °C and are extremely resistant to foods, cosmetics, and pharmaceuticals.

The use of carboxy-functional or anhydride-functional low molecular mass polyesters as curing agents is more recent. They are used to formulate white-pigmented can coatings. Blocked isocyanates (Desmodur AP stable, Bayer) cure high molecular mass epoxy resins on heating. The films have an extremely high hardness, good flexibility, and high resistance to chemicals.

Hot-Curing Epoxy Resin Esters. Solid epoxy resins with a molecular mass of ca. 2000 can be esterified with semidrying and/or nondrying fatty acids. These resins are used in combination with amino resins for stoving primers, can coatings, and industrial finishes. The importance of epoxy resin esters has, however, decreased in recent years.

Waterborne Paints and Coatings [2.132]. For ecological reasons waterborne industrial paints and coatings have become extremely important. The epoxy or hydroxyl groups of epoxy resins must be chemically modified to make them waterborne. Water-soluble epoxy paints can be produced by esterifying the resins (M_r 1000–2000) with a monocarboxylic fatty acid, followed by reaction with an anhydride. The resins are then dissolved in a water-soluble solvent (e.g., glycol ether or an alcohol) and made waterborne by neutralization with a monoamine. Such resins are cured with amino resins (e.g., hexamethoxymethylmelamine) at ca. 160 °C. The waterborne epoxy paints that are applied by anodic electrodeposition have a similar structure. Cathodically deposited coatings (as employed in the automotive industry) are generally produced by reacting low molecular mass epoxy resins (M_r ca. 1000) with a secondary amine, followed by neutralization with an organic acid. Blocked isocyanates are usually used for curing [2.133].

Waterborne coatings are increasingly used for can coatings. The binders are mainly based on epoxy resins. High molecular mass epoxy resins (M_r 3000–4000) can be made waterborne by graft polymerization with acrylates, methacrylates, and/or acrylic or methacrylic acid in the presence of styrene, and subsequent neutralization with amines. These waterborne polymers are normally cured with amino resins. All of the coatings exhibit good adhesion and flexibility and are mainly used for beer and beverage cans [2.134].

2.10.4.3. Radiation Curing

Cycloaliphatic epoxy resins are normally used for cationic radiation curing (see also Section 3.7). A strong Lewis or Brønsted acid is formed after photolysis of the initiator. This acid opens the epoxide ring, thus initiating the polymerization reaction. The chain length is thereby increased. Aryldiazonium salts are no longer used as initiators due to their poor dark storage stability and unsatisfactory film quality. Diarylhalonium salts are only employed to a small extent. Triarylsulfonium salts are the most important initiators. The special advantages of cationic cure are the possibility of coreaction with substances containing reactive hydrogen atoms (e.g., for flexibilization), insensitivity to oxygen during cure, and low shrinkage [2.135].

The systems are color-stable, weather resistant, flexible, and also adhere well to tin-free and low-tin sheet metals, as used in the manufacture of beverage and aerosol cans. Although arene–ferrocenium complexes even allow the cure of bisphenol A epoxy resins, provided postcuring takes place, these systems have not yet become important in surface protection applications. The reasons for this are inadequate color stability and weather resistance as well as the intense color of these films.

2.10.5. Toxikology

Liquid unmodified bisphenol A-, -A/F- and -F resins produce little or no irritation of skin or muscous membranes; they can however act as sensitizers and cause dermatological allergies.

Solid epoxy resins based on bisphenol A can be categorized as effectively non-toxic; they are not irritating and scarcely sensitizing. The same is true for the novolac epoxy resins although their liquid forms can have a sensitizing effect.

Reactive diluents and low molecular mass aliphatic epoxy resins are of low viscosity, show a noticeable vapor pressure and have to be handled with care. These substances produce medium to strong irritation of skin and mucous membranes; they are also sensitizers and can cause skin disorders.

Epichlorhydrin which is used for the synthesis of epoxy resins is a suspected carcinogen in animals. Thanks to the well developed manufacturing processes, practically all the current commercially available epoxy resins contain only the smallest traces of epichlorhydrin (Occupational Hygiene Advices for Manufacturing and Processing Plastics of CIBA, Company brochure).

2.11. Silicone Coatings

Polysiloxanes or silicones have been used in surface coatings since the earliest days of the silicone industry [2.136]. They are versatile materials with an enormous range of properties and physical forms [2.137]. They have a very low surface energy, high thermal stability, are UV transparent, and have a low glass transition temperature (T_g). These properties are utilized by the paint industry to provide excellent water repellency, resistance to weathering, UV radiation, and thermal cycling. Silicones used in paints can be conveniently classified into four types: silicone resins, blending resins, silicone organic copolymers, and paint additives.

Silicone Resins. Silicone resins used in paint applications are usually produced from mixtures of di- and trifunctional organosilane monomers (e.g., phenyl- and methyltrichlorosilane, methylphenyldichlorosilane, and dimethyldichlorosilane) by controlled hydrolysis and condensation. Typically, mixed chlorosilanes in xylene are added to water; hydrolysis produces a layer of concentrated hydrochloric acid which is then removed. The remaining solution of siloxane oligomers (M_r 2000–4000) is neutralized with sodium hydrogencarbonate and washed with water. At this stage the low-viscosity resin contains 3–6% silanol. Further condensation (bodying) and azeotropic distillation of water in the presence of catalysts such as zinc octanoate reduces the silanol level to 0.5–2% and yields high-viscosity network resins (M_r 200 000–500 000). Alternatively, alkoxysilanes (most commonly methoxysilanes) may be used as starting materials in which case alcohol, usually methanol, is liberated during hydrolysis, bodying, and curing. The final properties of the resin can vary greatly depending on the nature of the substituents, the degree of substitution, and the extent of the condensation reaction and hence the content of residual reactive groups [2.135]. Products vary from hard, brittle monomethyl flake resins to soft, flexible phenyl-containing resins dissolved in a solvent. Resins with a higher phenyl content are more thermoplastic, have greater heat and oxidation resistance [2.138], longer shelf life, and tend to be more compatible. A high methyl content confers faster cure rates, flexibility, hardness, and better retention of properties at low temperature. The level and type of substituents selected depend on the application requirements. For example, silicone resins formulated with white or colored pigments can withstand continuous service temperatures of 260–370 °C, whereas aluminum-filled paints made with phenyl-containing silicone resins resist temperatures up to 700 °C by forming protective ceramic frits [2.139]. Silicone-resin-based paints are used in extreme environments where organic paints are ineffective [2.140], e.g., boiler stacks, exhausts, ovens, furnaces, heat exchangers, hot plates, grills, air-conditioning units, and jet engines.

Silicone Organic Blends. Pure silicone resins are too soft and thermoplastic for general paint application but are frequently blended with traditional organics such as alkyd, acrylic, epoxy, and phenolic resins. The higher levels of reactive groups in unbodied silicones provide sites for copolymerization on curing; the low molecular mass (2000–4000) ensures good compatibility which is essential for optimum film-forming properties. The organic component can improve the hardness and drying properties of the silicone while the silicone adds gloss and color retention at elevated temperature. These properties are employed in exterior maintenance paints. Improvement in durability is generally proportional to the silicone content of the blend. This in turn is determined by performance – cost considerations and usually lies in the range 30–50%.

Silicone Organic Copolymer Resins. Simple blends of organic and silicone resins are unsuitable for many high-performance applications where mechanical properties and good weather resistance are of primary importance. They may become incompatible at the required levels and copolymerization is therefore preferred. At levels of 20–60% the silicone copolymers impart considerable improvements over unmodified organics, particularly in heat stability, chalking resistance, and gloss retention [2.141]. The silicone intermediates are usually low molecular mass oligomers that are rich in reactive hydroxyl or alkoxy groups and have a high level of organic substitution. They are copolymerized with polyester, alkyd, and acrylic resins containing 80–250% excess hydroxyl groups, usually in the presence of titanium catalysts [2.135]. If silanol-functional intermediates are used, the titanium catalysts promote copolymerization via –COH and –SiOH groups, thus eliminating the tendency for homopolymerization via SiOH groups. Choice of solvent and catalyst, concentration of reactants, and extent of reaction play an important role in the final composition and performance of the copolymer. Methoxy-functional silicone intermediates react with organic hydroxyl groups by elimination of methanol. Use of special catalysts, e.g., $Al(CH_3COCHCOCH_3)_3$ and $Ti[OCH(CH_3)_2]_4$, allows better control of the condensation reaction and lessens the yellowing caused by many catalyst systems at elevated temperature [2.142]. Materials made in this way (e.g., silicone polyesters) can contain as much as 80% silicone resin. Almost all types of organic coating have at some time been modified with silicone resins [2.143] although commercial success has been realized mainly with polyesters. A major user of silicone polyester paints is the coil coating industry where coatings containing > 50% silicone have retained their gloss and color for more than ten years [2.140]. Other applications include appliance finishes, exterior cookware coatings, and colored maintenance and architectural paints.

Curing. Curing cycles are primarily determined by the silicone content of the resin [2.139]. A typical cure schedule for a 100% silicone resin is 30 min at 260 °C. For 50–80% silicones, satisfactory cures are obtained in 15 min at 220 °C. For silicone-modified organic coatings, complete cure can be achieved using ambient or low-bake (< 180 °C) cure conditions.

Silicone Paint Additives. (See also Section 5.3.). Because of their very high surface activity, modified silicone fluids and resins are widely used in paint formulations to achieve special desirable effects [2.144], [2.145]. The only silicone additive likely to cause surface finish defects is an overdose of polydimethylsiloxane. New siloxane organic copolymers are compatible with organics and overcome problems with fish eyes (oval surface defects), cratering, and recoatability. They perform at very low concentrations (generally 0.01–0.5%) and are used to improve flowout, mar and scuff resistance, adhesion, leveling, and gloss. They can also control foaming [2.146], aid pigment dispersion, and impart special textured effects (e.g., hammer finishes). Silicone additives can be used in most paints including solvent-based, waterborne, high-solids, and powder coatings.

2.12. Urea, Benzoguanamine, and Melamine Resins for Coatings

Alkylated urea–, benzoguanamine– and melamine–formaldehyde resins represent a versatile group of cross-linking agents for hydroxy-, amide-, and carboxy-functional polymers. They are used in both waterborne and solventborne coatings, including high-solids industrial coating systems. These amino resins are almost always employed in combination with other flexibilizing functional polymers (backbone polymers) to form highly cross-linked networks.

The major application for amino resins is in baking systems; there is a small market for acid-catalyzed wood coatings that are cured at room temperature. Excellent properties can be obtained in both the formulated coating and the cured film. Advantages include excellent exterior durability, adhesion to metal, gloss and color retention, relatively low cost, and excellent stability. Amino resins have found widespread use in automotive primers and topcoats, appliance coatings, metal decorating, and most general industrial applications. They are usually modified by etherification with alkyl ether groups; in the paint industry the term alkylated (methylated, butylated, etc.) is commonly used for these modified resins.

The first amino resins used in coatings were partially butylated, polymeric urea–formaldehyde (UF) resins introduced in the late 1920s. Melamine–formaldehyde (MF) resins followed in 1935. In 1960 a range of fully alkylated, predominantly methylated, MF resins were introduced. These products are more monomeric in nature and supplied as 100% solids; hexakis(methoxymethyl)melamine (HMMM) is a typical example. Fully alkylated MF resins have gained a major market share in the United States accompanied by increasing penetration in Europe and Japan. Replacing a conventional partially alkylated amino resin with a fully methylated resin increases the solids content of a paint by 3–8%. Reasons for their success include higher application solids, improved hardness–flexibility relationship, better exterior durability, improved reaction with functional groups on the backbone polymer, stability in waterborne coatings, reduced emission of formaldehyde, and high reactivity under controlled conditions. The incorporation of cross-linking agents of this type provides the formulator with the capability to utilize low molecular mass backbone polymers.

Properties. Alkylated amino resins can be classified into two general classes: (1) polymeric, partially alkylated resins which have a lower solids content and (2) the more monomeric, fully and partially alkylated products which have a higher solids content.

Typical *polymeric, partially alkylated amino resins* are butylated or isobutylated condensates with an average degree of polymerization between 3 and 8 and a combined formaldehyde content of 1.4–1.8 per amino group [2.147]–[2.149]. The hydroxymethyl (methylol) groups are partially alkylated (degree of alkylation 40–80%).

The final resin has a solids content of ca. 50–65%, and a combination of functional sites of different reactivity [2.148]. The solvents most commonly used are aromatic hydrocarbons and alcohols.

The higher reactivity of partially alkylated amino resins was long attributed to the presence of free hydroxymethyl groups. Under weakly acidic cure conditions, these hydroxymethyl groups demethylolate to form alkoxymethyl groups adjacent to secondary amide groups which respond more favorably to general acid catalysis [2.149]–[2.152].

Since 1970 *fully alkylated monomeric melamine resins* based on hexakis(methoxymethyl)melamine (HMMM) and higher-solids methylated melamine resins have replaced conventional partially butylated amino resins in many applications. They have also made significant inroads into the UF and the benzoguanamine market. Although these resins are designated monomeric, they are of oligomeric nature, with a degree of polymerization between 1.4 and 3.0. The amount of combined formaldehyde is about 1.7–1.9 per amino group. The degree of alkylation for fully alkylated products is 90–95% of all available hydroxymethyl groups.

Fully alkylated amino resins require strong acid catalysis for fast and/or low-temperature cross-linking. Their catalysis mechanism is different from that of partially alkylated resins which respond to weak acid catalysts or general acid catalysis. A fully alkylated melamine resin catalyzed by a strong acid catalyst is a faster curing (cross-linking) agent than a partially butylated amino resin.

The ether linkage formed during the crosslinking reaction of amino formaldehyde resins with hydroxy functional polymers is sensitive to acid hydrolysis. [2.197] Exposure to acid rain is sufficient to damage high solids acrylic/melamine coatings. The use of either active CH [2.198] or NH functional polymers [2.199] creates linkages more stable to acid hydrolysis and permits the formulation of automotive clearcoats with improved acid resistance.

The catalysts which are predominately used are p-toluenesulfonic acid, dodecylbenzenesulfonic acid, dinonylnaphthalenedisulfonic acid, and their amine salts [2.152]. Compared to partially alkylated amino resins, fully alkylated resins have a lower tendency to undergo self-condensation and produce films which are hard and still more flexible.

Production. Although the patent literature cites many other amino and amide compounds, only urea, melamine, benzoguanamine, acrylamide, and glycoluril have found a market position as starting materials for the production of amino resins. Formaldehyde is the only aldehyde used on a commercial scale. Methanol, butanol, and isobutanol are mainly employed as alkylating alcohols.

The manufacturing process for partially alkylated amino formaldehyde resins consist of two consecutive reactions. The first reaction is the methylolation (hydroxymethylation) of an amino compound such as urea, melamine, or benzoguanamine with formaldehyde and can be carried out under basic or acidic conditions. Hydroxymethylation under acidic conditions leads to simultaneous alkylation and polymerization of the amino resin. The second reaction, alkylation (or etherification), is carried out under acidic conditions.

Conventional low-solids butylated and isobutylated amino resins are normally prepared in this fashion [2.153]. The reaction water formed during the hydroxymethylation or alkylation step can be removed by azeotropic distillation. The acids used for catalysis may be organic (e.g., formic, oxalic, or phthalic acid) or weak inorganic acids (e.g., phosphoric acid). The end point of the alkylation reaction can be controlled by compatibility testing of the resin with low-polarity solvents. The hydroxymethylation catalyst remains in the resin either in free form or as the salt and may influence the reactivity of the resin.

Monomeric partially and fully alkylated amino resins are prepared in two separate reaction steps. Hydroxymethylation is carried out under basic conditions to minimize self-condensation of the amino resin. The pH is then lowered by addition of a mineral acid and alkylation with methanol is carried out. To obtain monomeric amino resins, an excess of formaldehyde has to be used to assure a high level of hydroxymethylation and a low level of residual amide groups that would otherwise lead to polymer formation during the alkylation step. Since water removal by azeotropic distillation is not possible with methanol, a large excess of alcohol is required to achieve complete alkylation. After completion of alkylation the resin solution is neutralized; water, alcohol, and residual formaldehyde are removed by vacuum distillation. The salt formed during neutralization is removed by filtration [2.154].

The fully alkylated amino resin is characterized by its degree of alkylation and polymerization, and its residual hydroxymethyl and amino content, all of which can significantly affect physical and chemical properties.

Environmental Protection. The potential release of formaldehyde during application and cure is a serious problem in the handling of amino resins. Depending on the structure and the conditions of manufacture a resin can contain 0.25–3% free formaldehyde. The formaldehyde liberated from an amino resin may originate from three sources: unreacted formaldehyde, formaldehyde bound to the resin in the form of hydroxymethyl groups, or as a result of hydrolysis of the alkoxymethyl groups during cure and subsequent dehydroxymethylation [2.149].

Fully alkylated MF resins may contain $< 0.25\%$ free formaldehyde. Partially alkylated resins usually have a higher content ($\leq 3\%$). The free formaldehyde content of an amino resin is largely responsible for the formaldehyde released during the application process. This is only a small fraction of the total formaldehyde released. Hydrolysis of the amino resin and subsequent dehydroxymethylation during curing account for $> 80\%$ of the total formaldehyde release (see also p. 84).

Although resins with a high amino content have a higher content of free formaldehyde than fully alkylated resins, they release lower amounts of formaldehyde during cure. The lowest amount of formaldehyde during application and cure is released by alkylated glycoluril–formaldehyde resins [2.155].

Quality and Specifications. A typical specification for an amino resin contains information on solids, viscosity, color, density, and free formaldehyde content.

A combination of analytical techniques has to be used to fully characterize an amino resin: molar ratios of amino compound to aldehyde and alcohol, molecular mass, molecular mass distribution, and functional group content are important data.

Amino resins are offered at a solids content of 50–100%. For solutions a reliable nondestructive drying method is required before analysis [2.156]. Drying at $< 50\,°C$ in a thin film or under vacuum gives reproducible results without loss of resin components or degradation of the resin. This is especially critical for highly reactive resins. Solids methods which dry the resin at $100-110\,°C$ result in dehydroxymethylation of the resin and reduced solids content particularly when applied to polymeric butylated types.

Total amino resin content can be determined in the solid residue by the Kjeldahl nitrogen method or by CHN analysis. In melamine resins titration of melamine with perchloric acid is possible. The total level of free and combined formaldehyde can be determined by phosphoric acid decomposition and by distillation.

Free formaldehyde is readily determined by titration in ice water with sodium sulfite solution. Titration at higher temperature results in dehydroxymethylation. An indirect method for determining the structure of an amino resin consists of CHN analysis and ^{1}H-NMR to determine the CH_2:alkoxy ratio [2.157]–[2.159]. The hydroxymethyl content of amino resins can also be determined by reaction with iodine under basic conditions or by ^{13}C-NMR [2.160], [2.161].

Storage and Transportation. Amino resins should be stored in stainless steel or in lined carbon steel containers. The lining should be resistant to degradation or swelling by alcohols. Unlined steel vessels can be used under certain circumstances. The unlined vessel should have a desiccator on the gas-inlet line to remove water vapor from the gas that is used to replace the resin as it is pumped out of the tank. Methylated amino resins are preferably stored in lined drums or in fiber pack containers.

Amino resins should be stored at $20-27\,°C$. Lower temperatures can cause crystallization of hexakis(methoxymethyl)melamine and partially methylated melamine resins. Careful reheating is then required. Maximum storage temperature for fully alkylated melamine resins is 50–65 °C for 14–30 days. Amino resins with a lower degree of alkylation should not be stored at these temperatures. Fully alkylated amino resins are stable at 20–27 °C for more than 24 months. Butylated and partially methylated amino resins should not be stored for more than 12–18 months.

Since many amino resins contain solvents, local safety codes regarding flammable liquids should be observed. For bulk storage an inert gas atmosphere above the resin should be used [2.162].

Uses. The use of amino resins in packaging is regulated by the FDA under 21 CFR 175.105, adhesives; 21 CFR 175.300, resinous polymeric coatings; 21 CFR 176.170, paper and paperboard in contact with aqueous, fatty, and dry food.

Alkylated melamine–, urea–, and benzoguanamine–formaldehyde resins are the principal cross-linking agents in many industrially applied baked coatings. They are combined with acrylic, alkyd, epoxy, and polyester resins. The amide, hydroxyl, or carboxyl groups of these backbone polymers are used as functional sites for reaction with the amino resin.

The choice of amino component depends on the end use. Butylated UF resins are principally used in low-temperature curing applications, such as wood and general industrial coatings. Small amounts of these resins are also employed in container coatings in combination with epoxy resins. With the development of high-solids coatings, urea resins have been replaced in many applications with HMMM-type resins.

The market for benzoguanamine resins is small, and is mainly limited to container and appliance coatings because of its superior anticorrosion properties.

The largest market for amino resins is in the automotive topcoat area. Most of the currently used heat-cured automotive primers, primer surfacers, basecoats, and clearcoats utilize melamine cross-linking agents. In the United States most of the amino resins used in high-solids automotive topcoats are monomeric, mixed ether, fully alkylated (methylated/butylated) MF resins. Outside the United States lower-solids coatings are predominantly used, which are cross-linked with conventional low-solids, partially butylated MF resins. The level of cross-linker is ca. 25% in low-solids coatings and 35–40% in high-solids coatings.

The cure temperature for most amino-resin-based automotive coatings is 120–150 °C. In end-of-line repair applications the same coatings are cured with addition of acid catalyst at 80–90 °C.

Amino resins are also employed in waterborne systems used in industrial coatings (e.g., automotive spray primers and basecoats, can and coil coatings). Stable coatings can be formulated with hexakis(methoxymethyl)melamine or partially methylated melamine resins.

Commercial products and suppliers of melamine resins are listed below:

United States
Cytec   Cymel/Beetle
Monsanto   Resimene

Europe

Vianova	Austria	Viamine
Krems	Austria	Sakopal
Vianova	FRG	Maprenal, Resamine
BASF	FRG	Luwipal, Plastopal
Hendricks & Sommer	FRG	Heso Amin
Akzo Nobel	Netherlands	Setamine
DSM	Netherlands	Uramex
Dyno Cytec	Norway	Dyanomin-Cymel, UFR
Ciba-Geigy	Switzerland	Cibamine
Reichhold	Switzerland	Super Beckamine, Beckamine
BIP	UK	Beetle
Monsanto	UK	Resimene

Japan

Mitsui Cytec	Cymel
Mitsui Toatsu Chemical	U-Van
Dainippon Ink & Chemicals	Super Beckamine, Beckamine
Hitachi	Melan
Sanwa Chemical	Nickalac
Sumitomo Chemical	Suminal

Korea

Monsanto–Kumbo	Resimene

Economic Aspects. The growth of the amino resin market is closely related to the overall growth of industrial coatings. The development of new coating technologies (e.g., powder coatings and radiation curing) may depress the future expansion of this market. However, this is counteracted to some extent by increased use of melamine resins in high-solids coatings which is greatly influenced by the global trend to less polluting coatings with a lower volatile organic content.

The world market for amino resins used in coatings in 1990 was estimated to be 200 000–225 000 t [2.163]. The U.S. market is about 50 000 t, with the European market estimated to be of similar size. The Japanese market is approximately 33 000 t. A breakdown of the U.S. market for amino resins used in coating applications is as follows (10^3 t):

Automotive	13.5
General industrial	8.5
Wood	4.5
Paper, film, and foil	5.8
Appliances	1.8
Coil	4.8
Container	4.8
Other	4.0
Total	47.7

In the United States high-solids amino resins (predominantly methylated resins with a solids content of > 80%) enjoy a major market position. Both in Europe and Japan methylated resins are used only in applications such as waterborne coil and metal decorating, where superior performance is required. Conventional low-solids butylated amino resins are still the most important products of the industry in these countries.

Toxicology and Occupational Health. Most of the toxicologial and occupational problems in handling alkylated MF, UF, and BF resins are caused by formaldehyde release during application and cure [2.164]. Additionally, some release of formaldehyde can take place from cured objects.

The amount of formaldehyde released during the application process is related to the free formaldehyde content in the coating formulation. Loss of formaldehyde from hydroxymethyl groups during storage can raise the free formaldehyde content of a coating formulation.

The release of formaldehyde during the curing process is predominantly the result of dehydroxymethylation reactions. Some of the hydroxymethyl groups are formed during the curing process by hydrolysis of alkoxy groups from moisture in the air. For example, HMMM releases about 0.5–0.8 mol of formaldehyde during the cure process at normal humidity levels. High-imino melamine resins exhibit lower emissions of formaldehyde during cure. Formaldehyde release during cure can be almost eliminated with glycoluril–formaldehyde resin [2.155].

2.13. Phenolic Resins for Coatings [2.165]–[2.167]

Phenolic resins (i.e., condensation products of phenols and formaldehyde) are among the oldest synthetic binders, and their first use in paint technology dates back to the early 1920s. Their primary uses have constantly changed since then, and new classes of synthetic binders have become increasingly important.

Initially, phenolic resins attracted a great deal of interest because they appeared to be the first synthetic products that could be used as a substitute for natural resins (rosin, copal resins, shellac). Nowadays attention is mainly focused on performance and technical-economic competition between the widely differing groups of resins.

The main disadvantage of phenolic resins is their intrinsic yellow to brown color. As a result they cannot be used for colored and white paints. They can only be used for decorative coatings in a few cases (e.g., gold lacquers). Phenolic resins have favourable mechanical properties and a high chemical resistance. Paint systems that are optimally adjusted to the requirement profile can be developed by suitable formulations.

2.13.1. Resols

Resols are phenolic aldehyde resins that undergo self-cross-linking catalyzed with bases or basic salts. Their structure depends on the choice and molar ratios of the raw materials (e.g., phenols and cresols), the solvents, and the type and amount of catalyst used. Resols have free hydroxymethyl groups which can react on heating to form homocondensates. Heterocondensation with other reaction partners is also possible.

Phenolic resins are generally pale yellow to dark brown in color, occasionally of high intensity. In special cases intensely colored resols are used as coloring resins to produce defined shades in coatings (gold lacquers). The products are available in solid (solvent-free) and dissolved form. The choice of solvents is governed by the intended application; resols are generally dissolved in alcohols, glycol ethers, aromatic compounds, or mixtures thereof.

Resols used for coatings generally have a relatively low melting point (ca. 50 °C), with the result that solid resins may sinter at high ambient temperatures (ca. 30 °C).

Resols suitable for use in paints generally cannot be used as the sole binder because they produce very hard, brittle paint films after stoving. They must therefore be combined with other paint raw materials [e.g., epoxy resins, poly(vinyl butyrals and polyester resins)].

Resols as Sole Binders. Resols are not normally used as the sole binder because of their brittleness. There are cases, however, where flexibility is not important. Uses of phenolic resins as sole binders are restricted to coating rigid constructions (e.g., pipelines and reaction vessels). Several coats are often applied to obtain the desired layer thickness. The first coats are cross-linked at relatively low temperature (ca. 170 °C) to prevent stress formation within the film. The overall film structure only undergoes complete cross-linking when stoving the last layer. Relatively high stoving temperatures of up to 220 °C are required for optimum cross-linking.

Resols used alone as thermohardening binders produce coatings with outstanding resistance to chemicals and solvents. Cross-linking takes place primarily via the free hydroxymethyl groups of the resols. The disadvantage of the low flexibility of these coatings has to be borne in mind.

Coatings for Sheet-Metal Cans and Containers. By far the most important area of use is the coating of sheet-metal containers and cans for the storage and preservation of food and other products. Thin sheets of galvanised steel (tin plate), nickel-plated steel (tin-free steel), or aluminum alloys are generally used as substrate. Untreated steel sheet may, however, also be used (e.g., to produce containers and drums for chemicals), and are usually coated with pigmented or nonpigmented paint systems (varnishes).

In many cases the metal sheets must be coated before they are formed into containers. This demands extremely high flexibility of the cross-linked paint system. However, the coating material must also be extremely resistant to the container contents, must not discolor, and must, at least in the food-packaging sector, satisfy relevant legal provisions for example the Food and Drug Administrations in the USA and the

BGVV Empfehlung XL for Germany. Depending on the raw material, most as-supplied resols contain relatively large amounts of toxic substances (phenols, formaldehyde, etc.). They must therefore be properly labelled, stored, and handled. As a result of improvements in production methods and the choice of suitable raw materials, useable resols with greatly reduced levels of toxic substance have been developed and made available to the paint industry.

Resols are of limited use as sole binders on account of their brittleness; they are generally used in combination with plasticizing co-resins.

Combinations with Epoxy Resins. When resols are cross-linked with epoxy resins, the latter serve as plasticizing components. The hydroxymethyl groups of the resols are assumed to react with the hydroxyl groups of the epoxy resins to produce flexible coatings.

Epoxy resins with an epoxy equivalent weight of > 1500, a relatively high molar mass (ca. 8000), and a high glass transition temperature (ca. 75 °C) are generally employed. Depending on the required flexibility and resistance properties, the resol–epoxy resin mixing ratio is generally between 15:85 and 40:60 (relative to weight of solid resin). The viscosity of the mixture is adjusted to the required value (governed by the method of application) with suitable solvents. The coatings are generally provided with flow-improving additives (e.g., melamine resins) and lubricants (e.g., waxes). Typical application methods for this area of use include roller coating (normal coating, reverse roller coating), coil coating, and spray coating. In order to obtain the required cross-linking densities, the resol–epoxy resin paints must be stoved at relatively high temperature, normally > 180 °C. Production demands very short stoving times, a typical schedule being 12 min at 200 °C.

The general trend toward shorter production times and rationalisation of the production process is reflected in the "shock drying process", in which the paints are cured within a few seconds at up to 300 °C. The thickness of the cured film is normally only 2–10 µm; optimum flow behavior is therefore a prerequisite for a satisfactory pinhole-free film.

Catalysts are used to accelerate hardening. Phosphoric acid and organic phosphates are generally used in practice. These compounds also improve adhesion of the resol–epoxy resin to the metallic substrate.

Resol–epoxy resin mixtures can also be precondensed by refluxing the prediluted components under an inert atmosphere for several hours at ca. 100 °C. Although a substantial increase in the mean molecular mass is not normally observed, mixtures pretreated in this way exhibit considerably better flow properties; the susceptibility to external flow interference (craters) is also reduced.

Waterborne systems based on resol–epoxy resin precondensates are already at an advanced development stage. Carboxyl groups are introduced into the preformed resols and made water soluble by salt formation with amines. These systems offer a significant saving in solvent compared with that of conventional high-solvent, high-viscosity products.

Combinations with Poly(Vinyl Butyrals). Poly(vinyl butyral) resins with a high acetal content are used, like epoxy resins, to plasticize resols. The weight ratio of resols to poly(vinyl butyral) is generally between 95:5 and 70:30. The employed application methods and curing conditions are similar to those used in the resol–epoxy resin systems. The resistance of such paint films to organic solvents, amines, and other chemicals are better than that of the corresponding combinations with epoxy resins. In contrast to the latter, resol–poly(vinyl butyral) combinations are often used as pigmented coatings, but heat-resistant inert pigments have to be used. Certain shades (e.g., gold lacquers) can often be obtained more satisfactorily with such combinations on account of the darker intrinsic color produced on stoving.

Priming Coats. Resols in combination with poly(vinyl butyral) are used for acid-catalyzed priming coats that are cured at room temperature. These systems produce rapidly drying corrosion-resistant priming coats for metallic substrates. The mixing ratio of resol to poly(vinyl butyral) is generally 25:75 to 50:50. These priming coats (shop primers or wash primers) serve as a temporary corrosion protection in steel construction work. They also form an excellent adhesive layer for subsequent topcoats on substrates that are otherwise difficult to coat (e.g., aluminium or galvanised steel). A major use is therefore as a priming coat for rust protection in ships. Phosphoric acid is the preferred catalyst and, depending on the amount used, this coating is formulated as one-pack (acid and paint are premixed) or two-pack (acid and paint are supplied separately) systems. Corrosion-protection pigments are also employed; for ecological reasons zinc phosphates have replaced the previously used chromate-containing pigments. Highly volatile alcohols with small amounts of aromatic hydrocarbons as diluents are used.

Electrical Insulation Paints. Resols combined with poly(vinyl butyral), oil-free polyester resins, or poly(vinyl formal), as plasticizing components are used for this special application. The presence of small amounts of free phenols can increase the reactivity of these systems. These paints are used to coat wires, sheet metal for transformers and dynamos, as well as to impregnate electric motor windings. The paints are cross-linked at $100-300\,°C$.

Commercial products include Bakelite (Bakelite), Phenodur (Vianova Resins), Schenectady (Schenectady Chemicals), Uravar (DSM Resins), Varcum (Reichhold Chemicals).

2.13.2. Novolacs

If phenols and formaldehyde are reacted in the presence of acid catalysts, condensation products are obtained that do not contain reactive groups. These products can therefore neither react with themselves nor with other compounds and are termed novolacs.

Novolacs are solid at normal temperature and generally have a well-defined melting point. The largely inert behaviour of novolacs means that they can only be used for physically drying paints and coatings. They are extremely good soluble in polar solvents such as alcohols, ketones, and esters, but are insoluble in aromatic and aliphatic hydrocarbons (e.g., xylene, benzene). "Spirit varnishes" are novolacs dissolved in alcohol that are primarily used as unpigmented varnishes; they have an intrinsic yellow color. Their main uses include furniture polishes and rapidly drying coatings (e.g., for wood and toys) in industrial and consumer sectors. Novolacs have a high dilectric constant which means that they can also be used as electrical insulation paints.

The paint solutions dry rapidly and produce hard coatings that are largely resistant against water and hydrocarbons. Novolacs are readily compatible with many other binder classes. Addition of small amounts of alkyd resins or poly(vinyl butyral) resins improve the flexibility of the coatings.

Compared with phenolic resol resins, novolacs are of only minor importance as paint binders. Novolacs are, however, still very important as binders for basic dyes in printing inks.

Commercial products include Alnovol (Vianova Resins), Bakelite (Bakelite), Schenectady (Schenectady Chemicals).

2.13.3. Modified Phenolic Resins

Modified phenolic resins are condensation products of the resol type which contain other starting materials besides phenol and formaldehyde (e.g., acrylic monomers). The phenolic component itself is often modified (alkyl- or arylphenols). Modification with rosin is the most important one. The compatibility of phenolic resins with other binders can be substantially improved by modification. Rosin-modified phenolic resins may be combined with linseed oils and alkyd resins. Examples of use include putties, priming coats, rust protection paints, and colored topcoats.

Only a limited amount of resin can be added to white paints due to its intrinsic color. Addition of these hard resins increases the hardness and the gloss of the paint films, accelerates and improves the drying of oxidatively cross-linking alkyd resins, and optimises sanding properties and corrosion protection in putties. Modified phenolic resins have lost much of their importance because they have been replaced by more efficient binder systems (e.g., thermoplastic and cross-linkable acrylic resins, polyurethane systems). Rosin-modified phenolic resins are, however, still extremely important in the production of resins for printing inks.

Commercial products include Alresen, Durophen (Vianova Resins); Bakelite (Bakelite); Schenectady (Schenectady Chemicals); and Super Beckacite (Reichhold Chemicals).

2.14. Asphalt, Bitumen, and Pitch Coatings
[2.168]–[2.171]

2.14.1. Asphalt and Asphalt Combination Coatings

Composition. Only natural asphalts with small amounts of mineral constituents (asphaltites) can be used to produce paints. Of the asphaltites, only gilsonite with its moderately high softening point, particularly low mineral content ($< 1\%$), and high solubility in hydrocarbons is suitable. This natural product is a black, resinous, thermoplastic material with a relatively high molecular mass. Since gilsonite is very hard and brittle, it cannot be used as the sole film-forming substance and must be combined with resins and/or plasticizers. It can also be plasticized with softer bitumen or mineral oils.

Combination of gilsonite with drying vegetable oils (e.g., linseed or tung oil) produces high-grade asphalt combination paints, especially if the two binder components are heated (cooked) together.

Asphalt and asphalt combination paints are produced and applied only in dissolved form. Aromatic and aliphatic hydrocarbons of the normal boiling point range as common in the paint industry are suitable as solvents [e.g., mineral spirit (bp 140–200 °C), light solvent naphtha (bp 155–180 °C)]. Unfilled (unextended) asphalt or asphalt combination paints are rarely used nowadays.

Finely ground minerals such as talc (10–20 wt%) and mica, or aluminum powder (1–5 wt%) are usually incorporated into the asphalt to improve its film properties, in particular its mechanical, weather, and aging resistance. Antisettling agents such as montmorillonite–bentone (0.5–1.5 wt%) also have to be used.

On account of the relatively low inherent color of gilsonite, asphalt-based paints can be colored with inorganic pigments (e.g., iron oxide, chromium oxide, or aluminum).

Properties and Uses. Asphalt paints have good resistance to weathering, temperature fluctuation, condensation, aging, and also have good anticorrosion properties. They adhere well to metallic substrates, asphalt, and roofing felt.

Asphalt paints, particularly those based on combinations of gilsonite with drying oils, are used to protect exposed steel structures (e.g., bridges, lattice towers, penstocks) and are recommended for these purposes in DIN 55 928. Since they are also heat resistant, paints of this type are used for protection against corrosion in coking plants and blast furnace plants, as well as to protect the reinforcement bars in aerated concrete. Soft-formulated asphalt paints can be obtained by blending with mineral oils, soft bitumens, or drying (fatty) oils. These are used to protect and seal roofs.

2.14.2. Bitumen Coatings

Composition. Bitumen paints contain bituminous substances as binders. Bitumen is defined in DIN 55946 as "A low-volatile, darkly-colored mixture of organic substances obtained in the processing of mineral oils and petroleum, whose viscoelastic behavior changes with temperature. The typical properties of bitumen can be attributed to a colloidal system in which the disperse phase (asphaltenes) is stably distributed in a coherent phase of high-boiling oils (maltenes)".

Bitumens are thermoplastic, weather resistant, water resistant, absorb very little water, do not release any toxic substances in water, and are soluble in hydrocarbons.

Depending on the application, relatively hard distillation bitumen, oxidation bitumen, or high-vacuum bitumen (hard bitumen) are used for physically drying bitumen paints. Waterborne bitumen paints and thick coatings (bitumen emulsions) are produced from moderately hard distillation bitumen.

The properties of various types of bitumen should be taken into account when selecting the most suitable bitumen for a particular application. Oxidation bitumens have the widest range between the temperature at which they become brittle (fracture point, $\ll 0\,°C$) and the softening point. Their plasticity range is between 90 and 130 °C measured by the ring and ball method DIN 52011. The relatively hard distillation bitumens have plasticity ranges of ca. 60–70 °C with fracture points around 0 °C. The high-vacuum bitumens have plasticity ranges of ca. 80 °C with fracture points above 0 °C. Distillation bitumens and high-vacuum bitumens (hard bitumens) are substantially more extensible in the plasticity range than oxidation bitumens.

The distillation and high-vacuum bitumens have a comparatively limited plasticity range and higher fracture point in comparison with the oxidation bitumens but this is compensated for by the following advantages. Distillation bitumens and high-vacuum bitumens require less solvent (less environmental contamination) to obtain the same paint viscosity and equal hardness. Their solutions are stable (long shelf life), whereas solutions of oxidation bitumens tend to thicken (gel). Solutions of distillation bitumen and high-vacuum bitumen penetrate porous substrates more readily. After drying, they adhere better to a wide range of substrates. Their weather resistance is better and they absorb less water. Since they are more stable to UV radiation than oxidation bitumens, they have a better aging resistance.

Petroleum fractions and aromatic hydrocarbons with a boiling point range between ca. 130 and 200 °C are suitable solvents for bitumen solutions. Benzines (mineral spirit) are preferred on account of their less obtrusive smell and the fact that they are more environmentally friendly. On account of their toxic properties, chlorinated hydrocarbons are no longer used.

Waterborne bitumens may be anionic, cationic, or rarely nonionic emulsions, and are produced with the aid of emulsifiers and, if necessary, stabilizers conventionally used in the paint industry. The properties of bitumen paint (solutions and emulsions) (e.g., stability; mechanical, weather, water, and chemical resistance; corrosion protection properties) can be substantially improved by incorporating extenders. Suitable extenders include ground slate, ground limestone, chalk, talc, and mica (25–45 wt%, depending on extender type and required properties). The use of extenders requires the addition of antisettling agents. Long-term weather resistance can be considerably improved by incorporating soot.

Bitumen paints can only be colored to a slight extent (e.g., with iron oxide or aluminum powder) on account of their intrinsic color.

Properties and Uses. Thoroughly dried films of bitumen paints are resistant to water, salt solutions, dilute acids, and alkalis. Extended paints are weather resistant. On account of the thermoplasticity of bitumen, resistance decreases with increasing temperature. Bitumen paints are not resistant to organic solvents, fuels, oils, and greases. Optimally formulated bitumen paints are nontoxic and may therefore be used to protect surfaces that come into contact with drinking water.

On account of their favorable properties and comparatively low price, bitumen paints have been used to protect concrete structures (foundations, bridge abutments), felt roofs, sheet-metal roofs, drinking water reservoirs, silos, pipes made from fiber-reinforced cement, concrete, steel and cast iron, and in vehicle construction (wagon underframes, car underfloor protection).

Commercial products include Eurolan (Deitermann).

2.14.3. Bitumen Combination Coatings

The properties of bitumen paints (Section 2.14.2) can be favorably modified and adjusted to suit practical requirements by combination with other film-forming substances. For example, the thermoplasticity can be reduced and/or mechanical properties (e.g., hardness, extensibility) can be improved by adding polymers such as polyethylene, polypropylene, polyisobutene, and styrene–butadiene copolymers. The chemical resistance can also be improved; high-quality corrosion protection coatings can be obtained by combination with alkyd resins.

Bitumen is permanently compatible with only a few polymers and only when the latter are added in small amounts. Compatibility depends largely on the nature and origin of the bitumen.

Combinations of bitumen with reaction-curing paint binders (e.g., epoxy resins and polyurethanes) are not usually employed in solvent-based paints due to their unsatisfactory compatibility. Only with high epoxy resin contents can a certain compatibility be achieved by addition of solubilizing phenols and aromatic oils.

One and two-component waterproofing thick coatings, based on destillation or polymer bitumen (bitumen emulsions), have become an increasing importance for the protection of buildings in wet-duty areas like cellar, bathrooms e.g.

Those trowel-grade or sprayable materials can bridge cracks up to 5 mm and are leakproofed for soil moisture and pressurized water, depending on the layer.

Fillers like fibres, bentone or small styropore balls guarantee a good remaining thickness of the material after drying.

In the case of anionic two-component bitumen emulsions, the second component consists of a cement powder that braks the emulsion during the potlife time (time of application) and reacts with the water in excess of the emulsion.

Commercial products include ®Plastikol, ®Eurolan, ®Superflux (Deitermann).

2.14.4. Pitch Coatings

Because of it's cancerogenic potential, pitch coatings are to longer used in Middle- and Northeurope.

2.15. Silicate Coatings

2.15.1. Water Glass Coatings

The expression water glass paints is understood to mean coating systems based on the binder water glass (potassium, sodium, or occasionally, lithium water glass). The general expression silicate coatings has been adopted for these systems and is therefore used throughout Section 2.15.1.

History. The history of silicate paints began with the rediscovery of potassium and sodium water glass by J. N. VON FUCHS around 1818 [2.172]. The use of these compounds as a flame retardant coating was first described in 1823 [2.173]. FUCHS, J. SCHLOTTHAUER, and W. V. KAULBACH subsequently developed the "stereochromy" process, a new painting technique in which water glass replaced the previously used binders lime and casein for mural painting. A further significant development was made by A. W. KEIM in 1878 [2.174], who created the two-component technique, still used nowadays, involving the binder (fixative) and pigments. The modern one-component dispersion silicate paints were developed and introduced to the market in 1967 (Silin silicate paint, van Baerle & Co.; Kieselit, Henkel).

Properties. The special properties of silicate paints are imparted by the binder, generally potassium water glass. The paint coats harden physically by evaporation of water, as well as chemically by several reactions. These stabilization and consolidation processes are termed "silicification" [2.176]. The water glass reacts as follows with atmospheric carbon dioxide:

$m\ K_2O \cdot n\ SiO_2 \cdot x\ H_2O + m\ CO_2 \longrightarrow n\ SiO_2 \cdot x\ H_2O + m\ K_2CO_3$
Water glass $\qquad\qquad\qquad\qquad$ Silica gel

Reaction also takes place with the pigments and extenders of the paint. If the paint is applied to a mineral substrate, a third reaction occurs, e.g., with calcium ions.

$m\ K_2O \cdot n\ SiO_2 \cdot x\ H_2O + Ca^{++} + m\ CO_2 \longrightarrow CaSi_nO_{2n+1} \cdot XH_2O + m\ K_2CO_3$
Water glass $\qquad\qquad\qquad\qquad\qquad$ Metasilicate gel

The silica gel and metasilicate gel formed in these complex reactions contain a large amount of water, and are then converted into solid, amorphous silicate structures with only a low residual water content by evaporation of water [2.175]. Extremely finely divided potassium carbonate is formed as a hygroscopic byproduct which occurs as a transparent layer that is slowly dissolved by moisture. The sodium carbonate produced with sodium water glass leads to the formation of saltlike efflorescences. Potassium water glass is therefore mainly used for coatings.

Coatings formed from silicate paints differ from organic coating systems. Chemical bonds are formed by silicification in the paint layer and with the mineral substrate, and adhesion is therefore not purely physical. The paint layer has a silicate structure and is thus relatively permeable with a high degree of hardness. Evaporation of moisture is not hindered, peeling and flaking are therefore avoided. Since silicate paint layers have a high gas permeability they do not disturb the equilibrium with atmospheric carbon dioxide when applied to a lime plaster or mortar. Since the paint layer and the mineral substrate have a chemically similar structure they both exhibit the same expansion behavior under the action of heat and cold; thermal stress between the paint and substrate is therefore minimized.

The good resistance of silicate paints to weathering, environmental influences, and temperature is a result of their mineral, chemically resistant structure and the almost exclusive use of lightfast inorganic pigments. Silicate paints dry to give a matt surface.

Composition, Application, and Uses. Architectural and exterior coating systems may be either one- or two-component (DIN 18 363, Section 2.4.1). Two-component silicate paints may consist only of potassium water glass, pigments, and extenders that are resistant to potassium water glass. One-component dispersion silicate paints may consist of potassium water glass, pigments that are resistant to potassium water glass, hydrophobing (waterproofing) additives, and no more than 5 wt % of organic constituents, relative to the total amount of the coating material. This total amount is measured by the ignition loss (Technischer Arbeitskreis Dispersionsfarben, TAKD, resolution, Feb. 1984) [2.177].

In two-component silicate paints, the potassium water glass binder (fixative) and the powdered pigments, extenders, and auxiliaries are mixed (kneaded) in a 1:1 ratio by the user and homogenized after a few hours. The end product and reproducibility depend on the kneading time, degree of homogenization, temperature, and metering accuracy. On account of the high reactivity the paint should be applied within a few days of preparation. In contrast, one-component silicate paints are formulated ready for application in the paint factory [2.178], [2.179]. Depending on the paint manufacturer, water or a water–primer (fixative) mixture may be added by the user to optimize adaptation to the substrate. The permissible proportion of up to 5 wt % of organic constituents does not affect the advantageous properties of silicate paints. Indeed, they stabilize the system and improve water resistance. Storage stabilities of more than one year can be obtained and requirements with regard to the water uptake coefficient (DIN 18 558) are met [2.180].

On account of their properties, silicate paints are used as weather-resistant and wash-resistant coatings on mineral substrates in both exterior and interior applications. Substrates include lime, lime cement, cement renders (mortar groups I, II, and III), concrete, sandy limestone, as well as old silicate or mineral coatings. Old lime coats should be removed completely, and gypsum substrates should be treated only after special priming. A special water glass priming coat and one or two topcoats are required, depending on the substrate. On account of their silicate nature and the fact that they have the same color effect as historical lime paints, silicate paints are also used for restoring and maintaining historical monuments. Their high lightfastness, weather resistance, and water vapor permeability reduce the danger of coating damage even in structures and buildings that do not have adequate transverse insulation. Further special uses of silicate coating systems include corrosion protection (see Section 11.1) and flame retardant coatings on wood [2.181], [2.182].

Commercial Products. Commercial one-component products include Calsilit (Kabe Karl Bubenhofer); Caparol–Sylitol (Deutsche Amphibolin Werke); Color-Silikatfarbe (Sikkens); Granital/Biosil (Keim Farben); Kieselit (Henkel); and Silin (van Baerle). Two-component products include Purkristalat (Keim Farben) and Silin (van Baerle).

2.15.2. Alkyl Silicates [2.183]

Alkyl silicates may be regarded as esters of the hypothetical silicic acid, which may exist in monomeric or oligomeric form. They are also referred to as silicate esters or alkoxysilanes.

Hydrolysis of alkyl silicates produces silicic acids with varying degrees of condensation and ultimately silicon dioxide. The reaction time increases with increasing size of the alkyl group. Elevated temperature and high moisture content accelerate hydrolysis and condensation.

The resultant silica gel can be used as a binder for solid particles, i.e., for pigments utilized in the paint industry. Inorganic coatings with special properties can be formulated. The silanol groups formed during hydrolysis make the alkyl silicate binders highly reactive in contrast to silica sols formed from colloidal silicic acid. They differ from water glass paints (Section 2.15.1) in that alkali is not released during hydrolysis.

The properties of the hardened film (high surface hardness, temperature resistance, and electrical conductivity) account for their main use as binders for corrosion protection coatings containing zinc dust (see p. 243). Depending on the temperature sensitivity of the pigment used, they can withstand temperatures up to 1000 °C. The absence of organic polymers makes the paints extremely resistant to organic solvents, and the high degree of hardness confers excellent abrasion resistance.

Composition and Properties. The binders are produced from orthosilicate esters $Si(OR)_4$ or industrial mixtures of polysilicate esters. Although the methyl esters are readily hydrolyzed and have a very high silicon content (generally given as SiO_2 content), the ethyl esters are normally used for physiological and toxicological reasons. The less readily hydrolyzable esters of longer-chain alcohols are employed in special applications.

Various polymeric silica distributions are obtained depending on the production process. Although this distribution may possibly affect hardening behavior, investigations have shown that this is insignificant in practical applications.

The use of alkyl silicates in coatings is based on the following reaction:

$$Si(OR)_4 + 2\ H_2O \longrightarrow SiO_2 + 4\ ROH \qquad \text{where R = alkyl}$$

This overall equation can be subdivided into two steps: hydrolysis

$$Si(OR)_4 + 4\ H_2O \longrightarrow Si(OH)_4 + 4\ ROH$$

and condensation

$$Si(OH)_4 \longrightarrow SiO_2 + 2\ H_2O$$

Both reactions occur in parallel, but their rates can be independently influenced. Condensation can be suppressed by lowering the pH value to obtain hydrolyzates that remain stable for relatively long periods (6–12 months) without gelling.

It has been suggested that this occurs because protons associate with the silanols which then repel one another on account of their positive charge and thereby oppose condensation. If the pH value is raised, these charges are dissipated and the silanols can condense [2.184].

The gelling time depends on a range of parameters [2.185], the most important being the SiO_2 concentration, the degree of hydrolysis, the type of solvent, and storage temperature. (The degree of hydrolysis is the degree of conversion, complete conversion to SiO_2 and ethanol corresponding to 100% hydrolysis.)

In practice the ester is mixed with the required amount of water and a suitable solvent (e.g., alcohols, esters, ketones). To obtain 100% hydrolysis of 1 mol of an orthosilicate, 2 mol of water must be added. By maintaining a constant pH (for ethyl silicate ca. 2) the condensation reaction can be suppressed to such an extent that hydrolyzates of ethyl silicate containing 20% SiO_2 and with a degree of hydrolysis of 80–90% remain stable for about one year. If this equilibrium is disturbed (e.g., by shifting the pH value or evaporating the solvent), condensation commences and the solution solidifies after a short time. This mechanism is used in two-pack zinc dust paints. In the simplest case the binder is a hydrolyzate in alcoholic solution. Zinc dust and, if necessary, extenders, antisettling agents, colorants, etc. are then added. Addition of zinc consumes acid and the pH value increases.

$$Zn + 2\ HCl \rightarrow ZnCl_2 + H_2$$

Condensation therefore commences and the mixture becomes solid.

On application, the paint layer hardens within a short time due to evaporation of the solvent, progressive hydrolysis of ester groups after uptake of atmospheric moisture, reaction of silanol groups with the zinc and substrate, and condensation of silanol groups to form sparingly soluble, cross-linked silica gel [2.186]. Conditions are generally chosen such that the film is hard enough to allow transportation after 4 h and a topcoat after 24 h. Alkyl silicate zinc dust paints are particularly suitable for applications involving high atmospheric humidity and high temperature because these conditions promote hardening. This explains their widespread use for corrosion protection in maritime tropical regions.

Historical Development. The importance and advantages of ethyl silicate zinc dust paints were recognized and intensively investigated in the late 1940s [2.187], [2.188]. The fundamental work of LOPATA and KEITHLER [2.189] on the partial acid hydrolysis of tetraethyl orthosilicate and the use of the product as a zinc dust paint binder was followed by attempts to improve the application properties and quality of the coatings (e.g., by adding poly(vinyl butyral) [2.190], alkyl titanates [2.191], borate esters [2.192], or trialkyl phosphates [2.193].

Ethyl silicate was transesterified with higher alcohols to produce binders with a higher flash point [2.194], [2.195].

In addition to the general disadvantages associated with a two-pack system, ethyl silicate hydrolyzate also ages. The condensation reaction is merely slowed down and not stopped. As a result application properties such as viscosity and reactivity (and thus the hardening rate), alter during storage. The use of binders of varying ages can lead to considerable practical difficulties.

Continuous efforts have therefore been made to develop binders for one-pack paints which must be completely compatible with active pigments such as zinc dust. Mixtures of ethyl silicate and zinc can be stored for unlimited periods in the absence of water. If the hydrolysis and condensation reactions caused by atmospheric moisture can be sufficiently accelerated during application of the coating, a one-pack system can be produced.

Use of basic catalysts allows formulation of such systems because the reaction mechanism of hydrolysis and condensation proceeds differently in the alkaline range [2.184]; condensation occurs more rapidly.

The pronounced settling behavior, due to the high density of zinc and the low viscosity of the alkyl silicate, frequently resulted in solid precipitates and caused problems. Nowadays, however, one-pack zinc dust paints can be used even after several years because they contain special rheological additives (e.g., bentonite, pyrogenic silica). The application properties of these paints are similar to those of paint systems based on organic polymer binders. In addition to the general advantages associated with a one-pack system, these alkyl silicate paints do not exhibit age-dependent changes in their application properties (e.g., viscosity, brushability, and hardening behavior) because the degree of condensation does not alter.

All binders formed from ethyl silicate have a relatively low flash point due to the ethanol released during hydrolysis (flash point of ethanol 12 °C). The paint may have a somewhat higher flash point (15 °C) because its ethanol content is low. The use of higher alkyl groups yields systems with higher flash points.

Production. On account of the many possible variations in batch formulation and production procedure, only a general outline and instructions for producing one- and two-pack binders from ethyl silicate can be given here. The following batch formulation (in parts by weight) is typical for a two-pack binder containing 20 wt% SiO_2: ethyl polysilicate 500, solvents 440, and 1% hydrochloric acid 60. Alcohols such as ethanol and 2-propanol, often mixed with one another and with ketones, ethers, and esters, are used as solvents.

The solvents, dilute hydrochloric acid, and ethyl silicate are mixed and stirred and the temperature rises; an upper limit of 25–30 °C is often recommended. The second component of the two-pack paint is the pigment–extender mixture. The binder can be used after 1–3 d; the storage stability is limited and is generally specified by the manufacturer. In the above example it is ca. one year.

Binders with a fairly high flash point require a suitable solvent (e.g., propanol, butanol) and a silicate base that cannot release any alcohols of low flash point during partial hydrolysis (e.g., tetrapropyl or tetrabutyl silicate). The slower evaporation of such alcohols delays the initial hardening of these paints.

The amount of water present determines the possible extent of hydrolysis of the Si−OR groups (in the above example ca. 82%) and thus storage stability and reactivity. For example, the addition of ten additional parts by weight of water would reduce the storage time to < 25% of the original value. Conversely, a lower amount of water would give a product with a longer storage stability which, however, would react more slowly possibly resulting in uneconomically long drying times. Suitable catalysts include hydrochloric acid, other acids, and salts (e.g., zinc chloride or amine hydrochlorides). Their concentration determines the rate of the hydrolysis, but also affects properties such as storage stability of the binder, pot life, and paint drying.

The production of binders for one-pack zinc dust paints is simple: alkyl silicate, catalyst, and solvent are mixed and diluted to the desired concentration. The resulting mixtures are extremely sensitive to moisture; even small amounts of water in the solvents can lead to gelling. Appropriate equipment and trained personnel are required to avoid mishaps. Interactions between the constituents can greatly influence application properties and lead to failure. The choice of an effective antisettling agent is difficult due to the large difference between the densities of zinc powder and alkyl silicate. Most rheological additives are ineffective in highly polar alkyl silicate solutions.

In order to produce ready-for-use zinc dust paint, zinc dust is slowly added to the binder in a ratio of 4–4.5:1 while stirring. Binder additives (e.g., antisettling agents such as bentonite, pyrogenic silica) and zinc dust additives (e.g., extenders such as mica, talc, kaolin, or colored components such as chromium(III) oxide, titanium dioxide) can be used to adjust the application behavior and properties of the paints.

Commercial products containing ethyl silicate binders include Dynasil and Silester (Hüls), Ethylsilicate (Union Carbide), Ethylsilikat (Wacker), Silbond (Akzo).

Important manufacturers of ethyl silicate zinc dust paints include Ameron, Carboline, Hempel's Marine Paints, International Paint, Jotun, Mobil Chemical, and Sigma Coatings.

Transportation and Storage. Containers and drums with a polyethylene insert or resistant lining are suitable. The binders should be stored at as low a temperature as possible and under absolute exclusion of moisture. The storage time of two-pack systems varies and is specified by the manufacturer. Commercial products are stable for about 9 months when stored at 20 °C.

Properties of the Coatings and Uses. Impurities in steel (e.g., carbon) can act as local cells and cause corrosion. The iron is first oxidized to divalent ions. A metal with a lower electrode potential (e.g., zinc) is, however, preferentially oxidized. The electrons that are thereby released counteract oxidation of the iron:

$$Zn^0 \longrightarrow Zn^{2+} + 2\,e^-$$
$$2\,e^- + Fe^{2+} \longrightarrow Fe^0$$

Theoretically, oxidation of iron can only begin when all the metallic zinc has been oxidized. This process can occur only if there is sufficient electrical conductivity (i.e., contact) between zinc and iron. Thorough cleaning of the metallic substrate is therefore important to ensure the effectiveness of corrosion protection coatings.

On account of their higher electrical conductivity, silicate esters have clear advantages over organic polymers as binders. The resistivity of a one-pack silicate ester zinc dust paint is ca. $1 \times 10^6\ \Omega \cdot cm$, i.e., two to three orders of magnitude lower than that of commercially available purely organic systems with a comparable zinc content.

The zinc oxidation products form a hard cementlike layer in the silicic acid film. Resistance to diffusion is increased and attack of atmospheric oxygen on the substrate is thus slowed down (barrier protection).

Ethyl silicate binders are mainly used in applications involving high temperatures and corrosive influences. The principal application of ethyl silicate zinc dust coatings is the protection of iron and steel against corrosion; it is effective even at temperatures of up to 400 °C. Examples include ships, harbor construction work, bridges, steel structures exposed to atmospheres in industrial areas, tanks, containers, pipelines, and chimneys.

Ethyl silicate zinc dust paints are applied by air or airless spraying, often directly after jet cleaning of the substrate. They are seldom applied with a brush or roller. Layer thicknesses range from 15 to 20 µm (shop primers) through 50 to 70 µm (primers) to > 100 µm (single thick-layer application).

The contents of two-pack paints must be constantly stirred after mixing to prevent settling. The pot lives are ca. 8 h. With one-pack paints, layer thicknesses of up to 200 µm can be obtained in a single application. Pot life is unlimited provided moisture is excluded.

Silicic acid is not attacked by radiation, and silicate esters are therefore particularly suitable binders for coatings in radiation-hazard areas such as nuclear power stations.

Ethyl silicate zinc dust paints can be covered with many conventional topcoat systems [2.196]. This is necessary in an acid or alkaline environment to prevent the zinc from being attacked. Ethyl silicate is also suitable as a binder for inorganic pigments on account of its UV resistance.

3. Paint Systems

3.1. Solventborne Paints

3.1.1. General Information

Paints and coating materials normally consist of a physical mixture of binders, pigments, extenders, additives, and solvents. Depending on the method of application and area of use, the solids content may reach 80 wt%, the proportion of pigment may reach 60% of the solids content. The technologically most important component is the binder (or binder mixture). Binders may be classified as physically or chemically drying according to their film-forming mechanism.

Physically drying paints are solutions of thermoplastic polymers with molecular masses exceeding 20 000; on account of their low solubility they have a high solvent content (> 60%) and a low solids content. Chemically drying paints have a fairly low solvent content (30–60%) and a high solids content because the polymer network is formed by cross-linking of the binder (M_r ca. 800–10 000) usually at elevated temperatures or via radiation to form thermosetting coatings. Oxidatively drying paints contain allyl groups and reactive double bonds, and cross-link by absorbing oxygen and forming ether bridges.

Normally paints are also classified according to the nature of the principal binder and its associated film properties; e.g., alkyd, acrylic, polyester, nitrocellulose, epoxy, and oil-based paints (see Chap. 2). The method of application, surface properties, and intended use are also utilized for classification [3.1]. Since the beginning of the 1980s environmental requirements have become increasingly important for two main reasons, especially in the case of paints with low material transfer (application) efficiencies (see Section 3.1.3):

1) Avoidance of the use of toxic, carcinogenic, mutagenic, or teratogenic organic solvents
2) Drastic and in some cases legally imposed reduction of solvent contents (see also Section 3.1.3)

Legal regulations have resulted in a sharp rise in the solids content and a reduction in the solvent content of the most important paint systems. Among solventborne paints, high-solids paints have enjoyed the largest growth rate; in the Federal Republic of Germany production of this type of paint rose by 32% between 1988 and 1989, whereas the production of all solventborne paints increased by only 3% [3.2]. A sharp drop in high-solvent paints is to be expected in the United States and Europe

[3.3], [3.4]. The definitions of "low-" and "high-solids" paints differ. For example, the Verband der Deutschen Lackindustrie (Association of the German Paint Industry) defines "high-solids" paints as paints with a nonaqueous (i.e., organic) solvent content of less than 30 wt %, whereas the Environmental Protection Agency (EPA) in the United States defines "high-solids" paints as having volatile organic contents of < 2.8 pounds/gallon (336 g/L) [3.5]. High-solids paints are described in Section 3.2.

Typical examples of low-solids paints (solvent content usually > 60 wt %) are:

1) Metallic (-effect) base paints for mass production of automobiles and touch-up finishes
2) Thermoplastic coatings
3) Coatings for the electronics and optoelectronics industries

Of these three groups metallic basecoats are quantitatively the most important. Technical reasons for the high solvent content of these paint systems include the need for thin layer thicknesses (e.g., when applying paint with a spin coater). Secondly, binders such as polyimides for electrical insulation coatings and thermoplastic acrylic resins have low solubilities. Finally, in some cases (e.g., in metallic basecoats) a high solvent content is needed to optimize rheological behavior [3.6]. Rheological factors limit the reduction of the solvent content, particularly when coating vertical surfaces. Significant improvements can often only be achieved by replacing organic solvents with water. For example, the solventborne metal effect paints used in the automobile industry are increasingly being replaced by waterborne paint systems with a reduction in the solvent content in the paint to ca. 10% based on 25% solids [3.7] (see also Section 3.3).

3.1.2. Properties and Raw Materials

Solvents. Organic solvents are used to dissolve binders (resins) and additives and to disperse pigments and extenders in order to produce ready-for-use paints and coatings (see Chap. 14). The dissolution behavior (solvency) and interaction of organic solvents (e.g., ketones, glycol ethers, glycol ether esters, alcohols, esters, hydrocarbons, and terpenes) with binders have been investigated in detail [3.1]. The objectives of such studies are the optimum adjustment of processability to obtain the appropriate rheological behavior before and during application, and the maximization of the binder content in the paint formulation. The legally imposed requirement to employ toxicologically harmless solvents for the formulation of paints has stimulated the development of computer programs for optimizing solvent mixtures based on less harmful or innocuous solvents. This required a detailed knowledge of the interaction between binders and solvents [3.8]. Initial programs for calculating optimum solvent mixtures were developed by taking into account dispersive interactions (C_d), polar interactions (C_p), and interactions resulting from hydrogen bond formation (C_h) (solubility parameters) [3.9], [3.10]. Solvents were chosen in regard to toxicological factors [3.11] (see also Section 3.1.3). Conventional tests for checking solubility are still used.

As well as the solubility, the solvent release during and after application is another important criterion for solvent selection. Evaporation of the solvent mixture from the film is important not only for a smooth surface, but also for obtaining optimal, reproducible mechanical properties of the film. The boiling points or vapor pressures of the solvents do not provide sufficient information for formulating a paint system [3.12]. A model for calculating solvent evaporation also includes transport data of the solvents in the film in the form of diffusion parameters [3.13]. Evaporation behavior is also determined experimentally (e.g., according to ASTM D 3539.76).

In solventborne paints, small amounts of the organic solvents may remain in the resultant film after cross-linking. Solvent retention can reduce internal tension, linear shrinkage behavior, and the glass transition temperature [3.14].

The selection of solvents is additionally restricted by legal requirements for storage and transport based on the combustion and flammability behavior of the paint.

Binders and Film Properties. Physically drying solventborne paints contain cellulose esters, vinyl resins, thermoplastic acrylic resins, polyurethanes, rubber derivatives, and hydrocarbon resins as principal binders. Chemically drying solventborne paints that undergo oxidative drying are based on alkyd resins, epoxy esters, modified phenolic resins, and urethane oils. Cross-linkable paints play the most important role. They contain binders with unsaturated components (e.g., unsaturated polyester resins), ethoxy groups (e.g., epoxy resins), and hydroxyl groups (e.g., alkyd resins, polyesters, vinyl resins, and acrylic resins). A wide range of cross-linking agents are available for a variety of applications [3.11]. They include phenolic resins and amino resins. Isocyanates also play an important role as cross-linking agents. Hydroxyl-containing binders (e.g., polyester resins, acrylic resins) that can be cross-linked with polyfunctional isocyanates have been introduced as two-pack paints, particularly in the automobile and plastics industries [3.15], [3.16]. Advantages lie in the good weather resistance of the urethane groups, the low curing temperature, and the high gloss level of the coating.

Clear varnishes used in the automobile industry have to satisfy particularly stringent requirements: high gloss, color durability, and maintenance-free behavior for more than 5 years [3.17]. This is expressed in terms of permanently high water contact angles ($> 90°$) and a high gloss after several years' exposure of the paint in Florida. New acrylic resins have been developed that are copolymerized to about 50% with fluorine-containing monomers such as chlorotrifluoroethylene [3.18]. Acrylic resins produced by group transfer polymerization represent a new class of resins with a narrow molecular mass distribution which is advantageous for dissolution [3.19].

New cross-linking reactions for curing paints and coatings have recently been developed, but their technical implementation has not been very successful. The Michael addition of C−H-azide compounds to activated unsaturated double bonds is used for automotive repair and touch-up finishes [3.20]. Esters activated with alkoxy and amide groups have also been described [3.21]. Epoxy–amine, epoxy–carboxy, and epoxy–anhydride reactions are gaining importance in aliphatic epoxy resin cross-linking agents [3.22], [3.23]. However, films produced in this way do not exhibit any major advantages over those produced from polyurethane resins. Much research and development has been devoted to solventborne paints that produce a multilayer system after a single application; this leads to considerable savings in material and time. The phase behavior of the binder components and solvents plays an important role in these paint systems [3.24], [3.25].

3.1.3. Environmental Protection and Application Technology

The paint transfer efficiency is defined as the amount of paint on the object expressed as a percentage of the amount of paint used for application. A low theoretical paint transfer efficiency means high amounts of solvent emission and waste generation. Special attention must therefore be given to environmental aspects when using conventional spraying methods for producing smooth, high-quality surfaces (see Table 3.1).

The development of electrostatic spraying (see Section 8.3.2) has led to a sharp rise in paint yield and material transfer efficiency [3.26]. The use of high-speed rotating atomizers has also contributed to this and allows a very fine droplet spectrum to be produced. The droplet mist is directed onto the object by a guiding air jet; with a favorable charge to mass ratio, a high material transfer efficiency ($> 75\%$) is obtained. The level of transfer efficiency together with the evaporation losses of the solvent results in a source of emission that has led to the development of solvent and paint recovery units in plants with a particularly high throughput (e.g., in the automobile industry). Successful solvent recovery is of paramount importance in automobile factories [3.27].

New resins, solvents, additives, cross-linking agents, and other paint components are subjected to toxicological tests to determine any possible health-hazard effects before they are used in commercial formulations.

Table 3.1. Comparison of theoretical solvent emission data of various application methods

Application method	A Paint solids content, wt%	B Organic solvent content, % of solids content	C Transfer efficiency, %	D* Solvent emission, kg/kg film
Pouring, roller application	70	43	100	0.43
Pneumatic spraying				
Metallic basecoat	13	670	60	11.20
Topcoat	50	100	60	1.66
High-speed rotating atomizers and electro- static spraying				
Metallic basecoat	13	670	75	8.93
Two-pack varnish	60	67	75	0.85

*$D = B/C$.

3.2. Solvent-Free and Low-Solvent (High-Solids) Paints

Solvent-free and low-solvent paints are a development of conventional high-solvent paints and are formulated to comply with increasingly stringent environmental requirements. In this section only those liquid coating materials are discussed that can be applied by conventional methods (coating, spraying, rolling, dipping, and pouring). Solvent-free paints that are applied with new technologies and methods are described in Sections 3.4 (powder coatings) and 3.7 (radiation-curing systems). Waterborne paints are discussed in Section 3.3.

Low-solvent paints are commonly referred to as *high-solids paints* and are defined as systems with a solids content exceeding 85 wt %. In practice, however, paints with a solids content of 60–80 wt % are also termed high-solids paints, particularly if the corresponding conventional (high-solvent) system normally contains < 50 wt % nonvolatile components [3.28].

Low-solvent paints have existed for a long time in the form of oil-based paints and varnishes (Section 2.1). The more recent development of high-solvent paints satisfying strict quality requirements has largely displaced this type of coating materials. Extensive development of high-solids paints occurred only after regulations governing solvent emissions and air pollution came into force ("Rule 66" in California, and TA Luft in the Federal Republic of Germany). There are several reasons for developing solvent-free and low-solvent paints:

1) Reduction of atmospheric pollution
2) Savings in materials and transportation costs
3) Savings in energy costs involved in paint production and use
4) Time savings due to higher layer thicknesses per application cycle
5) Improvement in safety due to low-solvent content

The advantage of high-solids paints compared with more recent technologies lies in the fact that they can be produced and applied with conventional equipment and by well-known methods. Consequently potential users are primarily small and medium-sized paint shops, the maintenance and architectural sector, as well as the do-it-yourself sector.

3.2.1. Principles

One way of lowering the solvent content of a paint is to reduce the viscosity during application by physical or chemical methods. For example, dilution of solvent-containing paints can be minimized by hot spraying. The solids content of solvent-containing paints can be increased by using solvents with solubility parameters that largely correspond to those of the binder, by adding viscosity-reducing cosolvents, or by using additives. For example, the formation of hydrogen bonds between hydroxyl and carboxyl groups increases the viscosity, which however can be overcome by adding small amounts of alcohols.

Solids contents exceeding 70 wt% cannot be achieved in this way—modification of the binder is necessary. Thus, the binders used in high-solids systems have a much lower intrinsic viscosity than conventional high-solvent paints. This is usually achieved by reducing the mean number-average molecular mass or by using a narrow molecular mass distribution (reduction of the weight-average molecular mass). The use of esterification and transesterification catalysts in polycondensation reactions can contribute to a saving in solvents in this manner. Lead and tin compounds have proved particularly suitable for this purpose, but are not appropriate for all applications on account of their toxicity. Production of the resin in solution followed by concentration to a higher solids content is a further possibility. On account of the low solvent content in the formulation further measures may, however, be required because the binder is also responsible for flowability, antisagging on vertical surfaces, prevention of wrinkle formation, and controlling the drying behavior. In high-solvent paints these properties are partially determined by the solvent. Evaporation of the solvent after application produces a sharp rise in viscosity which prevents the paint from sagging on vertical surfaces and accelerates the initial drying. Solvents also permit the use of high-viscosity binders with high glass transition temperatures which undergo physical drying. These advantages are largely lost in low-solvent paints, and development is therefore concentrated on the synthesis of special binders with the following features:

1) Reduction of the number-average molecular mass
2) Narrow molecular mass distribution (reduction of the weight-average molecular mass)
3) Chemical modification of the binders
4) Use of nonvolatile reactive diluents
5) Use of metal compounds as cross-linking agents

In addition to the absence of physical drying, low-solvent paints require a longer time for cross-linking due to the lower velocity of polymerization of the binder. This has to be compensated by higher unsaturation or a higher proportion of functional groups in thermosetting paints. When larger amounts of curing agents are used, however, a denser network than in conventional systems may be produced, resulting in better chemical and weather resistance.

As well as the mean molecular mass and molecular mass distribution, the number and nature of the functional groups as well as the degree of branching of the polymer molecules are important. Starlike branched polymers exhibit weaker intermolecular interactions on account of steric factors. The rigidity of polymers containing olefinic double bonds interferes with the association of the chains. Both effects lower the viscosity.

The term reactive diluents denotes low molecular mass reactants that act as solvents with an extremely low volatility and participate in the cross-linking reaction. Reactive diluents have become especially important in one-pack systems because they can partially or wholly replace the solvent in high-solvent paints. Reactive diluents participate in the cross-linking reaction because their functional groups are identical or similar to those of the principal binder. They are completely incorporated into the polymer network during curing. Good compatibility with the principal binder in all cross-linking stages is necessary for continuous curing of the paint. If this is not the case, component segregation can occur, resulting in hazy films with reduced gloss. Resistance to chemicals and weathering as well as mechanical strength are also adversely affected. The most commonly used reactive diluents for oxidatively drying systems are unsaturated compounds. Polyfunctional methacrylate esters have replaced the more volatile monomers. Diacrylates or triacrylates of polyalcohols [3.29], dicyclopentenyloxyethyl methacrylate [3.30], [3.31], or their (co) polymers are recommended. Reaction products of dicyclopentadiene and butadiene also act as reactive diluents [3.32].

Low-viscosity metal compounds are also used in cross-linking reactions. For example, alcoholates of aluminum, titanium, and zirconium as well as their complexes with volatile ligands are primarily employed. The reaction involves substitution of the alcohol and formation of a coordination compound with the binder [3.33].

During evaporation of the solvent, sagging on vertical surfaces must be prevented by using thixotropic paints. These paints can be formulated with extenders and additives (e.g., smectite, montmorillonite, highly dispersed silica, hydrogenated castor oil, polyethylene waxes, and metallic soaps). These paint additives also act as antibleeding and antisettling agents. Modern urethane additives achieve the same effect without, however, affecting the gloss of the paint films [3.34]. A widely used alternative is thixotropic modification of the binder or combination of the binder with highly thixotropic resins (e.g., alkyd resins that have been reacted with polyamides).

3.2.2. Production and Uses

Low-solvent and solvent-free paints are not associated with specific classes of binders, and may be either one- or two-pack systems. In one-pack systems, hardening is effected by oxidative cross-linking with the aid of driers (see Section 2.6) or by thermoreactive curing agents such as etherified amino resins or "blocked" polyisocyanates. Two-pack systems are mixed immediately prior to application, and curing occurs as the result of a polyaddition reaction between the binder and a curing agent. Such paints can be classified according to the nature of the raw materials and binders.

Alkyd Resins. The high versatility and broad range of applications of alkyd resins (Section 2.6) have benefitted the development of high-solids paints. The commonly used one-pack paints contain high-oil, oxidatively drying alkyd resins, and have a solids content of 85–90 wt %. They are used as maintenance and architectural coatings. Unmodified alkyd resins require long drying times, tend to form wrinkles when applied in thick layers, and become yellow under the action of heat and in the dark, and they have therefore been superseded by modified alkyd resins. Low molecular mass alkyd resins with a narrow molecular mass distribution are reacted with modifiers such as trimethylolpropane diallyl ether to yield low-viscosity alkyd resins with improved drying properties [3.35]. Incorporation of 1,4,5,6,7,7-hexachlorobicyclo[2,2,1]-5-heptene-2,3-dicarboxylic anhydride [*115-27-5*] yields low-viscosity, satisfactorily drying binders that are recommended as flame-retardant marine and maintenance coatings.

Nondrying alkyd resins cross-linked with melamine resins at elevated temperature are used for industrial coatings. Resins containing synthetic fatty acids and resins based on glycidyl esters of Versatic Acid (Cardura E 10, Shell) are used in combination with low-viscosity melamine resins such as hexamethoxymethylmelamine. *p*-Toluenesulfonic acid and its amine salts are used as catalysts. On account of their high storage life, hydrophobic oxime esters of sulfonic acids are also employed.

Polyester Resins. Saturated polyesters (Section 2.7) of low molecular mass are formulated with hydroxymethylated, highly etherified melamine resins or blocked isocyanates to form one-pack paints, or with polyisocyanates to form two-pack paints. 2-Butanone oxime, caprolactam, and glycol ethers are used as blocking agents for isocyanates. Aliphatic polyisocyanates are preferred on account of their high resistance to yellowing. Linear or slightly branched oligoesters with terminal hydroxyl groups are usually used as resins. On account of their low molecular mass they require more curing agent than high-solvent binders and more powerful catalysis for sufficiently rapid cross-linking.

Given the possibilities for postcombustion of the solvent in industrial coating, the trend towards high-solids paints is, however, not so pronounced in this class of binders as it is, for example, in the case of maintenance paints and architectural coatings. High-quality paints with a solids content > 70 wt% are not widely available. Powder coatings are preferred as an alternative in the industrial sector.

Polyurethanes. The comments made concerning polyester paints also apply to polyurethanes (see also Section 2.9). Recently developed aliphatic isocyanates with a high yellowing resistance are increasingly used despite their high cost. Polyurethanes have already made large inroads into all application sectors [3.36]. Low-viscosity polyurethane oligomers can be combined with polyester, acrylic, and alkyd resins. They serve as modifiers for waterborne and low-solvent binders and improve the hardness and the flexibility of the paint film [3.37]. Two-pack polyurethane systems may contain as polyhydroxyl component both low-solvent or solvent-free oligoester diols and oligourethane diols. Although the latter are more expensive, they are more stable against hydrolysis.

Acrylic Resins. Low molecular mass acrylic resins (see also Section 2.5) may be obtained by using fairly large amounts of chain-transfer reagents (e.g., thiols). The resultant problems of smell have, however, led to the implementation of new polymerization techniques, such as group-transfer polymerization or dead-end polymerization. Bifunctional (telechelic) polymers [3.38] are obtained and can be used as binders for low-solvent paints.

Epoxy paints (see also Section 2.10) are usually applied as two-pack systems. Curing with amines is preferred in high-solids systems. Epoxy resins may also be used as coreactants or reactive diluents in low-solvent paints based on acrylic, polyester, or alkyd resins.

PVC plastisols, i.e., poly(vinyl chloride) dispersed in plasticizers, are used in special applications such as underseals for automobiles or as sealing compositions.

Commercial products (binders and crosslinkers for high solids paints) include Alkydal, Desmodur, Desmophen (Bayer); Beckopox, Macrynal, Maprenal, Viacryl, Vialkyd (Vianova Resins); Cythane (Cytec); Dynapol, Vestanat (Hüls); Johncryl (Johnson); K-Flex (King Industries); Synocure, Synolac (Cray Valley); Synthalat (Synthopol); Tolonate (Rhone Poulenc); Uracron, Uralac (DSM); Worléecryl and Worléekyd (Worlée).

3.3. Waterborne Paints

Waterborne (water-thinnable) paints were developed in the 1950s with the aim of replacing the common organic paint solvents by water, which has the obvious advantages of being noncombustible and nontoxic [3.39].

Waterborne paints provided the technological basis for electrodeposition paints (Section 3.8), in which negatively charged paint particles (anaphoresis, industrial introduction at the beginning of the 1960s), or positively charged paint particles (cataphoresis, industrial introduction at the end of the 1970s) are deposited from aqueous solution onto metallic substrates by application of an electrical field [3.40].

With the scarcity of raw material resources (e.g., petroleum) and the introduction of more stringent environmental legislation, waterborne paints were developed on a broader basis during the 1970s (e.g., as spraying and dipping paints). Among those coating materials that can be classed as environmentally friendly, waterborne paints have the widest potential as regards application, drying methods, and industrial uses.

3.3.1. Properties

In the production and application of waterborne paints, water is used as solvent or diluent. The physical properties of water and organic solvents differ (Table 3.2); some of these properties have to be taken into account when water is used as a paint solvent. The water molecules have a high dipole moment and associate with one another. This means that water has a high boiling point and high latent heat of evaporation despite its low molecular mass. This in turn results in fairly long evaporation times or in the need to supply energy in the form of heat to evaporate the water and dry the paint film. The high dipole moment of water is also responsible for its high surface tension. With substrates having a low critical surface tension (e.g., plastics or unsatisfactorily degreased metals) this leads to inadequate wetting, unsatisfactory edge covering, and crater formation. Critical surface tensions (at

Table 3.2. Physical properties of water, butanol, and xylene

Property	Water	Butanol	Xylene
Dipole moment, 10^{-30} C·m	6.17	5.54	2.07
Density, 10^{-3} g/cm^3	1000	811	874
bp, °C	100	117.7	140
Specific heat, J g^{-1} K^{-1}	4.187	2.998	1.666
Latent heat of vaporization (100 °C), J/g	2261	590	523
Evaporation number (diethyl ether = 1)	80	33	13.5
Latent heat of vaporization (liquid at 20 °C), J/g	2596	829	657
Surface tension (20 °C), mN/m	72.5	25.5	29.5

20 °C and in millinewtons per meter) of water and other substances follow:

Aliphatic hydrocarbons	18.0–28.0
Aromatic hydrocarbons	28.0–30.0
Ketones	22.5–26.6
Esters	21.2–28.5
Alcohols	21.4–35.1
Water	72.5
Poly(vinyl fluoride)	18.5
Oil	28.9
Polypropylene	29.0
Stainless steel	30.0
Steel	55.0

The high polarity of water is responsible not only for differences between the application behavior of waterborne paints and solvent-containing paints. It also means that the organic polymers used as binders for waterborne paints must have a different structure from those used in solventborne paints.

Waterborne paints differ according to the nature of their stabilization in water; the polymer molecules are dissolved in water or dispersed in water in the form of polymer dispersions or emulsion polymers. Recent developments include polymer particles formed in organic solvents and then emulsified in water with low or, more commonly, high molecular mass emulsifiers being used for internal or external emulsification. Internal emulsification denotes that part of the binder molecule functions as an emulsifying moiety, whereas for external emulsification separate emulsifiers are required.

Water-Soluble Binders. Water-soluble binders consist of relatively low molecular mass polymers ($M_r < 10\,000$) (e.g., alkyds, polyesters, polyacrylates, epoxies, and epoxy esters) whose individual molecules dissolve in water due to salt formation involving functional anionic or cationic groups.

Most water-soluble binders are anionic. They are made water soluble by neutralizing carboxyl groups with ammonia or volatile (generally secondary or tertiary) amines that evaporate during film formation. In cationic binders, salt formation usually occurs between the amine groups of the resin and organic acids (e.g., acetic or lactic acid). Salt formation with inorganic acids (e.g., phosphoric acid) is, however, also used. Typical applications include cationic electrodeposition paints.

The binder structure must be resistant to hydrolysis in aqueous, alkaline, or acid media. For alkyd or polyester resins the introduction of the carboxyl groups to be neutralized via full-ester bonds is advantageous; semi-ester bonds tend to undergo hydrolysis. Organic solvents and organic cross-linking agents (e.g., melamine resins) also influence viscosity stability on storage [3.41].

Water-soluble binders generally contain organic solvent ($\leq 10-15$ wt%) that originates from their production via polycondensation or polymerization reactions in an organic medium. They can, however, still be dissolved or diluted with water after neutralization. Anomalous viscosity behavior may, however, arise; it is characterized by a viscosity increase on dilution with water due to association of the binder molecules in water (Fig. 3.1) [3.42]. More modern binders are supplied as preneutralized solutions in water and cosolvent.

The organic cosolvents are mainly alcohols, glycol ethers, and other oxygen-containing solvents that are soluble or miscible with water.

Waterborne paints formed from water-soluble binders can undergo different forms of film formation and hardening, including drying at room temperature (e.g., polyacrylates, epoxy esters, and alkyds). The rates of initial drying and through drying are controlled by evaporation of water from the paint film (physical drying). Oxidatively drying, water-soluble binders (e.g., alkyds) also undergo chemical cross-linking that can be accelerated with driers.

Water-soluble binders (e.g., polyesters, polyacrylates) can also be chemically cured with cross-linking agents at elevated temperature (oven drying). These binders can be reacted with water-soluble or water-miscible amino resins (especially fully or partially methylated methoxymethylmelamine resins) or with water-soluble or water-miscible blocked polyisocyanates, which only react after the isocyanate groups have been unblocked.

On account of the viscosity anomaly, waterborne paints based on water-soluble binders have a relatively low solids content (ca. 30–40 wt%) and require relatively large amounts of organic cosolvents (up to 15%) to ensure water solubility and film formation. They also have the advantage of a broad drying spectrum (physical, oxidative, oven drying) and a wide range of application methods (dip coating, flow coating, spray coating, electrodeposition coating).

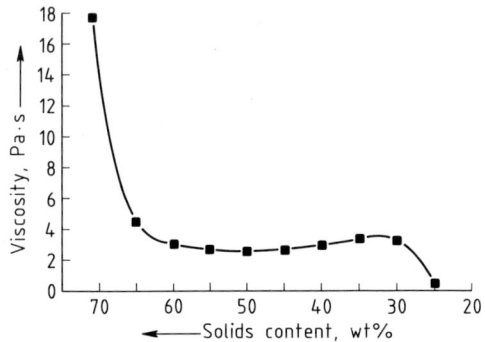

Figure 3.1. Viscosity behavior of a short-oil alkyd resin on dilution with water
Starting conditions: solids content 71.0 wt%; organic cosolvents 24.3 wt%; dimethylethanolamine 4.7 wt%

Paint films formed from some water-soluble binders tend to be water-sensitive due to their hydrophilic, solubilizing groups. They can be formulated to have a high gloss due to good pigment wetting and stabilization. They can also have a high level of corrosion protection which depends on the corrosion-inhibiting pigments used and the chemical nature of the binder. The latter determines adhesion to the substrate and diffusion of water and oxygen through the paint film.

Polymer Dispersions (Emulsion Polymers). Waterborne paints based on polymer dispersions (usually referred to as emulsion paints) are not water soluble. They are water-thinnable systems composed of dispersions of polymer particles in water (see Section 3.5). The particles consist of high molecular mass polymers (e.g., of styrene, butadiene, acrylate, or vinyl monomers) and are produced by emulsion polymerization. These waterborne paints also contain small amounts of organic solvents (< 5 wt%) that serve as film-forming (coalescing) agents that partially evaporate on drying.

The required application viscosity of waterborne emulsion paints is generally obtained by adding a small volume of water. The evaporation behavior of polymer dispersions is similar to that of conventional, solvent-based paints.

Film formation occurs via evaporation of water from the paint film and coalescence of the high molecular mass binders; the release of water takes place relatively quickly (physical drying). The minimum film-forming temperature depends on the chemical structure of the polymers. Film formation can be facilitated by the addition of organic solvents (e.g., alcohols, butylglycol ethers) or by the action of heat.

Alternatively, chemical cross-linking reactions can also be used to harden polymer dispersions (oxidative drying or oven drying). An example is the cross-linking of hydroxyl groups of the polymer dispersions with hexamethoxymethylmelamine resins.

Characteristic properties of waterborne emulsion paints are relatively high solids contents (50–60 wt%), rapid dilution with water without any viscosity anomaly, and a low content of organic solvents ($< 5\%$).

As a result of the particulate nature of the binders only a low gloss and, in some cases, only limited corrosion protection can be obtained. Waterborne paints based on dispersions can be applied by spraying, however they are of only limited use for electrostatic coating and dipping applications due to their rapid drying properties.

Hybrid Systems. Combinations of water-soluble and water-dispersed binders may be used to achieve synergistic effects (e.g., to control the amount of organic solvents or the application behavior). Polymer dispersions or powdered binders may be used as water-dispersible binders [3.43]. The use of polymer dispersions allows the solids content of water-soluble binders to be increased, the level of organic solvents to be reduced, and the physical drying time to be shortened. The use of water-soluble binders provides a broad application spectrum and yields paint films with well-balanced properties.

Internally and Externally Emulsified Binders. In recent developments polyurethane, polyacrylate, or polyester binders are synthesized in an organic medium and partially dissolved in water by salt-forming groups to form colloidal solutions. This involves internal emulsification. Alternatively, the polymers are dispersed and stabilized in the aqueous phase by adding low or high molecular mass emulsifiers (external emulsification). The disadvantages of water-soluble binders as regards low solids and high organic solvent contents, and polymer dispersions as regards film formation can be compensated in this way.

These binders form the basis for new industrial waterborne paints, an example being aqueous metallic base paints for automotive coatings and finishes. The binders of these paints are acrylic core–shell polymers [3.44], [3.45]. The core and shell have a different monomer composition, the shell has anionic groups which interact with water and stabilize the dispersion. The anionic shell allows pseudoplastic flow behavior which ensures parallel orientation of the aluminum pigments in the wet paint film. This orientation and the low solids content are responsible for the metallic gloss and high color flop (change in color observed on varying the viewing angle) of the base paints.

Generally, the binders have a low content of organic solvents (ca. 10%) and can be applied over a wide temperature and humidity range. Water and organic cosolvents used in clearcoats applied by wet-on-wet coating evaporate rapidly after predrying with convection or IR radiation.

3.3.2. Production and Application

Most of the pigments used in conventional paints and coatings are also suitable for waterborne systems. Some exceptions are lead chromates and molybdates, manganese lakes (Pigment Red 48 and 52), some perylenes (Pigment Red 223), benzimidazolone yellow (Pigment Yellow 151), and isoindolines (Pigment Yellow 13a) which all have limited alkali resistance.

Another limitation results from pigments that contain a high amount of water-soluble components and electrolytes which may reduce storage stability, cause coagulation of the binder emulsion or dissolved resins, and result in loss of adhesion or blistering during exposure to humidity and water. Examples are surface-treated organic and inorganic pigments (some phthalocyanine blues, carbon black, transparent iron oxides).

As a rule, pigments can easily be dispersed in water-soluble binders because the water-soluble binder molecules promote wetting and stabilization of the pigments. The dispersion and stabilization of pigments in polymer dispersions is, however, more difficult because the dispersions form a continuous phase in water and uniform distribution of the dispersed pigment particles in the paint film is hindered by coalescence of the polymer particles. To overcome such problems new dispersion additives have been developed in recent years.

Waterborne paints can be applied by a wide range of methods (e.g., by dipping electrodeposition, or spraying) followed by air, forced-air, or oven drying. The influence of climatic conditions (e.g., temperature and atmospheric humidity) on sprayability, substrate wetting, pouring and leveling behavior, as well as the rate of water evaporation from the wet paint film should, however, be taken into account. For example, the drying of waterborne paints proceeds more slowly at low temperature and/or high relative atmospheric humidity. These disadvantages can be remedied by suitable formulation of the paints, by heating the intake air to the spray cabins and the ventilation zone, and also by maintaining the atmospheric humidity within specific limits (e.g., 50–80%).

In order to reduce the tendency to "pinholing" that occurs with higher layer thicknesses, control of application parameters is advantageous. Examples include using suitable spraying equipment, feeding preheated air to the ventilation zone, using IR drying in the ventilation zone, and avoiding steep stoving curves.

Corrosion-resistant and wear-resistant materials (e.g., stainless steel and in special cases suitable plastics) have to be used for the equipment required to produce and apply waterborne paints. This applies to production vessels, storage and transportation vessels, to the feed system used for application (e.g., closed circuit tanks, pipelines, and pumps) all of which must be able to withstand chemical mechanical stress. The use of corrosion-resistant materials for the spraying equipment is also advantageous.

Waterborne paints should be protected against frost during storage. In general, storage areas do not have to satisfy special fire prevention regulations.

3.3.3. Uses and Environmental Aspects

Waterborne paints for industrial use only account for a small proportion of the total production of coatings, paints, and thinners in the Federal Republic of Germany (Table 3.3) (i.e., for the serial production of industrial and consumer goods such as machinery, tools, electrical equipment). Whereas the proportion of waterborne emulsion paints and plaster for decorative use (architectural paints, see Section 3.5) accounted for ca. 63% of the total production in 1996, the proportion of waterborne paints for industrial use accounted for 5,3%. However, industrial waterborne paints now have a broad application spectrum based on the development of new paint technologies. The most important use sectors include electrodeposition coating of automobile bodies and accessories; spraying and dipping primers, as well as one-layer coatings for engines, chassis, axles, and automobile accessories; sprayed fillers for automobile bodies (first industrial use in Europe 1981); waterborne paints for automobile bodies (1987); aqueous protective waxes for preservation of automobile body shells (1989); waterborne automotive clear coats (1992); aqueous powder slurries (1997); aqueous primers and one-layer coatings for metal furniture, electrical and electronics equipment, machines, machine parts, agricultural equipment and machinery, construction machinery; and aqueous primers and one-layer coatings for plastic parts (e.g., radio and television cabinets and office equipment).

The prospects for the increasing use of industrial waterborne paints lie in their economic advantages and in the possibility of reducing solvent emission during application to comply with legal requirements. Savings in organic solvents as diluents, savings in insurance premiums, lower energy consumption in spray cabins,

Table 3.3. Production of paints in the Federal Republic of Germany (Statistisches Bundesamt, BRD)

Production $t \times 10^3$	1987	1996
Waterborne coatings	66	94
Powder coatings	25	53
Dispersions and plaster (decorative)	526	963
Total production	1163	1783

ventilation zones, and drying ovens all contribute to the overall economy of waterborne coating materials.

Thanks to their low organic solvent content, the use of waterborne paints substantially reduces solvent emission from paint or spraying shops (spray cabin, ventilation zone, drying oven).

Waterborne paints can generally be classed as less toxicologically harmful than corresponding solvent-based paints. Nevertheless, lung-penetrating paint mists (aerosols) of water-based paints present a health hazard and appropriate protective measures (e.g., use of respirators) must be taken depending on the workplace concentration. Water-based paints are not, however, subject to such stringent regulations concerning hazardous substances as conventional, solvent-based paints.

3.4. Coating Powders

3.4.1. Introduction and Economic Importance

Coating of substrates (particularly metallic substrates) with particulate plastics is a well-established technology. The fluidized-bed process was developed in 1952. The electrostatic powder coating technique introduced in 1965 has, however, made a substantially greater impact [3.46]. These two processes are discussed in Section 8.3.5.

Other powder coating processes include blowing thermosetting plastics powders onto hot parts for electrical insulation, flocking polyethylene onto hot pipes to produce an external coating, and coating small, cold items in electrostatic fluidized beds, followed by thermal curing. Lances are also used for the internal coating of preheated vessels and pipes, vibration and rotation sintering, as well as flame spraying for repair work and large articles, and the minicoat and maxicoat processes for coating small items.

Thermosetting coating powders are applied mainly by electrostatic spraying and dominate the market. The annual growth rates worldwide are about 10%. Total world production in 1996, including "functional materials", was 565 000 t with a ca. value of ca. $ 2.6 \times 10^9$. This was subdivided according to geographical regions and countries as follows (10^3 t):

Europe	253
North America	140
Far East	122
Rest of the world	50
United States	122
Italy	73
Federal Republic of Germany	56
PR of China	35
United Kingdom	30
France	26
South Corea	24

In contrast to wet materials, almost no material is lost during application of coating powders (no solvents, no split-off products, powder overspray is recovered). These tonnages roughly correspond to two to three times the amount of ready-for-spraying, conventional industrial paint.

Table 3.4 summarizes the consumption of thermosetting binders used in various regions. The epoxy–polyester systems predominate over polyester–trisglycidyl isocyanurate and epoxy systems. The consumption of thermoplastic coating powders, particularly those used in the fluidized-bed and flocking processes, is substantially lower than that of thermosetting powders (Table 3.5). Their production does not show a comparably significant growth rate.

Table 3.4. Production of thermosetting coating powders in 1995 classified according to binder types (as percentages)

Binder types	Europe	North America	Far East
Epoxy	13	20	19
Epoxy–polyester	56	30	45
Polyester–trisglycidyl isocyanurate	24	20	15
Polyester–polyurethane	3	26	17
Others	4	4	4

Table 3.5. Estimated production (in 10^3 t) of thermoplastic coating powders in 1991 classified according to binder types

Binder types	Europe	World
Polyethylene (PE)	15	29
Poly(vinyl chloride) (PVC)	10	20
Polyamide (PA)	3	5
Ethylene–vinyl alcohol copolymer (EVOH)	3	4
Thermoplastic polyester (SP)	1	2
Total	32	60

3.4.2. Production (See also Section 7.2.3.).

Although other processes are described, thermosetting coating powders are produced exclusively by the extruder process. This method is employed because it is the cheapest and the most widely applicable and flexible; it allows large plants to be supplied with material of consistent quality. Production with extruders includes the following steps:

1) Testing of raw materials
2) Weighing of the premixture
3) Mixing in the premixer
4) Intermediate testing (monitoring the premix)
5) Homogenization and dispersion in the extruder (80–140 °C)
6) Cooling the extrudate to room temperature
7) Intermediate testing (monitoring the extrudate)
8) Flaking
9) Fine grinding in an air-classifying mill
10) Screening in a sieve
11) Checking specific properties of the coating powder

Since modification of the coating powder (e.g., tinting after production) is nearly impossible, all relevant process parameters (e.g., raw materials, machine settings) must be kept constant. This has the advantage for the user that changes such as thickening, settling, or floating cannot occur during application and storage. Large batch sizes are necessary because cleaning of the plant and equipment is time-consuming and expensive. An extremely high level of cleanliness must be maintained during production because different coating powders may be incompatible.

Thermoplastic coating powders can also be produced with extruders. Grinding is particularly important here and, depending on the binder and desired particle size distribution, can be carried out at ambient temperature [e.g., some polyethylene (PE) powders for the fluidized-bed process] or at low temperature using liquid nitrogen, etc., [e.g., polyester (SP) electrostatic powders]. Some poly(vinyl chloride) (PVC) powders are also produced in this way. The most efficient process for producing PVC powders is to mix the components in a high-speed mixer at 80–110 °C, followed by cooling in a second mixer, and screening.

Polyamide (PA) powders are generally precipitated from solution. Ethylene–vinyl alcohol copolymer (EVOH) powders are precipitated by hydrolyzing ethylene–vinyl acetate copolymer solutions.

3.4.3. Properties

The typical constituents of a coating powder are binders (resins, if necessary hardeners and accelerators), colorants (mainly pigments), additives, and in some cases extenders.

Table 3.6. Summary of coating powder binders

Use	Thermosets		Thermoplasts	
Mainly interior	EP–SP	epoxy resin–COOH polyester resin	PE	polyethylene LDPE, LLDPE, HDPE
	EP–DCD	epoxy resin–modified dicyanodiamide		
	EP	epoxy resin–phenolic hardener	PA	polyamide 11, 12
	EP	epoxy resin–imidazoline derivative	SP	polyester
	EP	expoy resin–anhydride adducts		
Interior and exterior	SP–TGIC	COOH polyester resin–trisglycidyl isocyanurate	EVOH	ethylene–vinyl alcohol copolymer
	SP–HAA	COOH polyester resin–hydroxyalkylamide hardener	PVC	poly(vinyl chloride)
	SP–PUR	OH polyester resin–isocyanate adduct		
	SP	COOH polyester resin–glycidylestermix		
	AY	glycidylacrylicresin–anhydride adducts		
	AY–PUR	OH acrylic resin–isocyanate adduct		
	AY–DDA	glycidylacrylic resin–dodecanedicarboxylic acid		

Binders. Thermosetting and thermoplastic binders are listed, together with their abbreviations, in Table 3.6. A broad range of raw materials is now available and it has become difficult to systematically classify binder properties. Outstanding properties are obtained with suitably selected systems and have contributed to increased use of coating powders.

The *thermosetting binders* are the most important. Their properties are summarized in Table 3.7. The quantitatively predominant EP, EP–SP, and SP–TGIC types cure by polyaddition without release of split-off products. SP–PUR and AY–PUR systems are cured by polycondensation and release some caprolactam. SP–HAA systems are also cured by polycondensation and release water as a product.

Thermosetting powders are mainly applied to cold workpieces by electrostatic spraying. Materials for interior use are generally produced from epoxy resins based on bisphenole A–epichlorohydrine and are cured with acidic polyester resins or modified (i.e., substituted or accelerated) dicyanodiamides. Materials for highly reactive systems are cured with phenolic hardeners (i.e., oligomers with phenolic terminal groups). Other curing agents include imidazoline derivatives (for matt coatings), and anhydrides and their adducts. The application of thermosetting powders to hot substrates by electrostatic spray or fluidized-bed processes is restricted to EP coating powders.

Coating powders for exterior use are mainly based on the SP–TGIC system. Of some importance is the SP–PUR system, i.e., the combination of hydroxy-functional polyester resins with caprolactam-blocked isocyanate adducts (primarily based on isophorone diisocyanate). The newly developed SP–HAA and SP-glycidyl estermix systems avoid the problems of TGIC as regards toxicology. Acrylic resins are of importance mainly in the United States (AY–PUR) and in Japan (AY–DDA). The AY–PUR system is used for automotive components and in domestic appliances. The PUR systems based on uretdione (a dimerization product of isocyanates) do not

Table 3.7. Typical properties of thermosetting powders[a]

Compound[b]	Stoving temperature, °C	Weight loss during stoving, wt%	60° gloss, %	Other properties
EP (cold electrostatic spraying)[d]	120–240	<0.1 organics	0–100[c]	depending on the hardener, excellent mechanical, chemical, electrical, anticorrosive properties, yellows if overstoved, marked loss of gloss and chalking under weathering; with coating of preheated parts and sufficient heat content of the parts, hardening takes place through the residual heat, organic pigmentation often restricted due to chemical reaction
EP (hot electrostatic spraying, fluidized bed, blowing)[d]	130–240	≤1 cracking products	70–100	
EP–SP	140–220	<0.1 organics	20–100[c]	excellent mechanical properties, almost no yellowing on overstoving, loss of gloss and chalking under weathering
SP–HAA	160–220	<1 water	20–100	excellent mechanical values, outstanding weather resistance, some yellowing on overstoving, layer thickness limited due to release of water
SP–TGIC	160–220	<0.1 organics	10–100[c]	excellent mechanical properties, no yellowing on overstoving, outstanding weather resistance
SP–PUR	180–220	2–4 caprolactam or <1 organics	0–100	good mechanical properties, some yellowing on overstoving, outstanding weather resistance, layer thickness limited due to caprolactam release SP-PUR systems based on Uretdiones only lose some organics on stoving
SP	170–220	<0.1 organics	70–100	SP cured with glycidylester mix hardeners behave similar to SP/TGIC
AY	180–220	<1 organics or 2–4 caprolactam	70–100	moderate mechanical properties, very good hardness and outstanding weather resistance, in the case of PUR hardening layer thickness limited due to caprolactam release AY cured with anhydride adducts forms very smooth films and can be cured at temperatures down to 140 °C

[a] The coating powders are applied by cold electrostatic spraying; EP resins can also be applied by hot electrostatic spraying, fluidized bed, and blowing. [b] For definition of abbreviations, see Table 3.6. [c] Smooth flow and low gloss are not obtained at low stoving temperatures. [d] Hot and cold denotes the temperature of the workpiece.

Table 3.8. Typical properties of thermoplastic powders

Binder*	Application method**	Preheating and stoving temperature [3.48], °C	Adhesive primer required [3.48], [3.49]	Weight loss during stoving, wt%	Other properties
PE	hot flocking, and hot fluidized bed	280–400	in some cases	≤1 paraffinic hydrocarbons	without primer (hot-melt adhesive) poor adhesion, soft, high viscosity, complicated shapes are difficult to coat, poor weather resistance
PA	hot fluidized bed	280–400	in some cases	0.2 laurolactam	primer required for good wet adhesion, hard, tough, good resistance to chemicals and detergents, good anticorrosive properties, average weather resistance, undergoes yellowing
PVC	cold electrostatic spray, hot fluidized bed	160–350	yes	1–6 plasticizer	extremely flexible, good weather resistance, undergoes yellowing, inexpensive
EVOH	hot fluidized bed	200–400	no	<0.1 organics	good adhesion, elastic, tough, good anticorrosive properties, good weather resistance
SP	hot electrostatic spray	240–400	no	≤1 organics	good adhesion, film formation possible in a few seconds, good weather resistance, very good elasticity, resistant to sterilization

* For definition of abbreviations, see Table 3.6. ** Hot and cold denote that the workpiece is preheated or at ambient temperature.

release reaction products and may be mentioned as a specialty. AY-systems cured with anhydride adducts are starting to be used as automotive OEM clear coat.

Typical properties of *thermoplastic binders* are summarized in Table 3.8. Application is mainly by fluidized bed but includes application by flocking. Electrostatic spraying is also used for finely ground powders. The main uses are in the interior sector. The PVC powders are the most widely used thermoplastic binders for exterior use.

Colorants. Coating powder colorants are almost exclusively pigments and must satisfy the following requirements:

1) Thermal stability at the curing temperature
2) Only a slight increase in binder viscosity

3) No reaction with other constituents of the formulation
4) Stability towards shearing forces during extrusion and grinding

Inorganic pigments (e.g., titanium dioxide, iron oxides, and chromium oxides) satisfy these criteria the best. In Europe the use of cadmium and lead pigments is declining and has stopped in some countries completely. Organic pigments are used in increasing volumes. Metallic and pearlescent effects are obtained by using flakes of aluminum or coated mica.

Other Constituents. The remaining constituents of coating powders impart specific properties (e.g., additives to adjust flow) and reduce cost (extenders such as calcium carbonate or barium sulfate).

3.4.4. Testing

DIN 55690 classifies and identifies coating powders according to the binder. Standards for testing coating powders in powder form are summarized in Table 3.9. Relevant European standards (CEN) have not yet been drafted. Quality aspects are described in detail in [3.50].

Table 3.9. Standards for testing coating powders in powder form

Parameter	ISO	DIN	AFNOR	BS	ASTM
Sampling	8130-9	53225/53226	T 30 – 048	3900-J 1	D 1898
Terms and Definitions	8130-13				
Characterization					
Compatibility	8130-12	55990-3	T 30 – 505		
Density (gas comparison method)	8130-2	55990-3		3483-B 8	D 3451
Density (liquid method)	8130-3	55990-8	T 30 – 506	3900-J 3	D 3451
Gel time	8130-6		T 30 – 502		D 3451
Sintering range		55990-7	T 30 – 507	3900-J 4	D 3451
Resistance to blocking	8130-8	55990-8	T 30 – 507	3900-J 4	D 3451
Chemical storage stability	8130-8	DIN-ISO 6713	T 30 – 210	3900-B 5	
Acid extract	6713				
Inclined plate flowtest	8130-11				
Application					
Particle size distribution	8130-1	55990-2	T 51-701	3900-J 2	D 3451
Lower ignition limit	8130-4	55990-6			
Free flow					D 3451
Fluidization	8130-5		T 30-501		
Flowability (fluidized)	8130-5		T 30-500		
Deposition efficiency	DIS 8130-10				D 3451
Stoving conditions		55990-4			
Loss on stoving	8130-7	55990-5	T 30-503	3900-J 5	D 3451

To check the film quality of a powder coating, the same standard methods as for liquid coatings are used. Minimum requirements for powder coatings have been laid down for certain areas of use in relevant standards (e.g., DIN 30 671 for pipes, DIN 30 677 for mountings and fittings, BS 6496 for individually coated structural elements) or by quality associations (e.g., Qualicoat or GSB, "Quality Association for Individually Coated Building Elements" relating to coating of aluminum and "Quality Association for Heavy Corrosion Protection" relating to the coating of valves and fittings).

3.4.5. Storage and Transportation

Coating powders are shipped in containers of varying size. They may, for example, be packaged in cardboard boxes each containing one or more polyethylene bags holding 20–25 kg. For larger customer off-takes, reusable metal (e.g., aluminum) containers with capacities for 400–1000 kg are used. Containers may be equipped with built-in fixtures such as a porous layer at the base for aerating and fluidizing the contents. However, containers without such fixtures are often preferred because they are easier to clean. Metal containers are increasingly being replaced by big bags that often contain a PE liner. Thermoplastic powders are often supplied in plastic drums or sacks.

Coating powders should be stored in a cool dry place. They should not be subjected to excessive pressure or temperature.

3.4.6. Environmental Aspects and Safety

Powder coating is the coating method that produces the least environmental pollution because of the following

1) The occurrence of paint waste is avoided
2) The separation of dust–air mixtures is simple
3) The coating powders are solvent-free and mostly cure by polyaddition

Emissions from coating powder spray cabins are generally $\ll$ 5 mg coating powder per cubic meter of exhaust air. Emissions from the driers are generally also extremely low. Other substances, apart from very small amounts of water ($\leq 1\%$), that are released in the stoving of coating powders are listed in Tables 3.7 and 3.8. On account of these very low emission levels, the use of coating powders is not limited by environmental protection legislation.

Coating powders that do not contain any hazardous components (e.g., lead pigments or TGIC) should generally be regarded as inert dusts and do not have toxic or fibrogenic effects. As a rule, the fraction of respirable dust that can penetrate the lungs is $\ll$ 10 wt%. Disposal of coating powder waste in landfills does not in general pollute groundwaters or surface waters, but is often not accepted by the authorities, even if deposed in a nondusty form.

Coating powders are combustible dusts and they can form explosive powder–air mixtures in certain concentration ranges. However, the danger risk is substantially lower than that for solvent-based paints for the following reasons:

1) When stored in bulk, coating powders are difficult to ignite and when heated generally only fuse together
2) Higher energy sparks are required to ignite powder–air mixtures than solvent–air mixtures
3) The glow temperature and the minimum ignition temperature of a powder–air mixture are very high (450–600 °C)
4) Powders without fines for fluidized-bed application are extremely difficult to ignite

The safe operation of coating powder plants and equipment is covered by national provisions and regulations. In the case of Europe, the ten-language brochures "Safe Powder Coating" published by CEPE (the European Committee of Paint Associations, Brussels) should be mentioned.

3.4.7. Uses

Thermosetting coating powders are widely used for industrial metal coating and are distributed among the various markets as indicated in Table 3.10. Uses range from manual application up to fully automatic, large-scale plants with an annual coating powder consumption of thousands of tons. Job coaters as a service industry account for a significant proportion of the coating powder consumption.

The main reasons for the steadily increasing use of thermosetting coating powders (see Section 3.4.1) are as follows:

1) Economy
2) Simple, easily automatable application
3) Ecological advantages
4) Good film properties, wide range of layer thicknesses
5) Improved coating powders and plants
6) New areas of application
7) Many years' experience with all sizes of coating plants

Table 3.10. Uses of thermosetting coating powders (1995) expressed as percentage of production

Use	Europe, %	Northamerica, %
Appliances	14	17
Automotive	15	16
Outdoor building elements	17	5
Garden furniture	7	8
General metal finishivp	47	54

The building and construction industry is the main user of coating powders, for both exterior and interior use. Coating powders have proved popular for coloring light metal façades, panels, and profiles on account of their outstanding weather resistance and robust surface. They also allow architects complete freedom of expression and use. New systems with extrem weatherability are introduced to the market.

In the interior-use sector, a diverse range of objects and parts can be powder coated. Examples include partition walls, ceilings, door and radiators. The coating of radiators is particularly interesting because it involves penetration of the coating powder into deep Faraday cages. The excellent impact resistance of most powder coating films is important in the building sector because the articles are roughly handled.

A further important use is the coating of office and garden furniture, machine parts, and electrical components. Applications in the furniture sector include the coating of metal rails and fittings.

The coating of steel chairs and garden furniture can be more easily automated with electrostatic powder coating than with conventional wet coating. Chairs have an unfavorable shape and so coating with wet paint can only be partially automated; generally a final coat is applied manually. With coating powders, however, coating can be performed fully automatically. The paint film has a high scratch resistance which considerably simplifies packaging.

The coating of cast is an important use in the engineering sector. Specially formulated coating powders are available for porous substrates to prevent bubble formation in the film.

In the electrical industry, the coating of switchboards with textured powders should be mentioned. Important requirements include satisfactory punching quality, recoatability with almost all paints, and constant visual properties.

Further important users are the household appliance industry and the automotive parts and manufacturing industry. The first large-scale plant for powder coating (for coating refrigerators and deep freezers) came into operation in 1973. Powder coating has since become widespread in the household appliances sector.

An extremely wide range of automotive parts and components on, in, or underneath automobiles are powder coated (e.g., trim, headlights, springs, transmission shafts, horns). By coating reflectors, smooth leveling is achieved on the rough, deep-drawn substrate, which eliminates the time-consuming, expensive polishing that is necessary with a liquid paint. Interesting multilayer coatings that combine powder and/or paint layers can be applied to wheels to obtain decorative effects. In the United States coating powders are used as primers or surfacers and blackouts on automotive bodies. In Europe use on automotive bodies has started including surfacers and clear coats.

Thermosetting coating powders are also used for storage furniture shelves, shopfitting, and wire goods. The coating of nonmetallic substrates (e.g., glass or ceramics) and highly decorative articles (e.g., spectacles and sanitary fittings) should also be mentioned.

Important areas of use of epoxy resin powders include the interior and exterior coating of pipes, fittings, fixtures, filter plates, and steel bars for reinforced concrete, as well as the electrical insulation and the encapsulation of electronic components.

Thermoplastic Coating Powders. *Polyethylene* is mainly used for the exterior coating of pipes. It is also employed for coating wire goods, especially in the domestic appliances sector.

Polyamide is extremely resistant to washing lyes and chemicals, provides good corrosion protection, and has a high impact strength and low degree of abrasion. Areas of use are therefore very varied, and include wire articles (e.g., dishwasher baskets) and metal furniture (e.g., for hospitals).

Poly(vinyl chloride) is mainly used for wire articles (e.g., fences, domestic and household appliances). The release of plasticizer during coating is a disadvantage.

The main use of *thermoplastic polyesters* is the coating of weld seams in cans.

3.5. Waterborne Dispersion Paints (Emulsion Paints) [3.52], [3.53]

The paints and coatings sector experienced a major revolution after World War II. In Germany the changeover to synthetic polymers [poly(vinyl acetate) and polyacrylates] took place during the war due to the lack of natural raw materials (vegetable oils for alkyd resins). Since the postwar demand for synthetic styrene–butadiene rubber fell in the United States, part of the production capacity was converted to styrene–butadiene latex for paints (latex paints). Dispersion paints, commonly referred to as emulsion paints are now the largest product group worldwide in the paint and coating industry. In Germany, for example, the total production of paints and coating materials in 1996 was 1.8×10^6 t, of which interior-use emulsion paints accounted for 450 000 t, exterior-use emulsion paints for 150 000 t, emulsion lacquers for 25 000 t, and dispersion-bound plasters for 180 000 t.

Binders consisting of plastics (polymer) dispersions can be used for matt to gloss coats, smooth paint-like products, and plasters (synthetic resin-bound plasters, DIN 18 550) containing coarse, structure-imparting aggregates.

Substrates such as plaster, concrete, cementbonded slabs, gypsum plaster, plaster board, wallpaper (wood-chip, glass textile, vinyl wallpaper, etc.), brickwork and sand–lime bricks, cement or concrete roofs, floors, sidewalks, wood and wood materials (e.g., chipboard and hardboard), and many other substrates can be coated with coating materials based on polymer dispersions. Many coating materials containing polymer dispersions that form films and dry at elevated temperature ($> 40\,°C$) are used for the industrial coating of plastic articles (e.g., audio equipment housings, dashboards and instrument panels in vehicles), metals, window frames, etc.

The advantage of waterborne coatings containing polymer dispersions is that they can be applied without stringent safety measures because water is their principal liquid component. Most of them dry physically, by evaporation of water, under ambient conditions and generally from $+5\,°C$ (minimum film-forming temperature, MFT) without producing any permanent odor. The MFT depends on the auxiliaries and residual monomers in the plastics dispersions. Emulsion paints are noncombustible in liquid form (easy storage and transportation); equipment, hands, and spillages can be washed and cleaned with water.

According to DIN 55 947, a plastics dispersion consists of polymer particles distributed in a liquid phase (water in the case of emulsion paints). Polymer dispersions may be fine (particle diameter 0.1–0.3 µm), medium (particle diameter 0.3–2 µm), or coarse (particle diameter 2–5 µm). Fine and medium dispersions are mainly used as binders for emulsion paints. If a particularly good penetration ability is required, very fine dispersions with particle sizes of 0.04–0.1 µm are also used. Polymer dispersions of relatively uniform particle size are termed monodisperse, while those with a nonuniform or broad particle size distribution are termed heterodisperse or polydisperse. Fine dispersions (0.1–0.3 µm) generally have a fairly high pigment binding capacity (i.e., at equal pigment volume concentration (PVC) the pigments are bound better than with coarser dispersions). Paints with improved application properties can be formulated with dispersions in the particle size range > 0.3 µm. Satisfactorily leveling and glossy coatings can be produced with fine, heterodisperse dispersions.

Emulsion paints consist of polymer dispersions as binders, pigments, extenders, and small amounts of auxiliaries (in some cases < 1%). Waterborne polymer dispersions are produced by emulsion polymerization; monomer droplets are polymerized in water-containing surfactants and protective colloids. The size and size distribution of the dispersed polymer particles can be controlled by adjusting the stirring rate in the polymerization reactor and by selecting appropriate protective colloids.

Polymers. The nature and amount of the employed dispersion greatly influences the quality of an emulsion paint. The nature of the plastics dispersion is governed by the monomers used for emulsion polymerization. The most important polymers for emulsion paints are listed below in order of increasing resistance to alkali (which may be of importance for use on exterior mineral substrates):

Vinyl acetate copolymers are used worldwide in polymer dispersions. Copolymers with dibutyl maleate are particularly elastic, while copolymers with vinyl esters of Versatic Acid have particularly good weather resistance for shade formulations. Copolymers with acrylate esters and vinyl acetate–ethylene–vinyl chloride terpolymers with a good cost–performance ratio are also important.

Vinyl propionate copolymers (e.g., with acrylate esters) are used in the same way as vinyl acetate copolymers.

Acrylate–methacrylate copolymers (pure acrylates) are used in particular for house paints and other emulsion paints with high weather resistance and for waterborne industrial paints.

Styrene–acrylate copolymers are widely used as binders for interior-use paints, plasters, and exterior-use paints with a relatively high PVC. Water absorption and elasticity decrease with increasing styrene content.

Styrene–butadiene polymers have a very low water uptake but tend to undergo chalking and are therefore used for formulating interior-use (latex) paints that can withstand repeated cleaning.

Commercial Products. Examples of commercial products are Acronal, Propiofane (BASF); Lipaton (Hüls); Mowilith (Hoechst); Plextol (Röhm); Primal (Rohm & Haas); Rhodopas (Rhône-Poulenc); and Vinnapas (Wacker).

Pigments. Many colored pigments are used in emulsion paints. Titanium dioxide is responsible for the hiding power, whiteness and, depending on the type of TiO_2 used (rutile or anatase), the chalking resistance of paints. Surface-treated rutile should be used for chalking-resistant paints because anatase reacts photocatalytically with oxygen and moisture resulting in radical reactions, decomposition of the binder at the TiO_2 interface, and chalking.

Extenders are used to adjust the properties of a coating (e.g., layer thickness and structure). They influence water uptake and permeability to water vapor, carbon dioxide, sulfur dioxide, and nitrogen oxides. Inorganic extenders are obtained from natural minerals by crushing, grinding, elutriation, and drying. The most important extenders are calcite, chalk, dolomite, kaolin, quartz, talc, mica, diatomaceous earth, and barytes. Synthetic inorganic extenders are produced from inorganic products by digestion, precipitation, or annealing. The most important synthetic extenders are precipitated calcium carbonate, barytes (blanc fixe), and silicates (Aerosil). Synthetic organic extenders include polymer fibers, as well as the recently introduced hollow polymer spheres which can be used to formulate paints with a particularly low density.

Auxiliaries. *Dispersants* ensure that the individual particles in the emulsion paint do not combine to form agglomerates. *Protective colloids and emulsifiers* are used during emulsion polymerization to ensure that small polymer spheres are formed in the aqueous phase but do not fuse together. They influence film formation of the emulsion paint and can cause foaming. Protective colloids include poly(vinyl alcohols) and cellulose ethers. Emulsifiers include anionic and nonionic surfactants.

Wetting agents are used to disperse the extenders and pigments. They include sodium polyphosphates, and sodium and ammonium salts of low molecular mass poly(acrylic acids). Polyphosphates tend to undergo secondary reactions; polyacrylates are therefore preferred from the point of view of storage stability, gloss, etc.

An excess of dispersion auxiliaries increases the water sensitivity of the coating films and thus reduces washfastness and weather resistance. The optimum amount and composition of dispersants is of decisive importance as regards the quality of each formulation.

During the production and transportation of emulsion paints the dispersants and emulsifiers cause foaming which leads to the undesirable formation of craters and pores in the coating film. *Defoamers* reduce the surface tension of the liquid to such an extent that the air bubbles in the foam collapse. They include silicone oils, waxes, fatty acids, if necessary on a solid carrier, e.g., on silicates. The defoamers are formulated in an auxiliary liquid (e.g., mineral oil, silicone oil, or glycol).

Preservatives prevent microbial growth in emulsion paints stored in drums or cans. Fungi and bacteria are present in the raw materials (especially water) and also pass during the production process into the emulsion paints. The most commonly used preservatives include formaldehyde-releasing agents, chloroacetamide, isothiazolinone, and heterocyclic nitrogen–sulfur compounds. For health reasons mercury compounds and organic tin compounds should not be used. At higher concentrations, preservatives have a fungicidal action and prevent fungal infestation of the coating and coated substrate. Coatings may also contain algicides.

Thickeners are used to modify the rheological properties of emulsion paints. Rheological properties include viscosity, consistency, application behavior, coating resistance, splashing during roller application, leveling, covering of butt surfaces and edges, pouring resistance, settling behavior, and storage stability (post-thickening).

Similar to all other auxiliaries, thickeners influence the water retention and open time, water resistance, washfastness, abrasion resistance, weather resistance, and gloss. The most important thickeners are methyl cellulose and hydroxyethyl cellulose. The viscosity rises with increasing polymerization of the cellulose ether. Synthetic thickeners based on poly(meth)acrylates, polyvinylpyrrolidone, and polyurethanes are also important. Inorganic thickeners include layer silicates of the montmorillonite type (e.g., aluminum silicate and magnesium silicate). They confer a better washfastness and abrasion resistance than the cellulose ethers and polyacrylates. However, they develop a lower water retention capacity (open time).

Film-forming auxiliaries are high-boiling solvents that promote film formation and reduce the minimum film-forming temperature. They act as temporary plasticizers until the end of film formation and then evaporate during the drying phase. Film-forming auxiliaries include butyl diglycol acetate, butyl diglycol, glycol ether, and glycol ether esters, which in some cases are also termed plasticizers. An example of one of the most important types is dibutyl phthalate.

Film formation is the physical "solidification" of an emulsion paint. Water evaporates from the wet coating film and the polymer particles approach one another until they come into contact. On further evaporation of water, capillary pressure is generated in the water-filled interstices between the polymer spheres, as well as between the polymer spheres and other solids (extenders, pigments). The polymer particles thus coalesce to produce a coating film. A precondition for this film formation is that the minimum film-forming temperature (MFT) is reached or exceeded during drying. Below the MFT an emulsion paint forms a noncoherent, fissured, mechanically unstable film. The MFT depends on the structure of the polymer binder, its glass transition temperature, and on the nature and amount of film-forming auxiliaries and other additives in the formulation. The MFT range may extend from -5 to $+50\,°C$; in the case of house paints and wall paints the range is $+5$ to $+10\,°C$.

Production. Emulsion paints are produced in paddle mixers, of which high-speed dispersers are by far the most widely used and effective. High shear forces are produced in these dispersers which account for their good dispersing action. With some viscous, coarsely filled emulsion paints, however, the shear forces can destroy the dispersion; beam agitators are more suitable in this case.

Properties. The most important properties of emulsion paints are their rheology (which affects the application properties); hiding power (extinction); pigment volume concentration (influences stress, water uptake, water vapor permeability, weather resistance); water vapor and water permeability (influences the moisture balance of the substrate and thus the durability and protection of the substrate against weathering); adhesion to the substrate (particularly with alternating moist–dry, moist–hot, and moist–frosty conditions); and the elasticity (in the case of crack-covering coatings and wood coatings).

Application. Emulsion paints can be applied by brush, roller, dipping, inundation, and spraying. A wide range of qualities of emulsion paints with tailor-made properties can be produced from the available binders, thickeners, and auxiliaries.

Types of Emulsion Paints. The most important types of emulsion paints are the following:

Emulsion lacquers have a pigment volume concentration of 15–20% and contain almost exclusively TiO_2 and finely particulate inorganic and organic colored pigments with which satisfactorily leveling paints with a uniformly high surface gloss can be formulated. Emulsion lacquers are produced mainly with fine, pure acrylate dispersions. Lacquers used for industrial coatings are also emulsion lacquers.

Exterior-use paints have a pigment volume concentration of 30–55%. They may be formulated with all of the aforementioned dispersions. Properties are adapted as appropriate to the substrates, climatic conditions, market circumstances, etc. The most important properties of exterior-use paints are their water and water vapor permeability (moisture balance), chalking resistance, and adhesion to the substrate. Special requirements apply in certain areas of application (e.g., CO_2 protection on reinforced concrete, fungal and algal resistance in moist surroundings, crack bridging on cracked substrates, and alkali resistance on new mineral substrates).

Interior-use paints have a pigment volume concentration of up to 80% depending on the required surface effects. The gloss, cleaning ability, mechanical strength, washfastness, and abrasion resistance decrease with increasing pigment volume concentration.

Tinting paints are used to tint and shade emulsion paints. They contain colored pigments and wetting agents to ensure good color development in white emulsion paints.

Synthetic resin-bound plasters contain, in addition to the components of an emulsion paint, coarse aggregates (sand, marble chippings, colored stones) to produce decorative effects.

3.6. Nonaqueous Dispersion Paints

The term nonaqueous dispersions can be applied to all particulate dispersions in any liquid medium other than water. It has, however, come to have a more restricted meaning, namely, particulate dispersions of organic polymers in organic liquids.

The upsurge of interest in this type of colloidal system followed the development in the 1950s of the technique known as dispersion polymerization [3.54]. This process provides a means of preparing nonaqueous polymer dispersions in a controlled manner. A wide range of such dispersions have been made, mainly by free-radical addition polymerization.

As with many technological advances, however, the development of stable, nonaqueous colloidal dispersions posed theoretical problems including the mechanism responsible for their stabilization. Electrostatic stabilization, operative in many aqueous systems, could not account for the stability of dispersions in low dielectric media and the concept of steric stabilization was proposed [3.55].

Steric Stabilization. Steric stabilization involves interactions between polymer chains located on the surface of particles [3.56]–[3.59]. Stabilizing repulsive forces are generated by the interaction of polymer chains located on approaching particles. These forces are related to the free energy changes accompanying the increase in the concentration of polymer segments during interpenetration or compression of the polymer chains. For practical purposes, sterically stabilized dispersions are achieved if:

1) Every particle in the dispersion carries a complete sheath of solvated polymer chains, each chain being anchored at one or a few points to the particle surface. Solvated denotes that the liquid phase of the dispersion would be a good solvent for the polymer chains if they were not anchored to the particle surface.
2) The chains are so strongly anchored to the particle surface that the sheath remains intact during encounters with other particles.
3) The sheath of polymer chains is thick enough to render the Van der Waals attraction between the particles insignificant at their point of closest approach. In general, all polymer chains are long enough to provide such a sheath [3.55].

Simple homopolymers or random copolymers that are solvated by the dispersion medium are only weakly adsorbed to the surface of the polymer particles and cannot therefore provide the strongly anchored sheath of solvated polymer chains necessary for steric stabilization [3.60]. Block and graft copolymers play a key role in the steric stabilization of nonaqueous dispersions. Block and graft copolymers contain long runs of identical monomers in the polymer chain (block) or in side chains (graft):

```
-AAABBBAAA-     Block copolymer
-AAAAAAAAA-    Graft copolymer
     B
     B
     B
     B
     B
     |
```

To provide strong adsorption (i.e., anchoring of a solvated polymer sheath) to the surface of the particle, at least one component of the block or graft copolymer must be solvated by the medium and at least one component must be insoluble in the medium and preferably compatible with, or identical to, the polymer particle to be dispersed.

Block and Graft Copolymer Stabilizers in Dispersion Polymerization. A sterically-stabilized, nonaqueous, polymer dispersion is made simply by heating a solution of a free radical initiator (e.g., azobisisobutyronitrile), an appropriate monomer, and a suitable block or graft copolymer in an organic liquid which is a nonsolvent for the polymer product and acts as a diluent for the dispersion. The block or graft copolymer acts as a stabilizer for the dispersed polymer obtained by free radical polymerization of the monomer. Although both block and graft polymers can act as stabilizers, graft polymers are normally used because they are more easily synthesized.

Although dispersion polymerization is similar to conventional aqueous emulsion polymerization, there are some crucial differences. In dispersion polymerization, the monomer is usually soluble in the organic liquid in which the polymer dispersion is to be made; a surfactant is therefore not necessary to maintain the monomer in emulsion form. Since the monomer is in solution, agitation of the polymerizing system is of secondary importance. Furthermore, in contrast to emulsion polymerization, the initiator fragments do not stabilize the resulting dispersion. The initiator to monomer ratio can therefore be used to control the molecular mass of the polymer without affecting the particle size of the dispersion [3.61].

In dispersion polymerization, sufficient graft copolymer must be present to satisfy the increase in particle surface area as polymerization proceeds. The graft copolymer can be either made in situ as a side reaction during dispersion polymerization, or it may be premade and added as required. The side reaction for the formation of the graft copolymer occurs in two ways:

1) Hydrogen abstraction grafting to a simple polymer dissolved in the diluent for the dispersion [3.62]
2) Copolymerization with a dissolved polymer carrying a copolymerizable residue (stabilizer precursor) [3.63]

In *hydrogen abstraction grafting*, the initiator abstracts a hydrogen atom from the main dissolved polymer. The graft copolymer is then formed by free-radical copolymerization of an appropriate monomer:

```
|              |              |
A              A              A
A  Hydrogen    A  Free-radical A
A  abstraction A  polymerization
A ──────────→  A· ───────────→ A ─ BBB·
A  by initiator A              A
A              A              A
A              A              A
|              |              |
```

For optimization of the hydrogen abstraction grafting process, the soluble polymer must be carefully chosen and should be used with an initiator that generates very reactive free radicals (e.g., benzoyl peroxide). Adventitious hydrogen abstraction and grafting can also occur. Many dispersions, in which steric stabilization is claimed to arise from the adsorption of simple soluble polymers, may thus actually be stabilized by graft copolymers formed in situ during the preparation of the dispersions [3.64].

Although the hydrogen abstraction process is a useful technique for many practical purposes, it is not easy to control. Control is considerably improved when a diluent-soluble polymer carrying one or a few copolymerizable groups, often known as *stabilizer precursor*, is used under conditions that minimize hydrogen abstraction grafting (e.g., use of initiators such as azobis-isobutyronitrile that produce moderately reactive free radicals).

Stabilizer precursors can be made by copolymerization of a monomer that contains a condensable group. The copolymer is then reacted with a monomer containing a complementary group. For example, lauryl methacrylate may be copolymerized with hydroxyethyl methacrylate. The resulting copolymer contains hydroxyl groups that can be esterified with methacrylic acid, thereby introducing copolymerizable vinyl groups:

$$\sim\sim\sim\text{COOCH}_2\text{CH}_2\text{OH} + \text{CH}_2=\overset{\text{CH}_3}{\underset{|}{\text{C}}}\text{COOH} \longrightarrow \sim\sim\sim\text{COOCH}_2\text{CH}_2\text{OCOC}\overset{\text{CH}_3}{\underset{|}{=}}\text{CH}_2$$

A more convenient way is to use the addition reaction of a carboxylic acid and a polymer containing an epoxide group [3.63]:

$$\sim\sim\sim\text{COOCH}_2\text{CH}\underset{\diagdown\diagup}{-}\text{CH}_2 + \text{CH}_2=\overset{\text{CH}_3}{\underset{|}{\text{C}}}\text{COOH} \longrightarrow \sim\sim\sim\text{COOCH}_2\underset{\underset{\text{OH}}{|}}{\text{CH}}\text{CH}_2\text{OCOC}\overset{\text{CH}_3}{\underset{|}{=}}\text{CH}_2$$

An alternative type of stabilizer precursor can be made by polymerizing a monomer with an initiator (e.g., azobiscyanovaleric acid) in the presence of a transfer agent (e.g., thioglycolic acid); both initiator and transfer agent have the same condensable group (in the above case a carboxylic acid group) [3.65], [3.66]. Under suitable conditions, > 95% of the resulting polymer chains carry a single terminal condensable group. When the polymer is reacted with a monomer containing an appropriate group (e.g., a carboxyl-terminated polymer with glycidyl methacrylate), a monofunctional stabilizer precursor is produced (i.e., it has a single terminal copolymerizable group).

The development of monofunctional precursors led to the production of graft copolymers that were fairly pure, in that they were relatively free from ungrafted polymer. Monofunctional precursors can also be made by condensation [3.66]. Indeed, graft copolymers made by the copolymerization of monomers with the reaction product of glycidyl methacrylate and the self-condensate of 12-hydroxystearic acid [3.67] have been used extensively in dispersion polymerizations, not only for technological applications, but also to study the mechanisms of dispersion polymerization and steric stabilization.

The emphasis so far has been placed on processes which, by their random nature, lead to polydispersity. For some fundamental studies and technological applications, block copolymers ionic initiation and "living polymer" techniques have been used [3.68], [3.69]. The growing polymer chains remain active until they are deliberately quenched. These processes yield polymers that are monodisperse with respect to molecular mass. Measurements made on dispersions, sterically-stabilized with almost monodisperse block copolymers, yield results which are more readily interpreted.

Stabilizer precursors can be used in conventional dispersion polymerization only because the starting reaction mixture is homogeneous. Some nonaqueous dispersions, however, are made by processes in which one or more of the reactants is/are insoluble in the liquid diluent (e.g., polycondensation) [3.70]. In these cases, preformed graft copolymers must be used because they function not only as stabilizers for the final dispersion, but also as dispersants or emulsifiers for the starting materials.

In order to maintain stability, the solvated sheath must remain anchored to the particle surface. In some circumstances (e.g., exposure to strong solvents which tends to swell the particles), anchoring can fail during preparation and use of the resulting dispersion. This can be avoided by including in the anchor component of the block or graft copolymer a group which is capable of coreacting with the polymer being formed in the polymerization or condensation process as the particles are being made [3.71]. The solvated sheath on the particles can then only be removed by totally disrupting the particles.

Characterization. The solvated sheath which provides the steric stabilization of nonaqueous dispersions has been extensively studied. Hydrodynamic techniques have been used to measure the thickness of the sheath [3.72], [3.73], low angle neutron scattering to measure the polymer segment distribution within the sheath [3.74], and compression experiments to measure the strength of the sheath–sheath interaction [3.57], [3.75]. The importance of the solvation power of the medium for steric stabilization has been demonstrated by flocculation experiments [3.76]. Flocculation is observed as the solvent approaches the "theta" condition for the unattached polymer chains, both by using solvent mixtures and by change of temperature. The theta condition is the state at which the configurational entropy losses caused by constraints on the polymer molecule are balanced by an increase in energy associated with polymer segment association.

Flocculation has also been observed when the diluent of the dispersion is a good solvent for the chains constituting the steric barrier. This is observed when free polymer, even when it is identical to that constituting the steric barrier, is dissolved in the diluent of the dispersion [3.77]. The magnitude of flocculation is a function of both the molecular mass and the concentration of the added polymer. The depletion flocculation theory suggests that this phenomenon is related primarily to the perturbation of the free polymer chains in solution rather than those in the solvated sheath [3.78].

The anchoring of the solvated sheath to the surface of the polymer particles has received relatively little attention. At least one component of a graft or block copolymer must be insoluble in the diluent for the dispersion. Preferably, the anchor component should be identical to or compatible with the polymer constituting the particles. There is some evidence that anchor polymer segments which are incompatible with particle polymers tend to clump at the surface of the particles [3.79], [3.80].

Similarly, the structure of polymer particles produced by dispersion polymerization has not been extensively studied. The kinetics of classic dispersion polymerization imply that particle nucleation is followed by gel-phase polymerization within the particles. This suggests that the resulting particles should have a fairly homogeneous structure [3.81], [3.82]. There is evidence, however, that in some circumstances (e.g., when the monomer is not a good solvent for the resulting polymer), polymerization occurs predominantly in solution. The polymer then precipitates out and is swept onto the surface of existing particles leading to a more heterogeneous structure [3.83]. Fluorescence quenching studies on dispersions made from labeled reactants have confirmed the heterogeneity of particles made by some dispersion polymerization procedures [3.84], [3.85].

Properties of Nonaqueous Dispersions. The stabilizing solvated sheath of nonaqueous dispersions is of limited dimensions. It therefore only has a significant effect on the effective volume fraction, and hence the rheology of the dispersions, when the dispersed particle size is very small and/or the volume fraction of the dispersion is very high. Dispersions of high molecular mass polymers have a much lower viscosity than their solutions at the same temperature.

Dispersions with uniform particle size have been used to model the rheological behavior of dispersions of noninteracting hard spheres [3.86]. Liquid dispersions with the appropriate particle size distribution have been made with a very high volume fraction [3.87].

Uses. Nonaqueous polymer dispersions have been used as the main film former in thermoplastic and, more sucessfully, in thermosetting automotive coating compositions [3.88]. The systems have the advantage of a higher solids content at application and hence lower emission of volatile organic solvents [3.89]. Cross-linked particles when swollen with solvent have been used as rheological control agents to improve the application of spray coatings in which the main film former is solventborne [3.90], [3.91]. Discrete polymer particles in a continuous matrix have also been used to improve the properties of the final paint films [3.92], [3.93]. Nonaqueous dispersions have proved convenient for the production of polymer in a form suitable for powder coating applications [3.94].

Nonaqueous dispersion polymer microgels have also found application as the main film former for certain types of exterior architectural coating where they give long term maintenance of mechanical properties leading to improved exterior durability on wooden substrates [3.94a].

More generally, nonaqueous dispersions have been used as components of polishes [3.95], adhesives [3.96], [3.97], fiber coatings [3.98], textile impregnating agents [3.99], printing inks [3.100], [3.101], and electrophotographic toners [3.102], [3.103].

Finally, the preparation of nonaqueous polymer dispersions has been used as a technique for polymer production. Process advantages such as the use of lower temperatures for polyamide and polyester synthesis can be obtained [3.70]. Products such as synthetic elastomers can be obtained in powder form [3.104]. The fine particulate form of dispersions of water-soluble polymers (e.g., polyacrylamide) offers a convenient route to aqueous solutions of polymeric flocculants and thickeners [3.105].

3.7. Radiation-Curing Systems

3.7.1. Introduction

Radiation curing is a technology which uses electromagnetic (mainly UV) or ionizing (mainly accelerated electrons) radiation to initiate a chain reaction in which mixtures of polyfunctional compounds are transformed into a cross-linked polymer network.

Resin Systems. Radiation-curable resin systems can be classified as follows:
1) Polyester–styrene mixtures
2) Products with methacrylic groups
3) Thiol–thiene mixtures
4) Prepolymers and monomers with acrylic functionalities
5) Epoxy resins

The polymerization of all these systems can be divided into initiation, propagation, and termination steps. The initiation step involves generation of a reactive species (free radical or acid). During irradiation with UV light, the reactive species are formed by chemical decomposition of a photoinitiator. In electron-beam (EB) curing, reactive species (radicals) are generated by interaction of accelerated electrons with organic compounds.

In systems 1–4 radical polymerization takes place. Decomposition of the photoinitiator (e.g., benzoin and its derivatives) leads to formation of free radicals which react with the carbon–carbon double bonds. In system 5 acidic reactive species are formed which react with cycloaliphatic epoxy compounds and vinyl ethers to form a cross-linked network. Sulfonium, and iodonium salts are used to initiate this cationic polymerization, a typical example is $Ph_3S^+PF_6^-$.

Polyester–styrene resins have a low raw material price and are used in wood finishing. Disadvantages are the presence of volatile styrene and the low cure speed. *Products with methacrylic groups* are used where line (cure) speed is less important, mainly in combination with acrylates in special end uses (e.g., in the electronics industry and in photopolymer printing plates) to improve adhesion or to obtain specific physical properties. *Thiol–thiene mixtures* can lead to very flexible cured films but their odor seriously limits their use.

Epoxy resins are recognized for outstanding adhesion to difficult substrates. They are used in clear decorative and protective topcoats for aluminum and steel cans and fixtures, plastic and paper products. Their low odor potential make them exceptionally well suited for food and beverage applications on plastic, paper and metal substrates.

Prepolymers and monomers bearing acrylic functionalities still dominate the radiation curing market and are discussed in more detail here.

3.7.2. Radiation-Curable Systems Based on Acrylates

A UV-curable acrylate system usually contains four main components: prepolymer(s), monomer(s), photoinitiator(s), and additives. A fully formulated EB-curable system is not merely a UV system without photoinitiator but the same principles apply and the same raw materials are used.

Prepolymers. A broad range of acrylated resins (oligomers) are commercially available. The film-forming properties depend on the oligomer system. One of the most common is the *acrylated epoxy* system. In *acrylated urethanes*, an isocyanate-functional prepolymer with a polyol backbone can be reacted with a hydroxy-functional monomer (e.g., hydroxyethyl or hydroxypropyl acrylate). Many different resins can be synthesized by varying the polyol backbone, the isocyanate type, and the hydroxy-functional monomer. *Polyester acrylates* are another example of commercially important prepolymers. *Acrylated acrylics* have an acrylic backbone with pendant acrylate functionality.

Selection of the prepolymer is guided by the desired film-forming properties. For example, acrylated epoxy resins typically provide hard, solvent-resistant films with fast cure at relatively low cost. Acrylated urethanes may be chosen for their excellent film flexibility and toughness. Aliphatic urethanes are used in preference to aromatic urethanes for stringent exterior weatherability requirements. The acrylated acrylics offer excellent nonyellowing properties when a soft film can be used.

Polyester acrylates are used in a wide range of applications from overprinting varnishes to lithographic (offset) inks.

Monomers. In radiation-curable systems, the solvent is replaced by reactive diluents (monomers) which are incorporated in the network during cross-linking. These monomers have two important functions: they reduce the viscosity of the mixture and they strongly influence the physical and chemical properties of the final coating. Monomers can be divided into three groups:

1) *Monofunctional monomers:* improve the flexibility of the end product (e.g., isobornyl acrylate).
2) *Difunctional monomers:* e.g., hexanediol diacrylate (HDDA), tripropylene glycol diacrylate (TPGDA).
3) *Tri- or tetrafunctional monomers:* increase the cross-linking density of the final coating.
   Examples include trimethylolpropane triacrylate (TMPTA), pentaerythritol tri- or tetraacrylate (PETA).

Additives can be extremely important in high-performance radiation-curable formulations. Pigments for inks or coatings, fillers for cost control or viscosity control, defoamers and wetting agents, flatting agents and slip aids can all play essential roles in UV/EB inks and coatings are used. Reactive additives such as surfactants or slip aids to avoid migration in the final coating.

Photoinitiators are essential for UV-curable systems and can be classified into two groups:

1) Photoinitiators that produce radicals by intermolecular abstraction of hydrogen (e.g., benzophenone, thioxanthone). These products require a co-initiator (e.g., tertiary amines, acrylated amines).
2) Photoinitiators that produce radicals by intramolecular photocleavage (e.g., benzoin ethers).

Radical polymerization is inhibited by oxygen, leading to tacky surfaces. This can be overcome in UV-curable formulations through the use of amines (co-initiator) and in EB formulations by working under an inert atmosphere.

Advantages. An important ecological advantage is that only reactive products are used; this means that there is no solvent recovery problem and no pollution. Energy consumption is low. Curing is performed at room temperature so heat-sensitive substrates can be used.

Perhaps the most important advantage is the economic aspect. Curing is done at high speed, manipulations and installation are easy, and only minimal floor space and manpower are needed. A further advantage is that the finished products are of high quality.

Disadvantages. Raw material costs are high but taking into account solvent recovery costs for solvent-based systems this is changing in favor of radiation curing. The products used are potentially hazardous (irritant) and special precautions must be taken for safe handling. Because of the extremely high reactivity and fast cure, there is a tendency for shrinkage leading to adhesion problems on certain substrates.

3.7.3. Equipment

Radiation curing requires only a fraction of a second exposure to the UV or EB source.

UV Curing. The UV lamps are usually medium-pressure mercury lamps or electrodeless gas-filled lamps. Cooling is necessary to avoid overheating of the lamp which would lead to decreased lifetime and changes in the spectral output.

Semielliptical reflectors are commonly used to concentrate the light and cure coatings applied to a moving flat substrate. Parabolic reflectors provide a parallel beam of light and are mainly used to cure coatings on uneven substrates.

EB Curing. Electron-beam accelerators (150–300 kV) used in finishing plants are of two types:

1) *Scanner type:* electrons are generated from a point cathode in high vacuum. They are then focused, accelerated, and scanned along the scan horn whose length is adjusted to the coating width.

2) *Linear cathode type:* electrons are emitted from suspended wire cathode(s). Contrary to the scanner-type accelerators, electrons are not focussed but are emitted over the total length of the wire(s).

3.7.4. Fields of Application

Radiation curing is used for special coatings, inks, adhesives, and in other special areas.

Coatings are either clear or pigmented. Coatings for paper include clearcoats for laminated paper on pressboard (wood-grain papers to simulate natural wood). Inks and overprint varnishes for the graphic art industry are a large, rapidly growing segment. High-gloss overprint varnishes for magazine covers, record jackets, and other consumer items are also often radiation-cured.

Plastic coatings for interior and exterior applications are also an important use segment. Important factors are cost, adhesion, weatherability, and availability of raw materials.

The coating of wood is an important use of radiation curing. Three-dimensional curing is very difficult and requires spray application, good control of film thickness, and well-engineered curing lines to assure complete cure of coated objects (e.g., chairs). UV-curable coatings are also used for cork in Spain and Portugal.

Adhesion to metal is sometimes difficult to achieve. However, due to the development of adhesion promoters industrial applications on metal exist. Electronics applications are important; photoresists (both wet and dry film), solder masks, potting compounds, and conformal coatings are products based on UV-curable materials.

Optical fibers have been coated with a protective layer of UV-curable materials for a long time.

Inks. Radiation-curable inks are applied to metal, paper, wood, and plastics. Lithographic (offset) and screen printing inks are the most important printing inks. Flexographic inks are already used on narrow web printing presses and are starting to find success in wide web applications. Presses have been modified to be able to print inks with a higher viscosity than conventional flexoinks.

Intaglio inks (special inks used to avoid counterfeiting of items such as bank notes) are also employed and ink-jet printing with UV-curable materials is starting.

Adhesives are well established, mostly in lamination. Some UV-curable pressure-sensitive adhesives are used in the automotive industry.

Other uses include the production of flexographic printing plates from UV-curable materials. An interesting application is three-dimensional modeling (stereolithography) in which a solid part is made from a vat of UV-curable liquor by the use of UV lasers controlled by computers using CAD/CAM software.

Release coatings (casting paper, caul sheet, labels) are manufactured by EB curing. Other new applications of radiation curing include optical applications (productions of lenses), composites, glasscoatings.

3.8. Electrodeposition Paints

Electrodeposition (electrophoretic) coating was developed in the 1960s as a method of priming electrically conducting bulk articles for industrial use [3.106]. A high application efficiency ($> 95\%$) can be achieved by using ultrafiltration techniques [3.107] and has promoted the use of electrodeposition coatings. In 1989 ca. 90% of all automobiles and many industrial goods were primed by electrodeposition. The market volume for electrodeposition paints in Europe in 1988 was estimated to be ca. 90 000 t/a, corresponding to a turnover of ca. 470×10^6 DM. The increase in sales in the Federal Republic of Germany in 1989 was ca. 8% versus 1988 [3.108].

Properties. Electrodeposition paints are aqueous dispersions of organic polymers which are used to apply organic coatings to electrically conducting substrates by passage of direct current [3.109]–[3.111]. When direct current flows between two electrodes, charged dispersion particles are deposited on one of the electrodes in a diffusion-controlled boundary layer after an induction time t_i. Coagulation of the particles occurs as a result of the change in pH value caused by the electrolysis of water [3.112]. If the surface charge of the dispersion particles is positive, coagulation occurs at the cathode [3.113]; if the charge is negative, coagulation occurs at the anode [3.114]. The deposited polymer particles consolidate due to electroosmosis and rapidly produce a high electrical resistance that slows down deposition [3.115]. This behavior results in a uniform coating, even of parts with complicated shapes (e.g., automobile bodies). The surfaces to be coated are pretreated with inorganic conversion layers (usually zinc phosphate) to improve adhesion of the electrocoat and corrosion protection [3.116].

The compositions of cathodic and anodic electrodeposition baths are given in Table 3.11. The main components are resins, pigments, extenders, some solvent, and neutralization agents. The latter are used to neutralize the charge on the resins so that they become water dispersible. The application data affect the plant and equipment technology (e.g., choice of construction materials, energy consumption, and rectifier design) [3.109], [3.117].

Electrodeposition tanks range in size from 0.2 to 400 m^3 depending on the objects being coated and the throughput rate. Tanks may be operated batchwise or continuously. The former are mainly used for truck bodies and small mass-produced parts, the latter for high throughputs as in car body priming. In addition to the coating tank, the most important parts of the unit are the counter electrodes, rectifier, circulating pumps, heat exchanger, filters, rinsing zones, and ultrafiltration units (see Section 8.3.3) [3.109].

Table 3.11. Typical bath and application data of electrodeposition paints

Parameter	Cathodic	Anodic
Solid (2 h at 130 °C), wt%	15 – 25	8 – 15
Ash (450 °C), wt%	1 – 5	2 – 6
Solvent, wt%	1 – 4	4 – 10
pH	4 – 7	7 – 9
Specific bath conductivity (20 °C), mS/cm	0.8 – 1.5	0.8 – 1.5
Voltage, V	250 – 450	150 – 350
Maximum current density, mA/cm^2	4 – 6	4 – 6
Bath temperature, °C	25 – 35	20 – 28
Coating time, s	120 – 240	120 – 240
Deposition equivalent, mg/C	30 – 40	60 – 80

In order to maintain constant bath composition and thus constant coating quality, two compensation methods are used for feeding the baths. In the first method the neutralizing agent released during the deposition period is used to disperse underneutralized, often one-pack feed material in the bath. The solids content of the bath is, however, usually kept constant by using fully neutralized, uniformly predispersed material. Excess neutralizing agent must then be removed through a circulation system at the counter electrodes with the help of ion-exchange membranes [3.118].

Binders. The non-water-soluble resins are converted into a soluble (i.e., ionized) state by using neutralizing agents (e.g., acetic acid). The first important, *anodic electrodeposition paints* were formulated with unsaturated oils that were functionalized with maleic anhydride. Reaction with maleic anhydride was followed by semi-esterification of the succinic anhydride derivatives [3.119]. These resins are cross-linked by oxidative radical polymerization via their residual double bonds. Specially developed polybutadiene oils subsequently replaced natural oils as raw materials. Unsaturated polyesters, saturated polyesters, and alkyd resins can be dispersed in water at an acid index of 40–80 mg KOH/100 g solids after neutralization with amines, and can be used as anodic electrodeposition paints. Acrylic copolymers form an important class of binders for one-coat, weather-resistant coatings. These electrodeposition paint binders can easily be synthesized by copolymerizing acrylic acid with unsaturated carboxylate esters and/or vinylogous compounds.

Cathodic electrodeposition paints important for corrosion-resistant priming coats are based on modified epoxy resins that contain amino groups. Dispersion is obtained by neutralizing the amino groups with organic acids [3.120], [3.121]. Acrylic resins modified with amino groups can also be used for cathodic one-coat paints. Polybutadienes can be modified for cathodic deposition (e.g., by amidation with diamines after reaction with maleic anhydride). Blocked isocyanates are mainly used as cross-linking agents [3.122], [3.123]. The properties of cathodic electrodeposited coatings are listed in Table 3.12.

Conventional *paint additives* (e.g., cross-linking catalysts, corrosion inhibitors, and surfactants) as well as pigments and extenders are used to complete the formu-

Table 3.12. Properties of cathodic electrodeposited coatings [3.124]

Property or test	Value
Stoving conditions	20 min, 165–185 °C
Layer thickness, mm	0.015–0.035
Corrosion protection on pretreated metal sheets	
Salt spray mist test (DIN 50021, 480 h), mm	0.3 (subsurface migration)
Alternating climatic test (VDA 621-415, 12 cycles), mm	1.8 (at the crack)
Draw behavior (DIN 53156), mm	4–8

lation of commercially employed electrodeposition paints. Dispersion mixtures for modifying film properties have also been described [3.124].

Environmental Protection. On account of the high paint transfer efficiency and low solvent content, environmental pollution in electrodeposition coating is much lower than with spray coating. The solvent contents of electrodeposition paints relative to the solids content are 5–10% in comparison with solventborne spray paints with a solvent content of 50–100%. The most important sources of emission are the electrodeposition tank, the rinsing zone, the stoving oven, and the drain of the ultrafiltrate [3.125]. Developments in new electrodeposition paints aim to reduce the organic volatile solvent content, to have fewer condensation products in the stoving ovens, and to have lead-free formulations [3.126].

Uses. Electrodeposition coatings are used primarily for corrosion protection of steel, galvanized steel, and aluminum (e.g., in automobiles, agricultural machinery, steel furniture, and appliances). New areas of application include coil coating primers [3.127], beverage can coatings [3.128], and photocurable electrodeposition paints as photoresists [3.129].

4. Pigments and Extenders

Pigments, dyes, and extenders are defined in DIN 55943 under the heading "colorants" (coloring materials). Further standards are DIN 55944 and 55949. Similar definitions are given in ASTM D16–84, ISO 4617, and ISO 4618. A *pigment* is defined as a substance that is insoluble in the application medium (e.g., a paint) and is used as a colorant or on account of its corrosion-inhibiting or magnetic properties. A *dye* is defined as a colorant that is soluble in the application medium.

Extenders are sometimes difficult to distinguish from pigments; they are defined as substances that are insoluble in the application medium and are used to increase the volume, obtain certain technical properties, and/or improve optical properties (see also Section 4.3.1).

Pigments, dyes, and extenders are incorporated in binders to obtain decorative effects (e.g., color, gloss, dulling) or functional effects (e.g., corrosion protection) on a surface coating. For optimization of these effects the colorants must be mixed in a suitable manner and in appropriate amounts with the relevant binder. The distribution of the primary pigment particles in the polymer matrix should be as uniform as possible.

4.1. Inorganic Pigments [4.1]–[4.9]

In this section the emphasis is on inorganic pigments for topcoats and finishing coats. Functional pigments (e.g., for corrosion protection) are discussed to some extent in more detail in Section 11.1. The particle size of inorganic pigments used in paints is 0.05–10 µm, that of transparent pigments 0.01–0.05 µm. The pigment volume concentration in gloss paints is 10–35 % but may reach >80 % in matt emulsion paints.

White Pigments. The most important white pigment for surface coatings is titanium dioxide [*13463-67-7*] with a worldwide consumption in 1996 of ca. 2.0×10^6 t (60 % of total consumption). Of the three known crystal modifications, only rutile and anatase are produced synthetically and utilized on a commercial scale. The brookite modification has no economic importance. The light scattering capacity of titanium dioxide is superior to all other white pigments due to its high refractive index (rutile $n = 2.80$, anatase $n = 2.55$); pigmented coatings, especially with rutile pigments, therefore have an extremely good hiding power. When combined with colored pigments, titanium dioxide exhibits the highest brightening power of all industrially produced white pigments. On account of these unique optical properties, its toxicological innocuousness, and its chemical resistance, titanium dioxide has largely replaced all other white pigments.

Doping the crystal lattice and coating the surface with metal oxides or hydroxides has reduced the photocatalytic polymer decomposition caused by titanium dioxide to such an extent that all stability requirements can be met. Since anatase is generally more photoactive than rutile, aftertreated and/or lattice-stabilized rutile pigments are preferred for exterior applications (exposure to sunlight).

Commercial products include Bayertitan (Bayer), Kemira (Kemira Pigments Oy), Hombitan (Sachtleben), Kronos (Kronos Titan), Tibras (Titanio do Brasil), Tiona (Millenium Chem. Inc.), Tioxide (Tioxide Group), Tipaque (Ishihara S. K.), Tipure (Du Pont), Tronox (Kerr McGee), and Rhoditan (Rhone Poulenc).

White pigments of historic importance or less economic importance than titanium dioxide include zinc sulfide, lithopone (mixtures of ZnS and $BaSO_4$, worldwide production in 1995 ca. 200 000 t), zinc oxide/zinc white (ZnO produced from metallic zinc by oxidation of metal vapors, worldwide production capacity ca. 35 000 t), and white lead (basic lead carbonate, $2\,PbCO_3 \cdot Pb(OH)_2$). On account of their special properties these pigments have been able to compete against titanium dioxide in some applications. Special properties of zinc sulfide and lithopone are their low absorption in the near UV range; they can therefore be used to color UV-curable paints. Lithopone-pigmented putties can easily be sanded and have favorable rheological properties. The luster of lithopone is used to advantage in wallpaper coatings. Zinc sulfide and white lead are sometimes used in coating materials because of their fungicidal properties and to neutralize acid to protect against corrosion. Also, the lower abrasivity in comparison to TiO_2 pigments can be of importance.

Iron Oxide Pigments. Iron oxide pigments may be natural or synthetic, the synthetic oxides have a wider color range and more uniform quality. Worldwide consumption of synthetic iron oxides in 1995 was 540 000 t; coating materials accounted for about 180 000 t. The color range includes the basic colors red (α-Fe_2O_3, hematite), black (Fe_3O_4, magnetite), yellow (α-FeOOH, goethite), and orange (γ-FeOOH, lepidocrocite). Mixtures of these colors are used as brown pigments; the γ-Fe_2O_3 (maghemite) normally used for magnetic recording purposes can, however, also be used to impart brown coloration. Manganese- and chromium-containing iron oxide phases are important as heat-stable brown pigments (e.g., for stoving enamels).

Iron oxide pigments are extremely important in coating materials due to their outstanding hiding power, excellent lightfastness and weather resistance, insolubility in water and organic solvents, resistance to alkalis, and their toxicological inertness. Only micronized pigments should be used in paints because they have better dispersion properties. Pigments surface-coated with metal oxides (e.g., Al_2O_3) frequently exhibit good flocculation stability.

Commercial products include Bayferrox (Bayer), Harcros (Harrison & Crosfield), Ferrofin (Laporte PLC), Mapico (Columbian Chem. Co.).

Chromium Oxide Pigments are composed of chromium(III) oxide [*1308-38-9*], Cr_2O_3, crystallized in a corundum lattice. This crystal structure impact has a certain degree of hardness and abrasivensss. On account of their excellent colorfastness and insolubility in solvents, acid, and alkali, chromium oxide pigments have successfully competed against more brilliant organic green pigments. Hydrated chromium oxide pigments are less important. Pure chromium oxide pigments are not toxic. About 25% of the consumption of chromium oxide pigments in the world (total production 70 000 t in 1995) was used for coating materials.

Manufacturers include Bayer (green chromium oxide), British Chrome and Chemicals Ltd. and Nikon Kagagu (chromium oxide).

Complex Inorganic Colour Pigments. This term usually refers to metal oxide pigments that are derived from spinel ($MgAl_2O_4$), rutile (TiO_2), hematite ($\alpha\text{-}Fe_2O_3$), or bixbyite ($\alpha\text{-}Mn_2O_3$) structures. Substitution of metal ions in the host lattice by other chromophoric metal ions opens up a broad color spectrum. The complex inorganic pigments generally exhibit outstanding lightfastness and resistance to weathering, heat, and chemicals. Heavy-metal ions can only exert their toxic effects in dissolved form. Since these pigments are sparingly soluble, they may be classified as toxicologically innocuous. Around 25% of the world production of ca. 25 000 t in 1995 was processed into coating materials.

Commercial products include Heucodur (Dr. H. Heubach), Irgacolor (Ciba-Geigy), Kerafast (Blythe Colours), lightfast pigments (Bayer), and Sicopal, Sicotan (BASF), other producers are Cerdec, Ferro Corp., Shephard Chem. Corp. and Ishihama SK.

Cadmium pigments are derived from the wurtzite lattice of cadmium sulfide (cadmium yellow). Partial substitution of cadmium by zinc shifts the color of cadmium yellow to greenish shades. Replacement of the sulfide ions with selenium yields orange to dark red pigments (cadmium red). Of all inorganic pigments, cadmium pigments exhibit the most brilliant shades; they have excellent hiding power and also a much higher heat stability than organic pigments and are completely stable against bleeding. Public discussion concerning environmental problems has lead to a sharp fall in the consumption of cadmium pigments in paints. Worldwide production capacity in 1995 is estimated at 3000 t. A new pigment class to substitute cadmium pigments is based on rare earth sulfides (trademark: Neolor/producer: Rhone Poulenc).

Commercial products include Languedomer (Société Languedocienne de Micron Couleur). Other manufacturers include Blythe Colours, Brown, Degussa, Ferro Corporation, Harshaw Chemical Company, and Reckitts Colours.

Bismuth Pigments. The development of pigments based on bismuth orthovanadate [*14059-33-7*], $BiVO_4$, is relatively recent. Bismuth pigments are also produced as the two-phase system $BiVO_4-Bi_2MoO_6$ to improve their colorfastness. The colors of this group of pigments are similar to these of cadmium yellow. Worldwide production in 1995 is estimated at 500 t.

Commercial products include Sicopal Yellow (BASF Lacke und Farben), and Irgacolor (Ciba-Geigy).

Chromate and molybdate pigments are generally mixed-phase pigments from the system $Pb(Cr, S, Mo)O_4$. They include chromium yellow (molybdenum-free) and molybdate orange and red. Variations in shades are possible by adjusting the crystal modification (orthorhombic or monoclinic) or proportion of free lead sulfate. Pigments from the system $PbCrO_4 \cdot PbO$ should also be classified in this group (chromium orange and chromium red). The chromate and molybdate pigments have a high tinctorial strength combined with good hiding power and brilliance. Depending on the nature and degree of stabilization, products are classified as standard, lightfast, or SO_2-resistant. Chromate and molybdate pigments are classed as harmful to health and must be identified. A considerable proportion ($>60\%$) of the total world production of chromate and molybdate pigments (90 000 t in 1996) is used in coating materials with declining tendency.

Commercial products include Heucotron (Dr. H. Heubach) and Sicomin (BASF). Other manufacturers include Ciba-Geigy, Capelle Frères, Belgium and Cookson Pigments, UK.

Ultramarine pigments are sodium aluminum silicates of the composition $Na_8Al_6Si_6O_{24} \cdot S_x$ (Na-rich) or $Na_{8-y}Al_{6-y}Si_{6+y}O \cdot S_x$ (Si-rich). Small sulfur-containing anions (e.g., S_3^- and/or S_2^-) are bound as chromophores in the interstices of the crystal lattice to equalize the charge. Depending on composition, blue, red, green, or violet pigments can be obtained. Ultramarine pigments have a high heat resistance but their universal use in paint systems is limited on account of their poor hiding power and limited weather resistance. Special types coated with silica must be used when resistance to acids is required. Ultramarine pigments are regarded as being physiologically innocuous. The worldwide ultramarine production capacity was ca. 20 000 t in 1995.

Manufacturers in Western Europe include Reckitts Colours, UK, Nubiola, Spain, and Prayon, Belgium.

Blue iron pigments have the general composition $M(I)Fe(III)(CN)_6 \cdot H_2O$, where $M(I)$ is Na, K, or NH_4. These pigments have a very high tinctorial strength, but are difficult to disperse on account of their fine state of division and tendency to agglomerate. Micronized types behave more favorably on dispersion. Thermal resistance (up to ca. 180 °C) is generally sufficient for stoving lacquers. Stability to dilute acid is good, although there are problems in the alkaline pH range. These pigments can be used in coatings for exterior use due to their lightfastness and weather resistance. Their main area of application is, however, the printing ink sector. Blue iron pigments are toxicologically innocuous. Total world production of blue iron pigments was ca. 30 000 t in 1991, a considerable proportion of which was used in paints and, in particular, printing inks.

Manufacturers include Degussa (FRG).

Nacreous (pearlescent) pigments and interference pigments generally consist of platelets with a high refractive index (at least in some regions in pigments with a layer structure) and a high transparency. The first such pigments were based on natural products such as fish silver, as well as synthetic products such as bismuth oxychloride, BiOCl, and basic lead carbonate, $Pb(OH)_2 \cdot 2\,PbCO_3$. Layered pigment structures (especially coated mica particles) now dominate, however. The mica surface is coated with a titanium dioxide layer of defined thickness. The photoactivity of the titanium dioxide for exterior applications can be lowered by suitable after treatment (see p. 144). Variations in the layer thickness produce a range of interference colors. Lowering the titanium dioxide content or additional coating with Fe_2O_3 or Cr_2O_3 (if necessary without a TiO_2 layer) results in further interesting color effects based on combination of interference and selective light absorption. These pigments can also exhibit metallic effects. Other layer structures (e.g., alternating coatings of mica with TiO_2 and Fe_2O_3) also produce metallic effects. Inorganic pigments based on mixed phases of platelets with the composition $Al_xMn_yFe_{2(x+y)}O_3$ ($x = 0-0.2$, $y = 0-0.06$) also create metallic or interference effects in coatings.

Special effect pigments based on mica are regarded as toxicologically harmless. Their economic importance has arisen steadily; production amounted to ca. 15 000 t in 1996.

Commercial products include Iriodin and Afflair (E. Merck), Merlin (Mearl Corporation), Paliochrome (BASF Lacke und Farben), and Flonac (Kemira Oy).

Metallic effect pigments consist of highly lustrous flakes of nonferrous metals (e.g., aluminum, copper) or alloys (e.g., brass). *Leafing types* undergo special organic surface treatment with fatty acids and collect at the surface of the coating film. *Nonleafing types* are coated with alkylamines and are uniformly distributed in the solid paint film. When the pigments are dispersed in aqueous media, care must be taken to prevent reaction with water on freshly created metal surfaces because this leads to graying; for the same reason care should also be taken when using chlorine-containing solvents. The importance of metallic effect pigments in the decorative sector has risen sharply in recent years (e.g., in the automobile topcoat and finish sector). Worldwide consumption was estimated to be ca. 20 000 t in 1995.

Manufacturers in Western Europe include Eckart-Werke (Fürth, FRG), Silberline Ltd. (UK), Wolstenholme Rink (UK), Alcan Toyo (France), and Schlenk (FRG).

Transparent Pigments. Transparent coatings are obtained when there are slight differences between the refractive indices of the pigment and binder. Pigment particle sizes that are substantially smaller than the wavelength of light must be used. Light scattering under these conditions is very low and light absorption predominates. Many transparent shades can be produced from colored inorganic pigments; examples include α- and γ-FeOOH (yellow), α-Fe_2O_3 (red), α-CrOOH (green), and $CoAl_2O_4$ (blue).

On account of their fine state of division, transparent pigments may cause dispersion and flocculation problems. They are used not only for their decorative effects — their high UV absorption and lightfastness are also exploited to protect the binder matrix and substrate (e.g., wood). Worldwide consumption was estimated to be ca. 4000 t in 1995.

Manufacturers in Western Europe include BASF (FRG), Blythe Colors (UK), and Cappelle Frères (Belgium).

4.2. Organic Pigments

Organic pigments are used to color and/or cover a substrate. They have gained importance because the use of some inorganic pigments containing heavy metals has been legally restricted.

Organic pigments have a high light absorption and a low scattering power, whereas inorganic pigments have a low light absorption and high scattering power. Combinations of organic and inorganic pigments are therefore often advantageous. Organic pigments are therefore often advantageous. Organic pigments also have a lower density and higher surface area than inorganic pigments; their color purity and tinting strength are often higher. Organic pigments tend to dissolve at high temperature and in binders and solvents (migration).

Organic pigments may be transparent or opaque. Transparent pigments are used in glazes and in combination with inorganic pigments including special-effect pigments.

The most important class of organic pigments in terms of production are the azo pigments. Other important pigment classes include metal-complex pigments (e.g., copper phthalocyanine) and higher polycyclic compounds (e.g., anthraquinone, quinacridone, isoindolinone, and perylene).

Organic pigments are synthesized and then dried to form powders. The powder consists of agglomerates of pigment particles. Breaking up and incorporation of the agglomerate in the paint (dispersion) are very important because they determine the appearance and film properties of the pigmented paint layer. These steps are described in detail in Chapter 7. Pigment dispersions for paint formulation are also commercially available.

Properties. *Lightfastness* and *weather resistance* are the most important properties of paint pigments. The organic pigments should be colorfast and stable towards radiation, heat, and atmospheric substances. Pigments may also be partially responsible for photochemical or thermal degradation of the binder which leads to deterioration of optical film properties (e.g., gloss).

Migration should not occur when organic pigments are used in paints that are applied in differently colored layers on top of each other. Resistance to migration depends on the pigment, binder, and solvent.

The pigment should be *heat stable* because it may be subject to increasing temperature during dispersion, stoving, or use.

Chemical resistance is required for various areas of use (e.g., in automotive finishes to resist car washes and battery acids and in washing machine coatings to resist detergent solutions). Chemical resistance is also of importance in paint formulation if acids or bases are employed as catalysts or if an aqueous basic or acidic binder is used. Since coatings are permeable to industrial gases (e.g., SO_2 and NO_x), the pigments must be resistant to these gases that can affect the colorfastness of the pigment.

The pigment must also be *compatible with the paint binder*. In chemically drying paints, the pigment may react with the reactive components of the paint system. This can lead to changes in the pigment or in the cross-linking stoichiometry. A change in the cross-linking density can affect the properties of the coating.

Uses in Paints. Most organic pigments on the market are suitable for use in paint systems. Apart from the properties mentioned above, the rheological character of the pigment is important because it affects both dispersion and the paint. Rheological behavior during dispersion determines how much pigment can be dispersed per batch. Favorable rheology (i.e., low viscosity) allows higher pigment and binder concentrations at equal application viscosity. The so-called "high-solids" coatings have a low solvent content and are therefore more environmentally acceptable.

Pigments can also influence other coating properties (e.g., film gloss) as a result of complex interactions between the pigment, binder, and solvent, and the method of application.

The choice of pigment depends on the required color, binder, and fastness. Pigment manufacturers have therefore produced pattern cards and information describing their properties and areas of application. The pigment should, however, be tested in the binder to be used and for the particular application.

Commercial Products include HELIOGEN, PALIOGEN, PALIOTOL, SICO (BASF); HELIOECHT, INDOFAST, PALOMAR, PERRINDO, QUINDO (Bayer); CINQUASIA, CROMOPHTAL, IRGALITE, IRGAZIN (Ciba Specialty Chemicals); DALAMAR, ENDUROPHTHAL, HANSA, HOSTAPERM, NAPHTHANIL, NOVOPERM, PERMANENT, WATCHUNG, (Clariant); CROMOFINE (Dainichi Seika); SUNBRITE, SUNFAST (Sun Chemical); MONASTRAL, MONOLITE (Zeneca); RUBICRON (Toyo Soda);

Commercially available pigment preparations containing dispersion resins or a dispersant are used when pigments are difficult to disperse in a given binder or if the dispersion stage of production is to be omitted. The pigment preparation is then simply stirred into the paint. The compatability of the dispersion resin or dispersant with the other paint components determines the area of application. Some preparations can be used for a wide variety of uses such as solventborne or waterborne systems. Others have very restricted application areas and can only be used with a few binders and solvents. Trade names and manufacturers include ENCELAC, LUCONYL, SICOFLUSH, SICOMIX (BASF); MICROLITH, UNISPERSE (Ciba Specialty Chemicals); COLANYL, FLEXONYL, HOSTATINT, PINTASOL, SANDOSPERSE (Clariant); SUNSPERSE, AQUATONE (Zeneca).

Solventborne Systems. The trend towards high solids in solventborne paint systems has made rheology an important factor. Viscosity is affected by the binder, solvent, and pigment. The effect of the pigment on viscosity is reduced by optimization of the particle size and by suitable surface treatment. This is particularly important for organic pigments which tend to have a higher viscosity than inorganic pigments because of their smaller particle size.

The pigment is dispersed in the binder and solvent which wet and stabilize the pigment. Destabilization of the pigment (e.g., reagglomeration, flocculation, and flotation) must not occur when the other paint components are added.

Aqueous Systems. Dispersion of organic pigments in an aqueous phase is difficult because of their hydrophobic character. Surfactants are therefore often used to reduce the surface tension of the water and thereby improve wetting. Suitable treatment of the surface of the pigments can also facilitate wetting. In addition to classical steric stabilization of the pigment, electrostatic stabilization is also possible due to the high dielectric constant of water. Special dispersion resins or surfactants are used to aid dispersion; they must be compatible with the binders and must not alter the paint properties. Commercial pigment preparations are also marketed which contain appropriate dispersion resins and surfactants.

Powder Coatings. Powder coatings do not contain solvents or water, they consist solely of binder and pigment. The pigment is incorporated into the reactive resin (e.g., with an extruder). The particle size required for the application is then obtained by grinding and sieving. Cross-linking of the paints is carried out at high temperature (> 150 °C), hence the heat stability of the organic pigments is very important.

Coil Coatings. Only a few high-grade organic pigments are suitable for use in coil coatings since they must be stable to stoving temperatures > 250 °C.

4.3. Extenders

4.3.1. Introduction

Definitions. According to DIN 55943 extenders are powdered substances that are practically insoluble in the application medium and are used to modify the volume, achieve or improve technical properties, and/or change optical properties. DIN 55945 states furthermore that in white extenders, the refractive index is generally below 1.7; in certain media an extender may also be a colorant (pigment).

Since the refractive indices of most binders are also < 1.7, extenders do not contribute significantly to the hiding power when they are used in an amount below the critical pigment volume concentration (*CPVC*). The hiding power of a dry film depends mainly on the amount of titanium dioxide present, a white pigment which has a high hiding power due to its high refractive index.

Above the *CPVC*, however, dry hiding becomes apparent. Air inclusions are formed in the micropores of the dry film, and consequently the hiding power depends on both the amount of titanium dioxide and the amount of trapped air.

If the refractive index of an extender is greater than 1.7 (e.g., zinc oxide), it can also be termed a pigment, though the transition is not clear-cut. The refractive indices of some important extenders and white pigments are as follows:

Water	1.33	Mica	1.58
Chalk	1.55 → Polymer 1.5–1.6	Barytes	1.64
		Zinc oxide	2.06
Kaolin	1.56	Zinc sulfide	2.37
Talc	1.57	Anatase (TiO_2)	2.55
Siliceous earth	1.55	Rutile (TiO_2)	2.75

Table 4.1. Classification of extenders used in coatings

Types of compound	Natural extenders (ISO 3262)*	Synthetic extenders
Oxides and hydroxides	aluminum oxide (DIN 55628) magnesium oxide	aluminum hydroxide (DIN 55628) magnesium hydroxide
Silica and silicates	quartz (DIN ISO 3262–13) cristobalite (DIN ISO 3262–14) kieselguhr (DIN 55920) talc (DIN ISO 3262–10 and 11) kaolin (DIN ISO 3262–8 and 9) siliceous earth (DIN 55920) mica (DIN ISO 3262–12) wollastonite feldspar plastorite slate	pyrogenic silica (DIN 55921) precipitated silica (DIN 55921) aluminum silicates (DIN 55921) calcined aluminum silicates (DIN 55631)
Carbonates	calcite (DIN 55918) chalk (DIN 55918) dolomite (DIN 55919) magnesite	precipitated calcium carbonate (DIN 55918)
Sulfates	barytes (DIN ISO 3262–2) gypsum	precipitated barium sulfate (DIN 55911)
Special extenders	cork flour, etc. rock wool	glass beads (DIN ISO 3262–15 and 17) glass fibers (DIN 60001) polymer fibers

* The DIN and ISO standards define supply specifications.

Classification. Extenders are mainly natural minerals that are converted into a usable form by working natural deposits, separating secondary constituents, and comminution (micronization). Synthetic products such as precipitated carbonates [e.g., calcium carbonicum praecipitatum (ccp), precipitated sulfates (e.g., blanc fixe), precipitated and pyrogenic silica, and silicates are used for optical brightening. Synthetic fibers (generally organic) are used for reinforcement. The classification of extenders used in surface coatings is summarized in Table 4.1.

Economic Aspects. The calcium carbonates (calcites, chalks, limestone powders, dolomites, precipitated calcium carbonates, and surface-treated carbonates) account for 80–90 % of the extenders used in Western Europe. In the Federal Republic of Germany calcium carbonates are more predominant than in the remainder of Western Europe (ca. 90 %); calcites with a high degree of whiteness and a mean particle size of 2–5 µm are most widely used. Next follow talc, kaolins (with calcined kaolins), silica, barytes, and mica.

Occupational Health. Some mineral extenders with extremely fine particles (dusts) are governed by special legal provisions. The most important regulations in Ger-

4. Pigments and Extenders

many, for example, are [14.10]:

1) Safety precautions (Unfallverhütungsvorschriften, UVV 26), "Protection against Noxious Mineral Dust" (VBG 119)
2) Technical rules for dangerous substances (Technische Regeln für Gefahrstoffe, TRGS), TRGA 508, (siliceous dust)
3) MAK values, TRGS 900, maximum workplace concentrations and biological tolerance values

Reference should also be made to the use of paints and coating materials for food containers and packaging [4.11].

4.3.2. Properties

Extenders were originally employed for increasing volume (i.e., filling) to reduce the cost of the end product. Sophisticated and modified processing techniques (e.g., micronization, surface treatment, and precipitation) have been used to modify and improve the extenders and have led to the discovery of other important properties (see Table 4.2).

Extenders included in Table 4.2 as well as their physical characteristic data are described in [4.12]. Measurement of these properties allows definition and comparison of the extenders. Such properties include:

chemical composition
particle shape and size distribution
density
tamped density and tamped volume
specific surface area
brightness (whiteness)
abrasion
pH
wettability, dispersibility
oil absorption

Chemical Composition. The extenders should be chemically inert in the paint. Secondary constituents (e.g., iron ions) may produce color changes. Sulfates, sulfides, alkali, alkaline-earth, and heavy-metal ions may cause problems due to their chemical reactivity and toxicity, and also affect abrasion and weather resistance.

Since these components mainly occur as water-soluble salts, they lead to problems in the paint film; for example, the paint film effloresces under the influence of humidity. This is important in exterior-use paints because it affects weather and corrosion resistance.

The extender should be acid-resistant (e.g., barytes, talc, mica, china clay, silica, silicates). Minerals with a platelet (lamellar) structure (e.g., talc, mica, and china clay) are particularly suitable for corrosion protection systems since they cover the surface better than particles of other shapes. Better covering makes the paint film less permeable to water vapor, thereby reducing bubble formation and improving film adhesion. New developments have shown that special sorts of calcites result in a remarkable corrosion resistance; this can be explained by the buffering effect when the paint film is exposed to harmful acid atmospheres [4.13]. Platelet-shaped extenders should not, however, be "contaminated" with differently shaped particles (e.g., talc with carbonates).

Particle Shape and Size Distribution (DIN 66111, 66116, 66117, 66141, 66143–66145). Particles may have the following shapes (figures denote the aspect ratio) [4.14]:

Spheres	1/1
Cubes	1/1
Parallelepiped	(1–4)/(1– < 1)
Platelets (lamellae)	$1/(\frac{1}{4}-\frac{1}{100})$
Fibers	$1/<\frac{1}{10}$

Extender producers try to retain the original particle shape during processing because the paint producer does not want to have the platelet shape of kaolin, talc, etc. destroyed during micronization.

Particle size and size distribution are important factors in determining the usability of the extender. The particle size spectrum produced by micronization corresponds to a Gaussian distribution. The size spectrum is numerically characterized by the mean particle size (d_{50}), the fines, an upper limit, and by the screen oversize.

A spectrum of particle sizes ensures optimum packing density. Interstices between individual particles can be filled with finer particles that prevent the penetration of binder. Both the manufacturer and user try to ensure that a specified product always has the same particle size distribution.

If, however, a certain extender grade does not have optimum packing density, it can be combined with fine extenders or extenders with different particle shapes.

Lamellar extenders improve coating properties, e.g., rheology (in particular thixotropy by talc and kaolin), matting (talc), dry hiding (talc, kaolin), and adhesion (talc, kaolin).

Density (ISO 787, Part 10). The densities of all mineral extenders lie in a relatively narrow range (ca. 2.5–2.8 g/cm³). The exception is barytes with a density of ca. 4.0 g/cm³.

Extenders with a lower density (< 2 g/cm³) aroused interest for volume packaging (i.e., packaging based on volume rather than weight) in the emulsion paint sector. However, paints produced with low-density extenders had completely different application properties. Recent developments have therefore been aimed at modifying emulsion paints to allow the use of more platelet-like extenders (e.g., talc or kaolin) in combination with the commonly used carbonates.

Tamped Density and Tamped Volume (ISO 787, Part II). The *tamped density* is the ratio of mass to volume, while the *tamped volume* is the ratio of volume to mass after tamping under specified conditions. These quantities provide information about the volume requirements of the products.

154   4. Pigments and Extenders

Table 4.2. Properties of extenders

Filler	Chemical composition	Occurrence	Particle shape	Density, g/cm^3	Oil number, g linseed oil/ 100 g extender	Brightness	Specific surface area, m^2/g	pH	Particle size distribution, μm[c]
Calcium carbonate	$CaCO_3$	natural	rhombohedral (calcitic)	2.7	14–21	75–95	0.5–3	9.0 ± 0.5	0.5–100 (300)
		natural	microcrystalline aragonite	2.7	18–22	70–90	0.5–3	9.0 ± 0.5	0.5–100 (300)
		synthetic		2.9	25–35	>95	7–10	9.5 ± 0.5	fiber length 1–3 fiber diameter 0.2–0.5
		synthetic	rhombohedral (calcitic)	2.7	35	>95	7–10	9.5 ± 0.5	fiber length 0.3–0.8 fiber diameter 0.2
Dolomite	$CaCO_3/MgCO_3$	natural	rhombohedral	2.8	16–25	85–92	0.5–3	9.8	0.5–100 (300)
Talc	Mg hydrosilicate	natural	platelet-like	2.7–2.8	30–55	75–>90	3–12	9.5 ± 0.2	0.2–80
China clay	Al silicate	natural	platelet-like	2.6	35–65	79–92	8–20	5 (natural) 7.5–9 (primary dispersion)	0.2–40
China clay	Al silicate	synthetic calcined	amorphous platelets	2.6	ca. 50	88–90	8–10	6.5 ± 0.5	0.2–40
Siliceous earth	Al silicate/quartz	natural	platelet-like corpuscular	2.6	ca. 45	81–90	10–15	7–8	0.1–20
Wollastonite	$CaSiO_3$	natural	fibrous	2.8–2.9	25–30	85–88	2–5	9–11	3–100
Pyrogenic silicia	SiO_2	synthetic	amorphous	2.2	280[a]	>90	100–400	4 ± 0.5	0.008–0.04 (average particle size)
Silicic acid	$SiO_2 \cdot nH_2O$	synthetic precalcined	amorphous	2.0	200–300	95–98	200–300 (700)	6.0–8.0	3–10 (secondary)[a]
Aluminum silicate	Al_2O_3/SiO_2	synthetic precalcined	amorphous	2.1	150–170	95–98	100–120	10.0–11.0	3–10 (secondary)[a]
Quartz	SiO_2	natural	trigonal	2.6	13–27	75–88	1–18 (90)	7–8	2.5–100 (250)

4.3. Extenders

	Formula	Origin	Shape	Density				pH	Particle size
Cristobalite	SiO_2	synthetic calcined	tetragonal	2.4	23–25	85–90	ca. 2	8–9	2.5–100 (250)
Mica	Al silicate	natural	platelet-like	2.7–2.9	35–75	60–85	3–12	7.5–9.5	1–100
Barytes	$BaSO_4$	natural	rhombic	4.0	8–10	60–90	1–3	8–9	0.8–50
Barytes	$BaSO_4$	synthetic	rhombic	4.0–4.2	13–20	92–95	1–4	6–7	0.3–10
Aluminum hydroxide	$Al(OH)_3$	synthetic	platelet-like	2.4	20–40	95–97	2–10	9±1	0.05–100 (250)
Gypsum	$CaSO_4$	natural	platelet-like	2.4–2.6	17–20	83–87	5000–6000[b]	7±0.5	2–40
Slate	$AlH(SiO_3)_2$	natural	platelet-like	2.7	30–35	28	8000[b]	9±0.5	30–200
Plastorite	Mg,Al,K silicate	natural	platelet-like	2.75	23–33	81–85	6000–14000[b]	7.5±0.5	5–250
Glass fibers	C-, E-Glass	synthetic	fibrous	2.5					fiber length 0.5–10 mm, fiber thickness 9–12
Glass beads	A-Glass	synthetic	solid, spherical	2.5	16			0.4–1.5	0–50
Glass beads	A-Glass	synthetic	hollow, spherical		0.2–0.4	16		0.4–1.5	0–50

[a] Oil adsorption (lubrication point according to ASTM D 281). [b] According to BLAINE, cm²/g. [c] This particle size is mostly used in paints, other particle sizes are also possible. [d] Consists of agglomerates.

Specific Surface Area (DIN EN ISO 3543, BET method). The activity of extenders (i.e., their influence on the properties of the end product) depends on their surface area and their interaction with, for example, the polymer matrix.

The specific surface area is measured by absorption methods. The BET (Brunauer, Emmet, and Teller) method is widely used for light-colored extenders.

Extenders can be classified as follows (BET surface area in m^2/g):

Inactive	< 10
Semiactive	10–50
Active	50–100
Highly active	> 100

Brightness and Yellowness (DIN 53163). Extenders should be chromatically neutral and should not cause discoloration or fading in the paint. Color, brightness, and yellowness are determined quantitatively (see Section 9.2.2).

Measurement is carried out by directing light on the surface of the test sample. The proportion of light reflected by the sample is determined with a spectrophotometer and compared with that reflected by a standard surface (usually barium sulfate). The brightness is the reflectance obtained with a R 457 filter ($\lambda = 457$ nm). The yellowness is obtained by subtracting the brightness from the reflectance measured with a R 57 filter ($\lambda = 570$ nm). Generally, measurement methods must be specified by the manufacturer to allow comparison.

Abrasion. The abrasion of extenders depends on their hardness and particle size. A high degree of abrasion is undesirable because it causes wear in machinery and contamination of the blended material with abrasion products. Contamination produces surface defects in the film. Einlehner's method is often used to investigate the influence of the extender on abrasion in machinery. An aqueous slurry of the extender is stirred for 1 h in a special apparatus. A metal sieve is placed in this slurry throughout this period and the weight loss of the sieve is then measured and converted to a specific surface area.

The abrasion of a mineral decreases with the particle size. Abrasion is low if a mineral does not contain any secondary abrasive constituents. Thus, for example, the abrasion of talc increases with the content of secondary constituents such as quartz and calcium carbonate. Abrasion can also be reduced by subjecting the material to surface treatment, normally with stearic acid or stearates.

pH Value (DIN ISO 787, Part 9). The pH value of an extender refers to the pH value of an aqueous suspension of the relevant extender. It is very important for paint manufacturers to know whether an extender contains acidic or alkaline constituents to predict its behavior in an existing formulation.

Wettability, Dispersibility (ISO 8780 Part 1–6, 8781 Part 1–3). The *dispersibility* of a powder is defined as its capacity for being dispersed. Dispersibility depends on the wettability of the substance and on the number and strength of the binding sites between the agglomerates.

The rate of color development and the decrease in granularity may serve as a measure of the dispersibility.

Dispersion is defined as the formation of smaller particles from powder agglomerates in a medium and the simultaneous wetting of the particles by the medium. In a fully dispersed material there is a statistically uniform distribution of the particles throughout the entire medium.

Grinding produces crystal fracture surfaces and increases the surface activity of extenders. Since the surface also bears hydroxyl groups, the ground extenders are generally hydrophilic.

Such extenders are consequently not wetted, or if so only poorly, by aliphatic hydrocarbon solvents. However, the binder takes over this function to some extent because it contains surfactants and wetting agents added by the formulator [4.15].

In nonaqueous systems the disadvantages of extender hydrophilicity can be overcome by surface treatment (stearic acid and stearates for carbonates, silanes for silicates). Such treatment is not necessary for aqueous systems and may even be damaging.

Wetting and micronization of the agglomerates play important roles in the dispersion process. The resultant dispersion should remain stable over a prolonged period, flocculation is undesirable.

Oil Absorption (DIN EN ISO 787, Part 5). The oil number is the amount of linseed oil that is absorbed by a pigment or extender sample under defined conditions. The oil number depends on the nature of the extender, its particle size distribution, packing density, and specific surface area.

Heterodisperse extenders have a smaller binder requirement than homodisperse extenders because their interstices can be filled with fine particles. In homodisperse systems the binder penetrates the interstices.

The wet point in waterborne paints is analogous to the oil number in solventborne paints. It represents the number of milliliters of water (containing 2% wetting agent) that are required to convert 100 g of extender to a compact mass.

4.3.3. Modification of Extenders

Precipitated Extenders. Precipitated extenders such as calcium carbonate (calcium carbonicum praecipitatum), barium sulfate (blanc fixe), and silica have been known for a fairly long time. When natural minerals are micronized, a minimum particle size limit is encountered. Lower particle sizes (< 1 µm) can only be obtained with considerable technical effort and expense. Extenders with finer particle sizes are produced synthetically by precipitation. The structure of one and the same extender can be modified by controlling the precipitation conditions. Particularly pure and thus bright extenders can be produced by a suitable choice of starting substances [4.12].

Surface-Treated Extenders. Some extenders tend to form agglomerates during storage, or absorb moisture or additives from the paint mixture. This can be prevented by surface treatment. Stearic acid and its salts are used in large amounts to coat calcium carbonates.

In subsequent developments the aim was to chemically bond the extender and substrate, with the surface-treatment agent acting as a bridging element. Silanes have proved particularly outstanding in this respect and a large number of products with widely varying organic groups (e.g., amino, mercapto, vinyl, and methacrylic groups) are available for this purpose. Such products are usually used to treat kaolins and other silicates.

5. Paint Additives

In addition to resins, solvents, and pigments, paints also contain additives. The additive content is typically between 0.01 and 1%. Paint additives are used to prevent defects in the coating (e.g., foam bubbles, poor leveling, flocculation, sedimentation) or to impart specific properties to the paint (e.g., better slip, flame retardance, UV stability) that are otherwise difficult to achieve.

Such products were formerly termed "auxiliaries" and were often only used to correct a paint batch that did not comply with the required specifications and showed some defects. Nowadays additives are already taken into consideration when a new paint formulation is created; they form an essential constituent of the coating. In view of the increasingly stringent quality and environmental requirements for the production and use of coatings, high-quality paint systems are almost always formulated with additives, and specific technologies (e.q., aqueous systems, powder coatings) require specific additive developments. Changes in legislation (reduction of aromatic solvents, VOC regulations, replacement of alkylphenolethoxylates) also make new additives necessary.

Many effects can be achieved with additives, and not all of their, in some cases, very specific uses can be discussed here. Information on paint additives and their mode of performance is given in the literature and company brochures. There is very little comprehensive literature on additives [5.1], [5.2]. The physical and chemical principles of paint production and application are helpful in understanding their mode of action [5.3], [5.4]. A large number of paint additives are listed according to areas of use in special tables [5.5]–[5.7].

Additives may be classified in the following groups:

1) Defoamers
2) Wetting and dispersing additives
3) Surface additives
4) Rheology additives
5) Driers and catalysts
6) Preservatives
7) Light stabilizers
8) Corrosion inhibitors

This subdivision is by no means exhaustive. Individual products cannot always be assigned to one of these groups because one additive may simultaneously influence a combination of paint properties. Additionally it is sometimes difficult to draw the line between additives and other paint ingredients (e.g., resins, pigments, extenders, and solvents) when these materials are used in an additive-like manner.

5.1. Defoamers

Although foam may occur as an interfering factor during paint production, most problems arise when it causes surface defects during the application process.

Liquid foams are a fine distribution of a gas (normally air) in a liquid. Thin films of liquid (the lamellae) separate the gas bubbles from one another and the gas–liquid interfacial area is quite high. Pure liquids do not foam; surface-active materials must be present in order to obtain stable foam bubbles.

Defoamers (antifoaming additives) are liquids with a low surface tension which have to satisfy three conditions:

1) They must be virtually insoluble in the medium to be defoamed
2) They must have a positive penetration coefficient E
3) They must have a positive spreading coefficient S

$$E = \sigma_L - \sigma_D + \sigma_{L/D} > 0$$
$$S = \sigma_L - \sigma_D - \sigma_{L/D} > 0$$

σ_L = surface tension of the liquid phase
σ_D = surface tension of the defoamer
$\sigma_{L/D}$ = interfacial tension between the liquid and the defoamer

If both E and S are positive, the defoamer penetrates into the foam lamella and spreads across the surface. This creates interfacial tension differences that destabilize the lamellae and cause the foam to collapse. In simple terms it can be said that defoamers act because of their controlled incompatibility with the paint system. If a defoamer is too compatible its defoaming effect is not sufficient, if it is too incompatible film defects occur (e.g., gloss reduction, formation of craters).

For *waterborne paint systems* (especially emulsions used for decorative purposes) defoamers based on mineral oils are often used. In addition to the mineral oil as carrier, these products contain finely dispersed hydrophobic particles (e.g., silica, metal stearates, polyureas) as defoaming components. A small amount of silicone is sometimes included to intensify the defoaming action. For high-quality waterborne coatings in industrial applications, defoamers are used that contain hydrophobic silicone oils as the principal defoaming component instead of mineral oils. They have a better defoaming effect, but are more expensive. In most cases silicone defoamers do not cause the gloss reduction that is often observed with mineral oil products.

Silicones are also the predominant defoamer components in *solventborne coatings*. Products with a correct balance of compatibility and incompatibility can be synthesized by selectively modifying the silicone backbone with polyether and/or alkyl chains.

$$(CH_3)_3Si-O-\left[\begin{array}{c} CH_3 \\ | \\ Si-O \\ | \\ R \\ | \\ O \\ | \\ \left[\begin{array}{c} CH_2 \\ | \\ CH-R \\ | \\ O \\ | \\ R \end{array}\right]_n \end{array}\right]_x \left[\begin{array}{c} CH_3 \\ | \\ Si-O \\ | \\ (CH_2)_m \\ | \\ CH_3 \end{array}\right]_y Si(CH_3)_3$$

Silicone-free defoamers based on other incompatible polymers (e.g., acrylates and acrylic copolymers) are also commercially available.

Commercial products include Agitan (Münzing); Airex, Foamex (Tego); BYK-024, -052, -066 (Byk); Colloid 681 F (Colloids); Dehydran, Foamaster, Nopco (Henkel); Disparlon OX-710 (Kusumoto); Dapro (Daniel); and Drewplus L-475 (Drew).

5.2. Wetting and Dispersing Additives

In the production of pigmented paints, the pigment particles must be distributed as uniformly and as finely as possible in the liquid phase (see Section 7.2.2). The pigment agglomerates must first be wetted by the binder solution. This process mainly depends on the chemical nature of the pigments and binders and can be accelerated by using *wetting additives*. Wetting additives are materials of low molecular mass with a typical polar–nonpolar surfactant structure; they reduce the interfacial tension between the binder solution and the pigment surface.

After the agglomerates have been broken down into smaller particles by impact and shear forces (grinding, milling), the pigment dispersion must be stabilized to avoid reformation of larger pigment clusters by flocculation. *Dispersing additives* are stabilizing substances that are adsorbed onto the pigment surface via pigment-affinic groups (anchor groups with a high affinity for the pigment surface) and establish repulsive forces between individual pigment particles. Stabilization is achieved either

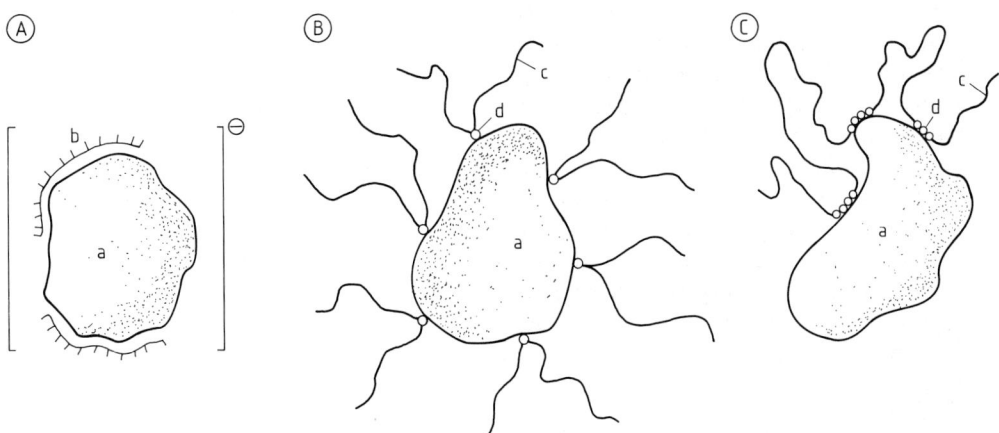

Figure 5.1. Stabilization of pigment dispersions
A) Electrostatic charge repulsion induced by polyelectrolytes; B) Steric hindrance through low molecular mass dispersing additives; C) Steric hindrance through polymeric dispersing additives
a) Pigment particle; b) Polyelectrolyte; c) Molecular structure causing steric hindrance;
d) Pigment-affinic group

via electrostatic charge repulsion (Fig. 5.1 A) or via steric hindrance due to molecular structures that project from the pigment surface into the binder solution (Fig. 5.1 B and C). The first mechanism is prevalent in waterborne emulsion systems, the latter predominates in solventborne paints. In coatings with water-soluble resins both mechanisms are equally important.

Good adsorption of the additive to the pigment surface is necessary for efficient stabilization. Problems may arise in this respect with many organic pigments because of their highly nonpolar surface. A new group of dispersing additives has therefore been developed recently. These polymeric wetting and dispersing additives can stabilize such difficult pigments by virtue of their macromolecular structure and the large number of pigment-affinic groups (Fig. 5.1 C). Such deflocculating wetting and dispersing additives are also very beneficial in highly filled *pigment concentrates*. As a rule, they strongly reduce viscosity thus allowing higher pigmentation levels.

Wetting and dispersing additives can also solve flooding and floating problems. Since most paints contain more than one pigment, the pigments often segregate in the paint film during drying. Nonuniform pigment distribution within the film surface is termed *floating* [formation of Bénard cells (Fig. 5.2 A) and streaks]. In *flooding* the surface is uniformly colored, concentration and thus shade differences occur only perpendicular to the surface; this phenomenon only becomes evident in the rub-out test (Fig. 5.2 B). In this test, after a short drying period part of the wet paint film is rubbed with the finger until almost dry (i.e., until it starts to become tacky). This treatment distributes the pigments evenly in the paint film and segregation is not possible. A color difference detected between the rubbed section and the untouched area indicates flooding.

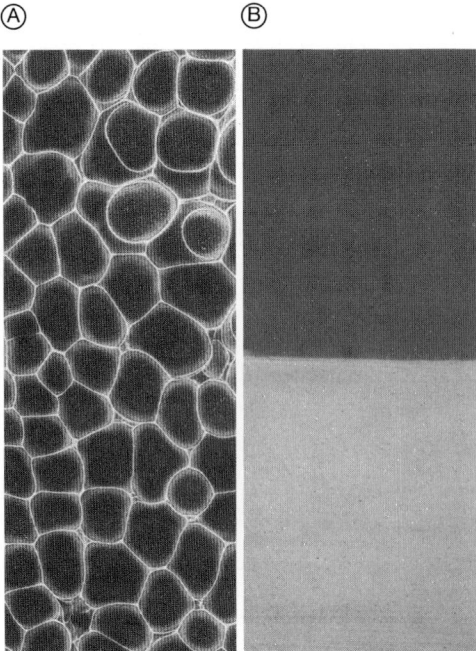

Figure 5.2. Uneven pigment distribution (color differences) due to flooding and floating
A) Bénard cells; B) Rub-out test performed on the lower part of the panel

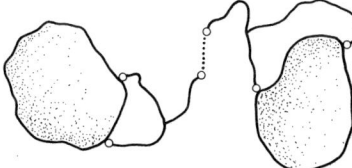

Figure 5.3. Controlled flocculation

Flooding and floating are caused by local eddies in the drying paint film. The pigment particles undergo eddy motion and if they differ in mobility, they can become separated from one another. The mobility of the pigments depends on density, size, and the strength of their interactions with the binder molecules. Additives can minimize mobility differences between different pigments by controlling these pigment–binder interactions and thereby prevent flooding and floating.

Another way of avoiding flooding and floating is to prevent the separation of the pigments by coflocculation. Additives that work in this way are known as controlled flocculating additives (Fig. 5.3). They form bridges between pigment particles and thus build up flocculates. Size and stability of the flocculates are controlled by the additive. This method is, however, not ideal for high-quality topcoats because flocculation may reduce gloss and impair other paint properties (e.g., hiding power, color strength, transparency). Controlled flocculation also changes the rheology of the paint system (see Section 5.6). Wetting and dispersing additives with such properties are often used in combination with other rheological additives. They enhance the action of the rheological additives, often synergistically, and problems such as sagging and settling can be overcome. In the case of settling, the presence of an additive layer on the pigment surface prevents the formation of hard sediment which would be difficult to stir in again. Instead any settled material formed is soft and easy to incorporate again. *Antisettling additives* generally increase the low shear viscosity to improve suspension of the pigment particles and avoid the formation of hard sediments.

Commercial products include Anti-Terra, Disperbyk (Byk); Borchigen ND (Borchers); Ser-Ad FA 601 (Servo); Solsperse (ICI); Surfynol (Air Products); Tamol, Triton (Rohm & Haas); and Texaphor (Henkel).

5.3. Surface Additives

Many surface defects can be explained by differences in interfacial tension. Poor substrate wetting, for example, must be expected if the paint has a higher surface tension than the substrate to be coated. When spray dust or solid dust particles fall onto a freshly coated surface, *craters* are formed if the deposited droplets or particles

5. Paint Additives

have a lower surface tension than the surrounding paint material. Craters are also formed if the surface to be coated is locally contaminated with substances having a very low surface tension (e.g., oils) and the surface tension of the paint is too high to wet these contaminated areas.

Surface tension differences may also develop within the paint itself: during drying the solvent evaporates and this change in composition also alters the surface tension. Even slight surface tension differences lead to the formation of Bénard cells which may result in visible surface defects such as *orange peel* and *air draught sensitivity*.

In general, surface tension differences lead to material transport in the liquid paint film from the region of lower surface tension to that of higher surface tension. This movement is responsible for the above-mentioned defects. Other phenomena such as fat edges, picture framing, and ghosting can be explained in a similar way.

Silicone additives (mainly organically modified methylalkyl polysiloxanes) lower the surface tension of coatings and minimize surface tension differences.

$$(CH_3)_3Si-O-\left[\begin{array}{c}CH_3\\|\\Si-O\\|\\Organic\\modification\end{array}\right]_x \left[\begin{array}{c}CH_3\\|\\Si-O\\|\\Alkyl\end{array}\right]_y Si(CH_3)_3$$

They are therefore ideal for solving the problems described above. Organic modification of the silicone (polyether and polyester chains, aromatic groups) serves to adjust the compatibility with the paint system. The alkyl groups have a strong influence on the surface tension: methyl groups give very low surface tension, longer alkyl chains give higher values. Special additives are available (fluoro surfactants, silicone surfactants) which are particularly effective in aqueous coatings to reduce surface tension.

Silicone additives also improve the *slip properties* of the dried coating which then exhibits improved blocking and scratch resistances. Also wax additives (wax emulsions and dispersions in water and organic solvents, or micronized waxes) are employed as surface additives. Besides giving better slip, they generally enhance the surface protection against mechanical damage (e.g., scratching, heel marking) and alter the "feel" of the surface ("soft-feel" effect). Depending on their particle size they also can contribute to the flatting effect.

Poor *leveling* is also considered a surface defect. The leveling properties of a coating depend on many factors. Silicones influence the surface structure by suppressing eddy motion during drying. Acrylate copolymers are also used for the same purpose. They are incompatible with the paint system and accumulate at the surface. They also have a stabilizing effect on the surface but do not lower the surface tension as strongly as silicones. Silicone and acrylate flow additives are also known as *surface flow control additives* (SFCA). Leveling also depends highly on paint rheology which can be modified by using special solvent blends. Finally it should be remembered that wetting and dispersing additives can also alter the rheology and thus influence leveling.

Commercial products include Baysilone (Bayer); Byk-306, -310 (Byk); Disparlon 1980 (Kusumoto); Paint Additive (Dow Corning); KP-321 (Shin-etsu); Perenol (Henkel); SF 69 (General Electric); Siliconöl AK 35 (Wacker); Silwet (Union Carbide); Slip-Ayd (Daniel); Tegoglide (Tego); Cerafak, Aquacer (Byk-Cera); Lanco Glidd, Lanco Wax (Langer); and Worlee Add 315 (Worlee).

5.4. Driers and Catalysts

Driers (siccatives) are used in paint systems that dry at ambient temperature by oxidation processes. They accelerate the drying process by catalyzing the autoxidation of the resin. Driers are in general organometallic compounds (metallic soaps of monocarboxylic acids with 8–11 carbon atoms), the metal being the active part. Cobalt and manganese (primary or surface driers), lead, calcium, zinc, zirconium, and barium (secondary or through driers) are mainly used. In practice, mixtures of metallic soaps are commonly used to obtain the optimum ratio of through drying to surface drying. Secondary driers cannot be used on their own, they always have to be combined with primary driers.

Driers can cause skin formation during paint storage, particularly if the can or container has been opened. Oximes or alkylphenols are added as *antiskinning additives*. They block the action of the driers in the can, but at the correct dosage do not prolong the drying time of the applied paint film due to their volatility.

The curing of coatings that are cross-linked by other chemical reactions can be accelerated with *catalysts*. Acid catalysts are the most important and are used for a large number of stoving enamels and force-dried, acid-curing wood paints. They are mostly sulfonic acids of widely varying structure, often blocked with amines to allow formulation of storage-stable paints. The use of a catalyst can lower the stoving time and/or stoving temperature to save energy or to permit the coating of temperature-sensitive substrates.

Catalysts also include accelerators for two-pack polyurethane paints (e.g., tin and zinc compounds, tertiary amines) and initiators for unsaturated polyester resins that act as radical-forming agents.

Commercial products include Additol XW 335 (Hoechst); Byk Catalysts (Byk); Cycat (Dyno Cyanamid); Dabco, Polycat (Air Products); K-cure, Nacure (King); Manosec Cobalt 6% (Manchem); Metatin Kat (Acima); Nuodex Cobalt 6% (Nuodex); and Troykyd Cobalt 6% (Troy).

5.5. Preservatives

Paints, the liquid paint as well as the dry film, are easily attacked by microorganisms and therefore biocides/fungicides are used as protective means. Microbial growth in the liquid paint may cause gassing, bad odour, discoloration and can finally render the paint totally unusable. This is mainly a problem in aqueous systems; in solventbased coatings the organic solvents effectively protect the paint against microogranisms. When the dry film is attacked by mildew and fungi, this first of all detoriates the optical appearance of the surface but also the mechanical properties of the film are degraded.

Preservatives have to be subdivided into in-can preservatives and in-film preservatives. *In-can preservatives* protect waterborne paint systems against contamination by microorganisms during production, transportation, and storage. *In-film preservation* is aimed at preventing the growth of bacteria, fungi, and algae on the applied paint film and is the more demanding task. A special area of use is the protection of wood against biodegradation by putrefactive fungi and insects. Antifouling additives for underwater coatings that are intended to prevent marine growth are also included in this category (see also Section 11.4).

Preservative measures are governed by the intended use of the coating. There are no universal additives on account of the large number of possible types of damage; combination products containing several active ingredients are available and often used. Organomercury compounds, chlorinated phenols, and organotin compounds were often used, but these environmentally harmful products are now being replaced more and more by metal-free organic substances, mainly nitrogen-containing heterocycles.

Commercial products include Mergal (Riedel de Haen); Metatin, Traetex (Acima); Nopcocide (Henkel); Nuodex Fungitrol (Nuodex); Preventol (Bayer); Proxel (ICI); and Troysan (Troy).

5.6. Rheology Additives

The rheology of a paint material can be described by its viscosity and the dependence of this viscosity on parameters such as shear rate, time, etc.

Newtonian liquids display no shear-rate dependence, their viscosity is constant over a wide range of shear rates. Only ideal liquids show this behavior. It is not found in coatings and it is also not desirable for coatings, because very low shear forces already will cause material flow leading to sedimentation and sagging (levelling, however, would be perfect in such a system).

Pseudoplastic flow behavior ("shear-thinning") is ideal for coating materials: viscosity is fairly high at low shear rates which avoids sedimentation and gives good anti-sag properties. At higher shear rates the viscosity is reduced, which allows easy handling and application of the material. Oftentimes it is observed that the flow behavior is further complicated by the fact that the viscosity does not only depend on the shear rate, but is also time-dependent: *thixotropic* materials do not show a constant viscosity for a given shear rate over time, but the viscosity decreases with time of shearing (Fig. 5.4). The measured viscosity of such materials depends on the shear history of the sample under test. In many systems with thixotropic or pseudoplastic flow behavior the occurance of a *yield point* is observed: a certain minimum shear force must be applied to the material before it will start to flow. If the applied shear rate is below this yield value, the material will not flow.

Rheological additives are employed to modify the flow behavior of coatings materials in order to get a more favorable rheology. In particular they are used to

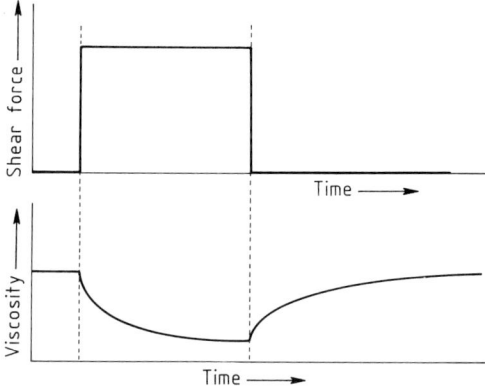

Figure 5.4. Thixotropy: viscosity is shear-force-dependent and time-dependent

create a pseudoplastic or thixotropic flow behavior in order to improve sag resistance and anti-sedimentation properties.

Thickeners, mainly cellulose derivatives (e.g., methyl cellulose, ethylhydroxypropyl cellulose) or polyacrylates, are generally used in emulsion paints. Recently polyurethane thickeners (associative thickeners) with more favorable leveling properties are also increasingly used.

A large number of rheological additives for solventborne systems are commercially available. Hydrogenated castor oils, pyrogenic silica, and modified montmorillonite clays (organoclays, e.g., bentonite) are preferred.

The rheological action of the above additives is based on the fact that they form three-dimensional networks in the paint. These lattice structures are destroyed by shear forces but are restored when the forces are removed. This recovery is not, however, immediate. The rising viscosity initially allows leveling of the surface but subsequently prevents sagging. This thixotropic behavior allows to adjust the balance between sagging and levelling.

Commercial products include Acrysol RM-4 (Rohm & Haas); Aerosil 200 (Degussa); Bentone, Thixatrol (NL); Talen 7200-20 (Kyoeisha); BYK-410 (Byk); and Tixogel (Südchemie).

5.7. Light Stabilizers

High-quality industrial coatings, especially automotive finishes, are subjected to severe weathering in exterior applications. In two-coat metallic coatings, exposure to UV light, oxygen, moisture, and atmospheric pollution causes decomposition of the polymer material in the automotive finishes. This decomposition results in loss of gloss, crack formation, color changes, and delamination phenomena [5.8].

High-energy UV light is particularly detrimental because each polymer material can be damaged particularly easily at one or more wavelengths in the UV range. Light stabilization is therefore essential [5.9].

Methods of Stabilization. Two stabilization methods have been adopted industrially [5.10], [5.11]:

1) Competitive UV absorption by UV absorbers in the wavelength range 290–350 nm
2) Trapping of the radicals formed during polymer degradation by radical scavengers (hindered amine light stabilizers, HALS)

In two-coat metallic paints, the basecoat is protected against color change and photochemical decomposition (which leads to delamination) by the filter effect of the UV absorber that is added to the clearcoat; UV absorbers cannot trap radicals. Hindered amine light stabilizers do not absorb in the UV region, but trap radicals already formed, and are mainly responsible for the gloss retention and prevention of crack formation in paints. Optimum protection against decomposition phenomena in the coating is achieved by using a combination of both stabilization methods.

UV Absorbers. Four different classes of UV absorbers are shown below:

Hydroxyphenylbenzotriazoles

Hydroxybenzophenones

Hydroxyphenyl-S-triazines

Oxalic anilides

The hydroxyphenylbenzotriazoles are the most important. They absorb the damaging UV light and rapidly convert it into harmless heat (keto enol tautomerism) [5.11].

The action of all UV absorbers depends on the Lambert–Beer law, and the absorption properties of the UV absorber. The further the absorption edge extends into the near UV region, the more UV light can be filtered out. Of the four UV absorber classes shown above, the hydroxyphenylbenzotriazoles have the broadest absorption band [5.8], [5.12]. In addition to thermal stability [5.12] and stability to extraction with water or organic solvents, photochemical stability is important [5.13]–[5.15]. Ultraviolet reflection spectroscopy can be used to establish whether the employed UV absorber is still effective, even after several years' external weathering [5.16].

Both the hydroxyphenylbenzotriazoles and hydroxyphenyl-s-triazines have a much higher photochemical resistance than oxalic anilides and hydroxybenzophenones [5.8], [5.12], [5.17].

Radical Scavengers (Sterically Hindered Amines). Two typical sterically hindered amines (HALS = Hindered Amine Light Stabilizer) follow:

$$\left[R-N \underbrace{}_{H_3C\ CH_3}^{H_3C\ CH_3} \!\!\!\!\!\!\!\!\! \overset{H}{} \!\!\! O-\overset{O}{\overset{\|}{C}}-(CH_2)_4 \right]_2$$

HALS I,   R = CH_3 $pK_b \approx 5.5$
HALS II, R = $O-i-C_6H_{17}$ $pK_b \approx 9.5$

The tetramethylpiperidine group is responsible for the stabilizing action. Different substituents on the nitrogen atom result in different pK_b values, which are important in the area of use of the products. HALS I, bis(1,2,2,6,6-pentamethyl-4-piperidinyl) ester of decanedioic acid [415526-26-7], is used in systems that are not catalyzed by strong acids (interaction of the acid with the basic nitrogen atom). HALS II, bis(2,2,6,6-tetramethyl-1-isooctyloxy-4-piperidinyl) ester of decanedioic acid [122586-52-1], was developed for acid-catalyzed systems because it does not undergo undesirable interactions with acid catalysts.

The mode of action (Densiov cycle) of HALS (as deduced from investigations on polyolefins) follows (P = polymer) [5.11], [5.17], [5.18]:

$$\text{>N-R} \xrightarrow[PO_2^-]{PO\cdot} \text{>N-H} \xrightarrow[PO_2^-]{\cdot OOH} \text{>NO}\cdot$$

$$\text{>NO}\cdot + P\cdot \longrightarrow \text{>NOP} \qquad \text{>NOP} + PO_2^- \longrightarrow \text{>NO}\cdot + POOP$$

The formation of nitroxyl radicals $\text{>NO}\cdot$ is essential for stabilization since the concentration of harmful peroxy radicals falls sharply in their presence [5.19].

Table 5.1. Outdoor exposure of a two-coat metallic coating[a]

Light stabilizer	20° gloss after n years Florida				
	$n=0$	$n=2$	$n=4$	$n=6$	$n=8$
Unstabilized[b]	93	45			
1% Benzotriazole I[c]	94	71	58		
1% Benzotriazole I and 1% HALS I	94	70	67	56	50

[a] Clearcoat, acrylic–melamine; basecoat, polyester–cellulose, acetobutyrate–melamine, silver metallic; bake: 130 °C, 30 min; exposure: Florida, 5° South, black box unheated; percentage of light stabilizer relative to binder solids; benzotriazole I, 2-(2H-benzotriazole-2-yl)-4,6-bis(1-methyl-1-phenylethyl)phenol [70321-86-7]. [b] Cracking after 2.25 years. [c] Cracking after 5.5 years.

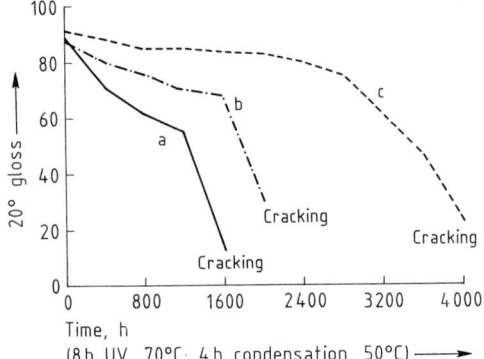

Figure 5.5. Accelerated weathering of an acid-catalyzed two-coat metallic coating in an UVCON apparatus (Atlas Corp.)
Clearcoat: high-solids acrylic–melamine; Basecoat: high-solids acrylic–melamine, silver metallic; Bake: 120 °C, 30 min; Benzotriazole II: 3-(2 H-benzotriazole-2-yl)-5-(1,1-dimethylethyl)-4-hydroxyoctyl benzenepropanoate [*84268-23-5*]
a) Unstabilized; b) 2.5% Benzotriazole II; c) 1.5% Benzotriazole II and 1% HALS II (percentage relative to binder solids)

Light stabilizers are tested under artificial weathering conditions (accelerated weathering) and under outdoor weathering (Florida, 5° South, black box, not heated) [5.17]. Figure 5.5 and Table 5.1 illustrate the influence of light protection agents on gloss retention and crack formation in two-coat metallic coatings.

5.8. Corrosion Inhibitors

To obtain a coating with good corrosion protection, *anticorrosion pigments* have to be used (e.g., red lead, zinc chromate, zinc phosphate) and/or the paint must act as a barrier against the aggressive media.

Corrosive inhibitors do not have pigment properties because they are soluble in the paint system and are not colored. They inhibit corrosion processes on ferrous surfaces (in the can and on the substrate to be coated). Typical examples are flash-rust inhibitors such as sodium nitrite and sodium benzoate for waterborne systems. Nitrogen-containing organic substances, tannin derivatives, and chelating compounds are still in industrial use. Some metal complexes of nitrogen-containing organics exhibit very good performance in combination with zinc phosphate.

Commercial products include Busan 11-M 1 (Buckman); Raybo 60 NoRust (Raybo); Ser-Ad FA 179 (Servo); and Sicorin RZ (BASF).

5.9. Use and Testing of Additives

Additives are generally used in very small amounts ($\leq 1\%$ of total formulation) and correct dosage is extremely important for optimum effectiveness and also to avoid undesirable side effects. The dosage has to be determined in test series.

Many additives can be incorporated relatively easily into the paint system during or after the last phase of production (let down). In some cases, however (e.g., with many rheological additives), certain incorporation conditions have to be observed. Wetting and dispersing additives, for example, always have to be present in the mill-base to achieve the desired results.

Ideally, the effectiveness of an additive should be checked in the complete formulation and under conditions as close as possible to those prevailing in practice. Checking a defoamer just in the binder system, for example, can only be regarded as a preliminary test, because the behavior in the complete formulation may differ substantially.

Many defects that are to be eliminated by the use of additives are also influenced by the substrate to be coated and the application method. Differences between additives can be established in simple laboratory tests. However the final composition of a specific formulation must take into account as many application parameters as possible (e.g., the state of substrate, application method, and drying conditions).

In most cases additives influence not just one property of the coating. They may also have undesirable or beneficial side effects. Additives are not "magical" products but need to be used rationally and carefully to provide the desired satisfactory results. As detailed and complete a knowledge as possible of the mechanism of action of the products, their possible effects and side effects, their limitations, and the underlying causes of paint defects are certainly helpful, but due to the complexity of paints and coatings empirical knowledge is indispensible.

6. Paint Removal

The nature, condition, and quality of the paint and substrate are important in paint removal. The paint binder plays a decisive role in paint dissolution; the substrate influences the choice of paint removal method. Various chemical and physical methods exist for removing paint from different substrates (metals, wood, and mineral substrates) [6.1], [6.2].

6.1. Paint Removal from Metals

6.1.1. Chemical Paint Removal

Paint layers can be stripped (dissolved) or degraded with chemicals [6.3]. Paint dissolution is performed with organic solvents, whose action is assisted by surfactants. Paint binders can be degraded with strong alkali or acid. Depending on their use, paint removers may also contain cosolvents, activators (acid or alkali), wetting agents, emulsifiers, evaporation retarders, corrosion inhibitors, and thickening agents. The efficiency of paint removers is improved by increasing the temperature; the paint removal time can be substantially shortened by raising the temperature of the bath (2–6 h at 20–95 °C).

The *advantages* of chemical paint removal are that it can be used for almost all types of paints, geometries, and heat-sensitive items. Investment costs are low and waste air does not cause serious environmental problems. The *disadvantages* are the relatively long removal time and formation of a paint slurry which leads to higher waste disposal costs.

Hot Alkaline Paint Removal Baths. Paint removal takes place in hot, aqueous, alkaline baths (> pH 13) at 50–95 °C. The alkalinity is adjusted by adding large amounts of alkali-metal hydroxides or organic amine or hydroxy compounds. The chemical bonds in the paint binders are hydrolyzed by the high alkalinity. Penetration into the paint layer and migration underneath the film is promoted by adding surfactants. Hot paint removers may, for example, consist of:

50–70% alkali-metal hydroxides (KOH/NaOH)
0–20% strongly basic organic hydroxy or amine compounds (e.g., alkanolamines)
5–20% high-boiling solvents (e.g., glycol ethers)

5–20 % organic acids (e.g., gluconic acid) and their alkali-metal salts
0–5 % surfactants

Strongly hydrolyzing, aggressive alkali lyes can be used for removing paint from steel whereas strong organic bases are used for light and nonferrous metals. The hot paint removal bath is overlaid with a layer of an organic compound of low volatility (e.g., paraffin oil) to prevent evaporation of water and active constituents. Alkaline hot paint removal agents are extremely economical and can be used for most paints.

Neutral paint strippers include halogen-free organic solvents (e.g., glycols, glycol ethers, 1-methyl-2-pyrrolidinone) which are generally used at 20–40 °C. In contrast to the alkaline paint removers, neutral paint strippers result in purely physical dissolution of the paint from the substrate. Their use is therefore restricted to removing physically drying paints.

Acid Paint Removers. Like the alkaline products, the acid paint removal agents (pH < 1) based on mineral acids result in chemical degradation of the paint binders. They are usually based on sulfuric acid and are covered with a protective layer of paraffin wax. Alkaline agents have a pronounced swelling action, whereas the acid agents result in disintegration of the binder. The acid products are used only in special cases (e.g., to remove epoxy and polyamide powder coatings from heat-sensitive substrates) because of their aggressive, corrosive properties; bath temperature is 20–50 °C.

Cold Paint Removal. Up to a few years ago, cold paint removal based on the use of dichloromethane [*75-09-2*] (methylene chloride) was the most widely used paint removal method. Use of this method is declining in favor of chemical or thermal methods due to concern about the environmental and occupational safety of halogenated hydrocarbons.

6.1.2. Thermal Paint Removal

Thermal paint removal methods exploit the fact that organic paint constituents can be readily pyrolyzed or combusted, as well as the high heat capacity of the organic material. They can be used to remove paint from steel and aluminum but not from heat-sensitive substrates (e.g., zinc and zinc alloys). The short removal times and small amount of waste ash formed are advantages, although this is offset by the high capital investment resulting from stringent safety regulations and the need for equipment to combust the resultant pyrolysis gases.

Low-Temperature Carbonization Method. Pyrolysis is performed for 3–9 h in closed furnaces at 380–420 °C in a controlled, low-oxygen atmosphere.

Fluidized-Bed Method. The fluidized bed consists of inert, finely granular, inorganic material (usually alumina) that is fluidized by injecting compressed air through a special fluidization floor. Organic paint material is oxidatively degraded by the entrained atmospheric oxygen at 400–480 °C in 20–60 min. This method is not suitable for some geometries (e.g., hollow items).

Salt Bath Method. The object from which the paint is to be removed is immersed in a salt melt heated to 300–500 °C. The short time required for paint removal (30–120 s) is the result of the spontaneous heat transfer and high oxidation potential of the salt melt (e.g., alkali nitrates).

6.1.3. Mechanical and Low-Temperature Paint Removal

High-Pressure Removal with Water. Paint can be removed from a metallic substrate (e.g., grids and skids) by the high kinetic energy of a high-pressure (70–100 MPa) water jet.

Blasting. In analogy with sandblasting of metal surfaces, paint can be removed from various substrates by blasting with air or water containing abrasive blasting media (e.g., sand, plastic granules, metal particles) and/or other additives (e.g., alkali salts). Paint is, for example, removed from aircraft (aluminum) with a water jet containing special additives.

Low-temperature paint removal exploits the shrinkage and embrittlement of paint layers that occurs after cooling in liquid nitrogen (-196 °C) for 1–3 min.

The mechanical methods and low-temperature paint removal are restricted to a few special areas of application. Chemical and thermal methods have their specific advantages and disadvantages, not only as regards paint removal but also with respect to environmental pollution by paint slurry, rinse water, effluent, and waste air.

6.2. Paint Removal from Wood and Mineral Substrates [6.4]

Paint strippers used to remove coatings from wood and mineral substrates may be either alkaline or based on organic solvents.

Alkaline paint strippers (e.g., alkali-metal hydroxides, sodium carbonate, sodium metasilicate, trisodium phosphate) can be manufactured in liquid, paste, or powder form. They are only suitable for removing hydrolyzable coating materials (e.g., alkyd resins and oil paints). Alkaline paint strippers may also be used to remove coatings on facades. Tests should be carried out to check that the coating is hydrolyzable.

The *solvent-containing paint strippers* may contain chlorinated hydrocarbons (dichloromethane) and cosolvents (e.g., alcohols and aromatic hydrocarbons). Systems that do not contain chlorinated hydrocarbons (CHC-free paint strippers) are, however, becoming increasingly important because environmental considerations demand reduction in the use of chlorinated hydrocarbons. The dichloromethane-containing formulations evaporate relatively quickly and subsequently enter the atmosphere. The CHC-free paint strippers are formulated with a combination of various slowly evaporating but more effective solvents. Typical solvents are high-boiling glycol ethers, dicarboxylic acid esters, N-methylpyrrolidone, alcohols, and ketones in combination with auxiliary substances. CHC-free paint strippers are suitable for removing facade coatings whereby many paint layers can be removed simultaneously. Removal takes a longer time than with CHC-containing paint strippers but this can be compensated for by using different application techniques. It is important that the paint stripper wastes are readily biodegradable.

7. Production Technology

7.1. Principles

A large number of paint formulations are produced in a limited number of production steps. The special know-how for paint production therefore involves skilful adaptation of the processing steps to the relevant production plant (scale-up, quality planning). Until recently, paint production was a traditional craft in which formulations were adapted in a large number of production stages. Quality testing was carried out at the end of the production process. Fluctuations in quality were rectified in time-consuming, and thus expensive, correction steps. With the transition to industrial production, quality planning and quality control systems were increasingly implemented in conjunction with the scale-up of unit operations to achieve the necessary quality specifications with a lower number of additional correction stages.

Many other factors besides the choice of suitable raw materials have to be taken into account if a paint formulation is to serve as a basis for commercial production of high-quality paints. This is because the quality and stability of the final product is determined not only by the choice and quantity of ingredients (pigments, film formers, solvents, additives, etc.) but also by how and in which order they are combined.

Source of Added Value. The main physical process involved in the production of coating materials is the homogeneous, irreversible mixing of the liquid components. In pigmented systems, complete wetting and a uniform, stabilized distribution of pigment particles in the liquid medium (resin solution or dispersion) are also important. The main reason why these apparently simple processes result in a relatively high added value is the crucial but difficult to achieve requirement for microhomogeneity with a particle size of ≤ 1 µm. Homogeneity or dispersity at this level is very important because of the interaction of the paints with the absorbed light with wavelengths in the visible region. High energy inputs are necessary to obtain microhomogeneity (or dispersity) due to the interaction of strong adhesive forces in this region.

Production Strategy. An all-embracing description of paint production processes cannot be given, mainly because paint manufacturers range from very small operations producing only a few hundred tonnes of paint annually up to large companies producing several hundred thousand tonnes annually. Large companies account for only about 20% of the worldwide market (ca. 30×10^6 t). A further difficulty is the multiplicity of formulations: large paint factories must hold a range of up to 20000 formulations to be able to serve their markets. In addition, national differences in environmental legislation make a uniform approach difficult.

Since customer orders differ not only as regards paint formulation but also in the amount ordered, manufacturers must have a corresponding "mix" of equipment (different sizes of mixers, tanks, production equipment, filling and packaging lines, etc.). Production according to customer orders (including just-in-time delivery) results in widely varying batch sizes, ranging from a few kilograms up to 20 or 50 t, but of the same, constant quality. A large paint factory therefore has a complex structure with regard to material flow and the provision of materials, equipment, and staff [7.11], e.g. up to 1500 raw materials and intermediates for 2000 various products.

In general the paint-making process can be subdivided into four main stages:

1) Preparation of the millbase (premixing)
2) Continuous dispersion
3) Completion of the formulation
4) Correction and adjustment of the final product

Two strategies are adopted for producing the millbase:

1) Ab initio production
2) Mixed production, using pastes

Ab initio production generally starts from raw materials, and pigment mixtures are dispersed directly in the batch. Mixed production starts from ready-prepared pigment pastes: only one kind of pigment is usually dispersed in the binders, and is ready-for-use. Paste series differ according to the nature of the film formers used.

Smaller factories generally buy in their film-forming materials and if necessary convert them into solutions. The major paint companies, on the other hand, produce considerable amounts of the film formers (resins) themselves and are therefore able to develop "tailor-made" film-forming agents that are optimized to the formulations, production conditions, application technique, and intended use of the resulting coating composition. Such companies can supply special, high-quality products.

Material Flow. Figure 7.1 illustrates the material flow in paint and coating production, starting with the reception and storage of the raw materials and finishing with the end product ready for dispatch. The type of raw material, its consistency, the amount consumed, and the packaging in the as-supplied state are decisive for material flow and metering. In large factories, large storage tanks for resin solutions and solvents have to be installed and equipped with the latest safety devices. The raw materials are usually delivered in tank trains. These bulk raw materials can be pumped and conveyed to the sites of use via pipes. A considerable number of other liquid, pasty, solid, or powder raw materials are supplied, palletized, and depalletized in containers, barrels, drums, or sacks. Open-air storage facilities and warehouses for free-standing stacking are provided for this purpose. The raw materials for 1, 2, 4, and in some cases 3, are delivered to the appropriate stores by road. The material flow then continues to the production facilities (9 and 10).

7.1. Principles

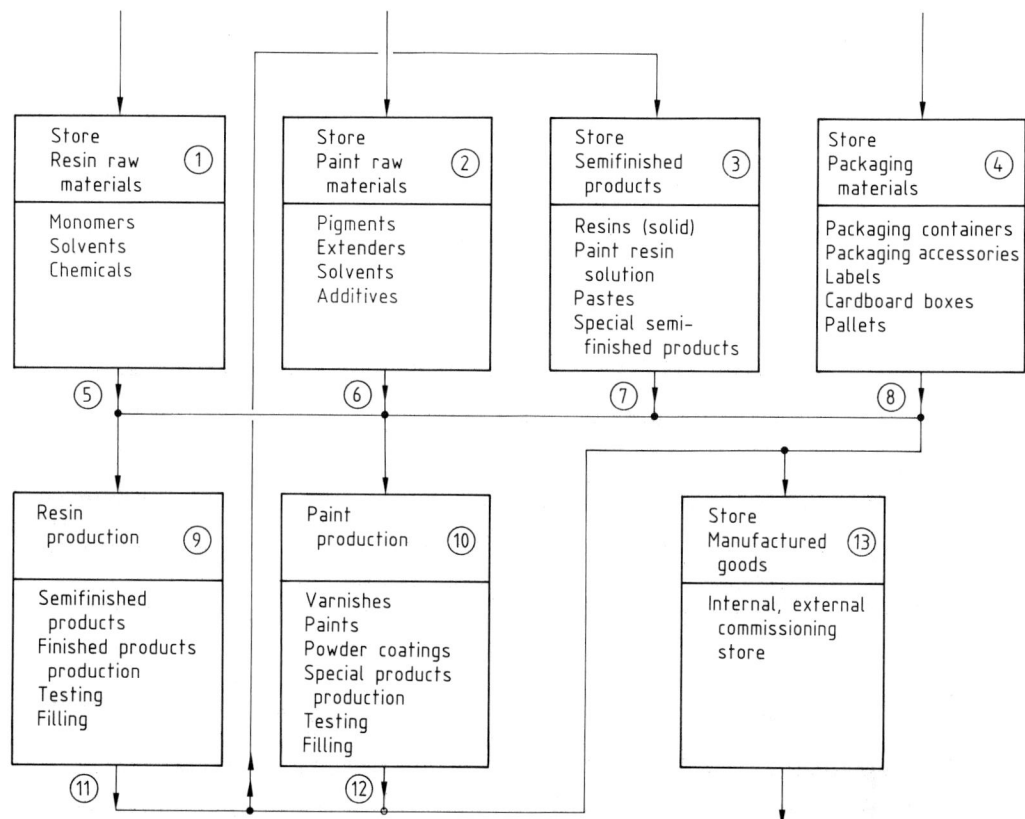

Figure 7.1. Material flow in paint production (for explanation of the numerals see text)

Four types of product are made (see Section 7.2):

1) Clear or colored varnishes (do not contain pigment)
2) Paints (contain pigment)
3) Powder coatings
4) Special products (e.g., putties, fillers, plasters)

The paint maker also produces semifinished products (e.g., pastes) for internal use. The semifinished products from 12 are stored in 3. Packaging materials flow via 4 to the filling areas at the end of resin (9) and paint (10) production. Many different packaging types, sizes, and designs are used. The end products may be dispensed into containers the size of a road tanker, into drums and barrels, or into cans with a capacity of as little as 50 mL. Finally, the ready-for-use material is sent to the storage area 13 before it is dispatched.

Production based on vertical material and process flow is most commonly used and will now be considered in a conventional factory (Fig. 7.2). Smaller factories operate on a horizontal concept, but the principles are the same.

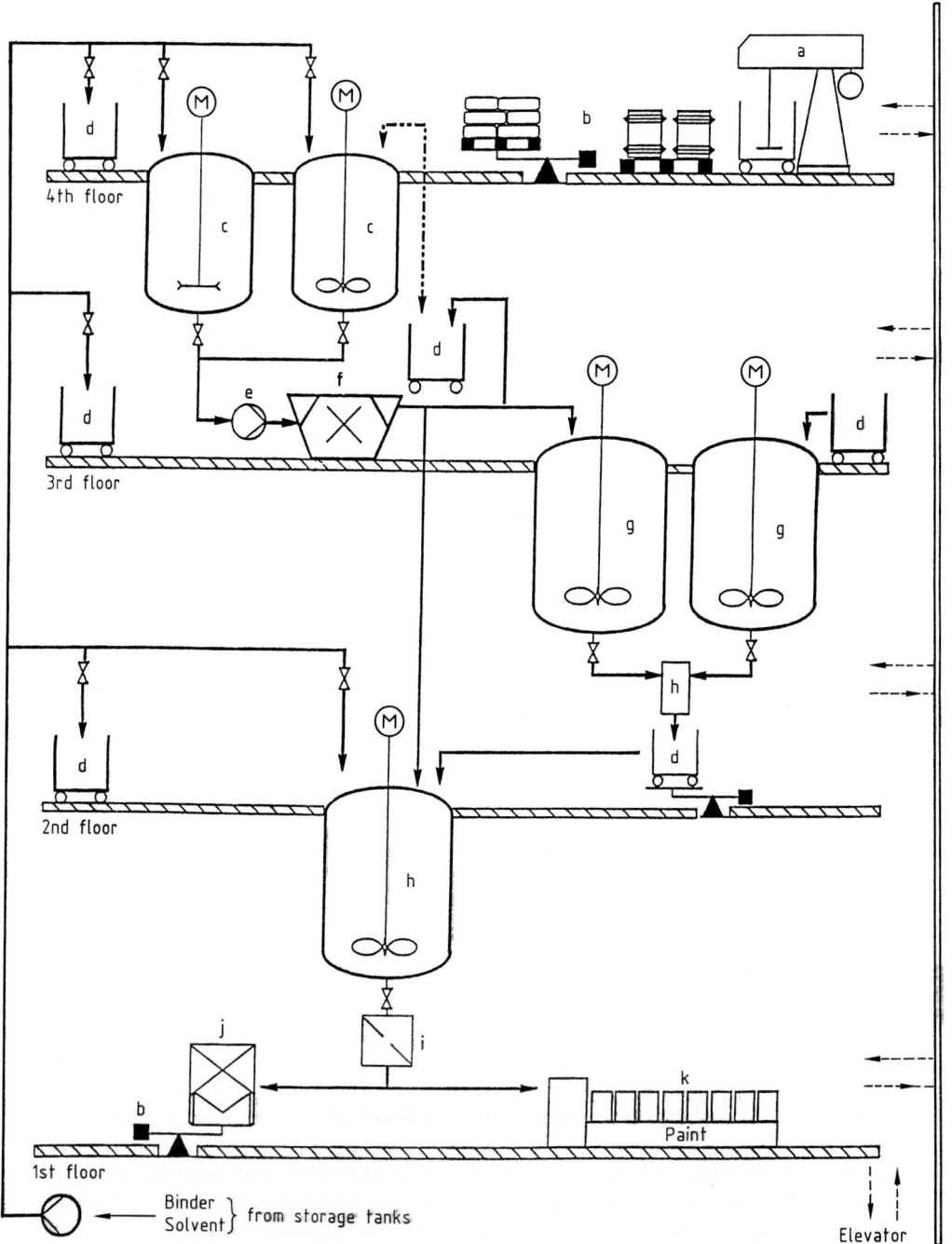

Figure 7.2. Flow sheet of a paint factory
a) Disperser; b) Scales; c) Premixer, dissolver; d) Movable container; e) Pump; f) Grinding mill; g) Paste mixer; h) Adjustment tank; i) Filter, sieve; j) Container; k) Packing machine

7.1. Principles

The material and process flow proceeds downward through four floors as follows:

Fourth Floor. The fourth floor is used for intermediate storage and metering of the raw materials required to produce the millbase. Solid raw materials (e.g., pigments and fillers) are stored on pallets in sacks, drums, barrels, or cartons. Liquid or pasty raw materials are kept in drums or cans. These materials are weighed (b) and charged into the premixers or dispersers (a, c). Solvents and solutions of film-forming materials are fed from the storage tanks into premixers or dissolvers (c) via pipelines and filling stations.

Third Floor. The mixture is dispersed using grinding mills (f). With mixed production the dispersed millbase is then fed as a single-pigment paste to paste mixers (g) and then to the adjustment tanks (h). In ab initio production the multipigment paste is sent directly from the grinding mills to the adjustment tanks (h).

Second Floor. A large number of adjustment tanks (h) of different sizes are used. The mixed pastes are metered from the paste mixers (g) into mobile containers and then added to the adjustment tanks. Formulation of the paste mixture or dispersion is completed in the adjustment tank by adding film-former solutions, solvents, and additives. The product is sampled, tested, and adjusted (e.g., colour, viscosity, other technically important properties).

First Floor. The end product is sieved or filtered and dispensed from the adjustment tank via filling lines into containers, drums, or small packages.

The material flow system is complex because the mixers and machines have to be changed in accordance with the product type and batch size. Equipment also has to be reserved for specific product groups with special shades or chemical properties. Only in this way can the effective cleaning of equipment and pipelines during batch changes be guaranteed.

Due to the high number of raw materials and intermediates logistics has become an important field of optimization. Various models of material storage and flow are installed in paint factories. The most common models are "storing in", what means that the most important materials are available inside the plant, or "batchwise precharging" of materials outside the plant. In the second model containers and big bags are used for the transport of the precharged solid materials from storage to production. Both models have economic and organizational advantages.

Filling is also centralized for the same reasons. Materials are conveyed in mobile containers or via pipelines. Cleaning inside pipelines can be optimized by using scraper (go-devil) techniques.

Very few product lines in the paint industry are mass-produced; white emulsion paints, plasters, rust protection paints, and underbody protection materials are, however, exceptions. Paint factories producing these bulk products are highly automated but extremely inflexible as regards the use of different raw materials and semifinished products and their proper scheduling. Enterprises that manufacture "tailor-made" products must have a wide range of equipment, and simple and flexible production methods.

Quality Assurance. Almost every aspect of the paint-making process is subjected to quality assurance measures. Depending on the type of product, this involves a number of correction stages and delay periods, particularly in the adjustment tanks.

In the most commonly employed correction strategy, the desired quality is approached and achieved in small steps. Correction in this context does not mean the elimination of defects, but the addition of small amounts of materials to hit the specification window. Quality assurance corresponding to the ISO Standards 9000–9004 (e.g., failure mode-and-effect analysis (FMEA) and statistical process control (SPC) [7.12]) has been introduced in modern paint factories.

Environmental Protection. The paint industry has adopted a two-pronged strategy to ensure environmental protection:

1) Apparatus and processes are enclosed as far as possible to minimize emissions
2) If enclosure is not possible, the sources of emission are selectively aspirated. Gaseous pollutants and vapors are collected and passed to central or decentralized disposal equipment [e.g., combustion plants (thermal or catalytic) or adsorption units], particulate impurities being retained in filters

7.2. Paint-Making Processes

The wide variety of paint formulations are reflected in the variety of paint-making processes. Figure 7.3 summarizes the essential features of these processes. Only a few special products (e.g., putties, plasters, or underbody protection materials) are not covered by this scheme; they are produced by special proprietary procedures (generally of the one-pot type).

7.2.1. Varnishes

Varnishes are usually colorless but may be slightly colored due to the intrinsic color of the binders, driers, or other additives. Colored transparent varnishes contain soluble dyes or a small amount of finely dispersed pigments. Matt varnishes may be opaque (milky), colorless, or transparent; after drying they have a matt to semi-gloss surface due to the presence of dulling (mainly inorganic) additives.

Liquid raw materials are mixed in a specific sequence in slow- to high-speed stirrers or mixers to give a homogeneous, streak-free mixture (see Section 7.3.3). In larger mixers and tanks (up to 50 000 L capacity) the mixing device is permanently installed in the vessels. Precondensates and wax solutions are prepared in heatable and coolable vessels equipped with stirrers.

Solids are usually dissolved in liquids using high-speed stirring equipment. If the formulation contains dissolving and nondissolving solvents (diluents), then the nondissolving solvents are often added first, followed by the solid raw materials (e.g., powders, granules, fibers), and then finally the "true" solvent. With this order of

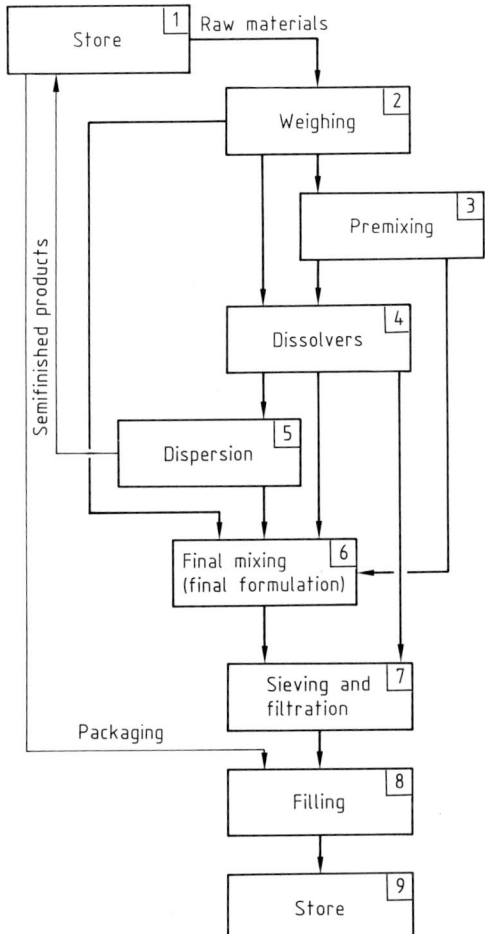

Figure 7.3. General flow diagram of the paint-making process

addition the surfaces of the solid particles are wetted more effectively, resulting in uniform penetration of the "dissolvers" into the solids. Agglomeration is thereby avoided and dissolution is speeded up. In Figure 7.3 the sequence of operations is 1, 2, 3, or 4, 7, 8, 9.

7.2.2. Paints

Dispersion. The liquid semifinished products (intermediates) used in paint making are prepared in a similar manner to varnishes (see Section 7.2.1). The most important stage in the production of pigment-containing paints is complete wetting and uniform distribution (i.e., dispersion) of the solid pigment particles in the liquid, film-forming material. To obtain optimal paint properties the pigment must be optimally dispersed in the liquid. Many pigments are surface-finished. An extremely fine grain size in the finished paint results in undesirable effects (e.g., gloss streaks).

Historically, to obtain homogeneous pigment wetting and dispersion, kneaders, rotor and stator mills, and roller mills came first. These were followed by ball mills, tank mills, attritors, and (open) sand (bead) mills. Closed grinding mills are now widely used to comply with increasingly stringent quality and environmental protection requirements. Fitted with an enclosure, a roller mill is still acceptable for some product lines (e.g., pasty printing inks). Paint production proceeds according to steps 1, 2, 4, 5, 6, 7, 8, and 9 in Figure 7.3.

Commercial coating materials seldom have the same composition as the ready-dispersed millbase. The proportion of film former in the final product is generally higher. Highly concentrated (high-solids) paint resins must be added to increase the proportion of film former to the required value. This is generally performed in formulation vessels equipped with appropriate stirrers. This stage can present serious problems because the widely differing solvent concentrations of the two substance streams may cause irreversible diffusion. The result can be a so called "pigment shock".

Since the solvent molecules are more mobile than the film-former molecules, they migrate down their concentration gradient from the millbase into the paint solution, resulting in a decrease in volume of the pigmented phase. The distance between the pigment particles is reduced to such an extent that, even in stabilized pigment dispersions, repulsive forces between individual pigment particles are overcome and flocculation occurs. This process (pigment shock) can be avoided by modifying the compositions of the millbase and the final formulation, or by suitable process technology (e.g., rapid homogenization during final formulation). Flocculation may also occur when solvents are added because they remove the stabilizing film-forming layers from the pigment surface.

7.2.3. Coating Powders

In order to produce coating powders (see Section 3.4), solid binders (e.g., epoxy, polyurethane, acrylic–polyester resins) are comminuted and thoroughly mixed with pigments, extenders, catalysts, hardeners, and additives (Fig. 7.4, a). The mixture then passes through a metering device into an extruder (b) where it is melted and homogenized at temperatures above the softening point of the binder and below the activation temperature of the catalyst. The extruded paste is rolled out to a thickness of 2–3 mm and transported on a cooling belt (c) or drum to a roll crusher (d) which comminutes the cold mass into chips. The chips are finely ground in an impact pulverizer or air separation mill (k). The powder is then sometimes screened (m) and classified to the desired particle size fineness. Oversize particles are reground [7.13]–[7.15]. In addition to the vertical plant design illustrated in Figure 7.4 production may also be performed horizontally.

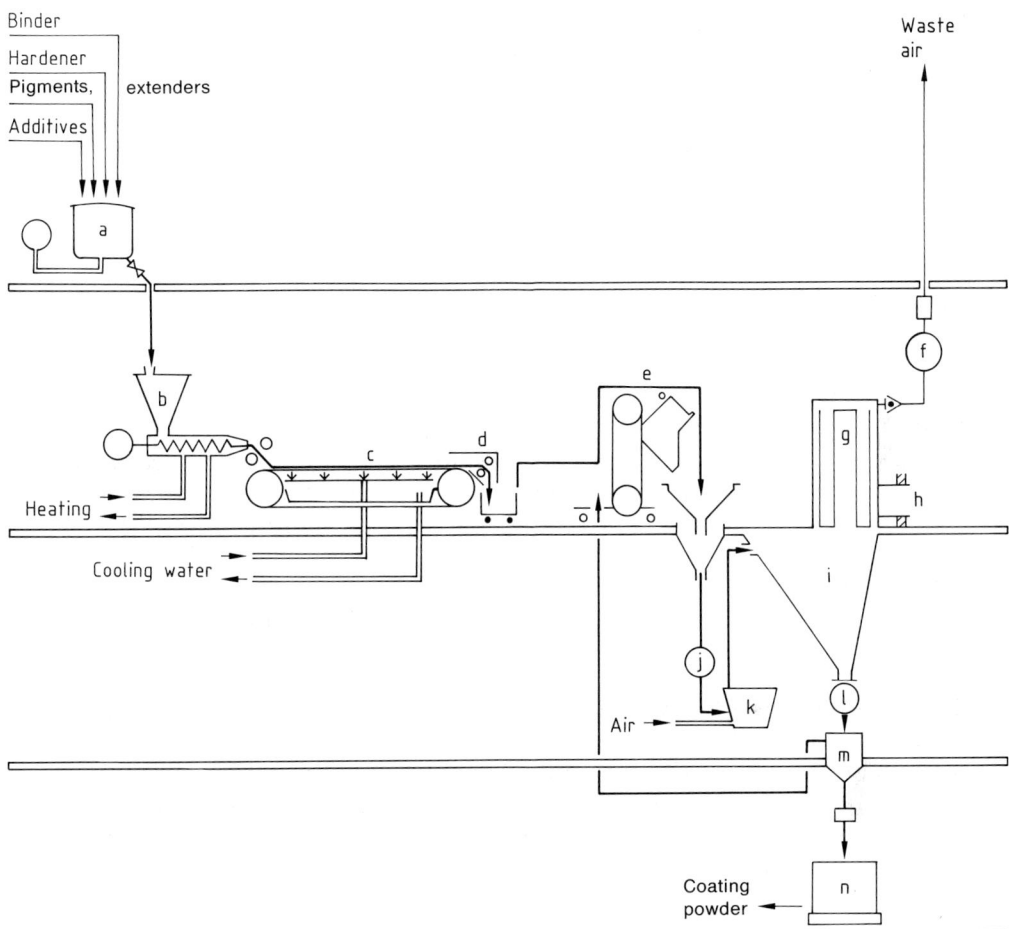

Figure 7.4. Powder coating production
a) Premixer; b) Extruder; c) Cooling belt; d) Roll crusher; e) Tipping device; f) Ventilator; g) Bag filter or cyclona; h) Explosion valve; i) Hopper; j) Cellular wheel sluice; k) Air separation mill; l) Cellular wheel sluice; m) Screening device; n) Packaging

7.3. Apparatus

7.3.1. Mixers

A wide variety of containers and tanks are used for the mixing process in the paint industry. Their materials of construction must be appropriate to the contents being mixed. Paints containing organic solvents are produced in steel or aluminum vessels.

Waterborne paints generally require stainless steel because of their high corrosion potential. The tanks may also be lined with enamel or plastics.

The stirrers are mainly of the flat-blade, disk, propeller, or turbine type [7.16]–[7.19]. Rotor–stator systems are also used as continuous mixers (Fig. 7.5). These machines are employed particularly if an emulsion or an emulsion-like product has to be produced (e.g., waterborne paints) [7.20]–[7.22].

7.3.2. Dissolvers

Dissolvers (Fig. 7.6), are mainly used in paint making for mixing, wetting and dispersing solids (pigments) in a resin solution [7.23]. They have a good shearing effect in the fluid and are even capable of breaking down coarse pigment agglomerates. The high shear forces required are produced by appropriate design and high speeds (up to 30 m/s at the stirrer tip). Power consumption may reach 50 kW/m^3. Dissolvers exist in single-shaft and multishaft versions. Vacuum dissolvers are used for product degassing.

7.3.3. Kneaders and Kneader Mixers

Various kneader designs (e.g., ram, internal, screw, paddle, Z-, arm, and planetary kneaders, Fig. 7.7) are used to produce highly viscous (pasty) paint concentrates (master batches), uncured rubber compositions, and putties. Kneaders have a high energy consumption because they are generally used to produce systems that have a low binder content or are difficult to wet [7.23].

Heatable and coolable kneaders are available and both types can be fitted with a vacuum device. They may operate batchwise or continuously and are equipped with screw conveyors.

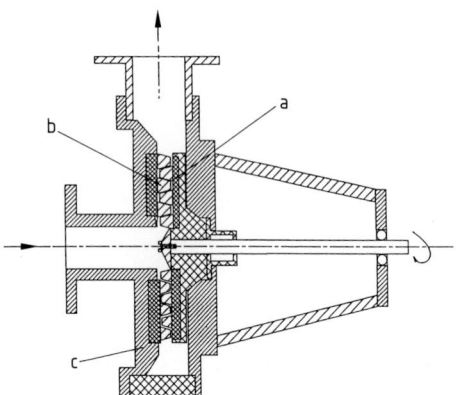

Figure 7.5. Continuous mixer for paint making (rotor–stator system)
a) Rotor; b) Stator; c) Housing

7.3. Apparatus 187

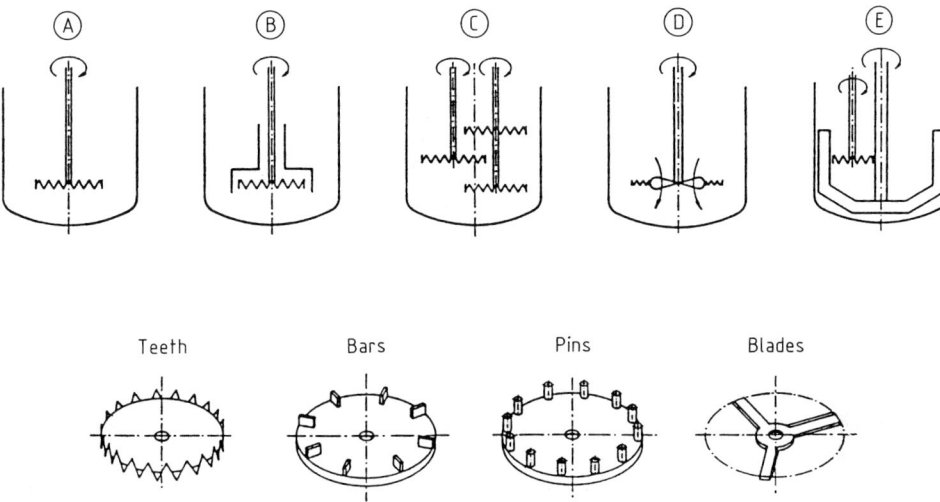

Figure 7.6. Types of dissolvers and dissolver disks
A) Simple dissolver; B) Dissolver with stator system; C) Double-shaft dissolver with three disks; D) Dissolver with disk for pumping and shearing; E) Dissolver with two drives for separate functions

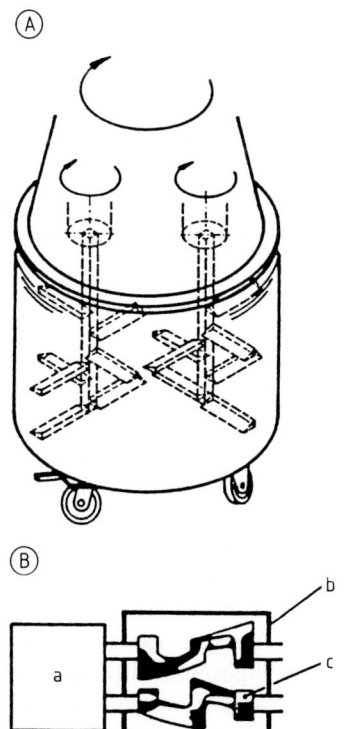

Figure 7.7. Planetary (A) and vat (B) kneaders
a) Drive; b) Vat; c) Kneader hock

7.3.4. Media Mills

On account of the widely varying production requirements (see Section 7.1), a wide range of mills are used for dispersion. *Ball mills,* rotating vessels filled with balls (diameter > 10 mm), were developed first. Ball mills are inefficient compared with modern units because they operate batchwise and large batches cannot be produced. In addition to safety problems, their high noise level is a disadvantage.

To overcome the safety problems associated with rotating ball mills, *tank mills* were developed which consist of a stationary vessel equipped with a slowly rotating stirrer. Use of grinding media less than 10 mm in size increased product quality, grinding efficiency, and thus output.

Increasing the stirring power, optimizing the geometry, and introducing an external forced circulation led to further advantages and the development of the *attritor* (attrition mill).

The transition from batch production to continuous production required the development of *sand mills.* Sand mills can be regarded as slender attritors with a forced product circulation. They are charged with product vertically from the bottom upward, gravity being mainly responsible for the separation of the grinding media. A sieve is installed to improve separation. The name of the mill is derived from the fact that Ottawa sand (particle size ca. 0.5–3 mm) is normally used as the grinding medium. The development of sand mills increased both product quality and productivity. Throughput was, however, restricted by the force of gravity. The viscosity of the paint pastes therefore had to be relatively low, which did not promote optimum dispersion. Another disadvantage is the open construction, which led to additional emission problems.

A further advance was the development of *enclosed agitator mills.* The horizontal design was intended to eliminate the influence of gravitational forces. Numerous design features have been optimized [7.18], [7.19], [7.22]–[7.32]. The most important developments are illustrated in Fig. 7.8.

The main disadvantage of these mills is the limited throughput due to the often extremely narrow devices used to separate the grinding media (rotating gaps, sieve). Since the free cross-sectional area of the separation devices can only be decreased within certain limits, small grinding media (< 1 mm) cannot be used to increase product quality [7.33].

An important step forward in wet mill technology came with the development of the *centrifugal vortex mill* (Fig. 7.9) [7.33], [7.34]. The major innovation in this mill is the separation of the active grinding zone from the passive separation and sealing zone. This results in the incorporation of large surface areas for separation (sieves). This advantage is essential for any substantial increase in output: the large sieve areas allow throughput performance to be considerably increased (by a factor of 10–20). With this type of machine it is possible to obtain a narrow dwell-time distribution by working in a recirculation or multipass mode. A fine, narrow particle size distribution can therefore be obtained more quickly. A recent development, which also combines the advantages of a large product throughput and the use of small grinding media, is the *double cylinder bead mill.* This mill has an internal grinding medium circulation to avoid compression of the grinding media [7.35]. The

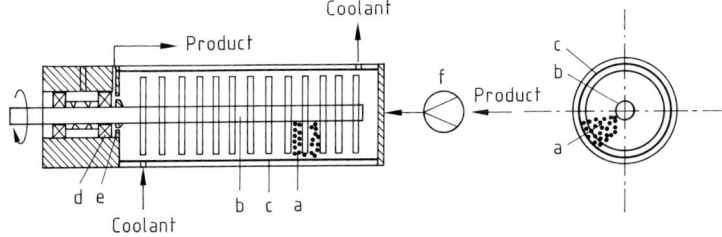

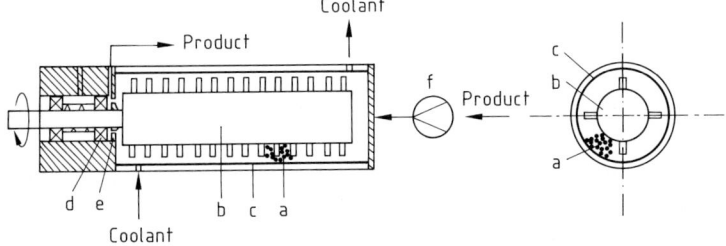

Figure 7.8. Conventioned closed grinding mills [7.4]
a) Grinding media; b) Rotor; c) Grinding chamber; d) Sealing system; e) Separator; f) Pump

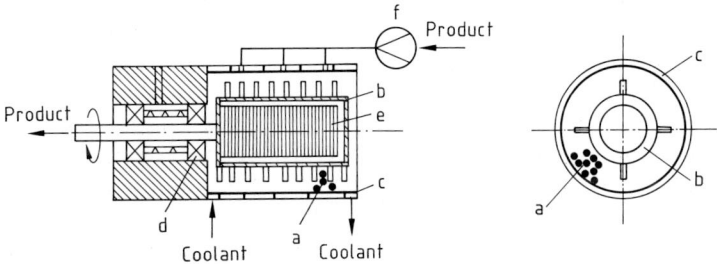

Figure 7.9. Centrifugal vortex mill
a) Grinding media; b) Rotor; c) Grinding chamber; d) Sealing system; e) Separator; f) Pump

construction of this type of machine allows a narrow dwell-time distribution and a narrow particle-size distribution to be otained by operation in the single-pass mode.

Dispersion Methods. The dispersion methods most commonly used in the paint industry are illustrated in Figure 7.11. The *single-pass batch procedure* (Fig. 7.11 A) is the most common method. Its disadvantage is the broad dwell time distribution and the resulting wide particle size distribution. It is also difficult to control, i.e., productivity falls sharply if the batch has to be treated again to improve its quality.

A *cascade of grinding mills* (Fig. 7.11 B) produces a considerably smaller particle size distribution. Control is, however, more complex than with the single-pass procedure.

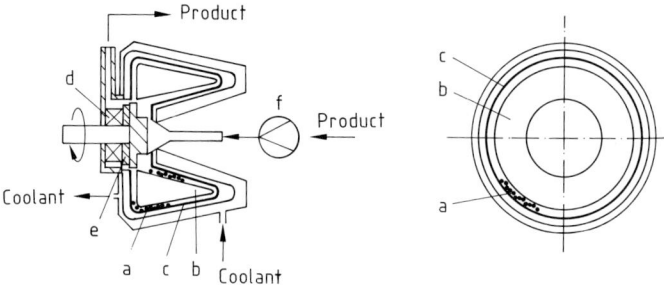

Figure 7.10 Double-cylinder bead mill
a) Grinding media; b) Rotor; Grinding chamber; d) Sealing system; e) Separator; f) Pump

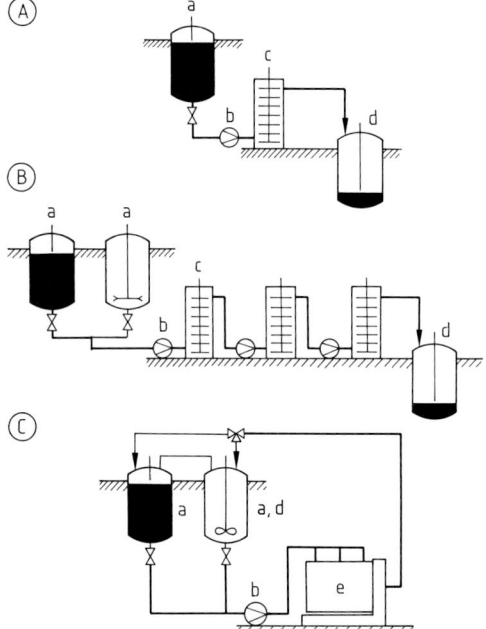

Figure 7.11. Dispersion methods
A) Passage; B) Cascade; C) Swing-batch, circulation
a) Mixer; b) Pump; c) Conventional mill; d) Let-down tank; e) Centrifugal vortex mill

The *swing-batch and circulation procedures* (Fig. 7.10) are suitable for machines with high product throughputs (e.g., the centrifugal vortex mill and double cylinder bead mill). Control is simple.

The swing-batch procedure involves two vessels whose contents are alternately passed through the agitator mill. The circulation procedure is even simpler to control because the product can be recirculated until the desired quality and properties have been obtained. Only one vessel is required per machine, which is an advantage with frequent changes of product. A disadvantage is the backmixing which occurs in the vessel, making the dwell time distribution less favorable than in the pendulum procedure. Above about 10 cycles or passes the difference is negligible and the particle size distributions in the two procedures are nearly comparable.

7.3.5. Roller Mills

Like agitator mills, roller mills are also used to disperse and homogenize highly viscous pigment pastes (Fig. 7.12). Roller mills are often also used for finishing (postdispersion or fine dispersion, e.g., in ink production). The fact that dirt and coarse particles float to the top in the feed nip is a beneficial side effect: the roller mill acts as a filter.

Roller mills may have one to five smooth rollers (rolls). Three-roller systems are most often used, the rolls being arranged in various ways (e.g., horizontal, inclined, or chair arrangement). In modern high-speed systems the axis of the third roll is generally fixed in position, while the first two rolls are hydraulically actuated. The rotational velocities are 40–100 min^{-1} for the front roll, 140–200 min^{-1} for the middle roll, and 300–600 min^{-1} for the third roll. Shear energy for dispersion depends on rotational speed differences, and the roll nip clearance (controlled by the compression force). In order to compensate for viscosity fluctuations the roller mills are generally designed so that they can be heated or cooled. Disadvantages of this type of will are the relatively low production rate and the normally open construction which results in high emission of solvent vapor.

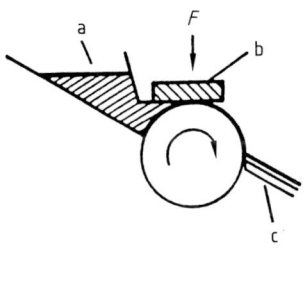

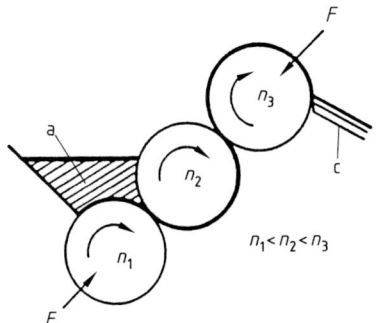

Figure 7.12.
Roller mills (one- and three-roller)
a) Filling; b) Compression bar; c) Scraper

7.3.6. Filter systems

Ensurance of the cleanliness of the end product is very important. Every batch therefore has to be filtered at least one before or during filling.

Sieves. To clean, paint must be screened. The simplest method containing undesirably coarse particles is to sieve it through metal (steel) or plastic fabrics of specific mesh size. Polyamide screen bags are employed to facilitate handling. Metal fabric is fastened or soldered onto cylindrical metal frames.

Vibrating screens are used for products (e.g., metallic effect paints) that are difficult to screen. The screens are vibrated horizontally or vertically and the material flows rapidly through the sieve.

Filters. In the production of paints inpurities sometimes occur which have to be eliminated by filtration. The simplest method is filtration through *bag filters* made of sintered synthetic micro fibres (Fig. 7.13). The bag filters can be installed in metal filter housings with closable, sealable covers. A perforated sheet-metal basket protects the filter bag against overloading.

If impurities cannot be removed with bag filters, *filter cartridges or candles* are used (Fig. 7.14). The filter cartridges are inserted in individual housings or, in large-scale equipment, up to 24 cartridges are inserted side by side or one above the other and held in place with quick-release mechanisms. Resin-bonded, sintered synthetic micro fibres and metals are used as filter material. The fineness of the filter fabric (5–100 µm) determines the filtration effect.

If serious problems arise (gel particles or turbidity, or surface demanging surfactance, especially in water-borne paints), layered filters such as plate and disk filters with or without adsorptive effects are used.

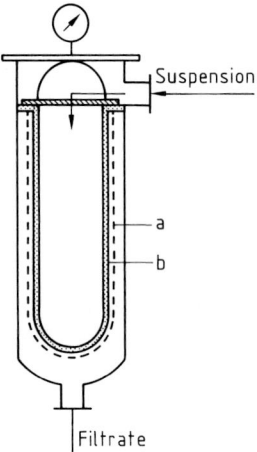

Figure 7.13. Bag filter system
a) Metal baket; b) Filter bag

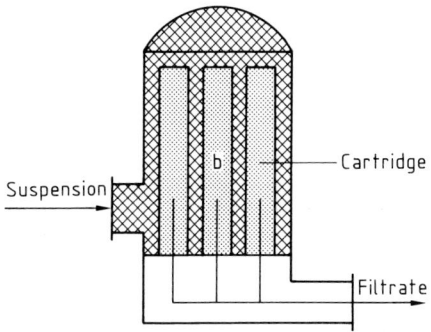

Figure 7.14 Cartridge filter system

An extremely time-consuming and expensive method of removing unwanted coarse material ("oversize") from paints and paint pastes, is *centrifugation*. This technique should only be used if filters and screens prove unsuccessful because it is often accompanied by a high product loss.

With the increasing intensification of quality assurance measures before (incoming inspection, FMEA) and during manufacture (cleanliness tests), the use of filter equipment is declining or limited to a protective function ("policing filtration").

8. Paint Application

8.1. Types of Substrate [8.1]

All solid substrates can be coated, including metals, plastics, wood, minerals, paper, and leather.

The surface structure of *metals*, their melting point, and susceptibility to the effects of weathering and chemicals (corrosion) can substantially influence the coating process and quality.

The coatability of *plastics* depends to a large extent on the material, design of the article, forming process, and the processing agents used for this purpose.

The coating of *wood* is strongly influenced by its structure and properties (elasticity, hardness, strength, shrinking tendency). Some wood constituents (e.g., pigments, resins, or tanning acids) can interfere with the coating process.

Coatings are used for different purposes:

1) Protection of an object against its environment (e.g., corrosion protection)
2) Improvement of appearance
3) Surface modification to meet special physical and functional requirements

The choice of coating technology and the quality of the coat applied to a given article primarily depends on the substrate, its geometry, design, and surface state.

8.2. Pretreatment of Substrate Surfaces

The surface of workpieces may be covered or contaminated with coherent or discontinuous layers of the following materials:

Lubricants	mineral oil products
	metallic soaps
	graphite
	molybdenum disulfide
Corrosion-inhibiting oils	petroleum sulfonates
	inhibitors
	mineral oil products
Fluxes	inorganic salts

Polishing pastes	mineral oils
	waxes
Scale, corrosion products	metal oxides
	metal salts
Finely divided, foreign substances	dust
	ash
	slag
	(from welding work)
Miscellaneous	dermal deposition from fingers/hands
	absorbed gas films
	old coatings
	adhering moisture

These foreign materials all have adverse effects on the wetting behavior, adhesion, and corrosion resistance of coating materials. The purpose of treatment prior to coating is therefore to remove foreign material from the workpiece surface. Another important function is to create a reproducible surface state that ensures long-term protection against corrosion subsurface migration and paint flaking at weak points or locally damaged areas of the paint film.

8.2.1. Pretreatment of Metallic Substrates

Many processes and techniques are used for pretreating metallic substrates (Table 8.1), [8.1]–[8.4].

8.2.1.1. Cleaning

Mechanical cleaning is mainly used to remove scale, rolling skin, rust, burr, and paint residues. Hammering, brushing, and lapping jelly are used in smaller workshops and on building and construction sites. Grinding is also widely employed for finishing weld seams and soldering points, smoothing and leveling file or tool marks, as well as for roughening material surfaces to improve paint adhesion. Blasting methods (sandblasting, compressed air, or water jet) remove impurities as a result of the kinetic energy of small hard particles. Abrasives may be metallic (e.g., spherical cast iron shot) or nonmetallic products (e.g., corundum). Blasted and abraded surfaces are extremely dusty and have to be cleaned by blowing, wiping, or washing.

Chemical Methods. Chemical methods include pickling, which can be used to remove scale and metal oxides. Sulfuric acid and hydrochloric acid are usually used to treat steel and zinc, while nitric acid, nitric acid – hydrofluoric acid, or sulfuric acid – hydrofluoric acid are used for aluminum. Although pickling in phosphoric acid is somewhat slower, it has the advantage that a thin conversion layer of the substrate phosphates is produced, which provides temporary corrosion protection

Table 8.1. Summary of pretreatment methods

Method	Material			Impurities						Temporary corrosion protection after treatment	Improved corrosion protection with coatings
	Fe	Al	Zn	Lubricants, corrosion protection oils	Scale, rolling skin	Dirt, shavings	Drawing agents	Corrosion products	Old paint layers		
Mechanical methods											
Steam and high-pressure water blasting	+	+	+	+	–	+	○	–	*	–	–
Jet blasting (e.g., wet jets)	+	+	+	–	++	+	○	++	+	– –	– –
Manual abrasive methods (e.g., brushing, grinding)	+	+	+	–	+	+	○	+	+	–	–
Wet chemical methods											
Solvent degreasing											
Steam	+	+	+	+	–	○	–	–	○	–	–
Dipping	+	+	+	+	–	–	–	–	○	– –	– –
Ultrasonic dipping	+	+	+	+	–	+	–	–	○	–	–
Alkaline degreasing											
Spraying	+	+	+	+	–	+	–	+	–	–	–
Dipping	+	+	+	+	–	○	–	+	–	–	–
Pickling											
Phosphoric acid	+	+	+	*	*	+	–	+	–	○	*
Sulfuric acid	+	–	–	*	+	+	○	+	○	– –	– –
Hydrochloric acid	+	–	–	*	+	+	○	+	○	– –	– –
Caustic soda	–	+	–	–	*	*	*	+	+	–	–
Phosphating											
Alkali phosphate	+	+	+	–	–	*	–	–	–	+	+
Zinc phosphate	+	+	+	–	–	–	–	–	–	+	++
Chromating	–	+	+	–	–	–	–	–	–	+	++
Adhesion promotors (wash primer)	+	+	+	–	–	–	–	–	–	++	+

The symbols have the following meanings:
++ Extremely satisfactory. + Satisfactory. ○ Moderately satisfactory. – Unsatisfactory. – – Particularly critical.
* Satisfactory only with special methods.

and is a good base for the subsequent coating. Previously, sulfuric acid was preferred for pickling because of its low vapor pressure; acid losses are therefore slight and environmental pollution is low. Nowadays the tendency is to use hydrochloric acid because it allows better cleaning of the metal surface (even slightly alloyed steels). Evaporation of hydrochloric acid from the pickling baths is limited by using self-contained or completely enclosed units. Organic inhibitors are generally added to pickling acids to limit the pickling action to the oxidic impurities and to minimize the dissolution of the base material. Addition of surfactants has a limited degreasing effect.

Pickling in aqueous caustic soda solution is widely used for cleaning aluminum workpieces on account of their amphoteric behavior. However, after alkaline pickling the article generally has to be treated with acid to remove loosely adhering layers of pickling sludge and to brighten the surface.

8.2.1.2. Degreasing

Processes using aqueous solutions or organic solvents have become extremely important for removing organic impurities. In special cases, salt melts and treatments at elevated temperature in the gas phase are also employed.

Organic Solvents. Chlorinated hydrocarbons are widely used. They remove oils and greases extremely effectively (generally in the vapor phase). Cleaning in immersion baths can be significantly improved by using ultrasound. Increasingly stringent environmental protection legislation will, however, greatly restrict the future use of chlorohydrocarbon solvents.

Aqueous Media. Degreasing may be performed with alkaline, neutral, or acidic aqueous media. Approximately neutral (mild alkaline) degreasing agents with fairly high concentrations of wetting agents and surfactants are mainly used. An advantage of these agents over alkaline or acidic media is a simpler and more economical effluent treatment. The degreasing agents hydrolyze animal and plant oils and grease. Non-hydrolyzable components (mineral oils and grease) are dissolved and dispersed by adding colloidal emulsifiers and wetting agents. These baths are operated at 60–80 °C and a pH of 8–9. The concentration of cleansing agents is between 1 g/L (spraying methods, pressure 0.15–0.25 MPa) and 50 g/L (dipping methods). Both stationary and flow-through baths are used.

8.2.1.3. Formation of Conversion Layers [8.1]–[8.5]

Conversion layers generally consist of inorganic compounds formed on the metal surface. They are used to increase the corrosion resistance and to improve the paint adhesion of the metal surface.

Industrially, phosphate layers are the most important and phosphating is used to treat steel, aluminum, and zinc. Chromating produces layers containing trivalent or hexavalent chromium compounds and is mainly used with aluminum and zinc.

Special oxide layers and inorganic–organic coatings are used for special purposes in strip treatment.

The surface weight of conversion layers is 0.05–5 g/m². With higher surface weights the flexibility of the layers decreases, which has an adverse effect on the flexural adhesion of the organic coating.

Phosphating Processes. The most important phosphating processes are alkali, zinc, and zinc–calcium phosphating. In alkali phosphating the layer-forming cation originates from the substrate, in zinc phosphating processes it originates from the phosphating solution.

Alkali phosphating (iron phosphating) is mainly used when corrosion protection does not have to satisfy stringent requirements. The solutions (pH 4–6) consist of acid alkali phosphates, free phosphoric acid, and small amounts of additives; oxidizing agents (e.g., chlorates, chromates, or nitrites), condensed phosphates (e.g., pyrophosphate or tripolyphosphate), and special activators (e.g., fluorides or molybdates). The first reaction is the pickling reaction which produces Fe^{2+} ions from the substrate (steel). These ions react with phosphate ions from the solution to form sparingly soluble iron phosphate that precipitates and adheres strongly to the metal surface. Zinc phosphate layers are formed in an analogous reaction sequence on zinc surfaces. Aluminum is usually treated with fluoride-containing solutions; thin, complex coatings are formed that contain aluminum, phosphate, and fluoride. The baths are adjusted to a concentration of 2–15 g/L. Contact with the surface may take place via spraying, flooding, or dipping. The bath temperature is normally 40–70 °C, but can be lowered to 25–35 °C with special bath compositions. Treatment times are 5–10 s (spraying of strip material) and 1–3 min (spraying or dipping of individual parts). Iron phosphating includes both thin-coating (0.2–0.4 g/m²) and thick-coating methods (0.6–1.0 g/m²). The color of the layers is blue-green, and in some cases also reddish iridescent. The surfaces become matter and grayer with increasing coating weight.

Zinc phosphating is primarily used for the surface treatment of steel and zinc as well as composites of these metals with aluminum. Aqueous phosphoric acid solutions (pH 2.0–3.6) containing dissolved acidic zinc phosphate, $Zn(H_2PO_4)_2$, are used.

The phosphate layers are gray in color (weight 1.2–6.0 g/m²) and consist of $Zn_3(PO_4)_2 \cdot 4 H_2O$ (hopeite), $Zn_2Fe(PO_4)_2 \cdot 4 H_2O$ (phosphophyllite), and $Zn_2Ca(PO_4)_2 \cdot 4 H_2O$ (scholzite). Layer formation is complete when the metal is completely covered with a phosphate layer, and the pickling action initiating layer formation has stopped. The treatment baths contain 0.4–5 g/L of zinc and 6–25 g/L of phosphate, calculated as P_2O_5.

The phosphating baths are usually used in automatic or semiautomatic dipping, spraying, or flooding plants at 45–70 °C; low-temperature processes operating at 25–35 °C also exist. Treatment times are 60–120 s (spraying process) and 3–5 min (dipping process). The iron phosphate sludge must be removed periodically or continuously from the bath.

Low-zinc processes were developed with the introduction of cathodic electrodeposition coating. In the normal zinc processes flat, sheetlike crystallites (mainly hopeite) are formed which may project from the surface. In the low-zinc process the layers mainly consist of phosphophyllite. They have a parallel orientation relative to the metal substrate and are more finely crystalline and compact than the hopeite layers. Very thin layers with a higher iron content are produced with nitrite-free low-zinc processes.

Chromating. In chromating, metal surfaces (mainly aluminum and magnesium) are brought into contact with aqueous acid solutions of chromium(VI) compounds and additives that activate and accelerate pickling. The pickling reaction converts acidic Cr(VI) into basic Cr(III); cations of the treated metal simultaneously accumulate in the liquid film on the metal surface leading to precipitation of a gel-like layer containing chromium(III), chromium(VI), cations of the treated metal, and other components. The final conversion layer is formed after aging and drying. Treatment can be carried out by spraying or dipping (6–120 s) at 25–60 °C. Effluent waste must be treated to remove Cr(VI); the most common method being reduction with sulfite to form Cr(III) followed by precipitation of chromium(III) hydroxide with milk of lime.

For the *yellow chromating* of aluminum, solutions containing chromium(VI) compounds as well as simple or complex fluorides and activators are used to accelerate layer formation. The pH value is 1.5–2.5 at total bath concentrations of 5–20 g/L. The conversion layers consist of oxides or hydrated oxides of trivalent and hexavalent chromium and aluminum. The color of the layer may range from colorless through yellowish iridescent to yellowish brown, corresponding to an increase in the surface weight from 0.1 to 3 g/m^2.

The essential constituents of the aqueous solutions for *green chromating* are chromic acid, fluorides, and phosphates. The pH value of the baths is slightly less than in the case of yellow chromating, and the bath concentration is normally 20–60 g/L. The conversion layers consist largely of chromium(III) phosphate and aluminum(III) phosphate, with small amounts of fluorides and hydrated oxides. The surface weight is 0.1–4.5 g/m^2, and the color ranges from iridescent green to deep green.

Aqueous, chromium-free acidic solutions have also been developed for aluminum materials that may contain complex fluorides of titanium and zirconium, phosphate, and special organic compounds. These solutions are applied by spraying or dipping (up to 60 °C) and produce thin, almost colorless conversion layers with a surface weight < 0.1 g/m^2.

Aqueous chromic acid solutions containing chlorides, simple and complex fluorides, sulfates and formates as activators, are used for *chromating zinc*. The total bath concentration is 5–30 g/L, the pH value 1.2–3.0. The bath temperature is in the range 25–50 °C, the process times are 5–120 s, the achievable surface weight is 0.1–3 g/m^2. The layer color changes with increasing surface weight from iridescent through yellow to brown or olive green.

Rinsing. A passivating rinse is necessary to exploit the quality-improving properties of conversion layers. The most important rinsing agents are dilute aqueous solutions of chromic acid, optionally with additional amounts of chromium(III). Equally good or only slightly worse results can be obtained with solutions free from chromic acid and containing polyvalent metal cations (e.g., chromium(III) or also organic components). The concentration of active substances in the rinsing baths is 100–250 mg/L. Rinsing is performed at 20–50 °C, treatment time ranges from a few seconds to about a minute. The surfaces should be sprayed with demineralized water as a last rinse to prevent crystallization of water-soluble salt residues.

8.2.2. Pretreatment of Plastics [8.1]–[8.3], [8.6]–[8.8]

Pretreatment of plastic surfaces is necessary for the following reasons:
1) To increase adhesion strength
2) To reduce the concentration of interfering constituents and mold release agents
3) To eliminate surface defects (e.g., bubbles)
4) To remove interfering foreign substances
5) To increase electrical conductivity

Many pretreatment techniques are used in practice (Table 8.2). The normal physical method used to improve the adhesive strength of the coating to the substrate is to slightly roughen the surface by solvent treatment, abrasion, or blasting. Some plastics (e.g., polyolefins) require special pretreatment methods; processes that modify the surface molecular layers of the plastic to increase their polarity have proved suitable (e.g., flaming, immersion in an oxidizing acid, immersion in a benzophenone solution with UV irradiation, corona treatment, plasma treatment).

Corona discharge is performed in a high-frequency alternating field (14–40 kHz) at 10–20 kV between two electrodes. The plastic surface is oxidized in a very short period (milliseconds). Plasma treatment is carried out under a moderate vacuum down to ca. 10 Pa. The advantage of this technique is the better penetration depth and the fact that it is also possible to treat shaped parts more easily. The plasma can also burn in gases (e.g., argon), whereby special effects (e.g., plasma polymerization) can be achieved.

Adhering processing additives (lubricants, release agents) can be removed by cleaning with solvents or aqueous surfactants. The solvent stability and solvent and water absorption of the plastic material should be taken into account. In order to reduce migration of constituents (shaping agents, plasticizers, dyes, organic pigments, stabilizers) during and after coating, preliminary tempering is often recommended (at the same time surface defects can also be detected).

Surface defects resulting from production (e.g., pores, bubbles, flow seams, and projecting fibers) are rectified by surface appearance enhancement. Deeper-lying defects are filled in and smoothed with putties.

Elimination of foreign substances (e.g., dirt particles and fibers) is very difficult due to the electrostatic charge of the plastics material. Alternative methods are wiping with a lint-free cloth wetted with water or a water–alcohol mixture or blowing with ionized compressed air. Plastics can also be made permanently antistatic by applying a dielectric coating.

Table 8.2. Pretreatment methods for plastics

Pretreatment methods		Use
Physical and mechanical methods		
Abrasion	dry, or wet dust cloth	removal of contamination (cleaning)
Blowing-off	oil- and water-free compressed air	
	ionized compressed air	+ reduction of electrostatic charge
Sanding		
Blasting		increase in adhesive strength, elimination of surface defects and foreign substances increase in adhesive strength, elimination or reduction of interfering constituents and adhering process auxiliaries, as well as foreign substances
Steam degreasing		
Washing	solvent, aqueous surfactant solutions	
Spraying	antistatically adjusted solutions	+ reduction of electrostatic charge
Dipping	conducting solutions	+ increase in electrical conductivity
Chemical methods		
Oxidative	flame treatment	increase in adhesive strength in special plastics, particularly polyolefins
	corona discharge, plasma treatment	
	oxidizing acids	
Cross-linking	benzophenone solution with UV irradiation	
Miscellaneous methods		
Tempering		elimination or reduction of surface defects, constituents, or process auxiliaries
Storage, aging		
Application of a dielectric layer		reduction of electrostatic charge

8.2.3. Pretreatment of Wood [8.1], [8.2]

Properly graded sanding with appropriate sandpapers is a prerequisite for a satisfactory wood surface. Industrially, sanding is performed on cylindrical abrasive-belt machines or automatic grinders, followed by brushing and suction to remove abrasive dust.

Surface pretreatment includes the following steps:

1) Removal of resins, e.g., by hydrolysis (wood soaps or soda solution) or dissolution (e.g., alcohol, acetone)
2) Removal of adhesive residues
3) Rectification (filling, patching) of processing and growth defects in the wood
4) Structuring by brushing, burning, sandblasting, embossing, or leaching
5) Staining with dyes dissolved in water or solvents, or with pigment dispersions

8.3. Application Methods

Many techniques have been developed for the industrial application of coatings. The individual industrial coating methods can, however, only be employed in limited areas if design and production sequence are not matched to the requirements of the coating technique. Adoption of more environmentally friendly coating methods is therefore often less a problem of investment, than a problem of the application limits of the relevant processes. Analysis of the criteria for choosing a coating method is a complex task. Of particular importance are the workpiece (design, material), the coating material, the number of workpieces and batch sizes, range of workpieces, requirements demanded of the coating (e.g., decorative appearance, corrosion protection), economic factors, legal provisions, and available facilities, premises, and equipment. Economic factors generally have top priority for choosing an application method on a commercial basis. Coating systems and processes are usually preferred that best satisfy the demands and requirements for thin coatings, high degree of material utilization, low energy costs, and good automation. Modern coating methods that best comply with these requirements are electrodeposition coating, electrostatic atomization, and electrostatic powder spraying.

8.3.1. Spraying (Atomization) [8.1]–[8.3], [8.9]

In conventional spraying, atomization is the result of external mechanical forces, i.e., the exchange of momentum between two free jets (air and paint). Atomization may be classified as compressed air atomization (air 0.02–0.7 MPa, paint 0.02–0.3 MPa), airless atomization (paint 8–40 MPa), air-assisted airless atomization, also termed airmix process (air 0.02–0.25 MPa, paint 2–8 MPa), and special technologies (Table 8.3).

In *compressed air (pneumatic) atomization*, compressed air flows through an annular gap in the head of the spray gun that is formed between a bore in the air cap and the concentric paint nozzle. Further air jets from air-cap bores regulate the jet shape and assist atomization. The expanding compressed air leaves the paint nozzle at high velocity. A low-pressure area is formed in the nozzle aperture which exerts a suction effect and assists outflow of the paint. The difference between the velocities of the compressed air and the exiting paint atomizes the paint into particles that are conveyed as spherical droplets in the free jet. In the high-pressure process (0.2–0.7 MPa) the exiting air jets can atomize the paint material extremely finely. The size of the liquid droplets varies from ca. 10 µm to 100 µm (depends on the liquid viscosity, amount of delivered paint, and air pressure). In the low-pressure process (0.02–0.2 MPa) atomization is correspondingly coarser (20–300 µm). Depending on the viscosity and throughput, the paint can be fed to the nozzle via a suction cup, a pressure cup, a flow cup, or pressure tank.

Table 8.3. Nonelectrostatic atomization methods

Advantages	Disadvantages	Examples of areas of use
Compressed air (pneumatic) atomization		
Universally employable	spraying experience required	large-scale series coating applications (automobile industry)
Simple to use	paint mists constitute a health hazard	
Uniform layer thicknesses		
Applicable to complicated workpiece geometries	ventilation required in enclosed spaces	repair and touch-up finishes
		industrial coatings (furniture, domestic appliances, etc.)
Small amounts can be applied	compressed air supply required	
Rapid change of colors	unsatisfactory material utilization	
Very good optical paint film quality	danger of film defects due to spray mist	
Suitable for special-effect paints		
Airless (hydraulic) atomization		
Very high operating speed	expensive apparatus	coating of large objects (shipbuilding, steel construction work, machines, lorries, etc.)
Low spray mist formation (low losses)	equipment parameters must be matched to the coating material	
Large film thicknesses in one application	limited number of spray jets due to overlapping	
Uniform surface and film thickness	amount can be regulated during application	
Suitable for high-viscosity paints	nozzle subject to high degree of wear	
Direct application from as-supplied drums and containers	danger of sagging with sensitive paints	
Substrate with deep pores can be wetted		
Airmix atomization		
Combines advantages of pneumatic and hydraulic atomization	paint pressure generator and compressed air supply required	industrial coating
Hot spraying		
High-viscosity paints can be applied	trained workforce required	industrial coating
Large film thicknesses	additional heaters necessary	
Low sagging tendency		
Quicker drying of the paint film		

Table 8.3. Continued

Advantages	Disadvantages	Examples of areas of use
Two-pack coating		
Advantageous for temperature-sensitive workpieces	expensive and complicated equipment	industrial coating, coating of wood and plastics
Hardening at room temperature (energy saving)	exact metering and mixing required (automatic monitoring)	protection of buildings and structures
Better paint film quality	safety measures required with isocyanate-containing systems	corrosion protection
High resistance to mechanical, chemical, and climatic influences	trained workforce	large equipment and apparatus (machines, aircraft, ships, commercial vehicles, etc.)
Low content of organic solvents		

In *airless (hydraulic) atomization* the paint is forced through a slit nozzle of hard metal under high pressure (8–40 MPa). On account of the high degree of turbulence, the paint stream disintegrates immediately after leaving the fluid tip. A similar atomization process occurs in spray cans where the paint pressure is produced by the propellant gas.

The combined *airmix process* operates at a lower paint pressure (2–8 MPa). Additional low-pressure air jets (0.02–0.25 MPa) from the air-cap bores impinge on the spray jet to mix and homogenize it. In addition to the atomizer and a compressed air generator (airless pump), the airmix unit therefore also requires compressed air for postatomization. Advantages over the airless method are the less sharply defined spray jet and the smaller droplet size. Compared with compressed air atomization, a low-mist coating is possible.

Hot spraying can be combined with all of the spraying methods described above and is used for large film thicknesses or highly viscous, high-solids paints (lower solvent consumption). The paint is heated to 50–80 °C in a heat exchanger. Immediately after atomization the heat content of paint droplets is transferred to the air and the workpiece. The droplets therefore cool and their viscosity rapidly increases; the risk of sagging at larger layer thicknesses is thus reduced.

In *two-pack paints* both the binder and the hardener have to be mixed before application. In paints with a short pot life the paint must be metered and mixed in the atomization equipment immediately prior to use. The two reactive components are normally mixed in static mixers after metering.

8.3.2. Electrostatic Atomization [8.1]–[8.3], [8.9]

In *purely electrostatic spraying*, the paint is atomized solely by electrostatic forces. In *electrostatically assisted spraying*, atomization takes place by the methods described in Section 8.3.1, with simultaneous or subsequent electrical charging.

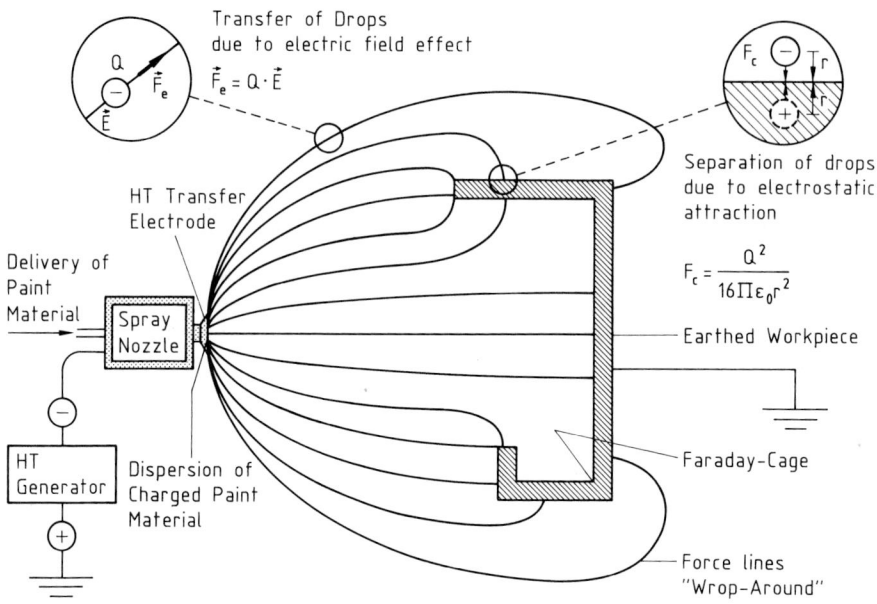

Figure 8.1. Fundamentals of the electrostatic coating process

In all electrostatic coating methods an electric field is applied between the atomization equipment and the workpiece (Fig. 8.1). The advantages and disadvantages of this technology are listed in Table 8.4. The paint is electrically charged by a concentrated electric field at a high-voltage electrode. In "lead charging" the nonatomized paint is charged by direct contact with the electrode. In "ionization charging" mechanically produced paint droplets are charged by attachment of ions from the air; the electrode serves as a corona tip that generates these ions.

In *purely electrostatic spraying* the paint is atomized solely by electric field forces. The paint flows as a thin film over a high-voltage sharp edge, where it is subjected to high field forces. The paint film breaks up into threads and then into charged droplets that follow the electric force lines to the workpiece. Only paints with a moderate viscosity and an electrical conductivity in the range 5×10^{-6} to 5×10^{-9} S/cm can be applied in this way. The best known designs are the electrostatic spray gap (AEG method), the electrostatic spray cone (diameter 70–250 mm, max. rotational speed 1500 min^{-1}), and the electrostatic spray disk (diameter 400–700 mm, max. rotational speed 3000 min^{-1}). Purely electrostatic spraying methods are only used in special cases on account of their limitations (workpiece geometries, type and throughput of the paint).

Electrostatically assisted atomization methods are more versatile than purely electrostatic methods because atomization takes place mechanically. The electric field serves only to charge the paint material and to transport the charged droplets to the workpiece. The following systems are used:

Table 8.4. Electrostatic atomization methods

Advantages	Disadvantages	Examples of areas of use
Purely electrostatic spraying methods		
Very high application efficiency	unsuitable for parts with complex shapes	housings and flat parts in the electrical industry (washing machines, refrigerators, switching cabinets)
Relatively low investment and operating costs	low paint throughput	
	limited choice of coating materials	
Low wear of structural parts	high consistency of paint data	
	low flexibility	
Electrostatically assisted rotation atomization		
High application efficiency	expensive plant and safety technology	large-scale series coating (automobile industry), industrial coating
Fine atomization		
Practically all paints can be applied	unsuitable for parts with complex shapes	
	excessive edge coating may occur	
Electrostatically assisted compressed air, airless, and airmix atomization		
As for nonelectrostatic methods but with higher application efficiency	coarser atomization otherwise as for rotation atomization	industrial mass-produced articles, handicrafts

1) Electrostatic high-speed turbo bells (diameter 30–80 mm, rotational speed 15 000–40 000 min^{-1}, voltage 50–120 kV)
2) Electrostatic high-speed disks (diameter 150–250 mm, rotational speed up to 20 000 min^{-1}, voltage 70–520 kV)
3) Electrostatically assisted atomization guns (charging by means of needle-shaped electrodes arranged directly in or on the paint nozzle, voltage 50–100 kV).

The extremely high electrical conductivity and high dielectric constant of waterborne paints should be taken into account. The paint supply system must be electrically insulated to prevent short circuiting. Ionization charging or, in the case of automatic equipment, the installation of an insulated paint supply system have proved of value.

8.3.3. Dipping [8.1]–[8.3], [8.10]

Dip coating is one of the simplest and oldest coating methods. In addition to dipping in solvent- or waterborne paints, electrodeposition has become important for large-scale series production (Table 8.5).

208   8. Paint Application

Table 8.5. Dipping methods

Advantages	Disadvantages	Examples of areas of use
Conventional dipping		
Simple process	special plant and equipment technology	priming or one-layer coating of mass-produced articles
Can easily be automated	unsuitable for many shades	
High application efficiency	ventilation and fire prevention measures required with solvent borne paints	
High economy		
Low wage costs		
High article throughputs	paint analysis control	
	edge coating often unsatisfactory	
	danger of sagging, dripping, paint splashing	
	foam and bubble formation	
Electrodeposition coating		
Complete and uniform paint film, also in cavities (wrap-around)	complex equipment	priming in the automobile industry
	unsuitable for more than one shade	priming of mass-produced articles
Very high material utilization		
No sagging or droplet formation	complex bath monitoring	
	highly trained workforce	
	high material costs	
Fully automatic operation		
High parts throughput		
Environmentally friendly (waterborne paints)		

Conventional Dipping. In conventional dipping the workpieces are immersed in the paint and then removed (Fig. 8.2). The liquid paint adheres to the surface and is then dried or stoved. Care should be taken to ensure that the workpieces do not float during dipping and that air bubbles do not become trapped. The speed at which the workpiece is removed from the bath must be selected so that excess paint adhering to the surface can run off. The draining and evaporation time must be sufficiently long to ensure satisfactory evaporation of the solvents (if necessary a hot air zone should be included for waterborne paints).

Electrodeposition. Electrodeposition paints are suspensions of binders and pigments in fully demineralized water with low concentrations (ca. 3 %) of organic solvents (see Section 3.8). Electrodeposition coating may be either anodic or cathodic.

In *anodic electrodeposition* the workpiece acts as the anode. This method is only used to a small extent. Disadvantages compared with cathodic electrodeposition are its poorer handling and corrosion protection. Advantages include the lower paint price and lower expenditure on plant technology.

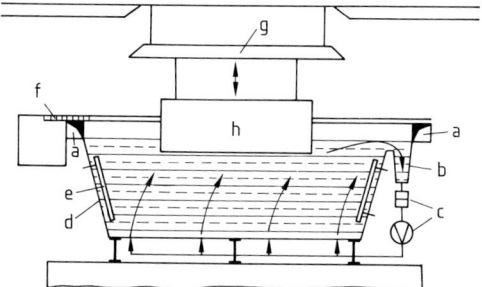

Figure 8.2. A conventional dipping unit
a) Edge suction; b) Overflow; c) Circulation system with pump, filter, piping, and nozzles; d) Dipping tank; e) Heating and cooling device; f) Cover (in the form of roller-type covers); g) Raising and lowering system; h) Workpiece

In *cathodic electrodeposition* the workpiece acts as the cathode; this method is more important than anodic electrodeposition. The binders consist largely of non-water-soluble epoxy resins, and to a lesser extent of acrylic resins (one-layer coatings). These resins are converted into a water-soluble (i.e., ionized) state by neutralization with organic acids (e.g., acetic acid):

$$R_2N-R + R'COOH \longrightarrow R_2N^+H-R + R'COO^-$$

Water is decomposed by electrolysis at the electrodes

$$2\,H_2O \longrightarrow O_2 + 4\,H^+ + 4\,e^-$$

and at the workpiece

$$2\,H_2O + 2\,e^- \longrightarrow H_2 + 2\,OH^-$$

Iron from the high-grade steel anodes is also oxidized at the anode and passes into solution. Hydroxyl ions are formed at the cathode and react with the solubilized resin causing reversal of the neutralization:

$$H_2O + 3\,H^+ + 4\,e^- \longrightarrow 2\,H_2 + OH^-$$

$$R_2N^+H-R + OH^- \longrightarrow R_2N-R + H_2O$$

The binder then coagulates and is deposited as an irregular, porous layer on the workpiece. It is converted into a uniform, sealed paint film by stoving.

Electrodeposition equipment and technology is expensive and is therefore practicable only for large-scale series coating (Fig. 8.3). In addition to the dipping tank and a storage tank, circulation systems for the paint and auxiliary materials, rinsing systems (ultrafiltrate, fully demineralized water), a regulated d.c. supply (200–400 V), and the workpiece transporting system (curent supply) also have to be installed. Since electrodeposition systems require temperature control, production tanks are equipped with heaters and chillers.

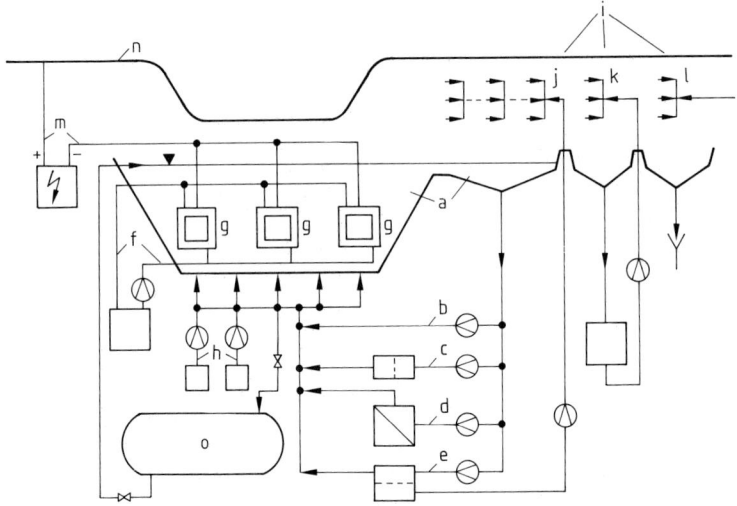

Figure 8.3. Cathodic electrodeposition coating unit
a) Dipping tank with overflow; b) Recycle circulation; c) Paint filter circulation; d) Paint cooling circulation; e) Ultrafiltration; f) Anolyte circulation; g) Anodes; h) Metering of paint; i) Rinsing system; j) Ultrafiltration rinsing; k) Recirculated material rinsing; l) Water rinsing; m) Power supply; n) Transportation system; o) Storage tank

8.3.4. Miscellaneous Wet Paint Coating Methods [8.1]–[8.3], [8.10]

Other wet paint coating methods are summarized in Table 8.6. Application with a *brush or roller* is now only used to a great extent in the handicrafts sector, do-it-yourself sector, or on building sites.

Flat workpieces (e.g., paper sheets, films, wooden boards, and panels) can be coated economically and quickly by rolling, pouring, or knife coating. These methods have become important because they are easily automated, have a high material yield, and are environmentally friendly. With *roller coating* the paint material is transferred from rotating rubber rollers to one or both surfaces of the workpiece (Fig. 8.4). In forward roller coating (layer thickness ≤ ca. 12 µm) the workpiece and paint application roller run in the same direction. With reverse roller coating (layer thickness 3–100 µm) they run in opposite directions.

The *pouring method* is commonly used in the wood and timber trade. Here the paint is pumped into a pouring head that has a paint outflow slit with adjustable lips (Fig. 8.5). Paint that does not come into contact with the workpiece is returned to the tank via a collecting channel. Since the paint is constantly circulated and is heated by the circulation pump, it must be cooled and must contain high-boiling solvents.

Knife coating is used in the paper and textile industry for coating continuous material. Knife coating can also be used to print and coat flat, two-dimensional

Table 8.6. Miscellaneous wet coating methods

Advantages	Disadvantages	Examples of areas of use
Brushing		
Simple equipment High paint yield No specially trained workforce required Universally applicable Good wetting of the substrate	highly labor-intensive (high wage costs) nonuniform film thicknesses danger of brush marks	steel superstructures lattice constructions handicrafts do-it-yourself
Roller application		
Fast and easy to master High paint yield Uniform film thicknesses	only suitable for smooth surfaces worse wetting of the substrate labor-intensive	steel superstructures handicrafts do-it-yourself
Wiping		
Fast application Uniform film thickness High paint yield	unsuitable for workpieces with complex shapes	exterior coating of pipes application of bitumen to pipelines wood coating
Rolling, printing, strip (coil) coating		
High degree of automation High paint yield High economy Very uniform coating	only suitable for flat surfaces (strips) high investment in plant and equipment limited potential uses	strip and panel coating (sheet metal, wood, films, paper, paperboard)
Pouring		
Very good material utilization Applicable for slightly curved parts Easily automated Uniform coating	special equipment not universally applicable	chipboard, paper, and cardboard coating
Troweling		
Easily automated Very low material losses	specially prepared materials only for flat, strip, or panel-type parts	wood panel coating
Flow coating		
Good material yield Easily automated	nonuniform film thicknesses danger of paint slurry formation	large, bulky articles (radiators, frames for commercial vehicles, etc.)
Centrifugation, drum application		
High paint yield Good economy	no special surface quality (nonuniform leveling, pressure points, etc.)	small mass-produced articles (hooks, eyelets, screws)

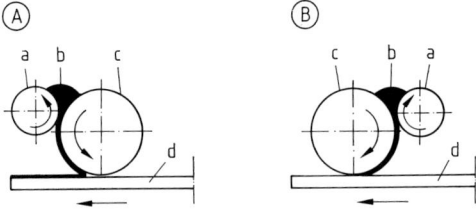

Figure 8.4. Roller coating
A) Reverse roller coating; B) Forward roller coating
a) Metering roller; b) Coating material; c) Paint roller; d) Workpiece

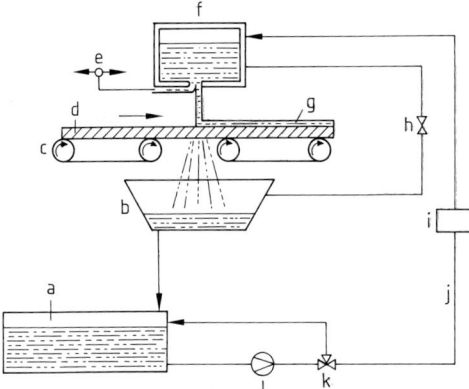

Figure 8.5. Pouring method
a) Paint tank; b) Drip pan; c) Transporting belt; d) Article being coated; e) Aperture regulation; f) Pouring head; g) Paint film; h) Pressure release value; i) Paint filter; j) Paint line; k) Valve for quantitative adjustment; l) Pump

workpieces. The coating is applied to the substrate which then passes under a doctor knife. The coating material is pressed onto the workpiece with the doctor knife (Fig. 8.6). The knife also removes excess coating material and smooths the surface.

Highly pigmented, pasty coatings and viscous putties and fillers are applied by *troweling*. The materials are applied with a pair of counterrotating rollers. Excess material is smoothed and "pressed" into the workpiece surface. This method can only be used for flat, striplike, or panel-shaped workpieces (e.g., wooden panels).

With *flow coating* gentle streams of paint are pumped over the workpieces via nozzles. Excess paint flows into a collecting trough and can be recirculated (Fig. 8.7).

Small items (e.g., hooks, eyelets, clasps, and buckles) can be coated by centrifugation or drum coating. The workpieces must not adhere to one another, nor must they become entangled. In *centrifugation* the workpieces are placed in a wire basket, and then immersed in a paint bath and centrifuged at a rotational speed of about 500 min^{-1}. With *drum coating* paint is fed into a rotating drum containing the workpieces. The paint feed can be achieved by adding special, low-viscosity paint or by spraying with spray guns.

Coil coating denotes the continuous coating of cold-rolled steel strip (including galvanized strip) or aluminum with organic polymers. In automated plants the metal strip is first cleaned and chemically pretreated. The strip is then roller-coated on one or both sides, with one or more coats of liquid, thermosetting or thermoplastic coating materials. The paint coat is dried in an oven after each application. Throughput rates are of the order of 1–3 m/s. Constant production conditions using appropriate control devices ensure high quality.

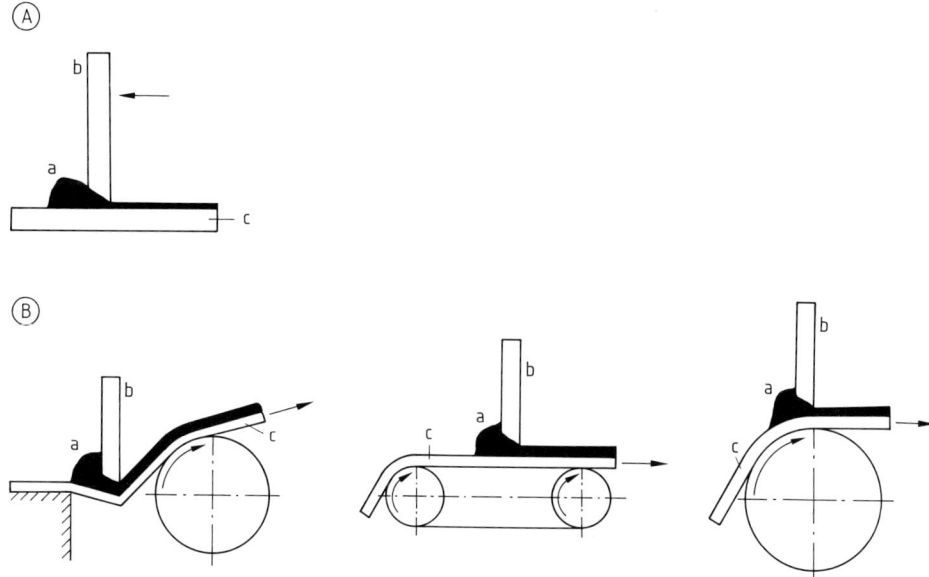

Figure 8.6. Knife coating
A) Coating of unit items (panels, plates, disks, etc.); B) Coating of semifinished articles (strips, sheets)
a) Coating material; b) Doctor knife; c) Substrate

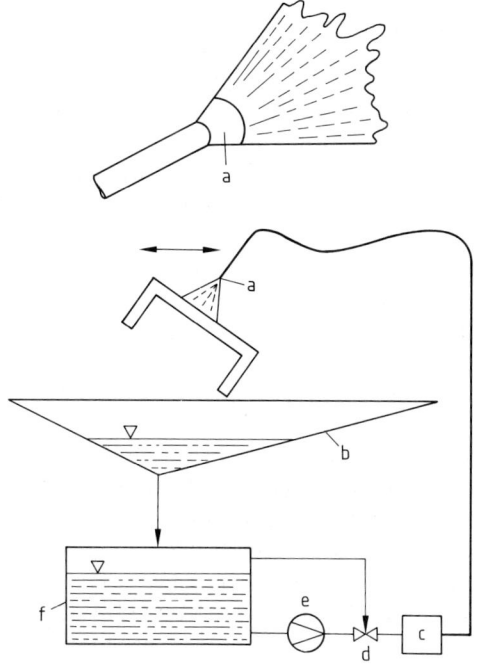

Figure 8.7. Flow-coating method
a) Paint nozzle; b) Drip pan; c) Filter; d) Paint adjustment valve; e) Pump; f) Paint tank

8.3.5. Powder Coating [8.2], [8.3]

In powder coatings the coating material is applied to the workpiece in the form of dry (i.e., solvent-free) thermoplastic or thermosetting powder. The powder particles are heated and melt to form a film. The thermoplastic powders melt and fuse on heating whereas the thermosetting powders also become chemically cross-linked. Two main application processes are used: electrostatic spraying and fluidized-bed coating.

Electrostatic Spraying. The principle of electrostatic spraying is simple. A coating powder is a dust with a particle size in the range 10–80 µm. When dispersed (fluidized) in air the powder flows in the same way as a liquid and is applied in spray cabins using special spray guns onto the cold or hot workpiece. The coating powder adheres to cold workpieces electrostatically, whereas on hot workpieces it adheres by fusion. The workpiece generally passes through a dryer. The coating powder overspray is suctioned off, separated from the air, screened, and reused.

Care should be taken to minimize the length of pipes for powder transport and to maximize the deposition efficiency in order to avoid problems caused by a shift in the particle size distribution. A sufficiently large number of guns, a coating powder cloud size matching the size of the workpiece, adjustment of the oscillating movement of the gun supports, gun triggering, and a continuous powder metering system with maximum accuracy are advantageous. Minimizing the powder content of a plant facilitates cleaning.

Coating powders are mainly sprayed with guns with negative corona charging. Guns with friction charging (positive charging by charge separation on polytetrafluoroethylene) also used in addition. A new generation of corona-charged guns are producing less free ions by inplementing special earthing devices and therefore better penetration into corners and cavities (Faraday cages). Higher deposition efficiencies are obtained by using slit-shaped nozzles and a triggered powder output that may be related to the shape of the object. Increasing the number of guns or reducing the gun output often improves coating application. Parts with complicated shapes, particularly those with Faraday cages, are extremely difficult to coat, at least with a constant layer thickness.

Good grounding (earthing) is necessary for satisfactory results. For safety reasons, earth leakage resistances of less than 1 MΩ (at 5 kV) are required. Good grounding can also be obtained via contactless, electrostatic processes.

A large proportion of coating powder plants operate partly or fully automatically. Many types of equipment are commercially available [8.11], [8.12]. An important cabin type is the compact unit with replaceable filters in which the air that entrains and removes nonadherent powder is fed to a filter housing, flange-mounted on the rear wall. In the filter the powder is separated from the air and concentrated. The air is then vented into the atmosphere or directly recycled to the working area. If the color has to be changed the cabin is cleaned and a new filter housing is mounted. If the unit operates with one filter housing only, the filter housing is cleaned.

Criteria for spray cabins, powder recovery systems, and powder preparation systems are mainly determined by the time required to change colors. A compact unit with one replaceable filter is used for one or a few colors. A compact unit with one or more replaceable filters or a filter carpet unit is used for several colors. For many colors a unit equipped with cyclones (or a filter carpet unit) or a compact unit with several replaceable filters should be used. If necessary, secondary colors can be applied without recovery or with liquid paints.

After the coating powder has been applied the workpiece can be passed directly to the dryer. An evaporation zone is not required as is the case for solvent-based paints. Since only the carrier medium air, and not a solvent, has to escape from the film, the coating powder can be heated much more quickly at the oven inlet than is the case with conventional solvent-based paints. Heating can be performed with IR zones or, after the coating powder has fused, with special blowing zones. Rapid heating results in short oven lengths and usually improves leveling and wetting. Since the stoving loss of coating powders is very low the ovens can be operated in a highly energy-efficient manner with a relatively low rate of exhaust air.

Fluidized-bed coating leads to substantially thicker layers than electrostatic spraying. The workpieces are preheated to 200–400 °C in an oven, dipped briefly (1–10 s) in vessels containing fluidized coating powder (particle size 40–200 µm). The excess powder is removed by vibration, shaking, or blowing and the workpiece is optionally postheated and then cooled with air or water. The capacity of the fluidization basins ranges from a few grams to several tonnes.

The layer thickness is determined by the preheating temperature, the heat content of the parts, the dipping time, and the coating powder. Parts with complicated shapes or with different substrate thicknesses are difficult to coat. The layer thicknesses obtained by the various application methods and the resultant use profiles are summarized in Table 8.7.

Table 8.7. Layer thicknesses and typical uses in powder coating

Application method	Temperature of workpiece	Typical layer thickness, µm	Uses
Electrostatic	cold	25–100	interior: decoration, corrosion protection
Electrostatic	cold	50–120	exterior: weather resistance, decoration, corrosion protection
Electrostatic	cold	50–250	resistance to chemicals, electrical insulation, corrosion protection
Electrostatic	hot	150–600	resistance to chemicals, electrical insulation, corrosion protection
Fluidized bed, flocking	hot	200– > 1000	resistance to chemicals, electrical insulation, corrosion protection

8.3.6. Coating of Plastics and Wood [8.2]

Most coating methods described in Sections 8.3.1–8.3.5 are also suitable for plastics and wood. The resistance to solvents and heat as well as the electrical conductivity of the plastics or wood must, however, be borne in mind.

Inmold coating (IMC) is an application technique used for plastics in which the coating material is applied during production of the plastics. The coating material is first injected (generally electrostatically) into the mold that is previously coated with a release agent. After the mold has been closed the plastic is then injected, foamed, and hardened. This method can be used for glass-fiber-reinforced plastics and polyurethane foams. A similar method is used with sheet molding compounds (SMC resin mats). After the resin mat has been inserted in the mold, the SMC is compressed and hardened. The mold tool is then opened and coating material is injected through the slit opening, compressed, and hardened.

8.4. Paint Curing Methods [8.1]–[8.8]

The performance of a coating depends on the chosen curing conditions. For example, mechanical and other properties associated with a given thermally cross-linking paint can only be guaranteed if curing is adequate. In undercured paint systems, adhesion to a subsequent layer may be adversely affected. If, however, the maximum curing temperature is exceeded the paint layer may become brittle or yellow. The curing time is also important.

Curing and hardening may be either physical or chemical. With a suitable combination of binders, both types of curing methods may proceed in parallel or overlap.

Physical Curing [8.1]. Physical curing occurs when polymers dissolved in organic solvents gradually cohere to form a solid film and then a network. Cohesion occurs solely as a result of solvent evaporation without chemical cross-linking. Such polymer films are generally reversible, i.e., they dissolve in the original solvent. Physically drying binders (e.g., nitrocellulose and its esters, vinyl resins, polystyrene, acrylate esters, chlorinated rubber, bitumen) are mostly chainlike or threadlike molecules with short side chains.

Chemical Curing [8.1]. During chemical curing film formation occurs as a result of formation of chemical bonds between the binder molecules. The binders become increasingly insoluble as cross-linking proceeds, and ultimately form irreversible (i.e., insoluble) thermosetting films. In solvent-containing systems physical curing also occurs simultaneously.

8.4. Paint Curing Methods

Curing methods may be divided into three groups:

1) Curing with a heat carrier (air)
2) Curing with radiation (IR, UV, electron beams, laser beams, plasma arc)
3) Curing by means of electrical processes (inductive curing, resistance, high-frequency and microwave curing)

Curing with Heat Carriers. Curing with a circulating hot air stream is the most important curing method. The heat causes the solvent vapors to evaporate from the coating film and they are removed by the air current. Heat transfer and thus film formation occur from the exterior to the interior. Paint curing is also possible with parts having a complicated shape. Since the whole workpiece has to be heated, long curing times and thus also large ovens are required. Heat consumption is relatively high because the workpiece, paint film, transporting device, parts of the conveying system, as well as the fresh air (maximum permissible concentration of combustible solvents in the oven is 0.8 vol%) all have to be heated.

Curing with Radiation. See also Section 3.7. Radiation curing methods have become increasingly important in the last few years. Paint curing proceeds more rapidly than in circulating air curing since the whole workpiece does not have to be heated. However, only large flat parts can be satisfactorily treated.

Microwaves (frequencies 3–600 GHz, wavelength 0.5 mm–10 cm) are generated in magnetrons and transmitted in hollow conductors. They exhibit wave effects such as interference which lead to localized concentrations of energy (wave peaks). Interaction between the polar material in the paint film and the electromagnetic alternating field is manifested macroscopically as a heating effect. Since microwaves are reflected by electrically conducting surfaces, this method can be used only for nonconducting substrates (plastics, wood, or paper).

Infrared (IR) radiation (wavelength 0.76 µm–1 mm) is absorbed, reflected, or transmitted by an object. In the paint film absorbed radiation is converted into heat. The paint film cures from underneath and the paint surface does not harden initially (solvents can, however, still escape without any problem). The wavelength and intensity of the IR radiation must be matched to the paint being hardened. Absorption behavior is determined by the pigment, pigment volume concentration, and binder. Longwave IR radiation (4.0 µm–1 mm) is absorbed by the pigments at the surface, while shortwave radiation (0.76–2 µm) can penetrate the paint. Normal thermal outputs are 5–25 kW/m^2, and may be up to 100 kW/m^2 if appropriate regulating equipment is used.

UV radiation with a wavelength range of 0.32–0.4 µm is used for curing paints. The UV radiation initiates photochemical reactions which lead to cross-linking. An added photoinitiator (sensitizer) decomposes in the paint into free radicals that initiate polymerization of the binder. Curing with UV radiation is of practical importance for hardening colorless polyester putties or primers, and offset and printing inks. The hardening times are of the order of a few seconds.

In pulsed radiation curing (PRC) exothermically reacting paints and printing inks are cured by UV pulses at a wavelength of ca. 197 nm. The pulses break the carbon double bonds of the binder because this wavelength range corresponds to their resonance frequency. A chain reaction starts throughout the whole layer, and curing takes place within a few seconds.

Electron beams are generated by applying an accelerating voltage (150 kV) to a thermionic cathode. An electron beam (ca. 6 mm diameter) is spread out into a curtain beam by a beam splitter. The electrons leave the beam distribution housing through a very thin metal sheet. When these electron beams strike binder monomers, they initiate polymerization in the paint film. Polymerization occurs in a fraction of a second and must be performed in a vacuum or in an inert gas atmosphere. The equipment must be screened to protect the operators.

Sharply localized curing can be achieved with *laser beams*. The carbon dioxide laser, which has a total beam output of 100 W/cm^2 at a wavelength of 10.6 µm, can be used for this purpose.

A *plasma arc* may also initiate cross-linking in a paint film. The high temperatures produced in the interior of the plasma arc are transmitted only to a small extent to the paint film.

Drying by Electrical Methods. In electrical methods electric current is directly converted into heat (resistance drying) in the workpiece or in the paint film.

In *inductive curing* an induction coil is located close to a metallic workpiece and generates eddy currents in the latter. The workpiece therefore becomes hot and the paint is heated from beneath the film; solvent loss or curing occurs.

In *high-frequency curing* the workpieces are arranged between two capacitor plates in a radiofrequency field (10^8–10^9 Hz). Molecules (dipoles) align themselves and are polarized in the alternating electric field. They therefore oscillate about their equilibrium position, resulting in heating.

9. Properties and Testing

Anyone wishing to test the quality of a paint or coating quickly realizes that only a few properties can be accurately scientifically defined. In many cases there is a good correlation between defined physical properties and the behavior of interest to the scientist or practitioner. In some cases, however, it is impossible to obtain such a correlation. A large number of laboratory testing methods have therefore been developed for paints and coatings that are intended to simulate in-use conditions. These testing methods are often similar but their results are not fully comparable. Standard manuals provide a good overview of available test methods [9.1]–[9.5]. In this chapter attention is focused on methods that are widely known and internationally standardized, or whose international standardization is in progress.

9.1. Properties of Coating Materials

Many properties of liquid paints can be measured with considerable accuracy. Samples must be homogeneous. They must also be sufficiently large to be representative for a given batch. Impurities and permanent material defects (e.g., skin formation, a hard sediment, or gelling of the paint) can be detected. The most common investigations to which a sample is subjected before starting the tests are described in ISO 1513. Only those tests that are important for paint storage, transportation, and application are described here.

Viscosity. Although paint viscosity can be accurately measured with viscometers, paint consistency is normally assessed with *flow cups*. The time in seconds required for a known volume of paint to flow out of the cup through a jet is measured. Paints of higher or lower viscosity can be matched by using cups with different jet diameters.

Measurement of the run-out time from flow cups has been adopted worldwide since this test can be performed anywhere (e.g., in the laboratory, during production, or on a building site). Nationally standardized sets of flow cups are normally used in major industrial countries and give similar but not identical results. An internationally standardized system of flow cups has been introduced (ISO 2431) to overcome this problem. Since temperature fluctuations greatly affect the viscosity measurement, the flow cups should only be used in conjunction with thermostated jackets.

Measurements should be made at 23 ± 0.5 °C. According to ISO 2431 flow cups should only be used for substances exhibiting Newtonian flow. However, they are also often employed for near-Newtonian paints where the flow behavior at the desired viscosity deviates only slightly from Newtonian behavior (e.g., when adjusting the application consistency by dilution).

Viscometers are being increasingly used for accurate measurements on modern industrial paints, especially waterborne paints. Rotating viscometers with a concentric cylindrical geometry (Searle system, Fig. 9.1) are advantageous for paints. Precise thermostatic control is easily achieved because the outer cylinder does not rotate. The drive and torque sensor are combined to form a single unit with the rotating inner cylinder.

The Couette system (Fig. 9.2) is another concentric cylinder system with the advantage that the drive and sensor are separate. The motor drives the outer cylinder; the inner, stationary cylinder is connected to the sensor.

Viscometers with cylinder and cone/plate geometries can also be employed. The cylinder viscometers are easier to use and provide more reproducible results. Cone and plate systems can be used to investigate the hardening behavior of paints. The system can easily be cleaned and only a small amount of sample is required. High velocity gradients can be achieved with small cone angles. The potential uses of cone and plate systems are limited for several reasons and they cannot be used with dispersions [9.6], [9.7].

Falling ball viscometers and capillary viscometers are not generally used for testing paints. Flow cups are, however, special capillary viscometers with a short capillary in which the force of gravity acts on the paint.

Other Rheological Properties. Paint brushability, sagging, and leveling are highly dependent on viscosity and are usually evaluated subjectively in an application test. *Brushability* (the ease with which a paint can be brushed) is evaluated by actual

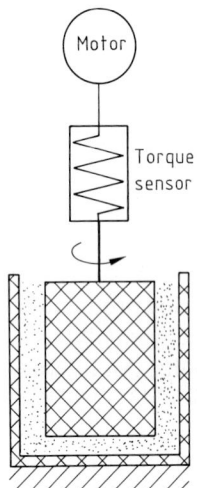

Figure 9.1. Rotating viscometer: Searle system

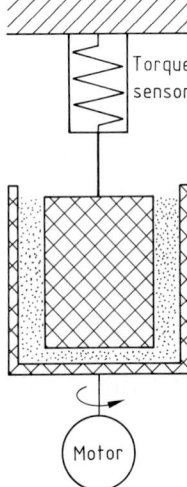

Figure 9.2. Rotating viscometer: Couette system

brushout of the paint by a painter. To evaluate the force required for brushing, the viscosity should be measured at a high shear rate of about $10\,000\text{ s}^{-1}$.

Sagging is the downward movement of a paint film that occurs between the time of application and setting. The viscosity has to be considered at a shear rate of at most 1 s^{-1} (low-shear viscometers) in order to assess sagging on vertical walls. The run-out rate depends on the square of the layer thickness. Irregularities in the layer thickness can cause undesirable sagging (curtaining or crawling).

Leveling is the measure of the ability of a paint to flow out after application (e.g., to obliterate brush marks). It is measured by comb tests. Theoretical correlations have been established between viscosity measurements and practical results. Surface irregularities can be sensed mechanically and visualized with modern methods of image analysis [9.8].

Pot Life. The pot life is the length of time that a paint can be used after necessary preparations for application have been made. The increase of flow time measured with a flow cup to twice the initial time is often used to assess the pot-life time.

Flash Point. The flash point of a liquid is a measure of the flammability of its vapors on application of an external flame. It is used to assess fire hazards. The flash point of paints may have to be measured to comply with legal requirements relating to the storage, transportation, and use of flammable products. According to ISO 1523, the flash point (closed cup) is the minimum temperature to which a product, confined in a closed cup, must be heated for the vapors emitted to ignite momentarily in the presence of a flame, when operated under standard conditions.

ISO 1523 states that the Abel, Abel–Pensky, and Pensky–Martens cups satisfy the necessary requirements.

ISO 3679 specifies another apparatus that provides similar results using a more rapid procedure and with a smaller test portion (2 mL).

ISO 9038 aims to test the combustibility of paints, and involves the use of a Cleveland cup (ISO 2592) or a similar device.

Content of Nonvolatile Matter. Nonvolatile matter is defined in ISO 3251 as the residue left when the product is heated at an elevated temperature for a definite period under prescribed test conditions. A 1 g sample is evenly distributed in a flat-bottomed metal or glass dish (diameter 75 mm) and weighed before and after heating. Six different heating conditions are recommended in ISO 3251. The nonvolatile content of a paint is not an absolute quantity, but depends upon the temperature and period of heating used for the test as well as upon the test portion. If the test method specified by ISO 3251 is used, only relative values are obtained due to solvent retention, thermal decomposition, and evaporation of low molecular mass constituents. The method is therefore primarily intended for testing subsequent deliveries of a given product.

Density. Paint density is usually expressed in grams per milliliter at a reference temperature of 20 °C.

Pyknometer Method (ISO 2811). A pyknometer is filled with paint of known mass. The density is calculated from the mass of the paint and the volume of the pyknometer. Cylindrical metal pyknometers and glass Gay–Lussac or Hubbard pyknometers are suitable. The Hubbard pyknometer is especially useful for highly viscous paints.

Immersed Body Method (DIN 53 217 T 3). The density is calculated from the buoyancy of spherical bodies immersed in the paint. This method is used for low- and medium-viscosity paints and is particularly suitable for production control.

Vibration Method (DIN 53 217 T 5). A U-shaped tube is clamped at both ends and filled with paint. When the tube is subjected to vibration its resonance frequency depends on the mass in a given tube volume and thus on the density of the paint. The tube is calibrated with two media of known density and the density of the sample is then calculated from the resonance frequency. This method is used for liquid and pasty coating materials [9.9].

Fineness of Grind (ISO 1524). The fineness of grind of paints is determined with a gauge graduated in micrometers. The reading obtained on the standard gauge under specific test conditions, indicates the depth of the groove of the gauge at which discrete solid particles in the paint are readily discernible.

9.2. Properties of Coatings

9.2.1. Films for Testing

Preparation of Test Samples. Almost all properties of coatings depend on the layer thickness of the film. To obtain reproducible comparable test results, measurements must be made on carefully prepared films with a defined layer thickness. Films may be either free or applied to a substrate. In some cases film properties are also measured on films with different thicknesses.

Many properties of coatings also depend on the substrate and the method of application. The substrate and method of application must therefore be described in detail when citing results.

Paint-making conditions should be reproduced in the laboratory when necessary. Often it is also useful to take into account the production boundary conditions (e.g., maximum and minimum temperature during stoving). If particularly good correlation of test results between a paint manufacturer and paint user is necessary, identical equipment (panels, spray units, etc.) must be used for sample preparation.

The use of standard panels according to ISO 1514 also has advantages. Standard panels are marketed by several companies. It is advantageous to use flat, inert substrates for optical and some mechanical tests if the reproduction of identical substrate surfaces presents serious difficulties (e.g., surfaces of wood or mineral substrates). For example, glass plates are used to test gloss and color; polished steel sheets are used to test flexibility.

In the laboratory, large-scale industrial plants for paint application are often simulated with simple coating equipment. The most commonly used devices for producing films on test panels are the film applicator, film caster, film spreader, and drawdown bar (Fig. 9.3). When operated at a constant rate these devices produce a uniform layer thickness on the test panels. Small machines are often used to coat test panels, in which either the film applicator or the test panel is uniformly moved by a motor-driven unit.

Combination of the drawdown technique with standard panels is quicker, universally applicable, and cheaper than simulating large-scale industrial application conditions, and is satisfactory for many tests. Drawdowns give more reproducible results than spray application.

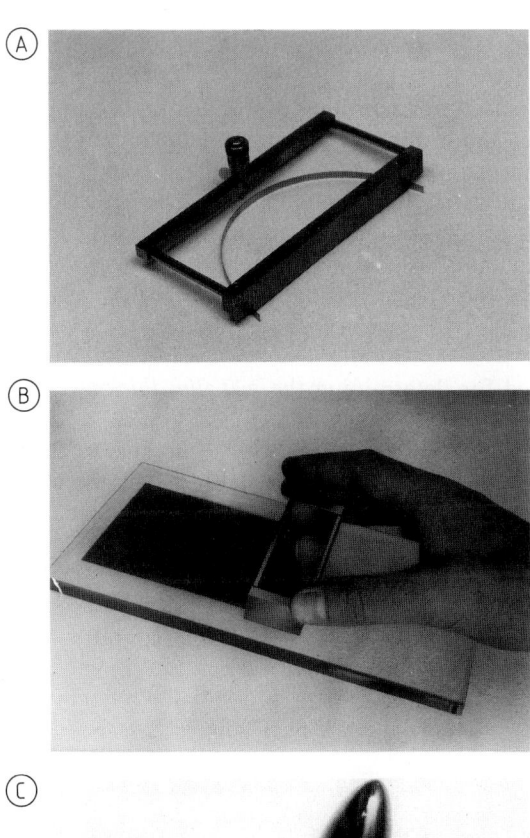

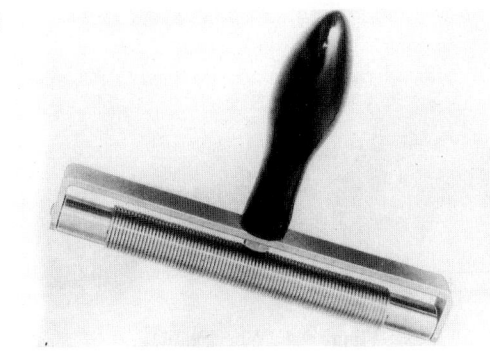

Figure 9.3. Devices for producing films on test panels (Courtesy of Erichsen GmbH + Co KG, Hemer, FRG) A) Film applicator with adjustable blade; B) Film applicator with four fixed blades of different clearance; C) Grooved-rod film applicator

Sample Conditioning. The physical and mechanical properties of paints and coatings generally depend on the environmental conditions used during testing; the most important variables are temperature and humidity. The degree to which each of these variables needs to be controlled is determined by the effect the variable has on the property being measured. Thus for the measurement of viscosity and density the temperature must be held constant within limits of $\pm 0.5\,°C$. Standard conditions recommended in ISO 3270 are $23 \pm 2\,°C$ and $50 \pm 5\%$ relative humidity. Paints that dry at ambient temperature should be dried under these standard conditions for the specified period of time. Paints that are dried at high temperature should be stored for two days under these conditions before being tested. Samples subjected to such conditions have the properties of a new coating. A coating that has been in use for a year exhibits different properties. Surprisingly, however, the properties of coatings are rarely tested after artificial aging.

Measurement of Film Thickness. An accurate definition of film thickness can only be given if the coating has even upper and lower surfaces and a defined density. In practice, neither the surface of the coating nor that of the substrate is even. Surface irregularities and density variations influence the results of each test method in a specific way. Results of different tests performed on the same sample may therefore differ substantially. Results of film thickness measurements therefore always have to be quoted together with details of the measurement method and instrument used [9.10], [9.11]. A survey of methods used to measure paint thickness is given in ISO 2808.

Two instruments are generally used for determining the *wet film thickness*: the Interchemical Wet Film Thickness Gauge (ASTM D 1212, Wheel Gauge) and the Comb Gauge. With the wheel gauge (Fig. 9.4) contact of the wet film and eccentric rim of the wheel is influenced by the surface tension of the paint. It is therefore necessary to observe the first and last contacts and calculate the mean value. The influence of surface tension is excluded if the comb gauge (Fig. 9.5) is used.

Many methods are used for measuring the *dry film thickness*. Thickness can be determined by mechanical contact instruments such as hand-held or stationary micrometers (ASTM D 1005). The apparatus comprises a dial comparator, dial

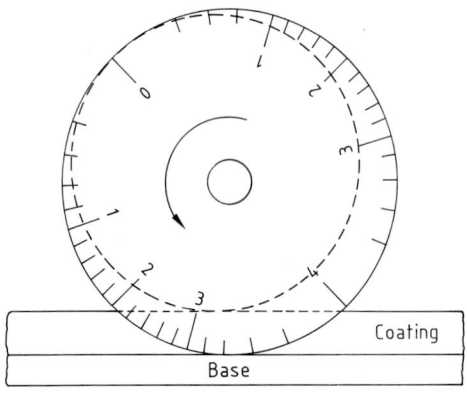

Figure 9.4. Wheel gauge

gauge, or suitable indicator for linear measurement, equipped with a mechanical contact foot and a mechanical, optical, or electronic indicator.

Two reference methods are recommended for painted surfaces: a method in which the surface profile is measured (ISO 4518) and a microscopy method (ASTM D 2691) for coatings on wood and other substrates. Other optical methods include the wedge-cut method and the profile-measuring microscope method.

Magnetic instruments based on the magnetic induction principle or on the permanent magnet pull-off principle are often used on magnetic metallic substrates (ISO 2178, ASTM D 1186). Instruments that operate on the eddy current principle are used for nonmagnetic metallic substrates (ISO 2360, ASTM E 376).

Two contact-free methods employ instruments based on the back scattering of β-particles (ISO 3543) or X-ray fluorescence (ISO 3497).

The uncertainty of film thickness measurement is usually 10% of the thickness or ca. 1.5 μm, whichever is the greater. The measurement uncertainty of reference methods is normally < 10% of the film thickness; microscopy methods have a minimum error of 0.8 μm.

Determination of Hiding Power. The hiding power is the ability of a paint to hide a surface over which it has been uniformly applied. The hiding power of light-colored paints can be determined to an accuracy of ca. ±10% by a simple method. Films are applied in different thicknesses to charts that are printed and varnished so as to produce an array of adjacent black and white areas. After drying, the surface is illuminated (1000–4000 lx) and visually examined. The layer thickness (in micrometers) at which contrast between the black and white areas is no longer visible is determined and is taken as a measure of the hiding power.

ISO 6504/1 describes a modification of this method in which the hiding power (V in m^2/L) is measured with a reflectometer; reproducibility is 2%.

The contrast ratio (opacity, hiding power) for a defined area covered by a given volume of paint is determined according to ISO 3905 using black and white charts, and according to ISO 3906 using a polyester film substrate. The contrast ratios of paints of the same type and color can be compared according to ISO 2814 [9.12].

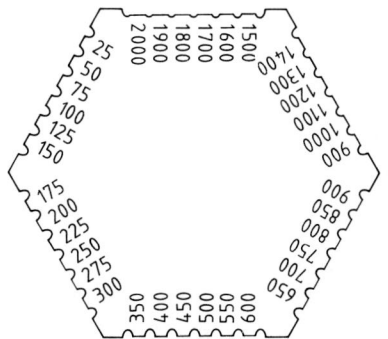

Figure 9.5. Comb gauge

Figure 9.6. Electromotive drying equipment (Courtesy of Erichsen GmbH + Co KG, Hemer, FRG)

Drying Time. Evaluation of the drying time is simple in the case of stoving materials, but is a major problem with all paints and varnishes that dry physically or chemically at ambient temperature.

Many expressions are used in connection with drying (e.g., dust-free, tack-free, surface-dry, dry-to-touch, dry-to-handle, dry-through, dry-to-recoat) but these terms do not have accurate definitions. However, equipment is available for measuring drying times in the laboratory under defined temperature and humidity conditions (Fig. 9.6). Internationally standardized test methods also exist:

Ballotini Method (ISO 1517). Small glass beads are poured onto a horizontal film surface. After 10 s the panel is inclined and brushed lightly. The coating is "surface-dry" if all the spheres can be brushed away.

Print-Free Test (ISO 3678). A square of polyamide gauze is pressed onto the surface. The surface is "print-free" if no imprint can be seen.

Pressure Test for Stackability (ISO 4622). Two panels are coated and dried; one panel is placed on top of the other and set under pressure. No damage to the coating must occur.

9.2.2. Optical Properties

The appearance of a coating involves not only color but also gloss, luster, and translucency.

Gloss. The term gloss is commonly used but is difficult to define [9.13]. The Commission Internationale de l'Éclairage (CIE) defines the gloss of a surface as "the mode of appearance by which reflected highlights of objects are perceived as superimposed on the surface due to the directionally selective properties of that surface."

Specular gloss denotes the degree of mirror reflection and is the primary visual gloss impression. Flat coated panels can be visually classified according to their specular gloss. The ISO 4628 system classifies specular gloss as high gloss, gloss, semigloss, semimatt, matt, and dull matt.

Arbitrary gloss scales were used initially. Little effort was made to establish perceptual gloss scales. "Psychromatic scaling of gloss" has recently been proposed and describes a method devised for the multidimensional assessment of gloss [9.14].

Although specimen panels can be classified according to the specular gloss criterion, the human eye recognizes additional gloss phenomena. These phenomena are generally described according to R. S. Hunter who, in 1939, proposed six types of gloss: specular gloss, sheen, contrast gloss/luster, absence of bloom/reflection haze, distinctness of image/image clarity, and directional/surface uniformity.

Specular gloss and the distinctness of image are of practical importance. In many paint uses maximum gloss is required. Humans experience gloss as particularly brilliant if images are reflected with clear definition from a plane surface. Various forms of black–white contrasts are formed on the paint surface for the visual evaluation of the distinctness of image (DOI).

Specular gloss and distinctness are also most accessible to objective measurement. In 1939 Hunter and Judd found that specular gloss measured at 60° in a reflectometer with specified light source and receptor apertures provided a useful classification of paint finishes according to glossy appearance [9.15]. The instrument was standardized as ASTM D 523, improved, and subsequently standardized as ISO 2813. The 60° reflectometer is now used worldwide as the standard instrument for measuring specular gloss because visual evaluation correlates well with the reflectometer values. The directionality can also be recorded with this instrument as the difference between the maximum and minimum readings when the instrument is rotated in its own plane [9.19].

The 60° reflectometer has been supplemented by the 20° and 85° reflectometers. The 20° instrument provides better differentiation in the high-gloss range, while the 85° instrument provides a better differentiation in the low-gloss range. However, values measured with these two reflectometer geometries do not always correlate satisfactorily with visual gloss evaluation.

In order to measure the distinctness of image, attempts were made to obtain a second reflectometer value at a larger receiver-side aperture angle and to combine the two reflectometer values in such a way that the measured values could be assigned to the relevant visual evaluations. Sophisticated instruments are now available, in which the reflected image of a grating projected onto the surface is analyzed by means of Fourier transformations.

Goniophotometers are used for more detailed laboratory investigations of the gloss indicatrix. A plane section is laid through the gloss indicatrix and can be measured (ASTM E 167) [9.16]–[9.22].

Color. The impression of color is produced by the absorbance of light by colorants (pigments) in the paint film. To ensure that the color of the coating is not altered by the color of the substrate, the coating must "hide" the latter. Visual color evaluation involves comparison of a sample with a standard (ISO 3668). Visual comparison of the color of paints requires an illumination of at least 2000 lx. The comparison should be made against a color-neutral (e.g., matt gray) background and the samples should be illuminated by diffuse daylight. This may be achieved with light falling onto the sample through a north-facing window (in the northern hemisphere), or in a color-matching booth with a light source giving a power distribution approximating to that of CIE Standard Illuminant D 65. If the standard and test panels contain different pigment mixtures, their colors may match under the standard light source but not under another source (metamerism).

Color comparison is most successful if the two paint films have the same or only slightly different gloss, or if the panels are viewed at an angle which minimizes gloss differences. Color differences are conveniently described according to the criteria of lightness, chroma, and hue in conjunction with the rating schemes given in ISO 4628/1. In the testing of coatings with special optical effects (e.g., metallic finishes) the flop effect also has to be included. Flop is the change in color observed when the angle of observance is varied. The tester should stand with his back to a north-facing window, hold the test panels in his outstretched arms at eye level, and tilt them slowly to give observation angles of 15°, 45°, and 75° [9.23], [9.24].

Colors can be defined with the CIELAB color space. All colors are arranged in a circle around a central vertical lightness axis. The center of the hue circle is considered to be neutral gray and saturation is quantified by the distance from the axis. All colors can be defined in the color space by color coordinates [9.25].

The three different color-response mechanisms in the human eye are the scientific basis for the measurement of color that is standardized in the CIE Standard Observer (1931). This system specifies color according to X, Y, Z tristimulus values and x, y, z chromaticity coordinates. Several alternative scales were subsequently developed to provide a better adaptation to the color differentiation ability of the human eye [9.26].

ISO 7724 describes methods for the instrumental determination of the color coordinates and color differences. The standard is based on the CIE 1976 (L^*, a^*, b^*) color space. ISO 7724 specifications are satisfied by many color measuring devices: tristimulus colorimeters, spectrophotometers, and abridged spectrophotometers. Spectrophotometric data are now preferred and have replaced tristimulus colorimeters. Measuring conditions 8/d or d/8 (with or without gloss trap), or 45/0 or 0/45 are used [9.26], [9.27]. Metallic automotive finishes are measured with a sensor having a variable angle geometry or with a two-angle instrument [45/0 for face tone (color observed when paints are observed at an angle close to the perpendicular) and a second viewing angle to evaluate the flop tone (color observed near the grazing angle)].

Spectrophotometers can measure the spectral reflection of a sample and represent it graphically by means of computer peripherals. Spectrophotometers are a useful aid in the analysis and synthesis of color samples. All colorimeters enable calculation of the total color difference with the CIELAB color difference formula.

Color measuring instruments are superior to the human eye as regards the ability to differentiate between chroma and hue. The trained eye is, however, superior to color measuring instruments in the evaluation of differences between full shade colors and dark colors. If paint has to satisfy stringent color limits, final pass–fail decisions are made on the basis of visual inspection; instrumental color measurement serves only as an aid [9.28].

Color values are generally only measured at one point on each sample panel. Colorimetric evaluation can, however, be improved if several measurements are made at different points on the surface and the results are statistically evaluated [9.29].

9.2.3. Mechanical Properties

Adhesion. The cross-cut test is the most commonly used procedure for assessing the resistance of a coating to separation from a substrate. A right-angle lattice pattern is cut into the coating and penetrates through to the substrate. Cutting was previously performed with a single cutting tool or, on site, with a multiple cutting tool (DIN 53151, ISO 2409). Sharp razor blades were also used. These tools are now being replaced by steel cutting blades which can be broken into sections. The tool may be mounted on a motor-driven apparatus to ensure more uniform cutting. A piece of adhesive tape is placed over the lattice pattern, firmly rubbed with a fingertip, and removed. A six-step classification is used to evaluate the results. This procedure should not, however, be regarded as a means of measuring adhesion [9.30].

The pull-off test for measuring adhesion is described in ISO 4624. Adhesion is assessed by measuring the minimum tensile stress necessary to detach or to rupture the coating in a direction perpendicular to the substrate. The result gives the minimum tensile stress required to break the weakest interface (adhesive failure) or the weakest compound (cohesive failure) of the test assembly. Mixed adhesive–cohesive failure may also occur.

Hardness. The hardness of a coating is examined by three types of test: the scratch test, the damping test, and the indentation test with a penetrating body. The results of one of these tests do not allow conclusions to be drawn as regards the behavior in another test [9.31].

Scratch Test. Many methods have been developed for testing scratch resistance. The method standardized in ISO 1518 employs a scratching needle with a hardened steel hemispherical tip of 1 mm diameter. The needle is moved along the surface of the coating under a specified load either in a hand-operated or in a motor-driven apparatus.

The coil coater method (ECCA-T 4, ASTM-D 3363) uses pencils of different hardnesses which have a "squared tip" and a circular cross section. The pencil is held at an angle of 45° to the coating surface. The hardness of the coating is equated with the hardness at which the pencil can press into and penetrate the paint film. Pencils that are too soft slip over the film without scratching it.

Damping Test. Pendulums according to KÖNIG [9.32] and PERSOZ (ISO 1522) or the Sward Rocker (ASTM D 2134) are used in the damping test. The damping time of a pendulum oscillating on the surface is taken as a measure of the hardness of the coating. With air-drying oil paints the hardness increases during drying. The pendulum oscillates for a longer time as the paint hardens and the internal damping decreases. The behavior is not found in modern stoving finishes. Surface hardness may be high but the internal damping may be large. A relationship between surface hardness and internal damping is no longer given in some synthetic resin binders.

Indentation Test. Indentation hardness is measured by two methods. A spherical or pyramidal indentor is applied to the paint surface for a specified period. The penetration depth under load is measured (ISO 6441, ASTM D 1474, Pfund hardness number). In the other method an indentor with a specified load is applied for a specified period to the paint surface and then removed. The depth of the impression is then measured.

According to ASTM D 1474 the Knoop hardness number is measured with the pyramidal Knoop diamond. According to ISO 2815 the Buchholz instrument is used. This method cannot detect elastic deformation of the coating that disappears immediately after the load is removed.

The Wallace microindentation tester employs a Vickers diamond pyramid indentor and electrical measurement of capacitance to determine the depth of indentation under load.

In modern instruments the load on the indentor is applied in very small steps and at varying rates, with the simultaneous measurement of the depth of penetration. The results are plotted as a curve showing the depth of penetration. The measurement can be continued with stepwise removal of the load [9.33].

Flexibility. Viscoelastic properties are sometimes measured on free paint films [9.34], [9.35]. However, the degree to which the coating allows bending of a substrate without cracking or peeling is more important. Flexibility depends on ambient temperature and humidity, as well as on the bending rate of the substrate. Both cylindrical and conical mandrels are quite often used to evaluate the flexibility of coatings in bending tests (ISO 1519, ISO 6860). Slow deformation tests such as the Erichsen cupping test (ISO 1520) also exist.

Rapid deformations are produced in impact tests (falling weight tests). A metal sheet coated on one side is deformed by a falling weight that has a hemispherical indentor. Deformation occurs directly or indirectly, i.e., the coated or the uncoated side of the sample is struck by the falling body and deformed (ISO 6272).

The viscoelastic properties of coatings depend on structural factors, particularly on the degree of chemical cross-linking, and can be accurately measured by monitoring the bending vibrations of thin, coated steel bars. The bars are subjected to forced oscillations whose resonance curves are evaluated with a computerized measuring system. The device can provide data on the flexibility of coatings after different curing processes. The method can be used to determine the optimum conditions for curing with heat [9.36], [9.37].

Abrasion Resistance. Abrasion (wear) resistance is a basic factor in the durability of a coating, and is determined by the interaction of the abrasive medium and the coating. Many types of abrasive media and treatment are used for testing e.g., fast and slow movement, small or large load, high and low temperature, constant or intermittent contact. The falling sand (abrasive) method, abrasive blast method, rotating disk method, rotating wheel method (Taber abraser, ISO 7784), and Gardner wet abrasion method (used with emulsion paints in DIN 53 778 T 2) are commonly used.

The stone impact resistance is an important property of automotive finishes. Several instruments (gravelometers) have been developed for testing this parameter. One or more projectiles are hurled or shot in the direction of the painted surface at a defined velocity [Verband der Automobilindustrie (VDA) Guideline 621-427] [9.38], [9.39].

9.2.4. Chemical Properties

Since coatings are generally used to protect substrates against corrosion, two factors must always be considered:

1) Are the coatings themselves resistant to aggressive media?
2) Can the coatings protect the substrate?

It should always be borne in mind that chemicals may have different effects depending on whether they are in liquid or vapor form.

Water Vapor and Water Resistance. The rate of water vapor transmission of free paint films is measured according to ISO 7783. The resistance of coated substrates to water condensation is, however, generally measured (e.g., according to DIN 50017). Water vapor condenses on coated substrates in airtight cabinets (capacity ca. 300 L) or in walk-in chambers. The coated substrates are subjected to 100% relative humidity at 40 °C for 8 h, followed by a 16-h period at 23 °C and 50% relative humidity.

The resistance to humid atmospheres containing sulfur dioxide is measured according to ISO 3231 in a 300-L cabinet, to which 0.2 or 1.0 L of sulfur dioxide is added before the cabinet is heated to 40 °C.

Resistance to humidity (continuous condensation) is measured according to ISO 6270 in the Cleveland condensation cabinet, in which the specimens form the roof of the cabinet. The cabinet contains a water bath set at 40 °C. The water vapor condenses on the coatings that are backed by porous or nonporous substrates which are maintained at 23 °C and 50% relative humidity. The water vapor diffuses through the porous substrate [9.40].

Corrosion Resistance. The determination of the resistance to neutral salt spray fog according to ISO 7253 has been adopted from ASTM B 117. This widely used method is regarded as the classic corrosion test. The specimen panels are artificially damaged with a scratch that exposes the substrate. The salt mist acts continuously on the specimen panels during the test. In filiform corrosion on steel and aluminum, thread-like infiltrations of the coatings start from edges or damaged sites. This phenomenon can be tested according to ISO 4623 on steel sheets by damaging a specimen with a scratch and immersing the specimen in salt solution or exposing it to a neutral salt spray fog for a short period (e.g., 1 h) and then exposing it for several days in a test cabinet maintained at $40 \pm 2\,°C$ and $80 \pm 5\%$ relative humidity.

Comparisons of the results of laboratory corrosion tests with damage sustained in use have revealed differences in corrosion and corrosion products. For example, the constant salt spray fog test leads to a different type of rust formation than that produced by natural weathering, with its alternating damp and dry periods. Test cycles have therefore been developed in which the individual stresses are applied in sequence. A good correlation with natural corrosion in automobiles was demonstrated in the VDA alternating test. The sample is subjected to salt spray fog for 24 h, followed by four 24-h cycles each consisting of 8 h at 40 °C and 100% relative humidity, followed by 16 h at 23 °C and 50% relative humidity, and two days of ambient conditions (VDA Guideline 621-415).

Since atmospheric corrosion is an electrochemical process, attempts have been made to develop electrochemical test methods that can predict the corrosion resistance of the coating–substrate system [9.41], [9.42].

The specimens are damaged with a scratch that exposes the metallic substrate, and are cathodically polarized. After several hours a loss of adhesion at the scratch can be observed, which can be correlated with natural corrosion.

Resistance to Liquids. Various methods are used to test the resistance of coatings to the action of liquids and pastes (e.g., mustard, soap solutions, ketchup). Standard methods are defined in ISO 2812. Coated specimen panels are partially immersed in the test liquid so that the changes in gloss, color, and swelling can also be evaluated at the liquid–air interface. Disks of absorbent material are immersed in the test liquid and placed on the coating. The test liquid or test paste is also dropped directly onto the coating. If the test medium is to be prevented from evaporating, the test surface is covered with a glass dish. Elevated temperature generally increases the aggressiveness of the test medium.

ISO 4628 describes a general system for evaluating the decomposition of coatings. The intensity, quantity, and size of common defects are classified on a numerical scale ranging from 0 to 5.

9.2.5. Weathering Tests

Coatings that are exposed to weathering undergo aging. Aging is defined as the sum of all irreversible chemical and physical processes that occur in the coating over the course of time. Aging is mainly caused by radiation, temperature, and moisture (rain and atmospheric humidity).

Solar radiation ($\lambda = 290-3000$ nm) is the primary cause of aging. Solar radiation causes heating of the coating that depends on the color of the surface. UV radiation initiates photochemical aging. Weathering produces changes in gloss and color, cracking, blistering, loss of adhesion, and loss of flexibility; these parameters are used to evaluate aging [9.43]–[9.48]. ISO 2810 gives guidelines on how to perform natural weathering tests. The manner and location in which the coating is exposed must be appropriate for the end use of the product under test. Flat specimen panels or structures are fastened to exposure racks at a defined angle. The exposure period should be one or more years. The intensity, quantity, and magnitude of defects (e.g., blistering, chalking, cracking) caused by natural weathering can be classified according to ISO 4628 (evaluation of degradation of paint coatings). Measurement of degradation of gloss and color is described in Section 8.2.2.

Commercial test institutes in Florida are often commissioned to carry out natural weathering tests. The weather conditions in Florida are preferred because the same aging phenomena occur as in other locations but more quickly. Exposure time can be reduced by a factor of two to four when compared with other locations. The test institutes in Florida also offer accelerated aging by means of exposure on black boxes and on heated black boxes.

In laboratories, coatings are artificially weathered in specially designed apparatus to simulate or measure the aging processes that occur during natural weathering. Artificial weathering involves a smaller number of parameters than natural weathering but can be controlled more uniformly and allows accelerated test conditions [9.49]–[9.51].

Generally valid correlations between aging processes during artificial and natural weathering cannot be expected because they are influenced by many factors. Clearly defined relationships can only be expected if the most important parameters are the same or their influence on the coatings is known [9.52], [9.53]. ISO 4892 and 11 341 specify filtered xenon arc radiation and other conditions used for the artificial weathering of coatings. The optical radiation source and its filter system is specified so as to produce a spectral distribution of the irradiance sufficiently similar to the global solar radiation defined in CIE Publication No. 85. The irradiance is measured inside the apparatus with a radiation meter. The temperature is measured with a black standard thermometer. The test panels are wetted by spraying or flooding with water. ISO 2809 and 787/15 also provide information on how to simulate aging processes occurring during natural weathering under a glass cover.

ASTM D 822 describes a standard procedure for operating light- and water-exposure apparatus (carbon-arc type) for testing coatings. ASTM G 53 and DIN 53 384 describe similar standard procedures for the light exposure or light and water exposure (fluorescent UV type) of nonmetallic materials [9.54].

10. Analysis

Modern coating materials are complex mixtures of binders, solvents, pigments, extenders, and additives. Most of these components consist of several constituents. Complete analysis of coating materials therefore requires comprehensive knowledge and the use of various analytical methods.

Surveys of the analysis of coating materials are given in [10.1]–[10.3]. The proceedings of conferences devoted specifically to paints (e.g., FATIPEC, International Conference on Organic Coatings, Science and Technology in Athens, Waterborne and Higher-Solids Coatings Symposium, the conference of the ACS Polymeric Materials Division, and ASTM meetings) are particularly important because they describe the use of analytical methods from the point of view of their suitability for investigating coating materials. Periodic literature reviews on the analysis of coating materials are also published in the journal *Analytical Chemistry*.

The analysis of coating materials is often employed in the investigation of complaints and substandard batches, or to evaluate competing products. Analysis also plays an important role in the assessment of raw materials, occupational safety and hygiene, and the emission of solvents and decomposition products during paint curing.

In the case of complaints, the binder and pigment (extender) composition of the individual layers have to be established to determine the origin of the paint material in doubtful cases. Establishing the cause of coating defects (e.g., inclusions, delamination, peeling) is also important. These and other analytical investigations are time-consuming and expensive. Optimal utilization of analytical resources therefore requires a clear definition of the problem in hand.

10.1. Analysis of Coating Materials

10.1.1. Separation of the Coating Material into Individual Components

For a complete analysis it is advantageous to separate the coating material into its components. The binder and pigment (extender) fractions can only be investigated in detail after separation and isolation. The solvent composition can usually be determined by gas chromatographic analysis of the complete coating material. The pigment (extender) fraction is centrifuged off after diluting the coating material with a suitable, volatile solvent (ethyl acetate, tetrahydrofuran, methyl ethyl ketone) [10.2]. Centrifuge speeds of 5000 min^{-1} are generally sufficient, however if finely divided pigments (extenders) or soot are present, a low-density, low-viscosity solvent (e.g., acetone) should be used and the rotational speed should be increased to 20000 min^{-1} [10.4]. To achieve quantitative separation the pigment sediment is repeatedly shaken with solvent and recentrifuged. Prior to analysis the combined binder containing supernatant fractions are dried in a drying cabinet. The centrifuged pigment is also dried in a drying cabinet before being analyzed.

Problems are often encountered if centrifugation is applied to waterborne systems, particularly if the binder is insoluble or only sparingly soluble in organic solvents. In these cases the coating material should be carefully dried (vacuum drying cabinet, freeze drying). A representative binder fraction is then obtained by exhaustive extraction with a suitable solvent (e.g., 1,2-dichlorobenzene, dimethylformamide and/or tetrahydrofuran) [10.5], [10.6]. For some microgel-containing waterborne systems the use of an isopropanol-water-mixture leads to a rather complete separation of pigment and binder, since the microgel remains in the supernatent.

10.1.2. Analysis of Binders

Modern coatings are expected to provide permanent protective action, outstanding mechanical–technological properties, and an attractive surface appearance. This requires binders consisting of special resin combinations (Chap. 2) and selected additives (Chap. 5).

Preliminary tests, color reactions, and spot tests [10.7] were formerly used to identify individual resins but are no longer important because they are not sufficiently specific and they do not provide quantitative results. They have been largely replaced by modern spectroscopic and chromatographic methods, which often require preliminary chemical workup of the sample.

Infrared and nuclear magnetic resonance spectroscopy are the most important spectroscopic methods for analyzing coating materials. Near infrared Fourier transform (NIRFT) Raman spectroscopy [10.8] also has great potential, particularly for aqueous systems. UV/VIS spectroscopy is used only in exceptional cases, e.g., to determine light protection agents (UV absorbers).

Infrared spectroscopy has the advantage of simple sample preparation and measurement; practically all types of samples (both as regards the state of aggregation and solubility) can be investigated with the aid of special measuring techniques. Infrared spectroscopy is frequently employed to obtain an overview of the binders and binder classes. A common procedure is to pour a few drops of the binder-containing supernatant from the separation (centrifugation) onto a NaCl or KBr crystal. A thin binder film is obtained after drying in the drying cabinet and IR absorption is then measured. The IR spectra provide qualitative and semiquantitative information about the binder composition. Comprehensive spectra catalogs of organic polymers [10.9] and commercially available coating materials [10.10] are available for comparison purposes. Modern IR spectrometers, particularly the Fourier transform infrared (FTIR) spectrometers, offer further possibilities. These instruments have short measurement times and high wavelength reproducibility; they can therefore accumulate data and provide a considerably better signal to noise ratio than conventional grating or prism instruments. IR spectroscopy can then be used to analyze extremely small sample amounts. Examples are diffuse reflectance FTIR spectroscopy (DRIFTS), investigations of sample surfaces by measurement with attenuated total reflection (ATR), or the analysis of local defects using an IR microscope [10.11]. A further advantage of FT spectrometers is the fact that the data are available in digital form. Spectroscopic data banks can therefore be compiled either on the spectrometer computer or on an external computer. Commercial data banks (e.g., Sadtler) and in-house, laboratory-specific data collections can be installed, the latter are often better suited to the particular interests of the laboratory. Data banks facilitate archiving, and also allow quick comparisons of measured and library spectra. Quantitative evaluations are also facilitated after appropriate calibration. The theoretical aspects of FTIR spectroscopy and analysis are described in [10.12]. The use of FT and conventional IR spectroscopy to investigate coatings and coating materials is described in [10.13]–[10.15].

Nuclear Magnetic Resonance Spectroscopy. Like IR spectroscopy, NMR spectroscopy requires little sample preparation, and provides extremely detailed information on the composition of many resins. The only limitation is that the sample must be soluble in a deuterated solvent (e.g., deuterated chloroform, tetrahydrofuran, dimethylformamide). Commercial pulse Fourier transform NMR spectrometers with superconducting magnets (field strength 4–14 Tesla) allow routine measurement of high-resolution ^{1}H- and ^{13}C-NMR spectra. Two-dimensional NMR techniques and other multipulse techniques (e.g., distortionless enhancement of polarization transfer, DEPT) can also be used [10.16]. These methods are employed to analyze complicated structures. ^{13}C-NMR spectroscopy is particularly suitable for the qualitative analysis of individual resins in binders, quantiative evaluations are more readily obtained by ^{1}H-NMR spectroscopy. Comprehensive information on NMR measurements and the assignment of the resonance lines are given in the literature, e.g., for branched polyesters [10.17], alkyd resins [10.18], polyacrylates [10.19], polyurethane elastomers [10.20], fatty acids [10.21], cycloaliphatic diisocyanates [10.22], and epoxy resins [10.23].

Chromatography. Liquid chromatography is the most important chromatographic method for the investigation of binders and resins. Special applications are also opening up for the recently developed technique of supercritical fluid chromatography (SFC) [10.24]. Methods of particular importance are size exclusion chromatrography (SEC) (also termed gel permeation chromatography, GPC) and reverse-phase high performance liquid chromatography (HPLC).

Gel permeation chromatography separates molecules according to their size. This technique is used to determine the molecular mass distribution of resins. A review of modern GPC methods for analyzing coating materials is given in [10.25]. Combination of GPC with modern spectroscopic techniques (in particular FTIR spectroscopy) is a valuable aid for identifying individual resins in binder systems; in some cases additives can also be identified, because their molecular mass is often substantially lower than that of the resin constituents. The coupling of liquid chromatography and FTIR spectroscopy is described in [10.26]. Off-line investigations of fractions from an analytical GPC run of a binder by FTIR measurements under diffuse reflection provide detailed information on the composition of the binder [10.27].

Reverse-phase HPLC (mainly octyl- or octadecyl-modified silica gels) separates molecules according to their partition coefficients. It is particularly suitable for characterizing relatively low molecular mass resins (e.g., epoxy, phenolic, and melamine resins).

The high resolution of modern columns and the high sensitivity of UV detection allow individual oligomers (including positional isomers and secondary compounds) to be separated and identified. The peak distribution pattern often permits identification of commercial products. Detailed HPLC investigations have been carried out on epoxy resins [10.28]. HPLC can also be applied to resins without UV-active groups by using mass detectors [10.29].

On-line coupling of pyrolysis, gas chromatography, and mass spectrometry is a quick and elegant method for the qualitative detection of monomer units in many resins (e.g., polyesters, polyurethanes, phenolic resins, and polyacrylates). Identification of comonomers of polyacrylates, including hydroxy-functional and carboxy-functional monomers, is facilitated if the sample is silylated before pyrolysis [10.30].

Chemical Workup. Chemical decomposition of resins followed by qualitative and quantitative analysis is still an important technique because, apart from NMR spectroscopy, none of the instrumental analytical methods provides reliable quantitative values. Chemical workup is also essential for low concentrations of resin building blocks that are often unknown; it simplifies and enables the desired substances to be concentrated. Qualitative and quantitative determination is carried out by instrumental methods.

Well-established methods of chemical workup are available for alkyd resins based on *o*-phthalic acid. Alkaline hydrolysis can be used for the quantitative determination of the dicarboxylic acid, fatty acid, and polyol fractions [10.31]. In the IUPAC method the carboxylic acids are determined by gas chromatography after transesterification with lithium methoxide; the polyols are determined by gas chromatography after aminolysis and acetylation [10.32]. Chemical workup methods (e.g., hydrolysis with alcoholic alkali, alkali fusion, aminolysis with hydrazine, and transesterification with sodium methoxide) for various resins are described in [10.33]. Alkaline hydrolysis has the disadvantage that it results in low fractions for polyunsaturated fatty acids and lower polyols. Transesterification with methanol or trimethylsulfonium hydroxide provides substantially better results for unsaturated fatty acids. This transesterification process was originally proposed for determining fatty acids in triglycerides [10.34] but can also be applied to alkyd resins and polyesters. After evaporation of methanol and silylation, the polyols from the reaction mixtures can then be determined by gas chromatography. This procedure yields more accurate values than with the previously mentioned methods, particularly for the lower polyols. In addition the amount of fatty acids and dicarboxylic acids can be determined with the help of gas chromatography by direct injection of the reaction mixture.

10.1.3. Analysis of Pigments and Extenders

The separated pigment (extender) is usually used as starting material for analysis (see Section 10.1.1). If quantitative separation is not possible (e.g., in emulsion paints) the inorganic pigment (extender) fraction can be obtained by ashing the nonvolatile fraction; for further details see [10.5].

The isolated pigment (extender) fraction is analyzed by various chemical and instrumental methods. Methods of elemental analysis are used for inorganic pigments (extenders). These include traditional chemical methods involving separation and gravimetric, tritrimetric, or polarographic determination of the elements. These methods are being replaced by instrumental methods such as atomic absorption spectroscopy (AAS), atomic emission spectroscopy (AES), and X-ray fluorescence analysis (XFA). A further valuable tool is IR spectroscopy, which provides characteristic spectra for many inorganic extenders and pigments (e.g., chalk, dolomite, kaolin, talc, and barium sulfate). The most elegant and informative method, but the most expensive as regards equipment, is X-ray diffraction [10.35], [10.36]. Sample preparation is simple and the method can be used on hardened coating materials; problems associated with isolating the extender–pigment fraction can therefore be avoided. The principal advantage, however, is that the substances (e.g., $BaSO_4$) can be identified directly and not indirectly via their elements (e.g., Ba and S). This is particularly advantageous if both silica and silicates (kaolin, talc) are present. Minerals such as dolomite, chalk, talc, and kaolin originating from different geographical locations have specific elemental compositions. Their spectra can therefore be used to identify the source of supply when investigating coating materials [10.37].

For many inorganic and organic substances, X-ray diffraction spectra recorded on powdered materials are commercially available on a data carrier (CD-ROM) [10.38].

The pigmentation of monochrome coating materials usually comprises several constituents: organic and/or inorganic pigments, titanium dioxide to improve the hiding power, and other inorganic extenders. The inorganic fraction is generally dissolved with acid and can then be analyzed by conventional chemical methods. For the investigation by atomic absorption or atomic emission spectroscopy a borax fusion is recommended as preparatory step. The organic pigments remain undissolved and are investigated by IR spectroscopy, photometry, X-ray diffraction, and separation methods such as thin layer chromatography (TLC) or HPLC. Modern FTIR spectroscopic methods including library searches; further developments in TLC and HPLC are discussed in [10.39]. A separation process for organic pigments has been developed which exploits solubility differences in various solvents (ranging from n-hexane to sulfuric acid) [10.40]. The extracted pigments are identified by their VIS spectra in the range 400–900 nm. The advantages and limitations of X-ray diffractometry in the analysis of organic pigments are discussed in [10.41].

The increasing use of special-effect paints, particularly in automobile finishes, places new requirements on analytical techniques. The previously described methods are largely unsuitable for identifying substances used to produce special effects. Aluminum pigments in metallic paints can be roughly classified by light microscopy. Since these pigments are often subjected to special pretreatment, the determination of foreign elements with X-ray microprobes or the determination of an organic agent used for surface treatment may be necessary for more accurate characterization. Light microscopy can also be used for initial characterization of nacreous pigments, more detailed information can be obtained with a microscope spectral photometer (e.g., UMSP 80, Zeiss) or transmission electron microscopy [10.42].

10.1.4. Analysis of Solvents

The solvent composition of coating materials can be determined in two ways. In the first method, the coating material is subjected directly to gas chromatography, if necessary after dilution with a suitable solvent. In the second method, the solvent fraction is separated from the coating material by vacuum distillation and then analyzed by gas chromatography or, in special cases, by IR spectroscopy. ASTM methods (D 3271 and D 3272) exist for both processes. The advantages and disadvantages of the two methods in the analysis of solventborne and waterborne systems are discussed in [10.2].

Direct gas chromatographic analysis with modern capillary column technology and reproducibly operating injection systems (e.g., modern autosamplers) gives highly reproducible, accurate results. This method is therefore used most widely. With complex solvent mixtures, unambiguous identification of the peaks is facilitated by simultaneously analyzing the sample on two columns with different polarities or by using GC–MS. For further information on gas chromatography, see [10.43].

10.1.5. Analysis of Additives

Qualitative and, in particular, quantitative analysis of coating additives is difficult due to their low concentrations and chemical diversity. A general outline for their investigation cannot be given, isolation and analysis depend on chemical structure. The liquid or hardened coating material is often extracted with a suitable solvent, followed by spectroscopic or chromatographic analysis. For example, plasticizers can be extracted very efficiently with pentane and detected by IR spectroscopy or gas chromatography. Supercritical fluid chromatography (SFC) is being increasingly used for the analysis of polymer additives [10.44]. Especially in combination with supercritical fluid extraction (SFE) as a relatively fast and efficient sample preparation method [10.45], it may prove to be of interest for the analysis of many additives used in coating materials.

10.2. Analysis of Coatings

Coatings are practically free of solvents and the binder is generally cross-linked (i.e., insoluble). These factors require special sample preparation and analytical methods, for a detailed discussion see [10.46]. The separation of binder and pigment (extender) fractions for further investigation is only possible with non-cross-linked (physically drying) binders. The coating film is soaked in a suitable solvent and the pigment is centrifuged off after dissolving the binder. Provided sufficient material is available, the isolated components can be analyzed by the methods described in Section 10.1. The above method cannot be used with cross-linked binders (e.g., two-pack systems, stoving finishes). The binder and pigment (extender) fractions in coating layers are most simply analyzed by means of IR spectroscopy. Different measurement techniques are available for this purpose, which require various degrees of preparative effort. The most important techniques are measurements on KBr pellets prepared from scratched off paint material, measurement of the coated surface with the method of attenuated total reflection (ATR), and measurements on cross sections of the coating with FTIR microspectroscopy [10.47].

Further methods for determining the binder structure of cross-linked systems include the use of pyrolysis gas chromatography or alkaline hydrolysis followed by analysis of the degradation products by gas chromatography. In multilayer coatings this may prove difficult because the materials have to be prepared from individual layers.

The inorganic pigment (extender) fraction can be obtained by ashing the isolated paint material and analyzing it by the methods described in Section 10.1.3.

The question of the cause of coating defects often arises in the investigation of coatings. Defects may be found at localized sites (specks, craters) and impair the appearance of the coating. They may also occur as planar defects that for example reduce adhesion to the substrate or between the individual layers. Defects are investigated with light microscopy, FTIR microspectroscopy, X-ray methods [10.48], and modern surface analysis methods such as time-of-flight secondary ion mass spectrometry (TOF-SIMS) [10.49], laser-microprobe mass analysis (LAMMA) [10.50], and X-ray photoelectron spectroscopy (XPS) [10.51]. Among the methods mentioned TOF-SIMS proved to be the method of choice for the analysis of craters and other defects caused by surface-active substances [10.52].

11. Uses

The application of paints to various substrates (e.g., metals, wood, plastics, and concrete) is the most widely used method of protecting materials against corrosion and degradation. It is also used to obtain properties that include gloss, color, completely smooth or textured surfaces, abrasion resistance, mar resistance, chemical resistance, and weather resistance. Normally, a combination of properties is required. Paint systems are therefore applied that generally consist of a primer, an intermediate coat, and a topcoat. These coats of paint together with the substrate surface and surface layers resulting from substrate preparation and pretreatment form the coating system. Only this complete coating system can provide the combination of properties required for the wide range of uses of organic coatings.

Most paints are supplied as liquids that are applied by different methods, using various types of equipment (see Chap. 8). The properties and uses of powder coatings are described in Section 3.4. Once applied, the wet paint film must dry to a hard solid film. Paints that dry at ambient temperature (air-drying paints) may be force-dried at temperatures up to 100 °C. Other types of paints require higher temperatures (120–220 °C) for film formation that involves reaction of two and more binder components. Thus, the paint formulator has to consider both the properties of the liquid film and those of the final dry film. Liquid film properties have to be considered during storage, application, and curing.

To obtain a properly formulated paint, testing has to be carried out in different stages (Chaps. 9 and 10): weathering and corrosion tests, application tests, field trials that test in-use behavior, and durability. The paints can only be used commercially when they have passed these tests.

11.1. Coating Systems for Corrosion Protection of Large Steel Constructions (Heavy-Duty Coatings) [11.1], [11.2]

Large steel constructions have vast metal surfaces which must be protected against corrosion to maintain their proper function. Such constructions include road and railroad bridges, electric pylon lines, radio and radar antennae, gas tanks, storage tanks (e.g., for oils, chemicals, cement, and grain), loading equipment (e.g., cranes, conveyors), mining and drilling constructions, as well as steelworks and chemical plants. Although these constructions are sometimes protected by inorganic coatings, about 90% are coated with paints based on organic binders.

Appropriate surface preparation is of utmost importance for a long service life. Best results are achieved by blasting, using grade 2 ½ or 3 according to the Swedish Standard SIS 055900 (equivalent to DIN 55928, part 4, FRG; BS 7079, part A 1, 1989, UK; SSPC-SP 5-SP 6, ASTM, USA; ISO 8501–8503). Although mill scale and old paint are sometimes removed by flame descaling, this method is less effective. Residues of rust and corrosion-promoting chemicals (e.g., salts) on the prepared surface lead to early rusting under the new paint and must therefore be removed completely.

In galvanized steel constructions, blasting is also necessary to provide a coatable surface. The sweep-blasting method is used in which the zinc surface is roughened without removing a significant amount of the zinc layer. When constructions made of aluminum alloys are coated, their surfaces must be blasted with iron-free blasting materials.

Heavy-duty coating systems generally comprise two primer coats and two topcoats, modern system sometimes consist of one primer coat and two topcoats. The total dry film thickness (DFT) of such anticorrosive paint systems is 150–200 μm, each layer has a minimum DFT of 40–50 μm. Due to their excellent adhesion, the first and second primers prevent corrosion of the metal surface. The pigments and extenders allow the primers to react with ions (Cl^- and SO_4^{2-}) that diffuse into the film from the atmosphere. The pigmented organic film also forms a barrier against humidity that may otherwise initiate a corrosive process.

Heavy-metal pigments (mainly lead pigments) and zinc chromates were used succesfully in earlier decades. These pigments are now being replaced by nontoxic pigments (see Section 11.3.1.).

The first and second topcoats build up the necessary dry film thickness and protect the entire coated construction against the adverse influence of the atmosphere.

Binders based on linseed oil and other oils have been used for many years in anticorrosive primers. Alkyd binders, especially those with high fatty acid contents, perform similarly. The main disadvantages of these binders is their limited chemical resistance and their slow drying.

Chlorinated rubber and poly(vinyl chloride) (PVC) resins allow the formulation of coatings with good chemical resistance. They are therefore used for steel constructions in chemical plants. Since they are not resistant to many organic solvents, they should not be used in oil refineries or plants handling solvents. The undesirable fact that these binders contain halogens in high amounts is responsible for their decreasing use. Overspray of chlorinated rubber and PVC paints and contaminated blasting materials produced after removing old paint cause severe problems in waste incineration plants (generation of hydrochloric acid), as well as in waste disposal areas (pollution of soil and water).

Epoxy resins cured with aminoamide resins or amine adducts are often used for large metal constructions. Paints based on these resins are normally applied in four layers. Epoxy coatings form films that are resistant to organic solvents and a wide range of chemicals. Epoxy coatings are currently used for the majority of steel and aluminum constructions, but are also suitable for use on other construction materials (e.g., concrete). They can protect buildings in chemical plants and nuclear power plants. Epoxy coatings are less susceptible to deterioration by radiation than other organic films, and are also resistant to decontaminating chemicals (usually aqueous-detergent solutions) used to remove radioactive dust from walls and other surfaces in nuclear power plants.

Heat-resistant coatings have silicone-resin binders. Pigments for such paints are zinc dust, flakes of aluminum or stainless steel, titanium dioxide, or silicon carbide. Such paints can withstand temperatures up to 600 °C.

Paints with inorganic binders are also used for corrosion protection of steel constructions. These paints are based on organic silicates which are soluble in mixtures of alcohols or other water-miscible solvents (see Section 2.15.2). Ethyl silicate is often used and mostly pigmented with zinc dust. Zinc-rich primers and single coats are available as one- or two-pack products. Zinc-rich ethyl silicate paints dry to form inorganic films that are very durable even under adverse atmospheric conditions, (e.g., onshore and at sea). These coatings have excellent resistance to oil, solvents, and mechanical impact, and are therefore used on drilling stations, oil rigs, and ships. Since zinc-rich silicate coatings are heat resistant, they are also used in hot areas of iron works, coal mines, and coking plants.

Heavy-duty coatings are often still applied manually with brushes or rollers that completely wet the metal surface; holes and pores are filled with paint. This is especially important when old, partially rusted constructions are repainted after sanding. Brushing and rolling, however, only allow a slow working speed. Larger surface areas must be painted with airless spraying equipment.

11.2. Automotive Paints

11.2.1. Car Body Paints

Cars are coated to achieve maximum, long-lasting corrosion resistance. Cars must also be given an optimum appearance that lasts for many years. Long-lasting color and gloss retention as well as resistance against cracking (especially in clearcoats of two-coat metallics) are therefore necessary. Topcoats of automobiles must withstand solar radiation and atmospheric pollution (e.g., acid rain and soot from oil combustion). Aggressive chemicals (e.g., road salts and cleaning agents containing detergents) can damage the coating if they come into contact with the car surface. Furthermore, small stones cause heavy impact on automobile surfaces and corrosion via chipping.

Large numbers of cars are manufactured on fast-running assembly lines. The paints must therefore be applied with highly efficient equipment, and must dry very quickly. The paint products are classified as primers, intermediate coats (also called fillers or surfacers), and topcoats (or finish). The primers and fillers are designated as the undercoating system.

Car paints are cured with heat in special oven lines. Electrodeposition coatings (used as anticorrosive primers) contain only small amounts of volatile organic compounds (VOC), whereas intermediate and topcoats release considerable amounts of VOCs. Intermediate coats based on waterborne resins have been developed to decrease VOC emission and are already being used in some automotive plants. Basecoats, as part of base–clear topcoat systems, contain very high amounts of volatile organic solvents. Waterborne basecoats were developed more recently to lower this source of solvent emission. Some car manufacturers are operating pilot lines with the aim of introducing waterborne basecoats into their production processes. Many car producers in the United States and Europe have already switched their topcoat lines over to waterborne basecoats [11.3].

Pretreatment. Various metals are used for manufacturing car body shells: steel, galvanized steel, aluminum alloys, and zinc-rich precoated steel. The surfaces of these metals are routinely contaminated with oils, drawing lubricants, dirt, and assembly residues (e.g., welding fumes). The body shells are pretreated to remove these contaminants and to obtain a well-defined, homogeneous surface that has the necessary properties for adhesion of primers. Pretreatment includes surface cleaning and formation of a phosphate conversion coat on the shell surface (see Section 8.2.1); six to nine discrete steps are involved using either spraying devices or baths. Continuous control of phosphating solutions ensures good results [11.1], [11.4].

Anticorrosive Primers. Anticorrosive primers are applied in dip tanks so that they reach all parts of the car body; dipping is a fast method of application. The standard method for application of primers is electrodeposition. Anodic electrodeposition paints were used when the electrocoating technique was first applied, but cathodic electrodeposition is now predominant because it provides better corrosion protection.

The binders for cathodic electrodeposition are epoxy resin combinations dispersed in water (see Section 3.8). Advantages of anticorrosive electrocoatings include excellent corrosion resistance at a dry film thickness of ca. 20–30 µm. Electrocoats are stoved at 165–185 °C to obtain films with the desired properties. The paint industry is now developing electrocoats that can be cured at lower temperatures (140–150 °C). Electrocoating produces a homogeneous film that covers the entire car body surface, including recesses and cavities.

Although the dry film thickness on the metal edges is somewhat lower, these areas are still efficiently protected against corrosion. The ultrafiltration technique results in a very high transfer effect and a uniform coating: paint solids from the bath are deposited on the metal surface without loss. Since electrodeposition paints have a low organic solvent content, air pollution is low. The dip tank contents are not flammable, which reduces insurance costs [11.5].

Intermediate Coats. Intermediate coats (fillers) are applied between the anticorrosive primers and the topcoat systems. They provide good filling and flowing layers which are normally smoothed by sanding. Oil-free polyesters are used as binders for fillers. They react with blocked isocyanates in 20 min at 165 °C. Their high flexibility gives the whole coating system a highly effective mechanical (stone chip) resistance.

Fillers are applied with electrostatic spraying devices (fast-rotating bells) to give dry film thicknesses of about 40 µm. Waterborne fillers with polyester–melamine binders (primer surfacers) have been developed to reduce the volatile organic content. They yield a film thickness of 30 µm after a prereaction time of 10 min at 100 °C and a reaction time of 20 min at 165 °C. The properties of the films are similar to those formed by solventborne paints. More recently, waterborne fillers based on blocked isocyanates have been developed. Field trials have shown that their mechanical resistance is very good.

Topcoat Systems. Topcoats form an important part of the protection system of the car body surface, but are much more important for decoration. The basic requirements for a car topcoat are:

1) Full, deep gloss (wet-look)
2) Highly brilliant metallic effects
3) Long-lasting resistance against weather and chemical influences
4) Easy to polish and repair

Topcoats based on nitrocellulose combinations with plasticizers and alkyd resins were used in the first decades of industrial car manufacturing. These were followed by thermosetting alkyd–melamine combinations, and later by thermosetting acrylics. The use of stoving enamels as thermosetting paints also accelerated production significantly. Although the properties of these coatings during application and in use were very good, their high content of volatile organic solvents had to be lowered to comply with legal restrictions.

The basecoat–clearcoat system is presently the most commonly used type of topcoat for cars because it is the standard application system for metallic colors. Today, about 70 % of all cars have metallic topcoats. The basecoat–clearcoat system consists of a colored layer (basecoat) which is overcoated after a short flash-off time with a protective layer of clearcoat. Both coats are cured together at 120–140 °C. The basecoat contains pigments which provide two types of finish: solid (straight) colors or metallic.

Solventborne metallic basecoats contain 15 to 30 % solids and 85 to 70 % volatile organic solvents. These solvents are not released into the atmosphere, but are converted to combustion gases in afterburners. To reduce emission of organic solvents from this source, waterborne basecoats have been developed.

Waterborne basecoats with higher solids contents are now available: metallic basecoats contain about 18 wt % solids and solid (straight) color basecoats 25–40 wt %. The solvent in waterborne paints is not pure water; about 15 % of organic solvents is still needed as a cosolvent for proper film formation. Metallic basecoats are applied at a DFT of 15 µm, solid color basecoats at a DFT of 20–25 µm.

Basecoats are sprayed in two layers. The first layer is sprayed electrostatically with high-speed rotation bells, the second layer is sprayed with compressed air to achieve proper orientation of the aluminum particles in metallic paints. The basecoat is then dried for 3–5 min in a warm air zone at 40–60 °C.

A final layer of clearcoat is applied with electrostatic high-speed rotation bells [11.3], [11.7] to protect the system against atmospheric influences, including wear and tear during use.

Alkyd–melamine clearcoats with an approximate solids content of 50% contain UV-absorbing agents to prevent deterioration in extreme climates.

Some car manufacturers use clearcoats with acrylic binders that are cured with aliphatic isocyanates. Their chemical and mechanical properties are better than those of alkyd–melamine clearcoats. Solid contents are as high as 58%.

Car Repair Paints [11.1]. Repair paints are used in considerable amounts for refinishing cars. Since repair shops cannot provide the same facilities as those of car manufacturers, repair paints are dried at ambient temperature or elevated temperature up to 80 °C (metal temperature). Alkyd repair paints and nitrocellulose paints were standard materials, but two-pack acrylate–isocyanate refinish paints are now more common. Their properties are similar to those of the original car coatings (long-lasting gloss and color, mechanical and fuel resistance). Car refinish paints are available in a wide range of colors, solids as well as metallics. They are often supplied to shops and retailers as mixing schemes.

Paint systems for car repair comprise anticorrosive primers, putties, intermediate coats, and topcoats; repair coatings applied to refinished cars have similar durabilities to those of the originally manufactured coating systems.

11.2.2. Other Automotive Coatings

The properties of coating systems used for car components differ considerably from those of systems used for exterior car surfaces. Color is not important (and is mainly black or gray), but anticorrosive properties similar to those of car body coatings are required. Since car components are produced in large numbers, coatings are commonly baked at high temperature to ensure a high reaction rate and rapid film formation.

Wheels are electrocoated; engine blocks are coated with heat-resistant, usually waterborne materials. Other parts (e.g., steering equipment and shock absorbers) are painted with two-pack, one-coat epoxy systems that are usually solventborne; use of waterborne systems is, however, increasing.

11.3. Paints Used for Commercial Transport Vehicles

11.3.1. Railroad Rolling Stock

Large numbers of railroad electric and diesel engines, passenger cars, and freight cars are in service. They are designed to function for at least 30 years with minimum maintenance of their coatings.

Coating systems for railroad rolling stock are used under severe working conditions. Engines and passenger cars are run at high speeds in many types of climates. They are frequently cleaned with strong chemical solutions to remove the heavy dirt adhering to the coating. Freight cars are also exposed to additional stress as a result of impact from mechanical handling during loading and unloading. Many transported goods (chemicals, fuels) attack the coating of freight cars. Coating systems for railroad rolling stock must be highly resistant in all respects. This is achieved by using two-pack coating systems (mainly epoxy but also polyurethane) and, more recently, coatings of acrylic resins which are applied as one-pack, waterborne dispersions.

Before applying the coatings, surfaces are pretreated by blasting. Longlife coatings require surface pretreatment according to Swedish Standard SIS 055900, grade 2½ (equivalent to DIN 55928, part 4, FRG; BS 7079, part A1, 1989, UK; SSPC-SP 5-SP 6, ASTM, USA; ISO 8501–8503).

Engines and Passenger Cars. The exteriors of engines, passenger cars, and similar vehicles (subway coaches) are normally painted in a three-coat system. An anticorrosive, two-pack epoxy primer is applied at a DFT of 80 μm. The curing agent is either an amine adduct or an aminoamide resin. Zinc chromate was used for many years as an effective anticorrosive pigment but has now been replaced by chromate-free pigments (e.g., zinc phosphates, barium metaborate, calcium borosilicates, and zinc phosphomolybdates) to avoid the risk of carcinogenicity.

Anticorrosive primers are followed by a 40–50 μm intermediate coat, based on two-pack polyurethane resins. This coat is sanded to yield a smooth surface.

Sometimes, the intermediate coat is overcoated with the topcoat in a wet-in-wet system. Normally, however, two-pack polyurethane topcoats are applied to the dried filler. These topcoats with aliphatic isocyanate hardeners have excellent gloss and color retention. The DFT of topcoats is 40–80 μm, depending on the hiding power of the topcoat.

Red, orange, and yellow pigments used in topcoats are now free of toxic lead and heavy metals to protect workers and the environment. Railroad companies expect a service life of 15 years before the topcoat has to be replaced (provided that spot repair is carried out when necessary). The whole coating system should not need to be renewed before a minimum service life of 30 years.

Although two-pack epoxy primers and polyurethane intermediate coats have high solids contents, they still contain significant amounts (20–30 wt%) of organic solvents. In polyurethane topcoats, the VOC is even higher. Anticorrosive, waterborne primers based on aqueous dispersions of two-pack epoxy resins and one-pack acrylic resins have been developed to decrease solvent emission. Waterborne, one-pack acrylic topcoats are also used. All of these waterborne paints contain 2–5% organic cosolvents that are required for film formation.

Field trials have shown that the expected durability of the new waterborne coating systems is equal to that of conventional solventborne paints. As a further advantage, the number of single coats can be reduced from three to two. Standard freight cars are painted with a three-coat alkyd system consisting of an anticorrosive primer, an intermediate coat, and a topcoat, total DFT is ca. 150 µm. Such systems are being replaced by single-coat systems of waterborne acrylic resins and acrylic copolymers to reduce solvent emission.

Freight Cars. Special freight cars transport dry and liquid goods that can contain aggressive chemicals which would destroy alkyd coatings. An *exterior coating system* based on two-pack epoxy resins is therefore used. It consists of the same anticorrosive primer employed for engines and passenger cars. This two-pack epoxy primer is applied to blasted steel, sometimes also stainless steel. The next layer is a two-pack intermediate epoxy coating with chemically inert pigmentation, DFT 40–60 µm.

Topcoats are also based on epoxy resins, their DFT is 40–50 µm. Chemical resistance requires pigments that are not affected by the transported goods; the available color range is therefore limited.

Although epoxy topcoats show chalking and loss of gloss after a short period of outdoor use, this does not affect their chemical resistance. Chalking does not cause a significant reduction of film thickness.

When goods such as salts or fertilizers are transported, the interior coatings must be chemically resistant. Chemical resistance is obtained by using the same system that is used for exterior coating. In the case of food transport (e.g., flour, sugar, or grain) the coating must also comply with relevant legal regulations; thresholds are stipulated to limit migration of coating ingredients into the transported goods. The same resin system as for the exterior coating may be used, but there are strong limitations for the use of pigments, plasticizers, and additives.

In the case of abrasive freight goods, the interior of the car must be lined with a thick coat system consisting of solvent-free two-pack polyurethane or epoxy material. Toughness combined with flexibility results in a film that is highly resistant to abrasion.

Application. The standard application method for most paints used on railroad vehicles is airless spraying. A combination of airless and compressed air spraying has been recently introduced, it is mainly used for applying waterborne topcoats.

11.3.2. Freight Containers

Hundreds of thousands of freight containers have been built worldwide since the early 1960s. They are exposed to many types of climate and are heavily stressed by wear and tear. Freight containers consist of steel or aluminum alloy frames, clad with sidewalls and roofs made of steel, stainless steel, or aluminum alloys. The walls and roofs may be coil-coated. Container frames are usually blasted to grade 2 ½ of the Swedish Standard SIS 055900.

Prime coating is either a two-pack, zinc-rich epoxy primer with a DFT of ca. 30–40 µm, or a chromate-free, two-pack epoxy primer. Topcoats based on PVC copolymerisates, chlorinated rubber, silicone-reinforced epoxy esters, or alkyds were used for many years. Now, the standard European topcoat is based on acrylic resins with terminal epoxy groups that are cross-linked with amine adducts or aminoamide resins.

11.3.3. Road Transport Vehicles

Coatings on road transport vehicles are expected to have a minimum service life of ten years. Paint systems must therefore be of high quality.

Trucks. Drivers' cabins are coated in the same way as cars (see Section 11.2.1). Truck chassis are especially prone to corrosion caused by stone impact and road salts. Chassis parts are supplied by special manufacturers who normally provide an anticorrosive prime coating.

Side and cross members are pretreated by blasting and phosphating. An electrodeposited primer coating follows and is sometimes immediately overcoated with a topcoat based on an air-drying alkyd or epoxy ester, or on an oven-drying alkyd–melamine resin combination. After assembly of the truck, a third layer (i.e., a second topcoat) is applied that serves as a supply finish. These finishes are mostly two-pack acrylic–aliphatic isocyanate topcoats. Water-based two-pack systems and powder coatings are the most recent developments.

Trailers. Trailer chassis are generally coated in the same way as truck chassis but with a two-coat system. The primer is either a one-pack alkyd paint or a two-pack epoxy paint that are cured at ambient temperature with amine adducts or isocyanates.

Truck bodies are often constructed of steel frames with aluminum side walls. The aluminum alloys are pretreated by application of 5–8 µm of a chromate-containing wash primer. The topcoat is a two-pack acrylic resin bearing hydroxyl groups and is cross-linked with aliphatic isocyanates.

Buses have a profiled steel framework that is clad on the sidewalls and roof with steel, galvanized steel, and aluminum alloys. The anticorrosive primer must therefore adhere well to all of these metals, and to glass-fiber-reinforced plastics.

Two-pack epoxy primers with chromate-free pigments show the best performance. They are applied with a DFT of ca. 80 µm, and simultaneously form the anticorrosive primer and a sandable filler. They can protect the metal substrates against corrosion.

The metals are joined by spot welds and rivets that may cause electric currents in the joining areas. Specially designed epoxy primers prevent corrosion in these areas. At ambient temperature epoxy primer fillers dry thoroughly in about 16 h. Since this is too long for assembly lines, they are normally cured in ovens (80 °C metal temperature).

Bus topcoats are mainly two-pack acrylic–isocyanate paints which dry very rapidly at ambient temperature or under low baking conditions. This is important because many buses, especially luxury coaches, are painted in multistripe color designs. The resulting coatings have excellent color and gloss retention under service conditions. This is not only important for long-lasting corrosion protection, but also for the image desired by the owners.

11.3.4. Aircraft Coatings

Aircraft coating systems are of an extremely high quality. Corrosion must be prevented to guarantee safe functioning of the aircraft. Aircraft manufacturers issue very comprehensive specifications to ensure the required coating properties. This demands extensive development work and careful paint formulation, because the coating properties must be obtained with a minimum dry film thickness to minimize weight. Aircraft coating systems must also be resistant to hot hydraulic fluids.

The aluminum alloys used for aircraft construction are pretreated by anodizing, chromating, application of wash primer (6–8 µm), or pickling.

Anticorrosive primer formulations are based on epoxy resins that react at ambient temperature with amine adducts, aminoamide resins, or isocyanates (two-pack primers). Zinc or strontium chromates are used as anticorrosive pigments. These electrolytically active pigments prevent dangerous filiform corrosion. Extensive development work is being carried out on chromate-free aircraft primers [11.8], [11.9].

Aircraft coatings normally consist of two coats. The topcoat used for civil aircraft is a high-gloss two-pack polyurethane product, cured with aliphatic isocyanates. Single-coat systems based on thermoplastic acrylic resins have been used for military aircraft, but are being replaced by polyurethane topcoats [11.10].

11.4. Marine Coatings [11.11]–[11.13]

The requirements of any coating system are determined by its in-use environment (service, construction, or maintenance). The sea is wet, salty, usually oxygenated, sometimes anaerobic, and full of living organisms which may colonize the surface of a vessel or platform [11.11].

During construction and maintenance cycles, coatings have to adapt to the environment without disrupting time schedules and productivity. They must also conform to increasingly stringent considerations of occupational safety and health, and to environmental constraints. While construction to the "block-stage" (stage at which the prefabricated block is transferred to the building dock) frequently takes place under cover, final assembly and painting take place under ambient conditions (-10 to $+40\,°C$, and 40–100% relative humidity). Marine paints must be able to tolerate this range of climate as well as application methods and skills. For maintenance, the advent of larger vessels, smaller and more specialized crews, and increased vessel activity, means fewer or no opportunities for in-service maintenance. Improved plant reliability means longer intervals between drydocking.

The requirements of marine coatings may therefore be considered under the following headings:

1) *Substrate*: usually hot-rolled mild steel, high-tensile steel, or aluminum
2) *Surface preparation*: usually centrifugal or pneumatic blasting with recycled shot or expendable grit blasting media
3) *Priming*: the dominant consideration is the effect on the efficiency and quality of new construction
4) *Design requirements*: examples are anticorrosive, antifouling, chemical resistance, impact and abrasion resistance, cosmetic qualities, friction and camouflage at various acoustic and electromagnetic wavelengths

11.4.1. Substrate, Surface Preparation, and Priming

Substrate. In marine construction the substrate is generally hot-rolled mild steel, with some high-tensile steel in highly stressed areas (apertures in container ships), and aluminum where there are weight penalties (e.g., topsides of passenger liners, ferries, and naval vessels).

Surface Preparation. Steel hot rolled at 800–900 °C acquires a tenacious oxide layer (mill scale) that is cathodic with respect to the steel to the extent of about 300 mV. In the presence of an electrolyte (seawater containing 3.5% salts, mainly sodium chloride) the steel would corrode and pit and roughen severely. The first process in new construction and refurbishment is therefore the complete removal of mill scale. A small amount of very light-gauge steel is prepared by acid pickling, but most steel for ship and off-shore construction is centrifugally or pneumatically blasted with steel shot that can be recycled or with expendable abrasive grit. Freedom from scale and soluble salt contamination are the main requirements, texture and profile are less important [11.13].

Priming. In new construction the freshly blasted steel surface is highly reactive and requires corrosion protection during construction. It immediately passes through an automatic shop or blast priming plant. The design criteria for shop primers are set principally by the ship-builder, with the consequence that their "fitness for purpose" is compromised. The main criteria are:

1) *Speed of Drying.* The primer is applied to the substrate on a horizontal moving conveyor; the plate must dry quickly to allow handling and stacking within a few minutes.
2) *Speed and Quality of Welding.* The primed plates are cut and welded by a number of techniques (plasma, submerged arc, inert gas, manual metal arc, laser). The primers must not interfere with the speed and quality of welding. The principal defects of the previous generation of materials was their high organic content, which yielded porous welds.
3) *Occupational Safety and Health.* Emission of toxic vapors must be minimized; hence, shop primers that form inorganic films (zinc silicate) are now mainly used.
4) *Lifetime.* Ship construction times have been dropping in recent years and much prefabrication is done undercover, so that anticorrosive protection (weathering) during construction may only be required for 3–9 months.
5) *Burning and Weld Spatter.* Limiting degradation of the primer in the heated zone and nonadhesion of weld spatter are important considerations for subsequent coating operations.

The above constraints mean that only a limited range of coating materials is suitable for marine construction:

1) Wash or etch primers based on poly(vinyl butyral) and phenolic resins pigmented with zinc tetroxychromate
2) Epoxy primers cured with amine adducts or polyamides, pigmented with zinc, zinc chromate and potassium chromate, strontium chromate, or, more recently zinc phosphate and calcium phosphate
3) Epoxy primers cured with amine adducts, containing inert pigments (usually iron oxide)
4) Epoxy primers, zinc-rich, cured with polyamines or polyamides
5) Epoxy primers cured with amine adducts, containing leafing aluminum flake pigmentation

These organic primers have, however, largely been discarded in favor of the "organic zinc silicate" shop primers (see also Section 2.15.2). These are low-zinc versions of zinc silicate anticorrosive primers. Low zinc concentrations are used to minimize volatilization of zinc metal and formation of zinc oxide aerosols at welding temperatures.

Dilution of the zinc with other pigments (extenders) requires careful compromise: the protective qualities of a conductive zinc film with continuous contact should be retained without increasing the emission of toxic vapors. Diluent pigments are therefore chosen for their heat resistance and contribution to the impermeability of the films; they include titanium dioxide, iron(III) oxide, talc, mica, and china clay.

Limitations in performance are several. Zinc is volatilized at weld lines and cut edges, reducing anticorrosive protection. In areas of marine and industrial pollution, thorough washing and/or grit sweeping is necessary to remove water-soluble corrosion products of zinc (sulfate, chloride, and ammonium salts) before overcoating. Performance is also reduced if the complete paint system requires chemical or impact and abrasion resistance. In such cases reblasting is necessary (particularly in chemical tanks, but also in ballast spaces and on the external hull) to completely remove the shop primer. Grit sweeping may also be used but is not as efficient. Maximum primer thickness must be strictly controlled to about 25 μm.

11.4.2. Ship Paint Systems

The painting areas of a ship are shown in Figure 11.1. The main requirement of marine coating systems is corrosion prevention. Detailed requirements vary with the particular internal or external area (e.g., chemical resistance in cargo tanks; resistance to seawater in ballast spaces; heat resistance in engine rooms; impact and abrasion resistance on boottoppings, external hulls, and decks; cosmetic qualities on superstructure and topsides).

Although still referred to as anticorrosives, marine coatings are technically barrier coatings that depend on ohmic resistance for their protective qualities. The ohmic resistance is maintained by impermeability to water and hydroxyl ions. The use of toxic anticorrosive pigments such as red lead, white lead, lead silicochromate, and zinc chromate has declined for safety and health reasons. Zinc (or calcium) phosphate is one of the few anticorrosive pigments still in common use. There has also been a move away from medium oil alkyds and tung oil/phenolic drying oils for primers and undercoats, and long oil alkyds for finishing. They are convenient maintenance materials, but reduced maintenance requirements have meant a move towards nonconvertible (chemically drying) epoxy and polyurethane coatings, and a decline in the use of convertible (physically drying) acrylic, vinyl, or chlorinated rubber coatings for topsides and boottoppings. Retention of color and gloss have been improved with rust-hiding, antistaining additives (e.g., calcium etidronate or other iron(III) sequestering agents).

For similar reasons in the boottopping area and on decks, where the most severe mechanical and corrosive problems occur, traditional drying oil–red lead compositions are being replaced by solvent-free or high-solids, impact- and abrasion-resistant epoxies with outstanding mechanical properties. External underwater hull coatings have changed more slowly. They originally consisted of mixtures of coal tar pitch, modified with natural bituminous materials (e.g., asphaltums and gilsonite) by cooking with lead and drying oils, pigmented with leafing aluminum flake. They had a high water and oxygen impermeability but deficient mechanical properties. Furthermore they are not resistant to cathodic protection, either by impressed current or sacrificial zinc or aluminum anodes.

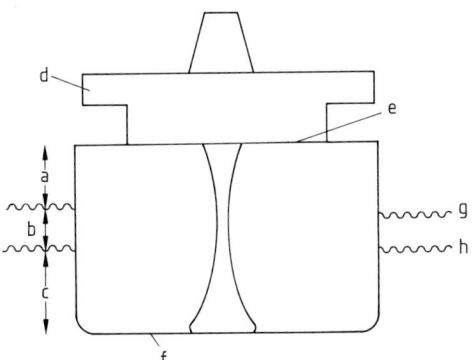

Figure 11.1. End-on view of a ship's hull showing painting areas
a) Topsides; b) Boottopping or "wind and water line" (may be exposed or immersed); c) Lower sides (always immersed); d) Superstructure; e) Decks; f) Bottom; g) Water level (heavy load); h) Water level (light load)

Coal tar pitches are less readily available than they used to be; their content of polynuclear aromatic hydrocarbons makes their safe use suspect. They have been replaced to some extent by the less effective petroleum bitumens whose mechanical properties are improved by combination with poly(vinyl chloride) and other chlorinated polymers as one-pack, nonconvertible coatings, or by combination with epoxy or polyurethanes in high-performance two-pack convertible coatings. Similar materials are preferred for coating ballast spaces and double bottoms in vessels, where economical systems are required that do not have to be attended throughout the lifetime of the vessel (15–25 years). Multiple coats with a total thickness of 250–400 μm are usually employed.

The environment in chemical tanks is amongst the severest to which marine coatings are subjected. Each tank may have to transport some 1500 bulk, liquid cargoes that include crude oil, refined gasoline, aviation spirit, diesel oil, solvents, vegetable oils, or wine. Inorganic cargoes are also carried in solution (e.g., alkalis and acids). New demands also appear such as methanol as a fuel and feedstock, and methyl *tert*-butyl ether as an additive for lead-free petrol.

No single type of coating is universally applicable, and nonpaint alternatives may also be used (e.g., rubber linings and stainless steel). Vessels that are not dedicated to a single type of cargo (parcel tankers) usually have a number of coating systems.

The tank coating has three purposes:

1) To prevent corrosion of the steel
2) To prevent contamination of the cargo
3) To facilitate cleaning and avoid cross-contamination of one cargo by another

Variants of six main paint systems are currently used (Table 11.1).

Zinc silicates are outstanding for nonreactive hydrocarbon cargoes. The remaining systems are all highly cross-linked and their advantages are generally a function of the cross-linking density and their specific chemical stability. Heat is of great

Table 11.1. The most important paint systems for marine tank coatings

Paint system	Advantages	Limitations
Zinc silicate	outstanding neutral solvent resistance	poor acid and alkali resistance poor film formation at low humidity
Epoxy cured with polyamide	tolerant to application conditions wide range of solvent resistance	poor resistance to polar solvents minimum application temperature 5 °C
Epoxy cured with amine adduct		
Epoxy cured with isocyanate	low temperature, elastic and adhesive	poor alkali resistance dangerous to human health
Epoxy–phenolic resin	outstanding overall resistance, including methanol	requires heat for ultimate cure
Polyurethane	excellent low-temperature curing	poor methanol alkali resistance and dangerous to human health

advantage in achieving adequate cross-linking density. Some systems require the carriage of heated cargo (e.g., palm oil) for up to six months after application. Heating, ventilation, and film thickness all need to be closely controlled during application. The steel surface temperature has to be maintained at a constant level. In Northern latitudes the outside of the vessel may be lagged with polyurethane foam and the tanks are ventilated with hot dehumidified air. Certain sequences of cargo have to be avoided since the coating systems swell and are stressed; for example, permeability to water (from damp cargo or steam cleaning) may increase following carriage of methanol or contact with tank-cleaning chemicals. Chemical tank coating is thus one of the most critical coating processes requiring careful coordination of coating formulation, testing, manufacture, surface preparation, and control and inspection of application. The paints are generally applied by airless spraying. Difficult geometrically complex areas and welds are usually cleaned, smoothed by mechanical abrasion, and coated with a brush [11.14]–[11.17].

11.4.3. Fouling and Antifouling

Fouling and antifouling are unique to the marine industry. In contrast to other coatings whose primary function is structural, preservation, the outer hull surface of a ship also has to be smooth to maximize ship speed and reduce fuel consumption [11.18]. Fouling with marine organisms, considered as biological roughness, can be immediate and drastically increase fuel consumption or reduce speed. Some 630×10^6 deadweight (tonnes) of marine transport burns 184×10^6 t of fuel at a cost of $\$ 18.4 \times 10^9$/a. Physical roughness would increase this by 10%, fouling by 30–40%. The improvement in fouling control and roughness of ships made since 1974 has been calculated to have led to worldwide annual fuel savings of $\$ 720 \times 10^6$; the extension of drydocking from 21 to 28 months has been calculated to have saved $\$ 800 \times 10^6$.

Much effort has been directed to the prevention of fouling [11.12], [11.19]. The major concern on wooden ships was the attack of wood-boring molluscs (*Teredo* and *Martesia*) or crustacea. Almost any sessile marine organism may be found as a fouling organism. Seriousness depends on size and frequency of occurrence; severity of settlement depends on geography and vessel itinerary. Because of the wide range of organisms (e.g., bacteria, diatoms, algae, barnacles, hydroids, molluscs, ascidians, and sponges) methods of prevention have to be extremely general. All antifoulings have so far depended on a small range of organic, metallic, or organometallic biocides. They include copper, water-soluble or water-reactive compounds, organoarsenical, organomercurial, organolead, and organotin compounds. The biocides function by slowly dissolving (leaching) in the seawater adjacent to the hull and killing the organism in question. There is a critical leaching rate for each biocide and for each organism. The objective of antifouling design is to select an optimum biocide or mixture of biocides and to control the release rate at just above the critical leaching rate for macrofouling (i.e., fouling that affects ship performance). Ideally the leaching rate should be linear or zero order. Because leaching has to continue for

3–5 years under adverse environments only the most potent biocides have been found useful (LD_{50} against target organisms may be 3 ng/mL). Control of release was at first poor because the choice of matrix or vehicle was limited to a small number of natural oils and resins with the appropriate permeability. Controlled-release media have, however, since been developed. Stable film-forming resins composed of trialkyl- and triaryltin esters of acrylic copolymers are used [11.20]–[11.22]. When exposed to water the film dissolves with zero-order kinetics and the surface becomes smoother in turbulent flow [11.23]. Self-polishing or ablating antifoulings based on tributyltin and copper(I) oxide can have antifouling lives of five years or more. Trade names include Intersmooth (International Paint, UK), Nautic (Hempel, Denmark), Seaflo (Chugoku, Japan), and Seamate (Jotun, Norway).

For reasons of safety, health, and environmental protection, organolead, -arsenic, and -mercury compounds are no longer used [11.24]. Use of organotin compounds is restricted to commercial vessels. Although nontoxic, nonpolluting, and nonhazardous solutions to the fouling problem are needed, alternative compounds to the tributyltin acrylates have proved extremely difficult to find. However, apart from their susceptibility to biocides one of the common features of all fouling organisms is adhesion. Attempts to prevent adhesion with antiadhesive coatings have shown recent success [11.25]–[11.27]. This approach requires an understanding of the physics and chemistry of bioadhesion. Hydrophobic fluoropolymers have proved ineffective as marine antiadhesive coatings. Silicone elastomers are adequate for most release purposes and physiologically benign, they have shown the most promise. Claims have also been made for nonadhesion with hydrophilic agents, but much work remains to be done in this area [11.28].

11.5. Coil Coating [11.1], [11.2]

Coil coating is a special application method in which coiled metal strips are unwound and then passed through pretreating, coating, and drying equipment before being finally rewound (see Section 8.3.4). The coated metal strips (0.2–2 mm thick) are supplied as coils, and are used to manufacture a wide variety of metal constructions, ranging from flat panels to complex geometrical forms. The organic coating film must have a high elasticity to allow the metal to be shaped without damaging the film at the edges.

Coil coating allows very efficient coating of large surface areas in a short time, the coated surfaces have a very high quality. The metal sheets are shaped to the required form after coating.

Coating coils of strips in a large coating facility, instead of in many small ones, results in optimization of the coating process and reduces pollution of the environment (lower emission of volatile organic solvents) [11.1].

Prior to coating, the metal surfaces must be cleaned and pretreated with aqueous solutions to form conversion coatings. The following metal surfaces can be coil-coated:

1) Steel strips with a width up to 1850 mm
2) Electrolytically galvanized metal bands
3) Hot-dipped galvanized steel bands (using the Sendzimir method)
4) Strips from aluminum alloys with a width up to 1650 mm

Coil coatings are normally applied by roller coating machines, but are sometimes sprayed. The topside of the metal band is normally painted with a primer (DFT ca. 5 µm) and a topcoat (DFT 20–22 µm). Zinc-rich coil coatings are applied with a DFT of 5–20 µm, plastisol films have a DFT of 80–400 µm. The reverse side of the coils is coated with backing coats based on binders, such as alkyd, polyester, and epoxy resins (DFT 8–10 µm).

After a very short flash-off time (3–10 s), the coated metal passes through a drying oven, which can be up to 50 m long. Drying time is 20–60 s at 180–260 °C, depending on the coating type.

Epoxy primers are used as general-purpose primers on steel, galvanized steel, and aluminum, and under all types of finishes. These primers give excellent corrosion resistance on steel. Acrylic primers have a corrosion resistance similar to that of epoxy primers, and are used primarily under fluorocarbon finishes because of their superior intercoat adhesion.

Polyester primers are widely used in Western Europe. Although their corrosion resistance (salt spray test) is not as good as that of the other primer types, their mechanical properties are superior.

The choice of binders for the topcoat depends on the end use of the coil-coated metals. Acrylic topcoats were developed earliest, and are widely used indoors for surfaces that are not exposed to water, chemicals, or mechanical stress. Polyester coil coatings provide a good general-purpose finish, and can be used for a wide range of applications (e.g., car interiors and accessories, caravan exteriors, domestic appliances). Tough, abrasion-resistant, and durable general-purpose finishes can be formulated with polyurethane and polyester resins in combination.

Topcoats based on fluorocarbon resins [poly(vinylidene fluoride)] produce extremely durable coatings (life to first maintenance > 20 years under most climatic conditions) with excellent color and gloss retention. Like silicone-modified resins they are used for applications where weather resistance is required (e.g., exterior panels on buildings). Because of high prices the importance of these systems is limited.

11.6. Coatings for Domestic Appliances [11.1]

Domestic appliances (e.g., refrigerators, deepfreezers, dishwashers, washing machines, and laundry dryers) are mainly made from blank sheet iron, or electrolytically or hot dip galvanized steel. Coatings must protect these construction materials against corrosion caused by food ingredients, household chemicals, and the humid atmosphere in a kitchen or laundry.

Domestic appliances are manufactured in large numbers necessitating adequate application and drying facilities and fast production lines. Stoving materials are used that cure in a short time at comparatively high temperatures (10 min at 160 °C or 30 min at 120 °C). Binders for these stoving paints are mainly self-cross-linking acrylic resins that cure by polycondensation to give a smooth homogeneous film with excellent mechanical and chemical resistance.

Stoved acrylic resin films are generally applied as single coats which requires careful pretreatment of the metal surfaces by cleaning and phosphating. The paints are applied by electrostatic spraying using rotating bells or disks.

Certain components of domestic appliances are also coated with electrodeposition paints. Two-pack polyurethanes are applied if the dimensions of the parts are too large for the electrodeposition bath. These paints dry at ambient temperature to give films with similar properties to stoving acrylic enamels. Solvent-free powder coatings are also used to coat domestic appliances. Nowadays powder coatings and precoated metal (coil coating) are gaining more importance.

11.7. Coatings for Packaging (Can Coatings) [11.1]

About 100×10^9 cans are produced annually worldwide for packing perishable food. Cans can be considered as a single material, consisting of a metal substrate with an organic lacquer.

The interior coating of cans is very important because it prevents the metal from reacting with the ingredients of the filling goods. The can exterior is painted to prevent corrosion, but also for decorative reasons. Coatings based on combinations of polyester or acrylic resins with melamine resins are used for can exteriors. Lacquers based on phenolic resins are especially resistant against aggressive can contents. Curing at 200 °C produces densely cross-linked films with high chemical resistance, but poor flexibility.

Combining phenolic resins with epoxy resins results in films with good chemical resistance and flexibility. Interior coatings based on these resin combinations are applied with a solids content of 33–43 % and cured at 200 °C in 80 s. Epoxy–phenolic resin lacquers are used in high amounts for can coating.

Interior can coatings based on PVC organosols are highly flexible and resistant. They are used mainly for cans which are heavily shaped during the manufacturing process.

Three-piece (body and two lids) cans are now welded instead of soldered. The welding seam is coated separately with epoxy–phenolic resins or PVC organosols that are cured at 260 °C for 10–20 s. The PVC organosols result in higher dry film thicknesses, whereas the epoxy–phenolic welding seam coatings have better chemical resistance.

Special filling goods and a long shelf life sometimes require an extremely high dry film thickness for the welding seam coating. Powder paints are then applied electrostatically to give a dry film thickness of ca. 50 μm.

Beverage cans are two-piece (body and lid) cans produced by the drawn and wall ironing (DWI) process. They require highly flexible coatings based on epoxy resins cross-linked with aminoplast resins.

For many years, paints used to coat cans contained considerable amounts of volatile organic solvents. Waterborne can coatings were developed to reduce solvent emissions and are used worldwide. Binders used in waterborne can coatings are modified epoxy resins (see Section 2.10). Acidic acrylate chains are grafted onto an epoxy molecule. After partially neutralizing with amines, the resins can be dispersed in water.

Both waterborne and solventborne can coatings must not affect the can contents, especially their taste. Components of coating films are not allowed to migrate into food, beverages, or other filling goods. In most countries, food packaging is subjected to legal regulations. The raw materials used to produce can coatings and the coatings themselves are strictly limited.

11.8. Furniture Coatings [11.1]

The long-term value of furniture depends to a high degree on its surface characteristics. Untreated or uncoated surfaces very quickly lose their good appearance and deteriorate under conditions of use. Different forms of wood are used for furniture (e.g., solid wood, veneer, plywood, particle board, and chipboard). Solid wood and veneer may originate from different types of trees with a wide range of properties partly due to their contents of resins and essential oils.

The natural humidity content of fresh wood has to be lowered to a maximum level of 10–15% before it is coated. Continuously varying atmospheric humidity in air leads to changes in the volume of the furniture wood. Furniture coatings must therefore have excellent film flexibility as well as film hardness, and resistance to abrasion and fluids (e.g., alcohols). Prior to coating, wood surfaces usually have to be smoothed by sanding, using putty, patinating, staining, or pore filling.

Resins and essential oils are often extracted with organic solvents to prevent the coating from cracking, discoloring, and developing other faults.

Automatic spraying, machine roller coating, and curtain coating are used for industrial application of furniture coatings. Drying is accelerated by air circulation and/or increased temperature up to 80 °C. IR drying ovens and tunnels are also used.

Furniture coatings can be clear or pigmented with many different color shades. The surface can be high gloss, semigloss, matt, or textured. Some coatings leave the wood pores open to give the wood surface a more natural appearance; the "bottom" of the pores must, however, be coated evenly to protect the wood surface completely. Application of a film that fills the wood pores results in a completely smooth surface coating.

For many years, nitrocellulose coating systems were preferred for indoor furniture and other wood parts. These systems are still used on low-price furniture, but are gradually being replaced by coatings based on polyurethanes and unsaturated polyesters. Most modern wood and furniture coatings are based on special acrylic resins and unsaturated polyester resins, cured by UV or electron-beam radiation [11.30], [11.31].

The properties of furniture coatings can be summarized as follows:

1) *Nitrocellulose Paints*. Cheap, high content of organic solvents, low solids content, dry at ambient temperature, not very resistant to chemicals or mechanical stress, dried film is soluble in organic solvents.
2) *Polyester Paints*. Binder based on unsaturated polyester resins (copolymerized with styrene), low emission of organic solvents, catalytic curing by organic peroxides or UV radiation, highly resistant to abrasion, alcohols, and other chemicals (cleaning agents).
3) *Polyurethane Paints*. Polyester, polyether, or acrylic resins containing hydroxyl groups are cured with isocyanate hardeners, medium to high solids contents, best overall properties in terms of mechanical and chemical resistance [11.32].

11.9. Coatings for Buildings

Coatings for buildings are solvent- or waterborne. They include coatings that can protect all materials used in building and construction work (e.g., wood, steel, light metals and alloys, plastics, concrete, plaster) against corrosion and decomposition; they can also give a decorative appearance.

Solventborne paints are mainly alkyd paints with mineral spirit as solvent. Paints based on synthetic resins and two-pack paints with hydrocarbon, ester, or ketone solvents are, however, also used. Gloss ranges from silk-matt to high gloss.

Waterborne systems are formulated with pure acrylates and have a solvent content below 10%, preferably glycols. The gloss is between silk-matt and full gloss.

Emulsion paints (see also Section 3.5) with a solvent content below 4% are based on vinyl or acrylic polymers and copolymers (e.g., with styrene). Gloss is between matt and silk-glossy.

Architectural paints are applied in situ with brushes or rollers in Europe, but spray application is widely used in the USA on concrete and wooden buildings. Paints on windows and doors are brush applied. Structural components (e.g., doors, windows, radiators) that receive primers or complete coating systems during production or in the workshop are dipped or sprayed.

Since the decorative effect is a very important factor in architectural paints, a large number of shades in various gloss gradations are available. Manufacturers' color cards, the German RAL-register (Deutsches Institut für Gütesicherung und Kennzeichnung), and the Swedish NCS (Natural Color System) shade charts provide a survey of available shades.

11.9.1. Exterior-Use Coatings

Exterior-use coatings must be weather-resistant: they must adhere to a wide range of substrates and retain their gloss and shade fastness for a number of years. Suitable binders and pigments must therefore be chosen [11.33].

Coatings for Mineral Substrates. Bricks, concrete, cement, and stone are the components used for the main construction of buildings. When the surface is new, most of these materials are very alkaline. Therefore the binders for primers and finishing coats must be particularly resistant to alkali because otherwise hydrolysis can occur in the present of moisture due to the high pH of the substrate. Another problem with concrete is carbonation, i.e., the chemical state of the concrete changes as a result of moisture ingress causing the steel reinforcing rods to rust.

The most important requirements and properties of exterior-use coatings on mineral substrates are the water permeability to water and water vapor and protection against carbon dioxide.

Aqueous systems such as silicate paints and vinyl or acrylic emulsion paints are suitable for opaque and semitransparent coatings. Silicate paints have a high degree of hardness and good permeability to water vapor and gas. Evaporation of moisture is not hindered, peeling and flaking are therefore avoided.

The most important properties of exterior-use emulsion paints are their rheology (good application) and the water and water vapor permeability which influences the moisture balance of the substrate and thus the durability and protection of the substrate.

Solventborne paints are based on synthetic resins (e.g., acrylate–styrene copolymers) which are soluble in aliphatic and/or aromatic hydrocarbons. The properties of these coatings are chiefly determined by the nature and proportion of the monomers.

Two-pack polyurethane coating materials with an aliphatic isocyanate as hardener can also be used.

Colorless, water-repellent impregnations are obtained with silicone resins or siloxanes in solvent and silicates in water [11.34].

Coatings for Wood. Wood is susceptible to attack from moisture, which is taken up through unprotected end grain, open joints etc. causing dimensional movement. Binder and pigments of the paint system must be formulated such that a sufficient film thickness reduces exchange of moisture and thus swelling and shrinkage of the wood.

Sunlight (mainly UV rays) breaks down the wood by depolymerization of the lignin. The UV transparency of the coating therefore has to be low. This is achieved by using UV absorbers in varnishes and transparent iron oxides in stains [11.35].

Fungi may grow when wood is open to moisture penetration. The mold grows quickly and disfigures the surface. It can lead to wet and dry rot when the moisture content exceeds 20%.

Timber used for structural purposes in buildings is impregnated with preservatives by the supplier, and can usually be overpainted without difficulty. Joinery (e.g., windows) is supplied either primed for painting or basecoated and preservative-treated for finishing with a varnish or woodstain.

Wooden structures can receive colorless, semitransparent, or opaque coatings. Long oil alkyd resins, mainly modified with soybean or talloil, are used as binders for solventborne products. Often, a small quantity of a thixotropic alkyd resins is added to improve brushability, to reduce sagging of the paint on vertical sufaces, and to obtain good edge covering.

Acrylid resin dispersions alone or mixed with alkyd resin emulsions are used for waterborne products.

Stains may be of the impregnating or coating type. Low-build stains (solid content < 30%) are used on nondimensionally stable wooden surfaces and on dimensionally stable structural parts as priming coats only. They show good penetration into the wood and contain preservatives against blue stain and wood-destroying fungi. High-build stains (solid content > 40%) are used as intermediate and topcoats to protect dimensionally stable structural parts direct weathering [11.36].

Coatings for Metals. Iron and steel are converted to their oxide form in the presence of oxgen and water. The oxide layer that forms on the surface is rust.

If steelwork is a chemical environment, the corrosion process will accelerate as the iron and steel reacts with the polluted atmosphere. In order to control the rate of corrosion the surface must be carefully prepared before applying and maintaining a suitable paint system.

Nonferrous metals (e.g., aluminum and zinc) are more corrosion resistant than iron and steel. They corrode in a different way and at a much slower rate. The corrosion products of aluminium and zinc are white salts which form on the surface. Aluminum and zinc are used to provide sacrifical protection to iron and steel in the form of galvanizing and metal sprays.

Choice of an appropiate surface treatment and a suitable primer are important because adhesion to the substrates presents difficulties [11.37]. Primers based on modified alkyd resins or two-pack epoxy-resins for derusted ferrous metals mainly contain zinc phosphate and zinc oxide as corrosion protection pigments. Nonferrous metals are first washed with an ammoniacal wetting agent before applying the primer that contains a binder based on synthetic resins (e.g., PVC copolymers, chlorinated rubber) which ensure good adhesion to the substrate. The same primer must be used on zinc or galvanized surfaces because the use of alkyd resins causes embrittlement [11.38] The primed surfaces are largely topcoated with alkyd resin systems.

Coatings for Plastics. The use of plastics in building and construction (e.g., rigid or unplasticized PVC, commonly used for replacement doors and windows) is increasing. Although not all of the many different grades of plastic can be painted, the trend toward renovating such surfaces. If painting is necessary, the surface is first cleaned and roughened.

A one-pack primer based on chlorinated rubber or chlorosulfonated polyethylene (as is conventional with galvanized surfaces) or a two-pack epoxy primer is then applied. The primer can be coated with a topcoat system that is usually formulated with medium or long oil alkyd resins [11.39].

11.9.2. Interior-Use Coatings

Wall Coatings. Plaster (a mix based on lime, cement, sand, and water) is applied to wall and ceiling surface. It subsequently hardens to give a smooth surface for decoration.

Depending on the required surface effects, interior-use paints have a $PVC \leqslant 80\%$. The mechanical strength and cleanability (wipefastness, washfastness, abrasion resistance) rather than weather resistance is, however, the prime concern (DIN 53778). Fungal contamination can easily occur in damp, moist areas and is prevented by adding fungicides (e.g., carbamates or imidazoles).

Various surface effects can be produced by varying the viscosity and adding coarse, possibly colored extenders or fibers. Two-pack systems based on polyurethane resins or epoxy resins are used for wall coatings that require a good resistance to agents used for chemical cleaning and decontamination.

Solid emulsions are relatively new. They are highly structured, white, interior-wall paints which do not drip or splash during application. The thixotropic character is achived by using a titanium chelate as a thickening agent.

Floor Coatings. Concrete floors are coated with low-solvent or solvent-free epoxy or acrylic resin materials that may be applied in any desired thickness. They are extremely resistant to abrasion, can be made slip resistant with sand, silicon carbide, or high-grade steel granulate, and are also resistant to mineral or vegetable oils and gasoline (used for warehouses and factory halls).

Pigmented, two-pack, waterborne epoxy resin coatings (garages) or one-pack waterborne acrylic resin emulsion paints are used for areas that receive less wear (e.g., cellars). Wooden parquet floors are coated with one- or two-pack polyurethane varnishes that can be applied by spraying or brushing. Acrylate-based waterborne parquet varnishes are also used because they are environmentally friendly.

Radiator coatings are intended to protect radiators against corrosion without, however, affecting their heating effects (DIN 55900). Primers based on special alkyd resins are generally applied industrially. They have to satisfy the usual requirements for preventing radiator corrosion during transportation and at the building site. The topcoats applied on site by rolling or inundation are based on medium oil alkyd resins.

Heating oil storage premises must be equipped with a collection trap so that any heating oil leaking from the tank cannot contaminate the soil. The interior of these premises must be painted with an officially approved coating material that is not dissolved or penetrated by heating oil. The coating must also cover cracks in the substrate. Multilayer systems based on waterborne acrylic resin dispersions are suitable for this purpose.

Fire Retardant Coatings. The flammability of combustible wood structures can be reduced in accordance with DIN 4102 by applying a fire retarding paint that forms an insulating layer [11.37]. Dispersion paints based on poly(vinyl acetate) with addition of ammonium phosphate, a nitrogen compound (e.g., melamine), and a carbon-forming agent (e.g., pentaerythritol) are suitable for this purpose. The thermal insulation is so good that ignition can be delayed by at least 10 min.

12. Environmental Protection and Toxicology

Paints and coating materials frequently contain substances that may be a hazard both to human health and to the environment. This applies particularly to organic solvents, to certain reactive binder constituents, to pigments containing heavy metals, and to some additives. Evaluation of the environmental properties of paints must take into consideration their effects on the atmosphere, water, and the soil, the potential danger to the user, the use of low-residue application techniques, and the suitability for use. The primary concern is to minimize adverse effects in all sectors.

12.1. Clean Air Measures

Organic solvents in paints constitute significant sources of atmospheric pollution. The direct effects of these substances and their mixtures (see Section 12.4), particularly the odor nuisance, should be taken into account in the vicinity of sources of solvent emissions. In the atmosphere the solvents gradually decompose or participate in chemical reactions under the influence of sunlight or traces of other substances present in the air. Photochemical decomposition in the presence of nitrogen oxides leads to formation of intermediates which are termed photooxidants on account of their oxidizing action. Ozone is regarded as the tracer for photooxidants. Even low concentrations of photooxidants harm plant life and may damage the human respiratory tract. Atmospheric pollution caused by photooxidants can occur particularly in the summer (summer smog), and was first observed in Los Angeles, where measures were adopted at an early stage to reduce pollution.

Solvent emissions can be reduced in three ways:

1) Use of low-solvent or solvent-free products
2) Application methods with low solvent emissions
3) Implementation of waste air treatment as a secondary measure in paint application facilities

In painting work performed by tradesmen, emissions can only be lowered by using low-solvent products.

Low-Solvent Coating Materials. A great deal of progress has recently been made in the development of low-solvent paints [12.1]. The use of water as a replacement for organic solvents is particularly important in this respect, although alternatives such as powder coatings or one-pack and two-pack high-solids paints have also led to significant improvements. Solvent emissions can be drastically reduced by using low-solvent coating materials. This is particularly apparent in automobile finishes: in a metallic effect finish based on conventional coating materials and solvent contents of up to 85%, the specific emission is ca. 180 g of solvent per square meter of car body. With an average car body surface of 90 m^2, ca. 16 kg of solvents are released per automobile. By using low-solvent coating materials (electrodeposition paints, waterborne primer surfacers, waterborne metallic basecoats, and a conventional solventborne clearcoat) solvent emissions in modern plants are now only 30–45 g/m^2. Waterborne clearcoats are currently under development, whereas powder clearcoats are used in the automobil production [12, 1 a]. Use of these systems should allow solvent emissions to be reduced further to ca. 20 g/m^2.

In other industrial application sectors there is an increasing tendency to employ low-solvent paints. In furniture production, for example, waterborne, UV-curing paints are superceding high-solvent nitrocellulose lacquers.

Stringent coating stability requirements on metal surfaces can be met by using powder coatings. Use of powder coatings is therefore increasing in new areas of application (e.g., for the colored coating of sanitary ware [12.2] and corrosion protection of truck tanks [12.3]).

The need for waste air treatment can generally be avoided by replacing conventional, high-solvent paints by low-solvent coatings. In some cases smaller waste air treatment plants may be necessary for specific demarcated areas where, for example, a solventborne layer is applied.

In the architectural paints sector, conventional, high-solvent paints based on alkyd and synthetic resins can largely be replaced by waterborne dispersions of acrylate polymers and copolymers, aqueous alkyd resin emulsion paints, and oil-based high-solids paints. Modern low-solvent coating materials not only decrease environmental pollution, but have also improved paint quality (e.g., resistance to weathering). In some countries low-solvent architectural paints are clearly labeled as such to assist workers and users. In the Federal Republic of Germany waterborne emulsion gloss paints containing $\leq$ 10 wt% solvent and high-solids paints containing $\leq$ 15 wt% solvent are identified by an "environment label", a symbol created for products having a favorable environmental compatibility. In Switzerland paints containing $\leq$ 5 wt% of solvents may be labeled as "solvent-free", and with $\leq$ 30 wt% as "low-solvent". Some states of the United States (e.g., California) stipulate maximum solvent limits for individual product groups.

Application Methods with Low Solvent Emission. The method used to apply the coating material also influences the level of solvent emissions as well as the amount of waste and water pollution. Application efficiency depends on the application method and shape of the article being coated. Application methods are described in Chapter 8. Spraying and dipping are most commonly employed for industrial coating. Spraying methods have the highest solvent emissions because overspray losses are unavoidable, and low-viscosity paints generally have to be applied. With conventional high-solvent paints the solvent content at the ready-to-spray viscosity value may be as high as 90 wt% but is usually 50–70 wt%. In conventional spraying with compressed air the application efficiency is particularly low (20–65%, depending on the shape of the part being coated). Higher application efficiencies are achieved with airless and electrostatic spraying methods, and may reach 90 % in the case of electrostatic spraying. The overspray of coating powders can be largely recovered and reused directly. Problems arise with changes in color which can, however, largely be solved. The highest application efficiencies for wet paints (nearly 100 wt%) are obtained with brushing, rolling, and pouring methods, which, however, can only be used on flat surfaces. Parts with hidden areas can be effectively coated by dipping methods.

Waste Air Treatment. Two main methods are used for treating waste air in coating and paint shops: afterburning with heat recovery and adsorption with solvent recovery. In order to reduce expenditure on waste air treatment, the amount of waste gas should be minimized by enclosing the paint application area and recycling circulating air. Water washers (scrubbers), fabric filters, and electrical separators are used to remove paint aerosols from the atmosphere. Recycling of circulating air is already current practice in modern automated paint shops (e.g., in the automobile industry). If the workforce is protected against relatively high solvent concentrations by respirators equipped with a fresh air supply, manual spraying zones can also be operated with recycled air. If such a concentration procedure is not possible, large-volume waste air streams with a low solvent content can be concentrated with a continuously operating adsorption wheel (e.g., with special activated carbon) and then treated [12.4]. The solvent from the waste air is adsorbed on one side of the rotating wheel and desorbed with a small air stream on the other side. The concentrated waste air (10 % of the original waste air stream) can then be purified either by an adsorption unit with solvent recovery, or by afterburning. The heat from the afterburning plant should be utilized; if this is not possible, newly developed thermal methods with internal heat utilization may be used [12.5].

Biological waste gas treatment methods are also suitable for purifying solvent-containing waste gases, especially slightly contaminated, large-volume waste gas streams; they have already been tested [12.6]. In biological methods organic substances are degraded by microorganisms on the surface of a wet filter layer or in a scrubber.

In various countries (e.g., the Federal Republic of Germany, the United States, Scandinavia, and Switzerland) regulations exist concerning the treatment of waste air from paint shops. Large paint shops (e.g., in the automobile industry) are covered by these regulations. In the Federal Republic of Germany the maximum permissible emissions from automobile paint shops are limited by the amount of solvent used per square meter of car body [12.7] (see p. 266). For automated spraying zones in other paint shops the emission of organic substances in the waste gas is restricted to a maximum of 150 mg/m^3.

12.2. Wastewater

In industrial paint application the principal sources of wastewater are spraying cabins, wet filters, and scrubbers. Further sources of wastewater are the cleaning of apparatus, equipment, vessels, tanks, and working areas, as well as the retentate produced in the ultrafiltration of electrodeposition paints. To reduce environmental pollution, attempts should be made to minimize the amount of wastewater. Wastewater from spraying cabins may be treated by coagulating the overspray and continuously extracting the paint slurry, as well as by reducing the amount of water used. Continuous methods for cleaning the circulating water are used both for solventborne and waterborne paints. Products based on alumina, metal hydroxides, and organic fatty acid derivatives are used as coagulating agents. These auxiliaries coat and envelop the paint particles.

With waterborne paints, stable dispersions or emulsions are sometimes formed in the circulating water. The coagulating agent also has to "break" these disperse systems. A very fine flocculant coagulate is often formed which has to be separated with special filters or a centrifuge, or converted into larger, more easily removable flakes by using a further coagulating agent. If the circulating water from the spray cabin has to be drained off due to high levels of contamination, it must be treated before being discharged into the sewage system or wastewater treatment plant. Treatment usually comprises flocculation, neutralization, and filtration. With certain water-soluble toxic substances (e.g., heavy-metal compounds), organic solvents, and additives, further purification steps may be necessary. Heavy metals can be precipitated. Methods used for organic solvents depend on their nature and concentration in the water; they include ultrafiltration, reverse osmosis, adsorption (e.g., on activated charcoal), biological purification, and, with high solvent concentrations, distillation [12.8]. The concentration of organic substances in the wastewater is described by the chemical oxygen demand (COD) and the biological oxygen demand within 5 days (BOD_5). Statutory requirements governing the preliminary purification of the wastewater can vary widely. They depend on purification facilities in the existing wastewater treatment plants. The effort and expense involved in wastewater pretreatment can be considerably reduced by avoiding the use of toxic substances (e.g., heavy metals). With waterborne paints particular attention should be paid to adequate removal of water-soluble organic substances (e.g., solvents) from the wastewater.

12.3. Solid Residues and Waste

Considerable amounts of solid residues and other waste are produced when paints are applied, particularly by spraying. The overspray is collected as a coagulated residue from the spray cabin water. Articles such as contaminated filters, paint residues, and empty containers also have to be disposed of. For ecological reasons minimization of waste production and reutilization should take precedence over disposal methods (incineration, landfill). In the Federal Republic of Germany, for example, this principle is laid down in waste control and emission legislation (Kreislaufwirtschafts- und Abfallgesetz, Bundes-Immissionsschutzgesetz). Up to now paint slurries were mainly disposed of in special landfills. On account of increasingly stringent requirements to prevent pollution of soil and groundwater, paint slurries will have to be disposed of in special refuse incinerators. This will inevitably lead to higher disposal costs; avoidance of waste and recycling will therefore be of economic advantage.

The production of paint residues and waste can be prevented or reduced by using coating methods with high application efficiencies (dipping, brushing, rolling, and pouring). Compared with the spraying technique, these methods generally also result in lower solvent emissions (see p. 266). The amount of overspray produced in spraying methods can be lowered by using electrostatic application procedures. In powder coating the overspray is trapped and separated in a dust-removal filter; it can then be directly recycled, if necessary after purification.

In the spray application of wet paints the overspray can also be recovered and recycled by various methods which have not, however, all been tested industrially [12.9]. The overspray can be recovered with rotating disks or circulating belts. Stable paints can be reused directly after conditioning (e.g., viscosity adjustment). In solventborne paints the disk or belt often has to be wetted with solvents, the waste air should therefore be treated to prevent high solvent emission. Some waterborne paints can also be recovered by this method. In paint recycling the overspray should not be entrained with the waste air from the spray cabin; certain preconditions should therefore be observed: airless spraying guns are most suitable and the parts to be coated should not be too large or have a complex three-dimensional structure (e.g., car bodies). The overspray can be recovered from the cabin circulation water if it can be coagulated without destroying its chemical structure. After mechanical dewatering with kneaders or mixers, purification, and work-up, small amounts of this material can be added to the new paint. A new develepment is the recycling of waterborne paints through ultrafiltration [12.10]. Modern paints based on waterborne binders fulfill the demands of ultrafiltration and common quality requirements so that direct addition to the new paint is possible. If direct recycling is not possible, the material can be separated by centrifugation into a binder solution and pigment concentrate that can be used as raw materials for paint production. Physically drying paints and stoving finishes are particularly suitable for recovery.

Paint coagulates can also be used as a binder constituent in the production of molded plastics and as a filler replacement in plastics dispersions. If used as a binder for molded plastics, the material must not be cross-linked; the coagulate is worked up into an aqueous dispersion which is used to wet or impregnate fiber mats that are then compressed. In order to produce fillers the paint coagulate first has to be dehydrated and dried, the material becomes completely cross-linked and can be ground into a powder. The powder is used as a filler for plastics dispersions (e.g., for underbody protection in automobiles and in sealing materials).

12.4. Toxicology

Many different substances are used in paints and coating materials as binders, pigments, solvents, and additives. Workers involved in painting and coating work are regularly exposed to volatile organic compounds, especially solvents. In spraying application methods the inhalation of all paint constituents in the form of aerosols should be borne in mind even if they are nonvolatile or of low volatility. Contact with the skin represents a further source of exposure to paint constituents, many of which can be absorbed through the skin.

With manual application by brushing or rolling the health hazards due to solvent exposure (aliphatic and aromatic hydrocarbons, esters, ketones, alcohols, and glycol ethers) are a major factor. Solvents are predominantly absorbed via the respiratory tract. Their toxic effects depend on the nature of the solvent, its concentration, and the length of exposure. Depending on the concentration, symptoms after acute exposure include irradiation of the mucous membranes (eyes and respiratory tract), vertigo, nausea, and vomiting; narcosis symptoms are also observed which are attributed to disturbances of the central nervous system. Chronic poisoning is initially undetectable, but may subsequently produce damage to organs specific for the solvent concerned.

The neurotoxic effects found in painters and coaters exposed to solvents are the subject of controversy. Some studies describe subjective symptoms such as fatigue, difficulty in concentrating, and short-term memory problems in workers employed in industrial paint and coatings application. These symptoms have not, however, been observed in painters employed in the architectural and exterior-use paints sectors who mainly use waterborne paints [12.11].

During surface treatment prior to paint application, abrasive dust and pyrolysis products produced during the removal of paints and solvents may also be inhaled. The dust produced from corrosion protection agents and some older colored paints is often contaminated with heavy metals. Chlorinated hydrocarbons are still used in paint strippers.

Frequent skin contact with paints and coating materials can cause skin disorders, particularly on the hands, in painters and coaters. The lipid-solubilizing properties of the organic solvents may cause or at least promote contact eczema. In particular, paints based on reactive resins (e.g., epoxy and polyester resins) may cause allergic skin disorders. Skin-sensitizing substances include residual monomers and reactive diluents (e.g., acrylates and epoxides) and paint additives (e.g., acid anhydrides, peroxides, amines, as well as cobalt and zirconium in driers, and formaldehyde and isothiazolinone in biocides). Some of these paint constituents are also skin irritants.

In spraying methods often employed for industrial paint application, workers are not only exposed to solvents, they may also inhale paint constituents in the form of aerosols. On account of their very small size, some aerosol components can reach and penetrate the lung virtually unhindered. Substances that have a particularly sensitizing and irritant action on the skin can thus also affect the respiratory tract. Isocyanates can have a sensitizing effect even at very low concentrations (1 $\mu L/m^3$) and can cause chronic bronchial asthma in particularly susceptible persons. A literature study carried out by a working group of the International Agency for Research on Cancer came to the conclusion that there is sufficient evidence for carcinogenicity due to the occupational exposure of painters [12.12]. Occupational exposure in paint manufacture cannot be assessed however.

Depending on the application method (brushing, spraying) and the paint used, technical and personal work safety measures should be adopted when applying paints and coatings. Technical measures include adequate supply of fresh air and removal of waste air (e.g., in special hoods), as well as the replacement of "hazardous" paints with less dangerous ones [12.13]. Many hazardous dangerous substances have maximum workplace concentrations (threshold limit values) which should be strictly observed and monitored. Adequate facial and skin protection must also be ensured (e.g., with masks and gloves).

13. Economic Aspects [13.1], [13.2]

Paints and varnishes (coatings) have two primary functions: protection and decoration. Other objectives include information, identification, safety, insulation, vapor barrier, nonskid surface, and control of temperature, light, and dust. A range of product categories with a wide variety of application is therefore available:

1) *Architectural (decorative) coatings* include exterior and interior house paints which are normally distributed through wholesale-retail channels and purchased by the general public, painters, building contractors, government agencies, etc.
2) *Product finishes* are coatings formulated specifically for original equipment manufacture (OEM) to satisfy application conditions and manufacturing requirements for a wide variety of industrial and consumer products, e.g., wood and metal furniture and fixtures; automotive and nonautomotive transportation, aircraft, machinery and equipment, appliances, electrical insulation, film, paper, foil, toys, and sports goods.
3) *Special-purpose coatings* are formulated for special applications or extreme environments and include automotive and machinery refinishing, high-performance maintenance, road markings, marine (bridge) maintenance, crafts, metallic and multicolored coatings.

The number of coatings producers worldwide was estimated at about 7500 in 1997. The total world coatings market was estimated to be ca. $ 55–60 \times 10^9$; the product market sectors were as follows:

Decorative	50%
Auto OEM	6%
Refinishes	5%
Can coatings	3%
Coil coatings	2%
Others including industrial paints	32%
Total production	23×10^6 t

World paint markets by region were in 1994

North America	29.4%
Western Europe	25.7%
Eastern Europe	12.9%
Japan	7.2%
Rest of Asia-Pacific	15.2%
South America	5.7%
Rest of world	3.9%
Total production	21×10^6 t

In 1996 the top ten paint companies accounted for about 60% of the total world market, by the year 2000 they could well account for more. This development is expected because of permanent streamlining of activities by larger companies through selective acquisitions and/or divestments.

Single sourcing as in the car industry (one supplier for one model), marine sector (direct availability at each ship yard), or canning industry (worldwide health and safety standards) is a key factor in this globalization. Recouping in international markets for expenditure in research and development for technically sophisticated, high added value products is the other reason for this evolution.

Internationalization and the generally high standard of technical products shifts the economic importance from countries to companies. Therefore simple national per capita consumption figures are no longer indicative of productivity and standard of living. This is independent of some standard parameters such as climatic, cultural, or other impacts.

The paint and coatings industry as a whole is considered as a mature industry. An overall growth of 2.5–3.0% is estimated for the 1990s assuming overall growth of the corresponding gross national product of + 3.5%. This is remarkable in view of the fact that improved techniques such as high-solids coatings and coating powders, reduction in overspray, recovery, and recycling have considerably increased the surface area covered by a given amount of paint.

Raw materials account for roughly half of the production costs; prices of many of them are linked directly or indirectly to the price of crude oil.

In the decorative market products are mainly waterborne and consumption is dominated by new construction work and maintenance. Higher growth rates are therefore expected in newly industrialized and developing economies. In the automotive sector paint supply (not necessarily production) follows the requirements of car producers. Major growth potential for packaging (food and drink cans) lies in developing countries. Industrial paint markets are characterized by replacement of solventborne paints by high-solids, waterborne, and powder coatings to reduce environmental pollution. Therefore, coil coating can also avoid the classical painting process.

14. Solvents

14.1. Definitions

Solvents are compounds that are generally liquid at room temperature and atmospheric pressure; they are able to dissolve other substances without chemically changing them. The liquid mixture formed on dissolving a substance (*solute*) in a solvent is termed a *solution*. The molecules of the solution components interact with one another. Solutions are obtained by mixing liquid, solid, or gaseous components with liquids, the liquid always being termed the solvent. When two liquid components are combined, it is arbitrary which of the two components is considered to be the solvent, and which the solute; the liquid component present in excess is usually termed the solvent. Accordingly, plasticizers that are used for flexibilization in plastics processing and paint production may also be regarded as solvents. Plasticizers differ from solvents, however, with regard to their technological significance. A good plasticizer should have a very low volatility and thus permanently affect the dissolved substance. An ideal solvent should, in contrast, have a high volatility, so that it can evaporate as rapidly as possible to leave the dissolved substance (e.g., in a paint film). The boundary between plasticizers and solvents is not clear cut—some high-boiling solvents of very low volatility exert a flexibilizing effect over a prolonged period.

A solvent should generally have the following properties [14.1]:

1) Clear and colorless
2) Volatile without leaving a residue
3) Good long-term resistance to chemicals
4) Neutral reaction
5) Slight or pleasant smell
6) Anhydrous
7) Constant physical properties according to the manufacturers' specification
8) Low toxicity
9) Biologically degradable
10) As inexpensive as possible

Inorganic substances (e.g., hydrogen sulfide, ammonia, sulfur dioxide, hydrogen fluoride, and hydrogen cyanide) that are used as solvents in special applications, mainly at low temperature or under pressure, will not be discussed here.

14.2. Physicochemical Principles

14.2.1. Theory of Solutions

During dissolution the solvent acts on the substance to be dissolved to increase its state of distribution. Dissolution results in the formation of real solutions, colloidal solutions, or dispersions depending on the size of the particles that interact with the solvent molecules.

In *real solutions* the diameter of the dissolved particles is ca. 0.1 nm, and is thus of the order of magnitude of the free molecules. Real solutions are formed by most inorganic and organic compounds of low molecular mass. They are clear, physically homogeneous liquids.

In *colloidal solutions* the diameter of the dissolved particles is ca. 10–100 nm. Colloidal solutions are generally clear to weakly opalescent liquids, but exhibit inhomogeneities as regards some physical properties (e.g., the Tyndall effect).

In *dispersions* the diameter of the particles is larger than in colloidal solutions. Dispersions are turbid to milky liquids consisting of at least two phases.

Intermolecular Forces. During the dissolution of a substance (A) in a solvent (B) the forces of attraction between the molecules of the pure components (K_{A-A} and K_{B-B}) are destroyed, and new forces are simultaneously formed between the solvent and substance molecules:

$$K_{A-A} + K_{B-B} \longrightarrow 2 K_{A-B}$$

A substance is generally readily soluble in a solvent if the forces of attraction in the pure substance are of the same order of magnitude as the forces of attraction in the pure solvent. A substance is generally insoluble in a solvent if the forces of attraction between its molecules are significantly higher or lower than in the pure solvent. In this case more energy is required to overcome the forces of attraction in the pure components than is released on formation of the solution. This is the explanation of the rule of thumb "Like dissolves like" (*similia similibus solvuntur*). The intermolecular forces of attraction differ—they are strongest in crystalline solids, weaker in amorphous solids and liquids, and weakest in gases. Intermolecular forces are classified according to their physical nature (Table 1) [14.2]–[14.5].

Ionic (Coulomb) Forces. Forces of attraction between ions of opposite charge are termed ionic or Coulomb forces. The force with which two ions 1 and 2 attract one another depends on their electrical charges e_1 and e_2 and the distance r between them:

$$K_{II} \approx -\frac{e_1 \cdot e_2}{r^2}$$

Table 1. Intermolecular forces

Type of force	Interaction between	Temperature dependence
Ionic	ions	weak
Ion–dipole	ions and dipoles	weak
Directional	permanent dipoles	strong
Induction	permanent and induced dipoles	weak
Dispersion	atomic dipoles	weak
Hydrogen bonds	groups of molecules	strong

Coulomb forces are responsible for the stability of ionic crystals (e.g., NaCl). When such a compound is dissolved in a polar solvent (dipole moment μ), dissociation and simultaneous solvation of the ions occur. The force of attraction between the ions is now inversely proportional to the dielectric constant of the solvent, and is thus reduced. New ion–dipole forces are formed as a result of the attraction of the permanent dipoles of the solvent by the ions:

$$K_{ID} \approx -\frac{e_1 \cdot \mu}{r^3}$$

The distance between the solvated ions in the solution generally changes only slightly with temperature and depends on the thermal expansion coefficient of the solution. The forces between the ions are therefore only slightly temperature dependent.

Dipole–Dipole Forces. Dipole–dipole (directional) forces are forces of attraction between molecules with a finite, permanent overall dipole moment. The forces of attraction resulting from the dissolution of a polar molecule (μ_1) in a polar solvent (μ_2) are given by [14.6]:

$$K_{DD} \approx -\frac{\mu_1 \cdot \mu_2}{r^4}$$

The distance between the dipoles depends largely on the position of the poles in the molecule (i.e., on steric molecular influences) and on thermal vibrational movements. The force of attraction between the dipoles accordingly decreases sharply with increasing temperature.

Induction forces are produced as a result of interactions between permanent dipoles and induced dipoles. The electric field of a molecular dipole leads to charge displacement in the neighboring molecule and thus to the induction of a dipole.

The magnitude of the induced dipole moment μ_{ind} depends on the magnitude of the permanent dipole moment μ and on the polarizability α of the second molecule [14.7]. Induction forces are only slightly temperature dependent:

$$K_{DDi} \approx -\frac{\alpha \cdot \mu^2}{r^7}$$

Dispersion (London–Van der Waals) Forces. Dispersion forces are formed by mutual induction of atomic dipoles due to the electromagnetic field between the nucleus and electrons of the atom.

Dispersion forces therefore depend on the displaceability of the electrons in the atoms, the polarizability α, and the availability of the electrons, i.e., on the ionization potential IP [14.8]:

$$K_{\text{Dis}} \approx -\frac{\alpha^2 \cdot IP}{r^7}$$

Since charge displacement is independent of the thermal motion of the particles, dependence of the dispersion forces on temperature is weak over the distance between the attracting partners. Dispersion forces act in all atoms and molecules.

Hydrogen Bonds [14.9]–[14.13]. Hydrogen bonding forces exist in substances that have hydroxyl or amino groups (e.g., water, alcohols, acids, glycols, and amines). These molecules act as hydrogen donors and thus form a bond with hydrogen acceptors (e.g., esters and ketones). Water, alcohols, and amines act both as hydrogen donors and acceptors. Very weak hydrogen bonds also exist in halogens and sulfur. Hydrogen bonds are highly dependent on the mutual orientation of the molecules and thus on the temperature.

Thermodynamic Principles [14.14]. Forces of attraction act between the molecules of the pure components and between the different molecules in the solution. If the forces of attraction in the solution are greater than those in the pure components, dissolution is accompanied by a decrease in the internal energy of the system. The process is exothermic and heat is released. If, however, the forces of attraction between the molecules of the pure components are greater than those in the solution, the internal energy of the system is increased with absorption of heat. In a closed system, this endothermic dissolution process is accompanied by cooling. In open systems heat is absorbed from the surroundings.

Most dissolution processes are endothermic and are thus promoted by a temperature increase: the solubility has a positive temperature coefficient. Exothermic dissolution processes have a negative temperature coefficient (i.e., the solubility decreases with rising temperature). Miscibility gaps frequently occur above certain temperatures (Fig. 1).

Why do endothermic dissolution processes occur spontaneously? The Gibbs–Helmholtz equation provides the answer:

$$\Delta G = \Delta H - T\Delta S$$

The reaction enthalpy ΔH is negative in exothermic processes, and positive in endothermic processes. The reaction entropy ΔS, which is a measure of the change in the state of disorder of the system, is always positive since the entropy of the solution (mixture) is always greater than that of the pure components.

The Gibbs free energy ΔG must be a negative quantity in all spontaneously occurring processes. In exothermic processes ΔH is negative: dissolution occurs spontaneously with release of heat, and is "enthalpy determined." In endothermic

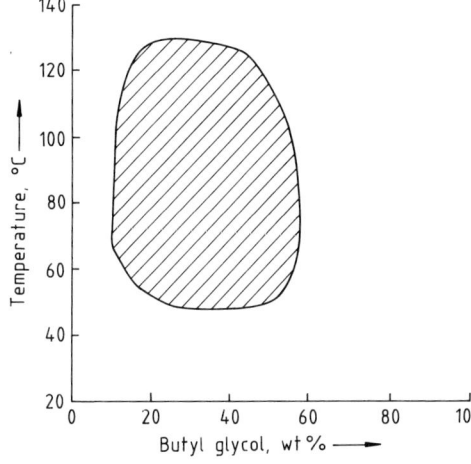

Figure 1. Miscibility gap for mixtures of butyl glycol and water at atmospheric pressure

processes ΔH is positive; these processes also proceed spontaneously, but the temperature of the system falls. On account of the large positive reaction entropy the negative contribution $- T\Delta S$ exceeds the positive reaction enthalpy ΔH, and ΔG is again negative. The process is "entropy determined." If, however, ΔH is positive and greater than $T\Delta S$, no dissolution takes place: the system remains heterogeneous.

Dilution processes are used very often, for example in the paint industry, and are always entropy-determined [14.15].

Systems with a negative temperature coefficient of solubility are mainly enthalpy-determined. The interaction between the partners is often due to weak hydrogen bonds, which become weaker than the forces between the pure components on increasing the temperature. The result is a decrease in solubility. Examples of enthalpy-determined systems with a negative temperature coefficient are summarized in Table 2.

The change in the total energy of the system can thus be negative or positive. This change in energy is termed heat of solution or mixing, or solution or mixing enthalpy [14.16], [14.17].

Systems that are miscible with one another without any change in temperature are termed athermic systems [14.18], [14.19]; they include the following solvent pairs:

Benzene – trichloromethane
Bromobenzene – chlorobenzene
Methanol – ethanol
Toluene – tetrachloromethane
Methyl acetate – ethyl acetate
Aniline – water

Negative heats of mixing (i.e., warming of the system) are found in the following solvent pairs:

Water – acetone
Water – methanol
Water – propanol
Trichloromethane – acetone

Table 2. Aqueous solution systems with a negative temperature coefficient of solubility

Solvent	Solubility in water	g/100 g
Methyl butyl ether	1.5 (10 °C)	0.9 (25 °C)
2-Methylfuran	18.2 (10 °C)	13.9 (25 °C)
Isoamyl alcohol	2.6 (22 °C)	2.2 (36 °C)
Ethyl acetate	10.0 (0 °C)	7.7 (25 °C)
Butyl glycol	∞ (40 °C)	11.0 (60 °C)

Positive heats of mixing (i.e., cooling of the system during mixing) occur with the following solvent pairs:

Trichloromethane–carbon disulfide
Acetone–carbon disulfide
Gasoline–carbon disulfide

Cohesive Energy Density and Solubility Parameters. As a result of attractive or cohesive forces the molecules in pure solvents have a cohesive energy that has to be expended in molecular separation processes (e.g., dilution, evaporation, or addition of another substance). The cohesive energy can be calculated from the enthalpy of vaporization ΔH_V and the work that is required to expand the vapor against the atmosphere (volume work) [14.20]. The cohesive energy per unit volume, i.e., the *cohesive energy density*, is defined as [14.16], [14.21], [14.22]:

$$e = \frac{(\Delta H_V - RT)}{V}$$

The cohesive energy density e is a specific temperature-dependent parameter for each solvent and represents a direct thermodynamic measure of the forces of attraction in solvents. It can be used to evaluate the solvency of the solvents if the cohesive energy densities of the substances to be dissolved are also known.

The cohesive energy density of a solvent A is altered when it is mixed with another solvent B. Two interaction pairs A–B are formed from each interaction pair A–A and B–B. The new cohesive energy density of the pair A–B in solvents of very low polarity being approximately equal to the geometric mean of the cohesive energy densities of the pure components [14.16], [14.21], [14.23]–[14.25]:

$$e_{AB} = (e_{AA} \cdot e_{BB})^{1/2}$$

The square root of the cohesive energy density is termed the *solubility parameter* δ. It is also a measure of the intermolecular forces in pure substances. Solvents with comparable solubility parameters have similar interaction forces and are therefore readily miscible and mutually soluble. In order to take account of not only the dispersion forces but also of the polar forces and hydrogen bonds, the solubility parameter is resolved into the following components [14.26], [14.27]:

$$\delta^2 = \delta_D^2 + \delta_P^2 + \delta_H^2$$

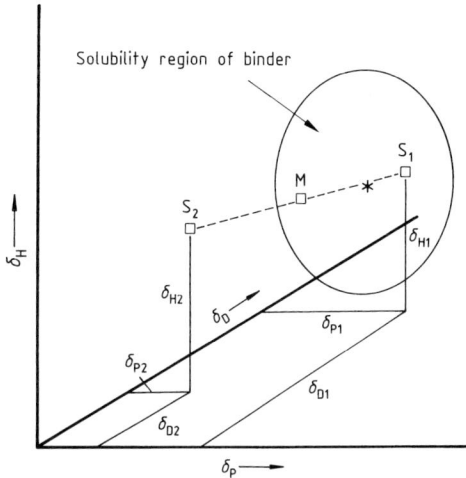

Figure 2. Solubility parameter diagram
S_1 = solvent; S_2 = non-solvent; M = solvent mixture

If the three solubility parameter components for dispersion forces δ_D, dipole forces δ_P, and hydrogen bonds δ_H are plotted on a three-dimensional space diagram (Fig. 2), a system is obtained in which a vector $\vec{\delta}$ is defined for each solvent. The vector describes the solvent's solubility and miscibility behavior [14.28], [14.29]. Solvents that lie close to one another in this space diagram (i.e., whose vector difference $\Delta\delta$ is small) have similar solution properties and often a similar chemical structure. Solvents that are far apart on the diagram differ greatly in their chemical and physical characteristics; they are generally immiscible [14.30]–[14.32]. The solubility parameters as well as their components are shown for some solvents in Table 15.

The predictions that can be made for solvent mixtures on the basis of the solubility diagram are not strictly valid because they involve thermodynamic simplifications and empirical parameters, and disregard temperature effects [14.29], [14.33]. A strong warning must therefore be given against the uncritical use of solubility parameters. Nevertheless, description of the solvents with the aid of the solubility parameter concept often provides useful information about their solvency and reveals similarities which can otherwise only be characterized empirically (e.g., latent solvents or dilutability).

Solvent Miscibility. As a rule solvents are readily miscible if the difference in their solubility parameters $\Delta\delta$ is $\leq 4-6$ units. Solubility is generally limited if the differences are larger (Table 3).

In the case of strongly polar (ionic) forces or hydrogen bonds, the solubility parameter concept is of limited applicability. For example, the following pairs of solvents are only partially miscible despite the small difference in their solubility parameters:

Nitromethane–ethyl glycol	3.5
Ethanol–methyl iodide	2.0
Decane–nitrobenzene	4.7
Ethanol–naphthalene	5.9

Table 3. Miscibility of solvents as a function of the difference in solubility parameters $\Delta\delta$

Solvent A	Solvent B	$\Delta\delta$, $(J/cm^3)^{1/2}$	Miscibility
Butyl acetate	methyl isobutyl ketone	0.2	good
Ethyl glycol acetate	xylene	0.2	good
Tetrachloroethylene	xylene	1.0	good
Ethanol	butanol	2.7	good
Methanol	dimethyl sulfoxide	3.3	good
Ethylene carbonate	dimethyl sulfoxide	3.7	good
Methyl glycol	hexane	7.2	poor
Cyclohexane	dimethyl sulfoxide	9.6	poor
Hexane	methanol	14.8	moderate

On the other hand, many solvent pairs are readily miscible although the difference between their solubility parameters is large:

Dimethyl sulfoxide – benzene	7.6
Nitromethane – tetrachloromethane	8.2
Dimethylformamide – hexane	9.9
Benzene – methanol	10.9
Ethanol – hexane	11.1
Methanol – tetrachloromethane	12.1
Ethylene carbonate – butyl acetate	12.7
Water – acetone	27.4
Water – tetrahydrofuran	28.7

In some cases the resolution of the solubility parameter into the three components δ_D, δ_P, and δ_H [14.31], [14.32], [14.34] is more suitable for explaining the experimental facts. For example, although the difference between the solubility parameters of the partially miscible solvent pair nitromethane – ethylene glycol is indeed small ($\Delta\delta = 3.5$), both solvents differ greatly as regards their solubility parameter δ_H ($\Delta\delta_H = 22.0$). Many empirical results (e.g., the good miscibility of benzene and methanol) cannot, however, be explained by this approach.

Solubility of Polymers [14.35]–[14.42]. Most polymers are readily soluble if their solubility parameters are comparable to those of the solvent in question. The upper limit for good solubility is a difference in the solubility parameters of 6 units (Table 4). The following selected systems are not, however, mutually soluble despite the small solubility parameter difference $\Delta\delta$:

Poly(vinyl chloride) – trichloromethane	1.6
Poly(vinyl chloride) – toluene	2.5
Polystyrene – heptane	2.9
Poly(vinyl acetate) – hexane	4.3
Poly(vinyl acetate) – diethyl ether	3.7
Cellulose – dimethyl sulfoxide	5.3

Furthermore, poly(vinyl acetate) is readily soluble in methanol and amylose is readily soluble in water, even though the differences in the solubility parameters are large ($\Delta\delta = 9.8$ and 22.5 units, respectively).

Table 4. Miscibility of polymer–solvent pairs as a function of the difference in solubility parameters $\Delta\delta$

Polymer	Solvent	$\Delta\delta$ (J · cm^3)$^{1/2}$	Miscibility
Polyacrylonitrile	ethylene carbonate	0.6	good
Polystyrene	ethyl acetate	0.8	good
Poly(vinyl chloride)	tetrahydrofuran	1.5	good
Polyisobutene	hexane	1.5	good
Polyacrylonitrile	dimethylformamide	4.9	good
Poly(vinyl acetate)	ethanol	7.0	moderate
Polyisobutene	ethanol	9.8	poor
Polystyrene	methanol	11.0	poor

14.2.2. Dipole Moment, Polarity, and Polarizability

In order to describe their solubility properties, solvents are often subdivided according to their polarity, i.e., polar and nonpolar solvents. Since the term polarity cannot be defined unambiguously in physical terms, such a classification is not meaningful. The term polarity includes parameters such as dipole moment, hydrogen bonding, polarizability, entropy, and enthalpy. The dipole moment μ of a substance is a molecular property resulting from the vector sum of bond-dipole moments. Highly symmetrical molecules (e.g., tetrachloromethane and benzene) accordingly have no dipole moment; other aromatic hydrocarbons and dioxane exhibit only very small dipole moments. Less symmetrical molecules with strong bond dipoles have dipole moments between 1.6 and 1.9 D (alcohols, esters, glycol ethers), glycols and ketones have higher values (2.3–2.9 D). The solvents with the largest dipole moments (3.7–5.0 D) are ethylene carbonate, nitropropane, dimethylformamide, and dimethyl sulfoxide (Table 15, p. 324).

The dissolution behavior of a solvent cannot be predicted solely on the basis of its dipole moment. For example, dioxane ($\mu = 0.4$ D) is a very good solvent and has a comparable solvency to dimethyl sulfoxide ($\mu = 4.0$ D).

Dipole–dipole and induction forces in solvents or solutions decrease with increasing molecular mass of the solvent [14.43]. Since this effect is not reflected in the dipole moment of the solvent, a polarizability parameter P is used to describe the dipole–dipole interaction forces [14.34]. This parameter can be calculated from the ionization potential IP, polarizability α, and dipole moment μ [14.44]:

$$P = \frac{2\,\mu^4}{3k \cdot T \cdot \varepsilon}$$

where $\varepsilon \approx \alpha \cdot IP$.

The dipole moment μ is also used in conjunction with the solubility parameter δ as a coordinate in solubility diagrams to take account of the influence of variations in polarity [14.44]–[14.46].

The polarizability α of an electrically neutral compound is a measure of the displaceability of charge carriers within the molecule. The greater the polarizability, the stronger are the dipoles induced by an external electromagnetic field; the magnitude of the polarizability and strength of the dispersion forces are thus related to one another.

14.2.3. Hydrogen Bond Parameters [14.47]

The strengths of hydrogen bonds in solvents have been divided into three classes [14.48]–[14.52]:

1) Solvents with weak hydrogen bonding (hydrocarbons, chlorinated hydrocarbons, nitro compounds, nitriles)
2) Solvents with moderately strong hydrogen bonding (ketones, esters, ethers, aniline)
3) Solvents with strong hydrogen bonding (alcohols, carboxylic acids, pyridine, water, glycols, amines)

This classification allows solvents to be described in terms of their solvency for polymers by means of solubility parameters, and also permits hydrogen bond forces to be taken into account. This is done by determining the solubility limits of the polymer in solvents belonging to each of the three classes. Three solubility parameter regions for a polymer are thus obtained. Attempts have also been made to use corrections to take account of the variations in solubility parameters caused by hydrogen bonding [14.53]–[14.56].

The hydrogen bond parameters γ provide a relatively accurate, simple characterization of solvents that can form hydrogen bonds [14.57], [14.58]. The parameter (Table 15) is determined from the shift of the oxygen–deuterium vibration frequency in the IR spectrum obtained when deuteromethanol is dissolved in the relevant solvent.

Solvents that undergo hydrogen bonding may act as proton donors or acceptors [14.9], [14.47]:

1) Proton donors (e.g., trichloromethane)
2) Proton acceptors (e.g., ketones, esters, ethers, aromatic hydrocarbons)
3) Combined proton donors and acceptors (e.g., alcohols, carboxylic acids, primary and secondary amines, water)
4) No hydrogen bonding (e.g., aliphatic hydrocarbons)

No hydrogen bonds exist in solvent mixtures comprising solely proton acceptors; hydrogen bonds are only formed in the presence of a proton donor and result in an increase in miscibility [14.47], [14.59].

14.2.4. Solvation

When a substance (solute) is dissolved in a solvent or solvent mixture, the forces of attraction between the solute molecules decrease because the solvent molecules penetrate between the solute molecules and finally surround them as a layer. This process, termed solvation, results in distribution of the solute in the solution at a molecular level. The strength of the solvation forces and the number of solvent molecules in the solvation layer depend on the solubility parameter, dipole moment, hydrogen bonding, polarizability, and the molecular size of the solvent and solute. The number of solvent molecules in a solvent–solute complex is determined by the degree of solvation β. Values for cellulose nitrate–ketone systems are given in Table 5 [14.60]. The degree of solvation β increases as the size of the solvent molecule decreases. It also increases with increasing solubility parameter δ. In the cellulose nitrate–ketone systems, hydrogen bond forces and dipole forces have little influence because both the binder and the solvent molecules are proton acceptors, and few hydrogen bonds are formed. Furthermore, the dipole moments of the solvents are roughly equal. In other cases, however, hydrogen bonding and dipole moments can greatly influence the degree of solvation.

14.2.5. Solvents, Latent Solvents, and Non-Solvents

True (active) solvents dissolve a given substance at room temperature; the solubility parameters of the solvent and the solute are similar.

Latent solvents cannot dissolve a substance by themselves. They can, however, be activated by adding a true solvent or a non-solvent. Latent solvents include solvents with extremely high or low solubility parameters or hydrogen bond parameters, such as alcohols, water, acetals, higher molecular mass esters, or ketones. In the solubility parameter space diagram, the latent solvents lie outside the solubility regions of the substances to be dissolved. On mixing with true solvents or non-solvents, mixed solvents are obtained whose solubility parameters fall in the region of the substances to be dissolved (Fig. 2).

Table 5. Connection between molecular mass of ketone solvents M, degree of solvation (for cellulose nitrate) β, solubility parameter δ, hydrogen bond parameter γ, and dipole moment μ

Solvent	M	β	δ	γ	μ
Methyl amyl ketone	114.18	1	8.5	7.7	2.7
Methyl propyl ketone	86.13	3	8.9	8.0	2.7
Methyl ethyl ketone	72.12	7	9.3	7.7	2.7
Acetone	58.08	12	10.0	9.7	2.9

Non-solvents are unable to dissolve the substance in question. Their solubility parameters and hydrogen bond parameters lie outside the solubility regions of the substances to be dissolved.

The expressions solvent, latent solvent, non-solvent only apply to a specific substance to be dissolved. A solvent can be a non-solvent for another substance and *vice versa*.

14.2.6. Dilution Ratio and Dilutability

If a non-solvent is added dropwise to a cellulose nitrate solution, the cellulose nitrate eventually separates as a precipitate or gel. The volumetric ratio of non-solvent to solvent that is still tolerated in the solution is termed the *dilution ratio*. The dilution ratio is an empirical, dimensionless quantity that provides information on the solvency of a solvent or solvent mixture. In order to determine the dilution ratio toluene or butanol is used as diluent (non-solvent).

Table 6 shows the dilution ratios for various diluents and solvents. As a rule the dilution ratio decreases with increasing boiling point of the solvent; however, the solvent of lowest molecular mass in a homologous series (e.g., methanol, methyl acetate, methyl glycol) often behaves differently.

Mixtures of two solvents frequently have a higher dilution ratio than the pure components: ether–alcohol and toluene–ethanol mixtures are better solvents for cellulose nitrate than any of the pure solvents; ethyl acetate and butyl acetate are

Table 6. Dilution ratios for cellulose nitrate

Solvent (volume ratio)	Diluent (non-solvent)			
	Toluene	Xylene	Aliphatic hydrocarbon	Butanol
Methyl acetate	3.0		0.9	
Ethyl acetate	3.4	3.3	1.0	8.4
Propyl acetate	3.0	2.0	1.2	
Butyl acetate	2.7	2.7	1.5	8.2
Ethyl glycol acetate	2.5	2.5	1.0	7.5
Acetone	4.5	3.9	0.6	7.0
Methyl ethyl ketone	4.5	3.3		
Methyl isobutyl ketone	3.5	3.7		
Cyclohexanone		7.5	1.0	
Methyl glycol	4.7	2.9	0.2	
Butyl glycol	4.0	3.2	2.3	
Methanol	2.5	2.0	0.3	
Butyl acetate/butanol (9/1)	3.5			
Butyl acetate/butanol (5/5)	5.5			
Methyl ethyl ketone/ethanol (85/25)	5.0			
Ethyl acetate/ethanol (1/1)	3.5			

better solvents for cellulose nitrate–synthetic resin combinations if they contain a proportion of ethanol or butanol, respectively. The solubility parameters of the mixed solvents are more comparable to those of the binder than those of the pure solvents.

The temperature dependence of the dilution ratio is shown in Table 7. The solvency of solvents consisting of small molecules increases with increasing temperature, whereas solvents consisting of large molecules behave in the opposite way. In other experiments, the dilution ratio generally decreases with increasing temperature, as has been demonstrated in cellulose nitrate solutions in butyl acetate, ethyl glycol, and methyl isobutyl ketone. The reason for this is that a cellulose nitrate solution adopts the gel state with a rise in temperature. The gel reverts reversibly to the liquid state when the temperature is lowered. However, for thermodynamic reasons the solvency of a solvent always increases with rising temperature. There is thus no meaningful connection between the dilution ratio and solvency of a solvent.

As expected, the dilution ratio falls with rising cellulose nitrate concentration in the solution (Table 8).

The dilution ratio reveals some similarities between the solvency of different solvents. However, most measurements are restricted to cellulose nitrate, and extrapolation of the results to other binders seldom provides correct results. Furthermore the dilution ratios, unlike solubility parameters, are unable to explain several phenomena (e.g., the enhancement of the solvency of a solvent produced on adding a non-solvent).

The apparent or theoretical *dilutability* denotes the dilution ratio that just results in a saturated solution. The true or practical dilutability predicts that this solution to which diluent has been added can still be applied satisfactorily, for example in the paint sector clear varnish films can form from these solutions after drying.

Table 7. Dependence of the dilution ratio on temperature (solute = cellulose nitrate; diluent = toluene)

Solvent	$-10\,°C$	$20\,°C$	$50\,°C$
Ethyl acetate	2.48	2.58	2.62
Butyl acetate	2.74	2.70	2.61
Amyl acetate	2.66	2.52	2.26
Octyl acetate	1.85	1.74	1.44

Table 8. Concentration dependence of the dilution ratio of cellulose nitrate solutions in butyl acetate

Cellulose nitrate, g	Butyl acetate, mL	Toluene, mL	Dilution ratio
1.0	20	75	3.75
2.0	20	71	3.55
3.0	20	67	3.35
4.0	20	63	3.15

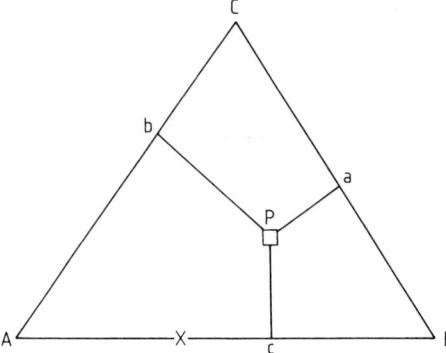

Figure 3. Triangular coordinates for determining the composition of a ternary mixture of solvents A, B, and C (for description, see text)

The limiting concentrations in a ternary system can be represented not only by a space diagram, but also by a three-coordinate system in the form of an equilateral triangle (Fig. 3).

The corners of the triangle represent the pure (100%) solvents A, B, and C. The point X on the side A–B gives the composition of a solvent mixture comprising B–X% of solvent A and A–X% of solvent B. The composition of a point P within the triangle can be determined by constructing perpendicular lines from P onto the sides of the triangle to obtain the points of intersection a, b, and c. The lengths P–a, P–b and P–c give the percentage contents of the solvents A, B, and C, respectively.

The sum of these lengths is taken as equal to 100%. Solubility regions can thus be determined in ternary systems by measuring the quantitative miscibility of the solvent A on dilution with solvents B and C or B–C mixtures.

14.2.7. Influence of Molecular Mass on Solubility

The solubility of chemically related compounds decreases with increasing molecular mass since the intermolecular forces of interaction increase. For example, benzene is completely miscible with ethanol, whereas anthracene and ethanol are only partially miscible. The influence of molecular mass on solubility is particularly evident in macromolecules. For example, alcohol, acetone, and acetic acid readily dissolve styrene, but not polystyrene; vinyl acetate dissolves in saturated hydrocarbons and ether, whereas poly(vinyl acetate) does not. Cellulose is insoluble in alcohols, poly(ethylene glycol) is insoluble in ethers, poly(vinyl chloride) is insoluble in vinyl chloride, and polyacrylonitrile is insoluble in acetonitrile, even though good solubility would be expected on account of the chemical relation between the polymers and monomers.

On account of their very large molecular mass, highly cross-linked polymers do not dissolve in solvents, even at elevated temperature. They are, however, swollen by solvents depending on the nature and density of the cross-linking sites [14.61], [14.62].

14.2.8. Dissolution and Solution Properties

Rate of Dissolution. Section 14.2.1 dealt with the thermodynamic equilibrium between the solvent and substance to be dissolved; the mixing time was disregarded. Dissolution proceeds, however, at a finite rate which depends on the specific surface area of the substance to be dissolved, its degree of crystallisation, its rate of diffusion into the solvent, and on the temperature. Often a substance is considered to be sparingly soluble if its rate of dissolution is low. This assumption is, however, incorrect [14.63].

Miscibility Gap. Some solvent pairs are miscible with one another in all proportions only above or below a critical dissolution temperature. The region in the concentration–temperature diagram in which both partners are immiscible with one another is termed a miscibility gap [14.16], [14.64], [14.65] (Figs. 1 and 4). Miscibility gaps may occur, if intermolecular forces are strongly temperature dependent. In the mixture triethylamine/water for example, the hydrogen bond N $\cdots$ H–O is weak. At temperatures higher than 17 °C the hydrogen bond will be destroyed and immiscibility occurs. Hexane and nitrobenzene are completely miscible only above 19.5 °C (Fig. 4A). Other systems with an upper critical dissolution temperature include:

Aniline–water (168 °C),
2-Chlorobenzoic acid–water (126.2 °C),
Water–phenol (68.8 °C)
Hexane–aniline (65.9 °C)
Cyclohexane–methanol (45.6 °C)

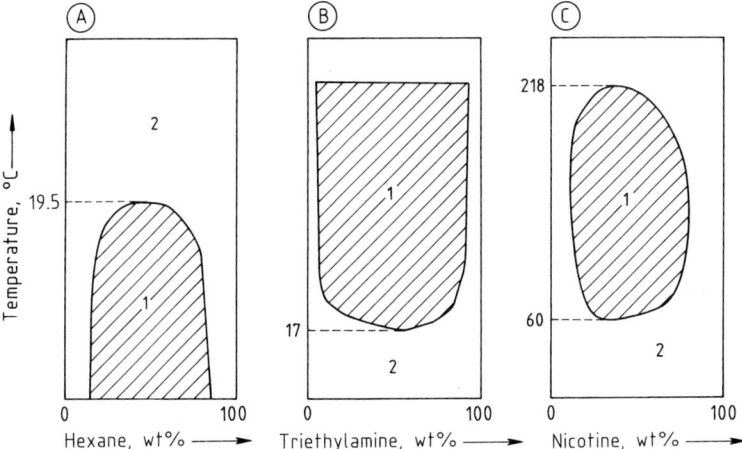

Figure 4. Two-phase solvent systems with miscibility gaps
A) Hexane–nitrobenzene; B) Triethylamine–water; C) Nicotine–water
1 = two-phase region; 2 = one-phase region

Systems with a lower critical dissolution temperature include:

Triethylamine–water (17 °C, Fig. 4 B)
Ethylpiperidine–water (7.5 °C)

Systems with an upper and lower critical dissolution temperature include nicotine–water (60 and 218 °C, Fig. 4C).

The miscibilities of various solvents with water are given in Table 16.

Physical Properties of Mixtures. The physical properties of an ideal mixture (e.g., vapor pressure) are obtained from the arithmetic mean of the properties of the individual components. In most mixtures, however, behavior is not ideal and the vapor pressure is greater or less than the sum of the values of the components. The reason for this is that the interaction forces in the pure components are different from those in the mixture. If the molecular forces of interaction in the pure components are lower than in the mixture, the vapor pressure of the mixture is lower than the sum of the pressures of the components (e.g., acetone–trichloromethane). In most cases, however, the forces of interaction in the pure components are higher than in the mixture, with the result that the vapor pressure increases on mixing (e.g., benzene–ethanol).

Similar variations in the densities and surface tensions of mixtures are also known. The density of a mixture of bromo- and chlorobenzene is equal to the arithmetic mean of the densities of the individual components, whereas the density of trichloromethane–diethyl ether is about 1.5% above the mean value. The surface tensions of the systems diethyl ether–benzene and benzene–carbon disulfide are equal to the calculated values. The surface tension of methanol–ethyl iodide is greater and that of benzene–ethanol and acetone–carbon disulfide are less than the calculated values.

Azeotropes. Molecular associations between the components of a mixture can result in systems that have a constant boiling point at a given concentration. The boiling point may be lower or higher than that of the individual components. Such mixtures are termed azeotropic mixtures or azeotropes. Low-boiling and high-boiling mixtures are denoted as minimum or maximum azeotropes, respectively. Examples of low-boiling azeotropes [14.66], [14.67] are benzene–water (*bp* 69.4 °C) and benzene–ethanol (*bp* 67.8 °C). Acetone and chloroform form a high-boiling azeotrope (*bp* 64.7 °C).

The composition of the liquid phase of the azeotrope is the same as that of the vapor phase with which it is in equilibrium. It depends on temperature and pressure, and azeotropically boiling mixtures are not necessarily azeotropically evaporating mixtures. However, azeotropically evaporating mixtures often form even at room temperature; their composition differs from that of the azeotropically boiling mixtures. Azeotrope formation can be suppressed by lowering the boiling point (vacuum distillation) and the components can then be separated if the vapor pressures of the pure components respond differently to a change in temperature. This is always the case if the heats of vaporization of the pure components differ widely.

The formation of low-boiling azeotropes is a fortunate phenomenon for the paint technologist because water and solvents then evaporate more rapidly than usual (azeotropes, see Table 17). However, these azeotropic mixtures also have several disadvantages:

1) The flash point is lower
2) The explosion limit is lower
3) The higher evaporation rate may have a negative effect on the paint surface
4) Paint flow may be poor because the azeotropic mixture constantly evaporates

Systems Containing More Than Two Components. As in binary systems, the behavior of systems containing more than two components can be understood on the basis of intermolecular forces and solubility parameters. Water and tetrachloromethane have widely differing solubility and hydrogen bond parameters, and are therefore immiscible. Added acetone dissolves partly in the aqueous phase due to hydrogen bond formation, and partly in the tetrachloromethane phase due to dispersion and induction forces. Twice as much acetone dissolves in the aqueous phase as in tetrachloromethane. On increasing the acetone concentration a homogeneous solution is obtained. The added solvent thus acts as a *solubilizer* for the two immiscible solvents.

Solvents having average solubility and hydrogen bond parameters are generally suitable as solubilizers, particularly ketones and glycol ethers. Butyl glycol, diglycol, and triglycol are often used because they contain hydrophilic and hydrophobic groups which effect miscibility between partners having widely differing solubility and hydrogen bond parameters. Very small amounts of a solubilizer are often sufficient to homogenize a heterogeneous solvent mixture. The two-phase system comprising hexane and methanol (mass ratio 2:1) is rendered homogeneous by adding 0.5 wt% dichloromethane.

Addition of solvents as solubilizers is important in the dissolution of polymers. Poly(vinyl acetate) is insoluble in pure ethanol, but dissolves fairly readily in ethanol containing 3% water. Cellulose triacetate is sparingly soluble in pure trichloromethane, but is readily soluble in trichloromethane containing 3% methanol. Many paint resins dissolve more readily in aromatic-containing naphtha fractions than in pure naphtha. Poly(vinyl chloride) is sparingly soluble in acetone and carbon disulfide, but is more readily soluble in a mixture of the two solvents.

14.3. Physical and Chemical Properties
see Tables 22 and 23.

14.3.1. Evaporation and Vaporization

Solvents are classified according to their boiling point ranges into:

1) Low boilers: $bp < 100\,°C$
2) Medium boilers: $bp\ 100-150\,°C$
3) High boilers: $bp > 150\,°C$

294   14. Solvents

The boiling point of a liquid is defined as the temperature at which the vapor pressure of the liquid reaches 101.3 kPa. Thermal energy is consumed in the evaporation of a liquid and is extracted from the surroundings, resulting in cooling.

A knowledge of the boiling point ranges of solvents is important in distillation and extraction processes. However, in the paint and adhesive sectors a knowledge of solvent volatility below the boiling point is more important. Below the boiling point the liquid is in equilibrium with its vapor, the vapor pressure p_s is < 101.3 kPa and is determined from the vapor pressure curve according to the Clausius–Clapeyron equation (integrated form):

$$\ln p_s = A - \frac{\Delta H_V}{RT}$$

where A is a general constant, ΔH_V the molar enthalpy of vaporization, R the gas constant, and T the absolute temperature. The higher the temperature, the greater the vapor pressure (Fig. 5); ΔH_V increases with increasing boiling point of the solvent according to the Pictet–Trouton rule:

$$\frac{\Delta H_V}{T_s} = \text{constant}$$

where T_s is the absolute boiling point.

There is no general correlation between the evaporation rates and boiling points of solvents. Solvent volatility generally decreases with increasing boiling point if the solvents are chemically related [14.68]. Solvents that tend to form hydrogen bonds (e.g., water, alcohols, amines) are less volatile than other solvents with the same boiling point since energy has to be supplied to rupture the hydrogen bonds before the transition to the vapor state occurs. The evaporation rate w of solvents is obtained empirically according to the Knudsen equation [14.69], [14.70]:

$$w = KP_V \sqrt{\frac{2\pi RT}{M}}$$

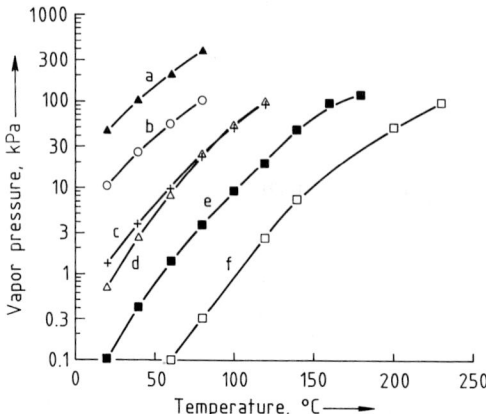

Figure 5. Vapor pressure curves of solvents
a) Dichloromethane; b) Ethyl acetate;
c) Butyl acetate; d) Butanol; e) Butyl glycol; f) Butyl diglycol

where the transmission coefficient $K \approx 1$, P_V is the vapor pressure, M is the molecular mass, and w is expressed in $g/cm^{-2} s^{-1}$.

The evaporation rate of solvents depends on:

1) Vapor pressure at the processing temperature
2) Specific heat
3) Enthalpy of vaporization
4) Degree of molecular association
5) Rate of heat supply
6) Surface tension
7) Molecular mass of the solvent
8) Atmospheric turbulence
9) Atmospheric humidity

Since these factors all depend on one another, it is impossible to give a theoretical prediction of the rate of evaporation [14.69]–[14.75]. In practice, the evaporation time of a given amount of solvent is determined experimentally under identical external conditions and compared with that of diethyl ether, or in some countries with that of butyl acetate.

Relative evaporation rates:

$$E \text{ (diethylether)} = \frac{t \text{ (test solvent)}}{t \text{ (diethylether)}}$$

t = evaporation time of the test solvent or diethylether in sec

$$E \text{ (butylacetate)} = \frac{t_{go} \text{ (butylacetate)}}{t_{go} \text{ (test solvent)}}$$

t_{go} = the time, where 90% of the test solvent or butylacetate evaporates in a given type of apparatus under strictly controlled conditions [14.75a]

The results are termed *evaporation numbers* (Table 9) and are referred to diethyl ether = 1. Evaporation numbers are listed in Table 23.

Solvents can be subdivided into four groups on the basis of their evaporation numbers:

1) High volatility: < 10
2) Moderate volatility: 10–35
3) Low volatility: 35–50
4) Very low volatility: > 50

A similar volatility classification of solvents is adopted in the United States with butyl acetate = 1 being chosen as the reference:

Class 1: rapid evaporation, > 3.0
Class 2: moderate evaporation, 0.8–3.0
Class 3: slow evaporation, < 0.8

If a substance with a relatively high molecular mass is dissolved in a solvent, its rate of evaporation decreases [14.76]–[14.79]. Solvents evaporate more rapidly from

14. Solvents

Table 9. Evaporation of glycol ethers (percentage evaporated as a function of time at 23 °C)

Solvent	Time, h										Evaporation number (diethyl ether = 1)
	10	15	20	25	50	100	150	200	300	400	
Ethyl glycol	40	58	77	93							43
Butyl glycol					30	60	92				119
Ethyl diglycol					0	3	11	18	34	51	1200
Butyl diglycol							0	3	6	9	> 1200

cellulose ester and cellulose ether solutions than from polymer or resin solutions. The volatility of a solvent in a polymer film depends on the difference between the solubility parameters of the solvent and the polymer. The polymer occupies a certain space in the solubility parameter diagram with coordinates δ_D, δ_P, and δ_H. Solvents that can dissolve the polymer are situated outside it (see Fig. 2). A solvent whose parameter lies close to the center of the polymer solubility space interacts strongly with the polymer; the solvent therefore evaporates slowly from the drying polymer film, and may even be retained. Solvents that interact weakly with the polymer are situated at the boundary of the polymer solubility space (i.e., in the transition region between solvents and non-solvents) and therefore evaporate more readily from the polymer film [14.32].

Solvents should not only evaporate rapidly after the application of a paint film, but should also promote the formation of a coating as follows:

1) Prevent blushing and absorption of condensed water vapor from the atmosphere, which is formed as a result of cooling due to solvent evaporation
2) Control the paint flow
3) Prevent precipitates and turbidity in the paint film
4) Prevent film shrinkage

A paint formulation therefore always contains several solvents: low boilers, medium boilers, and a small proportion of high boilers. The vapor pressure of the individual solvents in the mixture is described by Raoult's law:

$$p_i = f_i \cdot x_i \cdot p_i^\circ$$

where p_i is the partial pressure of component i, x_i the mole fraction of component i, and p_i° the vapor pressure of the pure component. The activity coefficient f_i, which may be larger or smaller than 1, takes account of the real behavior of the solvent (i.e., deviations from the ideal state).

14.3.2. Hygroscopicity

Some solvents, particularly those containing hydroxyl groups (e.g., alcohols and glycol ethers) are hygroscopic: they absorb moisture from the atmosphere until an

Table 10. Hygroscopicity of glycol ethers expressed as equilibrium water content (wt %) of solutions in contact with air of different relative atmospheric humidities (101.3 kPa, 25 °C)

Solvent	Relative atmospheric humidity, %						
	40	50	60	70	80	90	100
Ethyl glycol	5	8	12	20	31	48	77
Butyl glycol	4	7	11	18	27	37	49
Ethyl diglycol	6	8	14	20	30	40	53
Butyl diglycol	3	6	9.5	17	25	37	49
Ethylene glycol	18	26	37	47	60	74	89

equilibrium is reached. The equilibrium water content of the solvent depends on the relative atmospheric humidity and temperature. The hygroscopic properties of glycol ethers have been investigated in detail (Table 10).

14.3.3. Density and Refractive Index

In addition to the boiling curve, the density and refractive index are used to assess and, with some restrictions, to determine solvent purity.

The *density* of a solvent is generally measured at 20 °C and referred to the density of water at 4 °C (relative density d_4^{20}). The densities of most organic solvents decrease with increasing temperature (Fig. 6) and are less than that of water, but halogenated hydrocarbons are denser than water. The relative densities (d_4^{20}) of homologous

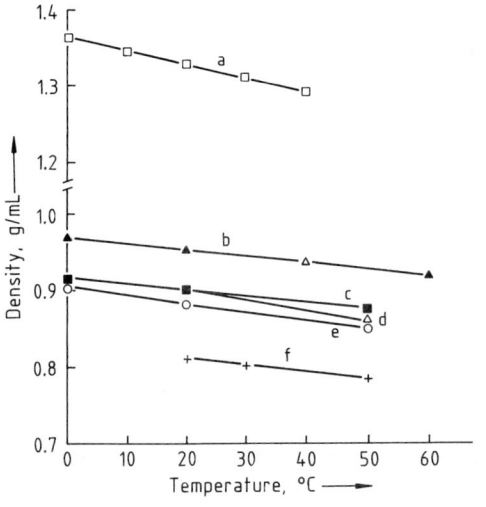

Figure 6. Temperature dependence of solvent densities a) Dichloromethane; b) Butyl diglycol; c) Butyl glycol; d) Ethyl acetate; e) Butyl acetate; f) Butanol

esters and glycol ethers decrease with increasing molecular mass, whereas those of ketones and alcohols increase:

Esters		Alcohols	
Methyl acetate	0.934	Methanol	0.791
Ethyl acetate	0.901	Ethanol	0.789
Propyl acetate	0.886	Propanol	0.804
Butyl acetate	0.881	Butanol	0.810
Amyl acetate	0.876	Amyl alcohol	0.815
Glycol Ethers		*Ketones*	
Methyl glycol	0.966	Acetone	0.792
Ethyl glycol	0.931	Methyl ethyl ketone	0.805
Propyl glycol	0.911	Methyl propyl ketone	0.807
Butyl glycol	0.902	Amyl methyl ketone	0.816

The *refractive index* n_D is measured in a refractometer with a sodium vapor lamp (Na-D lines, 589.0 and 589.6 nm). The value of the refractive index [14.45], [14.46], [14.80] is largely determined by the hydrocarbon skeleton of the substance in question. Aliphatic esters, ketones, and alcohols have refractive indices between 1.32 and 1.42. In homologous series the refractive index increases with increasing length of the carbon chain, and decreases with increasing branching. Cycloaliphatic and aromatic structures increase the refractive index (n_D^{20}), as does the incidence of functional groups:

Methyl acetate	1.3610	Cyclohexanol	1.4667
Ethyl acetate	1.3718	Benzyl alcohol	1.5390
Propyl acetate	1.3844	Methylbenzyl alcohol	1.5270
Isopropyl acetate	1.3773	Toluene	1.4955
Butyl acetate	1.3961	*p*-Xylene	1.4956
Isobutyl acetate	1.3898	Tetrahydronaphthalene	1.5443
Methanol	1.3290		
Ethanol	1.3619	Ethyl glycol	1.4075
Propanol	1.3859	Ethylene glycol	1.4310
Isopropyl alcohol	1.3772	Diethylene glycol	1.4460
Butanol	1.3994	Dichloromethane	1.4234
Isobutanol	1.3960		

The refractive index decreases with increasing temperature (Fig. 7).

14.3.4. Viscosity and Surface Tension

The *viscosities* of homologous series of solvents (Table 22) increase with increasing molecular mass. Solvents bearing hydroxyl groups have higher viscosities due to the formation of hydrogen bonds. The solvent viscosity greatly influences solution viscosity (see Section 14.7.1). The viscosity decreases with increasing temperature (Fig. 8).

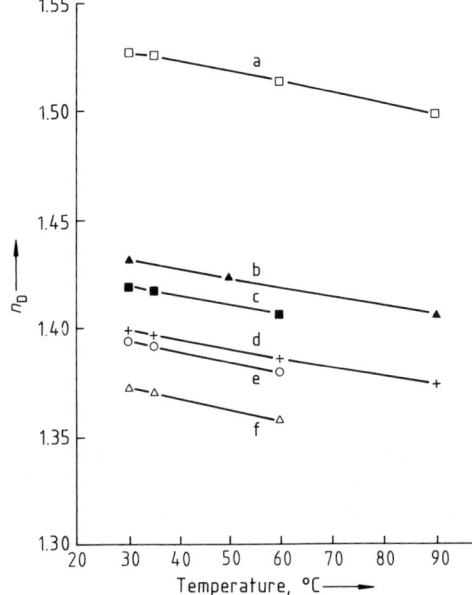

Figure 7. Temperature dependence of the refractive index of solvents
a) Methylbenzyl alcohol; b) Butyl diglycol; c) Butyl glycol; d) Butanol; e) Butyl acetate; f) Ethyl acetate

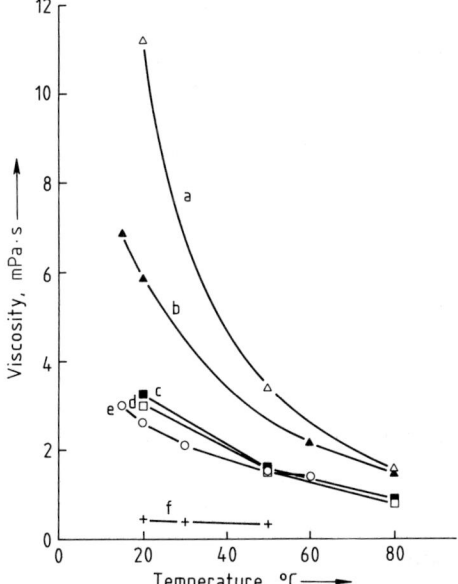

Figure 8. Temperature dependence of solvent viscosities a) Methylbenzyl alcohol; b) Butyl diglycol; c) Butyl glycol; d) Butanol; e) Isophorone; f) Ethyl acetate

The surface tension of paint solvents is of importance for the rate of evaporation, for the formation of the coating surface, and as for the wetting of the substrates, extenders, and pigments.

The *surface tension* of a solvent is related to the cohesive energy density and internal pressure of the liquid. A relationship can be derived between the solubility parameter δ (which is defined as the square root of the cohesive energy density) and the surface tension γ_o [14.81]. For polar and nonpolar solvents the following relationship applies:

$$\delta = 2.1 \, K \left(\frac{\gamma_o}{V^{1/3}} \right)^a$$

where V is the molar volume and K and a are constants ($K \approx 3.6$, $a \approx 0.56$). This empirical relationship is extremely satisfactory (Table 18, p. 328).

14.3.5. Vapor Density

The vapor density is the mass of solvent vapor per cubic meter of air (kg/m³) that is in equilibrium with the liquid at 101.3 kPa. The vapor density thus corresponds to the solvent content in the atmosphere at saturation and is temperature dependent. The relative vapor density of solvents d_S is referred to the density of air and can be calculated according to:

$$d_S = \frac{M_S}{M_{air}}$$

where M_S is the molecular mass of the solvent, M_{air} is the mean molecular mass of air ($= 28.95$ g/mol).

In the ideal case, the relative vapor density is temperature independent. Relative vapor densities of some solvents follow:

Toluene	3.18
Xylene	3.67
Ethanol	1.59
Butanol	2.56
Ethyl acetate	3.04
Butyl acetate	4.01
Methyl ethyl ketone	2.49
Ethyl glycol	3.11
Dichloromethane	2.93

14.3.6. Thermal and Electrical Data

The dielectric constant and thermal conductivity decrease with increasing temperature, whereas the specific heat increases. The thermal conductivities, cubic expansion coefficients, dielectric constants, and electrical conductivities of various solvents are listed in Table 11. Critical data of solvents and the technical use of supercritical liquids are described in [14.82].

Table 11. Thermal and electrical properties of solvents

Solvent	Thermal conductivity (20 °C), $Wm^{-1}K^{-1}$	Cubic expansion coefficient, $K^{-1} \times 10^{-3}$	Dielectric constant (20 °C)	Electrical conductivity, S/cm
Cyclohexane			2.01	1.9×10^{-14}
Xylene		0.9	2.3	10^{-5}
Dichloromethane	0.159	1.3	7.0	4.3×10^{-11}
Tetrachloroethylene	0.16	1.08	2.5	5.6×10^{-4}
Ethanol	0.171	1.1	22.4	1.4×10^{-9}
Butanol		0.79	18.2	9.1×10^{-9}
Butyl acetate	0.14 (30 °C)	1.21	4.5	5.2×10^{-11}
Ethyl glycol acetate	0.143 (25 °C)	1.12	9.0	1.2×10^{-8}
Methyl ethyl ketone		1.27	18.4	3.6×10^{-9}
Methyl isobutyl ketone		1.15	13.1	3×10^{-9}
Butyl glycol	0.162 (30 °C)	0.92	9.2	4.3×10^{-7}
Butyl diglycol	0.139	0.85	11.0	2×10^{-8}
Diethyl ether	0.132		4.34	4×10^{-13}
Tetrahydrofuran	0.14	1.29	7.6	0.5×10^{-8}

14.3.7. Flash point, Ignition Temperature, and Ignition Limits

The temperature at which a solvent vapor–air mixture is ignited by a naked flame is termed the *flash point* of the solvent. The flash points of solvents increase with decreasing vapor pressure and thus with increasing molecular mass and boiling point. They form the basis for their classification according to flammability (danger classes, transport regulations, Section 14.5.3). Flash points of some solvents follow:

Methyl acetate	−13 °C	Acetone	−19 °C
Ethyl acetate	−12 °C	Methyl ethyl ketone	−14 °C
Propyl acetate	12 °C	Methyl propyl ketone	22 °C
Butyl acetate	22 °C	Methyl butyl ketone	23 °C
Amyl acetate	34 °C	Methyl isobutyl ketone	15 °C
Ethyl glycol acetate	52 °C	Methyl amyl ketone	49 °C

The absolute flash point T_F of a solvent is related to its absolute boiling point T_S according to the following empirical equation [14.83]:

$$T_F = 0.736 \, T_S$$

Flash points calculated according to this equation are sufficiently accurate (deviation ± 5 K) in the case of hydrocarbons, ketones, and esters. The equation is, however, unsatisfactory in the case of solvents containing hydroxyl groups such as alcohols, glycols, and glycol ethers (deviation up to ± 30 K).

A relationship between the flash point and vapor pressure p has also been derived:

$$T_F = a \log p + b$$

where a is a general constant and b is a substance-specific constant.

For safety reasons numerous attempts have been made to raise the flash point of solvent mixtures [14.84]. Azeotropes having a lower flash point than those of the pure components can be formed on mixing solvents; use of solvent mixtures to raise the flash point is therefore extremely problematic. The flash point of toluene can, for example, be significantly raised above 21 °C by adding chlorinated hydrocarbons [14.85], [14.86]. However, the flash point is first lowered instead of raised if dichloromethane is added. Non-flammability is achieved only above a minimum addition of dichloromethane.

The flash point of a solvent mixture is not identical to that of its most flammable component. When solvents with widely differing hydrogen bond parameters are mixed (e.g., alcohol–hydrocarbon), the flash point is significantly reduced. On the other hand, the flash point of a mixture of chemically related solvents lies between those of the individual components [14.87]–[14.89]. Methods have been developed for calculating the flash point of solvent mixtures and solutions, activity coefficients are used to account for nonideal behavior [14.90].

Flash point determinations on mixtures of xylene with polar solvents led to the following conclusions [14.91], [14.92]:

1) Small quantities of solvents containing carbonyl groups (esters, ketones) reduce the flash point, larger amounts increase it again
2) The flash point is raised by adding 4 wt% of butanol or 8 wt% of isobutanol
3) Small additions of water do not significantly influence the flash point
4) Addition of ethanol sharply reduces the flash point; ethyl glycol acts in the same way, although to a lesser extent

In order to evaluate the flammability of water-thinnable paints, a knowledge of the increase in the flash points of water-miscible solvents after addition of water is important. These mixtures no longer ignite above a specific ratio of water to organic solvent (following results at 25 °C):

1) Isopropyl alcohol–water mixtures are still combustible in a 40/60 ratio, but not combustible in a 30/70 ratio
2) Propanol–water mixtures are still combustible in a 70/30 ratio, but not combustible in a 60/40 ratio

Flashpoints of solvent–water mixtures are given in Table 12.

A solvent vapor is ignited not only by a naked flame, but can spontaneously ignite when the solvent vapor–air mixture has reached the *ignition temperature*. A distinc-

Table 12. Flash points (°C) of solvent–water mixtures

Flash point	Solvent to water ratio			
	1:0	9:1	1:1	1:9
Ethanol	10	19	25	51
Isopropyl alcohol	12	19	24	41
Ethyl glycol	46	57	72	
Methyl ethyl ketone	−9	−7	−6	5–9
Butyl glycol	60	> 65		

tion is made between the gas ignition temperature and drop ignition temperature, depending on whether the measurement is made by determining the gas temperature or by allowing the solvent to drop onto a hot surface of known temperature [14.93]. The value of the ignition temperature is used to group solvents into temperature classes (ignition groups). For example, VDE Regulation 0171 specifies five groups:

T5 100–135 °C   T2 300–450 °C
T4 135–200 °C   T1   > 450 °C
T3 200–300 °C

An explosion is a self-propagating, particularly rapid combustion process that is initiated by an ignition and proceeds without external energy or air. The explosion of a solvent vapor–air mixture is possible only within a specific solvent concentration range, defined by the lower and upper ignition limits (explosion limits) which are specified in volume percent or grams per cubic meter of solvent in the solvent vapor–air mixture at 101,3 kPa.

The temperatures at which the concentrations of solvent vapor in equilibrium with the liquid reach the lower or upper ignition limit are obtained from the vapor pressure curves of the solvents; these temperatures are termed the lower and upper explosion points. The lower explosion points roughly correspond to the flash points of the solvents [14.94]–[14.98].

14.3.8. Heats of Combustion and Calorific Values

The combustion of organic solvents releases energy since organic compounds generally have higher energy contents than their combustion products. The combustion energy, also termed heat of combustion or gross calorific value, is given in kilojoules per mole or kilojoules per kilogram.

The gross calorific value refers to liquid water of reaction, whereas the net calorific value refers to the water vapor that is formed. The two calorific values thus differ by the contribution of the enthalpy of vaporization of the water formed during combustion.

The heats of combustion of alcohols, esters, and glycol ethers are very much smaller than those of hydrocarbons (Table 19, p. 329). Electronegative substituents generally reduce the combustion energy in the order $NH_2 > Cl > OH > SH$. The larger the number of substituents in the molecule, the lower the combustion energy. A knowledge of the calorific values of solvents is of great industrial importance because they decisively influence the size and cost of combustion and solvent incineration plants [14.99], [14.100].

14.3.9. Chemical Properties

A high chemical resistance is an important prerequisite for the use of a liquid as a solvent. Aliphatic and aromatic hydrocarbons are chemically inert and thus satisfy this requirement extremely well.

Alcohols too are chemically very resistant but they react with alkali metals, alkaline-earth metals, and aluminum to form salts. Under certain conditions alcohols can be converted into carboxylic acids by powerful oxidizing agents. They are, however, stable toward atmospheric oxygen. On account of their reaction with isocyanates to form urethanes, solvents containing hydroxyl groups (e.g., alcohols, glycols, and glycol ethers) must not be used as solvents for polyurethane paints [14.101], [14.102].

On prolonged storage most ethers and glycol ethers form peroxides with atmospheric oxygen. In the case of glycol ethers this leads to acid formation due to oxidation which can be prevented by adding stabilizers.

Esters and ketones are chemically very resistant under normal conditions, especially in the paint industry. However, it must be remembered that esters can be hydrolyzed to form alcohols and acids. For ethyl acetate the equilibrium constant K at 40 °C is 2.51, and at 100 °C is 2.56.

$$K = \frac{[\text{Ethyl acetate}] \cdot [\text{H}_2\text{O}]}{[\text{Acetic acid}] \cdot [\text{Ethanol}]}$$

In a neutral medium hydrolysis begins extremely slowly, but is autocatalyzed by the acetic acid product. The hydrolysis kinetics proceed according to:

$$\frac{dx}{dt} = k(a-x)\sqrt{x}$$

where k is the reaction constant, a is the initial concentration of the ester, and x is the acid concentration. For example, when 44 g of ethyl acetate is dissolved in 1 L of water at 20 °C, about 15 g is hydrolyzed after 130 days. Under acid or base catalysis the reaction rate is 10^5-fold or 10^8-fold higher, respectively.

The hydrolysis rate of the ester depends on its chemical structure. The rates of alkaline hydrolysis of isobutyl and butyl acetates are greater at room temperature than that of *sec*-butyl acetate; *tert*-butyl acetate is even more stable under alkaline conditions (iso $\approx$ n > sec $\gg$ tert). The rates of acid hydrolysis of the butyl acetates decrease in the order iso > n > tert > sec.

Chlorinated hydrocarbons can release hydrogen chloride in the presence of bases or metals, but stabilizers added by the manufacturer ensure high chemical resistance. Highly stable solvents are nitro compounds, dimethylformamide, and dimethyl sulfoxide, which, however, often become dark on prolonged storage in air.

Before using a solvent it is advisable to become acquainted with the chemical properties so that interactions can be anticipated and explained. A comprehensive literature exists on solvent effects in chemical reactions [14.2], [14.102], [14.103].

14.4. Toxicology and Occupational Health

14.4.1. Toxicology [14.103 a]

Acute and Chronic Toxicity. Solvents act with different intensities on human, animal, and plant organisms. Their effects depend on the amount of solvent and the exposure time. Under short-term exposure to high solvent doses acute damage may occur, whereas the absorption of smaller amounts over a longer period leads to chronic damage and sensitization. The chronic effects are more dangerous since they are accompanied by an acquired tolerance, with the result that they are often not detected early enough.

In order to evaluate the solvent toxicities, their LD_{50} values (oral administration to rats) are listed in Table 23. Since most cases of solvent poisoning are caused by inhalation of solvent vapors, the LC_{50} values are also given.

Inhaled solvent vapors pass via the lungs and blood circulation into the body, where they accumulate in tissues with high lipid content (e.g., nerves, brain, bone marrow, adipose tissue, liver, and kidneys) [14.104]. The cells can either be damaged by the solvents or by their decomposition products. Solvents can also pass into the body via cutaneous or, more rarely, gastrointestinal absorption [14.105], [14.106].

Symptoms of acute solvent poisoning include dizziness, drowsiness, headache, loss of consciousness, and narcotic effects which are attributed to disturbances of the central nervous system. Chronic poisoning is initially undetectable, but subsequently causes damage to organs that are specific for each solvent [14.107]–[14.110].

Solvents have two effects on the skin: (1) they dissolve the natural fatty layer, the skin therefore cracks and microorganisms and dirt particles can penetrate more easily and cause infection; (2) solvents can act directly to cause inflammation or blistering.

The following solvents are absorbed very readily through the skin and pass into the body: aniline, benzene, butyl glycol, butyl glycol acetate, dimethylacetamide, dimethylformamide, dioxane, ethyl glycol, ethyl glycol acetate, ethylbenzene, isopropyl glycol, carbon disulfide, methanol, methyl glycol, methyl glycol acetate, methylcyclohexane-2-one, 4-methyl-2-pentanol, nitrobenzene, nitrotoluene, isopropylbenzene, 1,1,2,2-tetrachloroethane, and tetrachloromethane.

After sensitization of the skin or respiratory system, allergic reactions may occur but vary widely, depending on the individual susceptibility. Observance of the MAK or TLV values is no guarantee against the occurrence of such reactions [14.111]. Turpentine oil is the only solvent that has attracted attention because of its ability to trigger hypersensitivity reactions of an allergic nature. Investigations have shown that some liquid products have carcinogenic, mutagenic and reproduction toxic (including teratogenic, embryotoxic) properties.

The carcinogenic, mutagenic and reproduction toxic properties in the EC classification are, in the jargon used, referred to as "cmr" properties [14.103a]:

- c: carcinogenic → causing cancer
- m: mutagenic → altering genetic material
- r: reproduction toxic → hazardous for reproduction

The EC classification of "cmr" substances uses categories 1, 2 and 3 (see also TRGS 905).

- Category 1: Substances which are known to have this property (c, m or r) in humans.
- Category 2: Substances which have shown this property to date unambiguously only in animal experiments.
  It must be assumed that this is applicable to the human situation.
- Category 3: Substances for which there is cause to suspect that they have this property.
  Information for a satisfactory assessment is available.

The classifications of the German "MAK Committee" in respect of these properties in some cases differ considerable from the EC categories, which are binding in the final analysis. In future, all the national classifications will lose significance, irrespective of the European country.

Carcinogenicity. The central principle for protection when handling carcinogenic substances in section 6 of the german Gefahrstoffverordnung (hazardous substance regulations) is, after the replacement requirement, the requirement to minimize exposure. For substances in categories 1 and 2, no MAK value is fixed because it is not possible to indicate any concentrations which can be regarded as safe. If use of these substances is necessary in industry, special protection and monitoring measures are needed (see the german Technische Richtkonzentrationen). Particular care is also necessary when handling substances in category 3. The substances in category 3 are divided into subgroups 3a (substances whose toxicity has been extensively investigated) and 3b (substances not yet adequately investigated).

Examples of substances in categories 1 to 3:

Category 1: Benzene
Category 2: Hydrazine, butadiene, 1,2-dichloroethane, hexamethylphosphoric-acidtriamide, 2-nitropropane
Category 3: Aniline, chloroform, dioxane, ethyl chloride, methylene chloride, pentachloroethane, 1,1,2,2-tetrachloroethane, 1,1,2-trichloroethane, tetrachloroethene, tetrachloromethane, trichloroethene.

Classification by the "MAK Committee":
The "MAK Committee" assigns carcinogenic substances to groups III A and III B. The unambiguously carcinogenic substances are in turn divided into two subgroups, similar to the subdivision made by the EC Commission:

III A: Products unambiguously proved to be carcinogenic
III A 1: Substances which, as shown by experience, may cause malignant tumours in humans
III A 2: Substances which have as yet clearly been shown to be carcinogenic only in animal experiments
III B: Substances for which there is cause to suspect a carcinogenic potential.

The use of III A and III B results from the divisions in the MAK list:

I	Significance and use of the MAK values
II	List of substances
III	*Carcinogenic products*
IV	Sensitizing products
V	Dusts and smokes (suspended particles)
VI	Particular products.

Relation between the classifications of the MAK Committee and the EC:

Carcinogenic products

MAK classification	EC classification
III A 1	Category 1
III A 2	Category 2
III B	Category 3a and 3b

Mutagenicity. Mutagenic means causing damage to male and female germ cells, resulting in genetic alterations in the progeny. This damage may take the form of genetic mutations, or alterations in the structure and number of chromosomes.

Examples of substances in categories 1 to 3 (EC classification):

Category 1: No substance has yet been included in category 1
Category 2: e.g. Hexamethylphosphoric-acid-triamide, ethylene oxide, ethylenimine, diethyl sulphate, acrylamide
Category 3: e.g. 4,6-Dinitro-o-cresol

Reproduction toxicity. Reproduction toxicity defined in the EC classification embraces two independent properties:

1. Substances which impair development of the unborn child (impaired development; Symbol R_E)
2. Substances which impair fertility (Symbol R_F)

As described above, the EC classification makes use of the usual three categories for each part.

The term impaired development embraces not just the occurrence of anatomical malformations (teratogenic effects); on the contrary, it also includes growth retardation without changes in organs as well as impairments of mental development.

Examples of development-impairing substances in the EC classification:

Category R_E 1: Lead chromate, lead acetate
Category R_E 2: Ethylglycol, methylglycol, ethylglycol acetate, methylglycol acetate, dimethylformamide

In contrast to the EC, the German MAK Committee takes only the embryotoxic effect of substances into account. Whereas the classification of the MAK Committee only provides information on the inhalational exposure at the relevant limit in air

(MAK), the classification of the EC Commission follows the generally valid scheme without reference to a limit for the workplace.

According to the MAK classification, the embryotoxic effect is divided into 4 pregnancy groups:

Group A: Certain proof of an embryotoxic effect. Harm cannot be precluded even below the MAK.
No solvent has yet been put in group A.

Group B: Embryotoxicity probable even below the MAK. e.g. 2-Methoxypropanol, 2-methoxypropyl acetate, methyl chloride, chloroform

Group C: There is no risk of embryotoxicity below the MAK. e.g. Ethanol, 1,1,1-trichloroethane, 1,1-dichloroethene, tri- and tetrachloroethylene, n-hexane di(2-ethylhexyl) phthalate, 2-butoxy-ethanol, 2-butoxyethyl acetate, 1-methoxy-2-propanol, 1-methoxypropyl acetate, ethylene glycol, butyldiglycol, isoamyl alcohol, 2-isopropoxyethanol, isobutanol, THF, toluene, cyclohexanone, dimethylacetamide, styrene

Group D: Assessment not yet possible.

Relation between MAK and EC classifications:

Embryotoxic substances MAK classification	Development-impairing substances EC classification
Reference point = MAK	Reference point = oral intake
Group A	R_E 1
Group B	R_E 2 and 3
Group C	no classification
Group D	–

Fertility-impairing substances (EC classification):

All substances classified as impairing fertility (R_F 1 or 2) are at present also classified as impairing development (R_E), e.g.

Category R_R 1: Lead acetate
Category R_F 2: Ethylglycol, ethylglycol acetate, methylglycol, methylglycol acetate

In connection with the detection and assessment of dangers due to hazardous substances at the workplace (TRGS 440) and finding possible substitutes ("replacement requirement"), account must be taken of the industrial regulations of the 600 series, including in particular TRGS 609 ("substitutes, replacement processes and restrictions on use of methyl- and ethylglycol and their acetates") and TRGS 612 ("substitutes, replacement processes and restrictions on use of dichloromethane in paint removers").

14.4.2. Occupational Health

Personal Safety Precautions. When working with solvents or solvent-containing preparations, contact with the skin and mucous membranes should be avoided. Protective goggles and gloves should be worn and the skin protected with skin cream. Wet articles of clothing should be changed immediately. Inhalation of solvent vapors should be avoided. The guidelines and codes of conduct published by the industrial and trade unions should be observed.

Odor Threshold. Most solvents have a characteristic odor. Human perceptibility and sensitivity to solvent vapors depends on habituation, which varies markedly from one person to another. Odors that are regarded as pleasant in small concentrations may be considered intolerable at high doses and under constant exposure. Other vapors that are initially considered objectionable may subsequently be regarded as tolerable [14.112]. It is therefore impossible to give objective rules for determining when an odor becomes objectionable. Odor intensity is subdivided into four levels:

1) Imperceptible
2) Weakly perceptible
3) Moderately perceptible
4) Highly perceptible

The odor threshold is the vapor concentration in a cubic meter of air (ppm) that is just still perceptible, [14.113]. The odor threshold values of some solvents are given in Table 23.

When handling solvents, the odor can be regarded as a preliminary warning sign but cannot replace necessary safety measures. Relatively high solvent concentrations may irritate the mucous membranes. It must be pointed out, however, that some mildly smelling solvents may present more of a health hazard than other, strongly perceptible products.

MAK Values. The toxicity of solvent vapors at the workplace has been investigated in animal experiments and by observing and monitoring humans. The MAK (Maximale Arbeitsplatzkonzentration) is the maximum permissible concentration of a substance in the atmosphere at the workplace that is generally not injurious to the health of the employees and is not regarded as intolerable by the latter, even after repeated and long-term exposure (generally 8-h exposure per day). As a rule, MAK values are average values over a period of a working day or a working shift. Since the actual concentration of the working substances in the inhaled air frequently fluctuates, uper exposure peaks are specified in the TRGS 900 section 2.3.

Technical equipment at the workplace should be designed and dimensioned so that the MAK values of the individual solvents are not exceeded.

No MAK values can be specified for a number of carcinogenic and mutagenic substances. For these substances the Technische Richtkonzentrationen TRK (lowest technically feasible levels) have been set to minimize the risk of a health hazard at the workplace. The TRK value for benzene and 2-nitropropane is 5 ppm (TRGS 102).

Since the MAK values apply only to pure substances in the workplace atmosphere, an evaluation index I_{MAK} has been defined in TRGS 403 according to the following formula for mixtures of solvent that occur in the majority of cases at the workplace:

$$I_{MAK} = \frac{C_1}{MAK_1} + \frac{C_2}{MAK_2} + \cdots \frac{C_N}{MAK_N} = \sum_{i=1}^{N} \frac{C_i}{MAK_i} = \sum_{i=1}^{N} I_i$$

where $C_1, C_2, \ldots, C_N$ are the average concentrations of a working day or working shift of the substances $I = 1, 2, \ldots, N$ with $MAK_1, MAK_2, \ldots, MAK_N$ that goes with it.

TLV Values. In the United States, the equivalent of the MAK value is the threshold limit value (TLV). The TLV value is that concentration of a substance in the air to which virtually all workers can be exposed daily without any harmful effects. This value is subdivided into TLV-TWA (time-weighted average concentration), TLV-STEL (short-term exposure limit), and TLV-C (ceiling limit). TLV-TWA applies to a normal 8-h working day or a 40-h week, TLV-STEL is the maximum concentration for an exposure time of 15 min, and TLV-C is the concentration that should at no time be exceeded. The TLV values published by the ACGIH are recommendations, whereas the PEL values (permissible exposure limit) specified by OSHA (Occupational Safety and Health Administration) are enforced by law. The safety limits are recommended to OSHA by NIOSH (National Institute of Occupational Safety and Health). Solvent TLV values for the USA and several European countries are listed in Table 13.

Other Limiting Values. In Scandinavia, solvent-containing products are identified with a YL value (yrkeshygieniskt luft behov = industrial hygiene air requirement). This value states how many cubic meters of air are required to dilute the amount of solvent contained in 1 L of product to such an extent that the concentration is below the TLV value. The YL value can be calculated for each solvent by multiplying the YL factor by the percentage content of the respective solvent in the liquid. A PWA

Table 13. Threshold limit values of Germany and the USA (ppm, 1996)

Solvent	Germany (MAK)	USA (TLV-PEL)
Cyclohexane	300	300
Diethyl ether	400	400
Tetrahydrofuran	200	200
Ethyl acetate	400	400
Pentyl acetate	100	100
Methyl ethyl ketone	200	200
Methyl isobutyl ketone	100	100

factor (**p**aint technology **w**ork hygiene **a**ir requirement) is allocated to the solvents on the basis of their YL value. The PWA number is in turn used as a basis for calculating the potential health hazard of a paint, and serves to specify the protective measures and working clothes that are to be used when applying paints (Swedish Work Safety Office Publication No. 463 and State Factory Inspectorate Publication No. 464, of 3rd August 1982).

A harmfulness factor has been proposed to quantify the harmfulness of solvents [14.114]–[14.116]:

$$\text{Harmfulness factor} = \frac{10\,000}{\text{TLV} \cdot \text{relative evaporation number}}$$

The harmfulness factor takes into account not only the actual health-damaging action of the solvent, but also the time for a health-damaging concentration to accumulate if the product leaks or is split.

14.5. Environmental and Legal Aspects

14.5.1. Environmental Protection

Being volatile substances, all solvents inevitably evaporate and pass as harmful substances into the atmosphere [14.117]–[14.120]. They may also affect waterways, lakes, rivers, ground water, and soil. Anthropogenic organic emissions total ca. 12 million t/a in the EC; of this, 30% stems from the use of solvents and 45% from the transport and traffic sector (Corinair Study 1990, Inventaire européen des émissions).

United States. The first environmental protection measures were implemented in particularly threatened, highly industrialized regions with a high density of motorized traffic. In the Los Angeles district of California, exposure of certain organic solvent vapors to solar radiation can lead to the formation of photo smog [14.121]–[14.123]. In the United States, Public Law 84–159 of 1955 passed by the Federal Departments of Health, Education, and Welfare was the basis for local clean air directives. In Rule 66 and Rule 442 (district of Los Angeles), aromatic hydrocarbons (xylene, tetrahydronaphthalene, toluene, and ethylbenzene) and branched-chain ketones (e.g., isophorone, mesityl oxide, methyl isobutyl ketone, methyl isopropyl ketone, and diacetone alcohol) are classified as photochemically reactive, smog-producing solvents. Aliphatic hydrocarbons and nitro compounds are considered to be photochemically nonreactive. Efforts are being made to reduce the solvent content of paints and use photochemically nonreactive solvents or water. Clean air measures in the United States have become more stringent as a result of further laws passed by the EPA, including the Air Pollution Control Act 1962, Clean Air Act 1963, Air

Quality Act 1967, National Environmental Policy Act 1969/1970/1975, Clean Air Act 1970 and 1990 (CAA) [14.124]–[14.127]. The implementation of these acts in the federal states is governed by executive provisions (e.g., New York State Rule 187). Since 1989, emissions have been regulated and reduced in stages by restricting the use of solvent-containing coatings in certain areas of application; for example, specific directives apply to the use of paints in automobile repair workshops, air-drying industrial paints, wood paints, swimming-pool paints, and marine paints. Special regulations apply to new industrial sites (e.g., new automobile factories).

Europe. An anti-emission law (Immissionsschutzgesetz) was passed by the state of North Rhine–Westphalia, Germany, in 1968. The purpose of this law and various executive directives was to improve living conditions and the atmosphere in the Ruhr district. In 1974, the individual laws of German states (Länder) were unified in the form of the Bundesimmissionsschutzgesetz (Federal Anti-Emission Law) and the Technische Anleitung zur Reinhaltung der Luft, TA Luft (Clean Air Regulations) of 27 Feb., 1986, which has since become known throughout Europe [14.128]–[14.131].

The following laws and directives currently exist in other European countries:

United Kingdom. Environmental Protection Act with Air Pollution Control APC (1990) [14.132].

Italy. Law 203 of the Ministry for the Environment (1988) including guidelines (1990) that restrict solvent emissions in certain application sectors (e.g., the automobile industry).

The Netherlands. Guideline KWS 2000 to reduce aromatic hydrocarbon emissions by ca. 50% of the 1981 level by the year 2000.

France. Guideline to reduce emissions of volatile organic compounds (VOC) by ca. 30% by the year 2000.

Austria. Anti-emission law to reduce the use of solvents, particularly aromatic hydrocarbons [14.133].

European legislation is standardized by a guideline on Integrated Pollution Prevention and Control (IPPC) which restricts and controls emissions produced during the erection of new technical plants and the expansion of existing plants. The guideline will come into effect shortly.

For solvents a european legislation will be standardized by the Solvent Emissions Directive that is currently beeing drafted and will become effective probably in 1998. Its aim is a reduction of the ozone content in the troposphere by stepwise limitation of emissions of volatile organic compounds beginning with a reduction of 30% until 1999 in comparison to 1990. In the long term there are plans to reduce the emissions up to 70–80%. Affected by the Solvent Emissions Directive are among others the paint and automotive, printing ink, metal degreasing, wood impregnation, chemical dry cleaning and pharmazeutical industries [14.134].

For the different solvents POCP values (POCP = photochemical ozone creation potential) were determined which are listed below:

Perchloroethylene	1	Trichloroethylene	7
Dichloromethane	1	Methanol	10
Methyl acetate	3	Isopropyl alcohol	15

Diacetone alcohol	20	Ethyl glycol acetate	60
Acetone	20	Butyl glycol acetate	60
Ethyl acetate	20	Diethyl ether	60
Ethanol	25	Ethoxypropyl acetate	65
Cyclohexane	25	Tetrahydrofuran	70
Methoxypropyl acetate	30	Ethyl glycol	75
Isobutyl acetate	35	Butyl glycol	75
Methylcyclohexane	35	Isophorone	80
Isobutanol	40	Methoxypropanol	80
Methyl ethyl ketone	40	Diisobutyl ketone	80
Hexane	40	Ethoxypropanol	85
Butyl acetate	45	*p*-Xylene	90
Propanol	45	1,3-Diethylbenzene	95
Toluene	55	Tetramethylbenzene	110
Butanol	55	Trimethylbenzene	120
sec-Butanol	55	Diethylmethylbenzene	120

In the TA Luft organic substances (including solvents) are divided into two groups of which each group is subdivided into three classes. The following concentration limits must not be exceeded in emissions:

Non carcinogenic substances:

Class I: Substances with a mass flow ≥ 0.1 kg/h: 20 mg/m^3
Class II: Substances with a mass flow ≥ 2 kg/h: 100 mg/m^3
Class III: Substances with a mass flow ≥ 3 kg/h: 150 mg/m^3

Carcinogenic substances:

Class I: Substances with a mass flow ≥ 0.5 g/h: 0.1 mg/m^3
Class II: Substances with a mass flow ≥ 5 g/h: 1 mg/m^3
Class III: Substances with a mass flow ≥ 25 kg/h: 5 mg/m^3

If non-carcinogenic organic compounds of several classes are present, the mass concentration in the waste gas must not exceed 150 mg/m^3 with a total mass flow ≥ 3 kg/h.

The classes contain the following solvents (non carcinogenic):

Class I. Aniline, 1,4-dioxane, nitrobenzene, phenol, 1,1,2,2-tetrachloroethane, tetrachloromethane, trichloromethane, furfurol.

Class II. Butyl glycol, chlorobenzene, cyclohexanone, dimethylformamide, diisobutyl ketone, ethyl glycol, ethylbenzene, furfuryl alcohol, isopropylbenzene, carbon disulfide, methyl glycol, methyl acetate, methyl cyclohexanones, methyl formate, styrene, tetrachloroethylene, tetrahydrofuran, toluene, 1,1,1-trichloroethane, trimethylbenzenes, xylenes.

Class III. Acetone, alkyl alcohols, methyl ethyl ketone, butyl acetate, dibutyl ether, diethyl ether, diisopropyl ether, dimethyl ether, ethyl acetate, ethylene glycol, diacetone alcohol, methyl isobutyl ketone, *N*-methylpyrrolidone, paraffin hydrocarbons (except methane), pinenes.

Organic substances not listed above should be allocated to the classes containing substances that they most closely resemble as regards their environmental effects. The toxicity, degradability, degradation products, and their odor intensity should be borne in mind.

The degradation of solvents by microorganisms in water, wastewater, effluent, and in clarification plants differs. In Germany solvents have been therefore classified as non, weakly, slightly, moderately, or strongly water-polluting [14.140]. In general, the discharge of solvents into water and waterways should be avoided to prevent contamination of rivers and ground water. The majority of solvents in wastewater and treatment plants are degradable if handled properly [14.141]. Detailed investigations have been carried out to assess the toxicity of solvents toward fish, which have been embodied OECD test standards [14.142], [14.143].

14.5.2. Laws Concerning Dangerous Substances

The EC directives and the wording of the dangerous substances directive are being continually updated. In order to assess the potential danger of a solvent the current wording of the directive should therefore be consulted.

On the basis of the EC directives the Verordnung über gefährliche Stoffe, GefStoffV (Directive on Dangerous Substances) of August 26, 1986 was issued in Germany (Bundesgesetzblatt I, p. 1470) in the wording of June 5, 1991 (Bundesgesetzblatt I, p. 1–1218).

Classification criteria such as danger of explosion, fire promoting, highly flammable, readily flammable, flammable, highly toxic, toxic, harmful, corrosive, irritant, carcinogenic, reproduction toxic, and mutagenic are defined in this directive. Instructions are given concerning the packaging and labelling of substances and preparations. The directive also contains instructions on the handling of dangerous substances (protective measures, prohibition of use, official regulations) and on health monitoring (preventive medicine, health data file). Individual paragraphs deal with legislation for the protection of young workers, pregnant women, and nursing mothers, and regulations governing trades, businesses, and working hours. Appendices I–VI include regulations concerning classification and identification; directives covering carcinogenic substances, aliphatic chlorinated hydrocarbons, lead and antifouling paints; provisions on working in rooms and containers; prescribed medical examinations; and a list of substances together with legally binding information on their labelling.

The German Chemikaliengesetz, ChemG (Chemicals Law) was passed on September 18, 1980 (Bundesgesetzblatt I, p. 1718) and deals with protection against dangerous substances. Newly developed substances must be registered with the authorities and subjected to toxicological testing. Similar regulations exist in other countries. Toxicological and ecotoxic data are also being compiled for chemicals that already exist and are listed in the European Inventory of Existing Chemical Substances (EINECS).

The Toxic Substances Control Act (TSCA) in the United States has been effective since 1977 and is similar to the German Chemicals Law. The TSCA specifies that chemicals must be assessed as regards their risks and registered before production and use.

The Technische Regeln für Gefahrstoffe, TRGS (Technical Regulations for Dangerous Substances) specify safety, industrial medicine, hygiene, and ergonomic requirements for introducing and handling dangerous substances.

The following regulations are of particular importance for solvents: recommended concentration (TRK) for dangerous substances (TRGS 102, limits e.g., benzene, 2-nitropropane); TRGS 150 covering direct skin contact with dangerous substances; TRGS 402 for the determination and evaluation of the concentrations of dangerous substances in the atmosphere of working areas; TRGS 403 for the measurement of mixtures of substances in the workplace atmosphere.

14.5.3. Fire Hazard

Most organic solvents are readily volatile and combustible, and their vapors form explosive mixtures with air. The fire hazard of solvents depends on their volatility and flash point (Section 14.3.7).

In Germany, the Verordnung über brennbare Flüssigkeiten, VbF (Directive on Combustible Liquids) divides solvents into danger classes:

Class A I: water-insoluble, flash point < 21 °C
Class A II: water-insoluble, flash point 21–55 °C
Class A III: water-insoluble, flash point 55–100 °C
Class B: flash point < 21 °C in the case of liquids that are miscible in all proportions with water at 15 °C

In the United States substances are subdivided according to their flash points:

1) Flammable
Class I: flash point < 100 °F (38 °C)
Class IA: flash point < 73 °F (22.8 °C), bp < 100 °F (38 °C)
Class IB: flash point < 73 °C (22.8 °C), bp > 100 °F (38 °C)
Class IC: flash point > 73 °F, bp < 100 °F

2) Combustible, flash point > 100 °F
Class II: flash point > 100 °F and < 140 °F (60 °C)
Class III: flash point > 140 °F (60 °C)
Class III A: flash point > 140 °F and < 200 °F (93.3 °C)
Class III B: flash point > 200 °F (93.3 °C)

Safety regulations concerning storage and transport are specified for each danger class.

Combustible liquids are also classified according to their flash point in other national and international regulations, e.g., ADN, ADNR, ARD, GGVS, IATA, IMDG, IMO, and RID.

The following regulations must be observed when combustible solvents are stored or handled in closed spaces:

1) Prohibition of naked flames
2) Prohibition of smoking
3) Provision of adequate ventilation
4) Provision of fire-extinguishing equipment
5) Protection against spark formation due to static charges, percussion, or impact
6) Provisions covering the installation of electrical plant and equipment in explosion-harzard workshops (e.g., VDE 0165)

7) Provisions covering explosion-proof electrical equipment and material (e.g., Part 1 – VDE 0171)
8) The Technische Regeln für Gefahrstoffe TRGS 514 (storage of highly toxic and toxic substances) and TRGS 515 (storage of fire promoting substances)

Storage and filling vessels must be earthed and connected to one another so as to ensure electrical conduction during solvent transfer to avoid spark formation due to electrostatic charges. The ignition groups of the solvents should be taken into account when installing and wiring electrical fittings and equipment.

14.5.4. Waste

Recycling and Waste-Gas Purification. Solvent recycling is becoming increasingly important for environmental reasons. The following methods may be employed to recover a solvent from a solvent vapor–air mixture but are still uneconomical due to their high costs [14.147]–[14.158]:

1) Condensation on cold surfaces
2) Adsorption on a solid adsorbent, e.g., activated charcoal
3) Absorption in high-boiling liquids
4) Membrane processes [14.159], [14.160]

Removal of solvent from a solvent vapor–air mixture is nowadays even more economically efficient than recovery methods, and can be effected by [14.161]–[14.164]:

1) Thermal incineration
2) Catalytic incineration
3) Biofilters (biosolve method)
4) Biowashers
5) Membrane reactors

On account of the rapid poisoning of the catalyst in the catalytic incineration method, the thermal incineration has been most widely employed. The combustion heat is used, for example, to preheat the air in paint shop dryers [14.165], [14.166].

The recycling of paint coagulates and slurries has been intensively investigated [14.167], [14.168].

Waste Disposal [14.169], [14.170]. Solvent residues, solvent waste, and solvent-containing preparations should be stored only in air- and liquid-tight containers in special dumps.

Waste treatment in Germany is regulated by provisions (Kreislaufwirtschafts- und Abfallgesetz, TA Luft, Verordnung über Verwertungs- und Beseitigungsnachweise, Verordnung über Abfallwirtschaftskonzepte und -bilanzen, Abfallverbringungsgesetz) which require product manufacturers and processors to describe residues in terms of their chemical and physical properties. They also cover the transportation of waste, proof of and permission for waste disposal, prohibition of mixing wastes, and assigning appropriate treatment techniques to specific types of waste (14.171–14.174).

The requirements regarding handling of waste is based on its characterization by the European waste classification catalog as hazardous or "normal" in conjunction with the corresponding European waste regulations. This has been implemented in German law with the Closed Cycle Waste Management Act (Kreislaufwirtschafts- und Abfallgesetz) and the related regulations cited above.

According to the German Technical Waste Control Regulations (TA Abfall) all waste containing solvent which cannot be thermally recovered must be disposed of thermally in special waste incinerators equipped with flue-gas cleaning systems. Residues are then stored underground or in special waste dumps. In most cases, however, thermal recovery is possible. Under the German Bundesimmissionsschutzgesetz and Kreislaufwirtschafts- und Abfallgesetz, prevention and recovery take precedence over disposal. Disposal is only permissible or called for if prevention and recovery are technically or economically unfeasible, or if disposal is more environmentally compatible.

The problem of wastewater is closely coupled with the processing of solvent-containing paints, adhesives, and other preparations.

In Germany uniform standards were introduced with the sixth amendment to the Water Management Law (Wasserhaushaltsgesetz, WHG). The state of the art now applies to all pollution measures. However, a material tightening over against the previously recognized technology regulations was not (yet) connected to this. Concrete standards are set by the Waste Water Management Act (Abwasserverordnung) and the related appendixes, as well as by a few more stringement regulations, depending upon the source of the waste water § 7a of the WHG in conjunction with the Abwasserverordnung governs discharge of sewage into bodies of water. Moreover, paragraphs 19a to 1 present guidelines for handling substances potentially hazardous to water, wherein waste water is not considered a substance hazardous to water.

Paragraphs 19g to 1 present standards for handling substances hazardous to water. The substances are compiled in a catalog. They are subdivided into classes (Wassergefährdungsklassen, WGK) according to their water-hazard potential:

The substances are compiled in a catalog. They are subdivided into classes (Wassergefährdungsklassen, WGK) according to their water-hazard potential:

WGK 0: Generally not water hazardous, e.g., ethanol, acetone
WGK 1: Slightly water hazardous, e.g., aliphatic compounds, alcohols, ketones, ethers, esters
WGK 2: Water hazardous, e.g., aromatic compounds, dichloromethane, carbon disulfide
WGK 3: Highly water hazardous, e.g., tetrachloroethylene

The water hazard class of mixtures is determined according to data obtained from the mixture. If such data do not exist, the component of the highest water hazard class is decisive.

14.6. Purification and Analysis

Purification. Distillation is most commonly used for purifying solvents. Solvents with different vapor pressures can be separated from one another by fractional distillation. Azeotropic mixtures can be separated by extractive or azeotropic distillation (e.g., addition of benzene to a water–ethanol mixture), by chemical reaction of a component (e.g., addition of acetic anhydride to an ethanol–ethyl acetate mixture), or by altering the pressure during distillation.

Further methods for purifying solvents include freezing out water from solvents that are partially miscible with water, extracting water-soluble constituents from water-immiscible solvents by shaking with water, and the use of adsorbents (e.g., activated charcoal).

Analysis. Solvent purity is assessed by means of gas chromatography [14.175]–[14.181], physical properties, water content, evaporation residue, and acid, saponification, and hydroxyl numbers [14.182]. Color and smell are also evaluated.

Standardized tests are employed in the analysis of solvents (see Table 20).
Further special analytical methods are discussed in [14.183]–[14.186].

14.7. Uses

14.7.1. Solvents in Paints (See also chapters 1–13)

Composition of Paint-Solvent Mixtures. The composition of a paint-solvent mixture is governed by the application conditions, drying temperature, and drying time of the paint. The solvent mixture in a paint that undergoes physical drying at room temperature contains ca. 45% low boilers, ca. 45% medium boilers, and ca. 10% high boilers.

True solvents and latent solvents are present in such a ratio that the paint dries to give a clear film without haze. Low boilers accelerate drying, whereas medium boilers and high boilers are used to produce a flawless surface. Oven-drying paints, stoving enamels, and coil coatings are applied at relatively high temperature and contain a large amount of high boilers and only a small amount of readily volatile solvents, if at all, because they may cause the paint to "boil" during stoving.

The nature of the solvents in the mixture also depends on the type of binder. In order to obtain rapid drying combined with low solvent retention, the solvent mixture should be formulated so that its solubility and hydrogen-bonding parameters lie at the boundary of the binder solubility range. On the other hand, the parameters of the solvent mixture should be similar to those of the binder to ensure satisfactory flow. Finding a sensible compromise is difficult and requires a great deal of experimental effort. It is advantageous if the non-solvents which, according to the solubility parameter concept, accelerate drying, are more volatile than the solvents, which remain behind and improve the flow [14.32].

The volatility and dissolution properties of the solvent mixture should be adjusted [14.187] so that its parameters move from the solubility boundary to the solubility center of the binder during evaporation. It must, however, be borne in mind that the solids concentration increases during evaporation of the solvent and that the paint temperature increases or decreases, thus altering the solubility range of the binders. Numerous studies have been published on the evaporation of solvent mixtures from paint films [14.188].

Paint Viscosity. The viscosity of a paint depends on:

1) The nature of the binder
2) The solvent composition
3) The binder concentration
4) The pigment content
5) The temperature

In homologous solvent series the paint viscosity generally increases with increasing molecular mass of the solvent. Since the solvency power of a solvent decreases with increasing molecular mass, a relationship between solvency power and paint vicosity seems likely. Comparison of solvents from different homologous series shows, however, that the viscosity is not generally dependent on the solvency power [14.189], [14.190]. The viscosity of a binder solution is determined by diverse binder–solvent interactions [14.191]–[14.195]. It is also influenced by the intrinsic solvent viscosity, the degree of uncoiling of the binder, the molecular mass of the binder, hydrogen bonding between binder and solvent molecules, as well as solvation and hydrogen bonding between binder molecules and between solvent molecules. Accordingly, the viscosity of a paint is also not generally at its lowest when the solubility parameter values of the solvent mixture coincide with those at the center of a binder-solubility region [14.32]. A binder is most strongly uncoiled (i.e., has its maximum volume) in the center of its solubility region. Solvents that have the solubility parameters of the binder solubility center therefore often form paints with a particularly high viscosity [14.196], [14.197]. In order to obtain low-viscosity paints, solvent additives are used to displace the solubility parameters of the solvent mixture toward the boundary of the binder solubility region; however, turbidity or a sharp increase in viscosity may be expected if this boundary is exceeded. The viscosity of paints that contain alcohol solvents or binders with hydroxyl groups can be reduced by adding small amounts of non-solvents (e.g., white spirit); on dilution viscosity-increasing hydrogen bonds are apparently ruptured. Relationships between viscosity and the hydrogen bond parameters of binders and solvents are described in [14.198].

The viscosity η of a paint is reduced on raising the temperature according to the Arrhenius equation,

$$\frac{1}{\eta} = A e^{-E/RT}$$

in which A and E are material constants [14.199]. Resultant sagging phenomena of the paint from vertical surfaces must be prevented by using solvents that form hydrogen bonds and by making the paint thixotropic.

Paints have widely differing viscosities that depend on their application conditions; low-viscosity paints are processed by dipping and spraying methods, while high-viscosity paints are used in casting, rolling, and hot-spraying methods. The correct choice of solvents serves to optimize the paint properties [14.200], [14.201].

Solvents in High-Solids and Waterborne Paints. In high-solids (low-solvent) paints, small amounts of auxiliary solvents are used to reduce the viscosity, as well as to optimize degassing and the flow properties [14.202]. Butyl acetate and butanol are mainly used to reduce the viscosity; in combination with glycol ethers and glycol ether acetates these solvents also improve flow properties and degassing. The viscosities (flow time in seconds, DIN-4 cup) of pigmented, high-solids paints after addition of 10% of a solvent are listed below [14.203]:

Butyl acetate	78	BUTOXYL	133
Butanol	80	Ethylbutanol	138
Xylene	92	Cyclohexanone	141
Butyl glycol	119	Butyl glycol acetate	148
SOLVESSO 100	126	Diacetone alcohol	169

Solvents for high-solids paints [14.204]–[14.206] cannot be selected according to the solubility parameter concept with the necessary degree of certainty. This is because the high-solids binders have a low mean molecular mass and are thus soluble in virtually all solvents with the exception of white spirit. Consequently, a boundary cannot be specified for the binder solubility region in the solubility parameter–hydrogen bonding parameter diagram and the influence of the solvents on the binder–solubility interactions cannot be estimated with sufficient accuracy [14.207].

In general it may be said that solvents of low intrinsic viscosity strongly reduce the viscosity of high-solids paints [14.208]. High-boiling solvents with a high solvency must also be used to obtain good flow properties [14.209].

Waterborne paints contain auxiliary solvents as solubilizers in amounts of 2–15%, depending on the binder. These solubilizers are water-miscible solvents or solvents that become water-miscible in all proportions in the presence of the binder [14.210], [14.211]. The most important are listed below:

Glycol Ethers [14.212], [14.213]. Isopropyl glycol, propyl glycol, butyl glycol, isobutyl glycol, butyl diglycol, 1-methoxy-2-propanol, 1-ethoxy-2-propanol, 1-isopropoxy-2-propanol, 1-propoxy-2-propanol, 1-butoxy-2-propanol.

Alcohols [14.214], [14.215]. Ethanol, propanol, isopropyl alcohol, butanol, isobutanol, *sec*-butanol, *tert*-butanol.

Butanol by itself is not miscible in all proportions with water, but its water miscibility is unlimited in the presence of paint binders. Butanol is an extremely effective solvent in waterborne paints, although it has the disadvantage of a somewhat more pungent smell than glycol ethers. The auxiliary solvents in waterborne paints promote solubilization of the binder and water, reduce the viscosity maximum that occurs on dilution with water, and yield smooth-flowing, flawless paint surfaces [14.216]–[14.226].

Auxiliary solvents and film-forming auxiliaries are also used as flow promoting agents in aqueous dispersions (emulsion paints). Propylene glycol acts not only as a solvent, but is hygroscopic and thus ensures a sufficiently high water content in the coating until a smooth surface has formed.

Pigment Wetting. The wetting of pigments is influenced both by the binder and the solvent. Good pigment wetting is important for pigment grinding in the finished paint product. The solubility parameters and, more commonly, hydrogen bond parameters of the binders and solvents influence pigment wetting [14.196], [14.197], [14.227], with the result that good or bad dispersibility, flocculation, and leafing effects may occur [14.32].

Blushing, Gloss, and Flow Properties. When the solvents evaporate from a paint the latter cools. At high atmospheric humidity water droplets condense if the temperature of the paint surface is below the dew point. This water is absorbed and homogeneously distributed in paints that contains solvents that are able to absorb water (e.g., ethanol or glycol ether). If the paint does not contain such solvents the water remains on the surface as a visible white haze (blushing). Blushing disappears if the paint contains solvents that form volatile azeotropes with water (e.g., aromatic hydrocarbons or butanol).

The gloss of paint coats is significantly enhanced if true solvents (high boilers) are the last solvents to evaporate in the paint formulation. Glycol ethers, in particular, improve the gloss due to their effect on the flow properties of the paints. A paint should dry to form a smooth, flat film without any surface structure as a result of the coalescence of the paint particles. Unsatisfactory paint flow leads to surface defects known as orange-peel effect, honeycomb structure, and fish eyes. These defects may be attributed to physical factors (e.g., to a change in the surface tension of the paint during solvent evaporation) and associated eddy formation in the paint film [14.191], [14.228]–[14.231]. Rapid evaporation occurs preferentially at the film surface causing the surface tension to increase more markedly than in the interior of the film. The resultant eddies in the film must be prevented by using slowly evaporating solvents with a good solvency for the binder. Additives that reduce the surface tension, e.g., wetting agents or silicone oils, also have a beneficial effect.

Mechanical Properties and Solvent Retention. Solvents significantly influence the mechanical properties of a paint for the following reasons:

1) Solvents influence the molecular structure of the film by aligning the binder molecules or preventing ordering
2) Solvents react to some extent with reactive, multicomponent lacquers containing a binder component and thus exert an internal flexibilizing effect
3) Solvents are retained by the paint film and exert an external flexibilizing effect

Glycol ethers have a flexibilizing action on saturated polyester resins that crosslink with melamine–formaldehyde resins; some functional groups in the melamine resin are apparently blocked by the glycol ether. The flexibilization of paint films by retention of solvents occurs in copolymer paints and has been investigated by gas chromatography [14.232] and radiotracer methods [14.233]. Dry, hardened paint films absorb solvents to varying degrees [14.234], [14.235]. Dichloromethane swells almost all paint films which is the reason for its good paint-removing action. The swelling action of aromatic hydrocarbons on polyester stoving enamels cross-linked with melamine resin depends on the monomer composition of the saturated polyester resin [14.62] and the degree of cross-linking after stoving [14.236].

Thick-Coat Paints and Multicoat Paints. Paints for anticorrosion systems that are applied by airless methods are prepared on the basis of, for example, vinyl chloride copolymers. Thick, high-solids coats can be obtained in a few stages with short drying times between coats. The solvents should deaerate the paint quickly and produce rapid drying and good flow. These properties can only be obtained by accurately coordinating the solvent components. In the case of multicoat paints the solvent composition of the top coat must be selected such that the primer is not swollen by inwardly migrating solvents. Mild true solvents (e.g., alcohols and glycol ethers), diluents, and non-solvents should be used for this.

14.7.2. Solvents in Paint Removers

Solvents used for paint removal are able to dissolve or considerably swell physically drying binders (e.g., vinyl chloride copolymers, cellulose nitrate, polyacrylates) and chemically cross-linked coatings (e.g., oil-based paints, dried alkyd resins, cross-linked polyester–melamine resins, cross-linked epoxy and isocyanate coatings) [14.237]. A combination of dichloromethane with low-boiling ketones or esters is particularly suitable. Small amounts of high-boiling solvents with a low volatility (e.g., tetrahydronaphthalene, solvent naphta, methyl benzyl alcohol, or benzyl alcohol) are added to these mixtures to retard evaporation and increase the solvency. Modern paint removers do not contain chlorinated hydrocarbons, they are formulated on the basis of high boilers (e.g., dimethylformamide, dimethyl sulfoxide, propylene carbonate, and N-methylpyrrolidone) in combination with alcohols and aromatics, or consist of aqueous, frequently alkaline or acidic systems.

14.7.3. Solvents in Printing Inks

Solvents in printing inks must readily dissolve the binder or resin and must not attack the printing rollers [14.238]. Acidic or sulfur-containing solvents must not be used with copper rollers. Aliphatic and aromatic hydrocarbons that attack rubber are unsuitable for offset inks. Alcohols, glycol ethers, and glycol ether esters are most suitable for gravure and flexographic printing.

The volatility of the solvents must be adapted to the application process, the drying process, and the drying time during printing. Solvents are rarely used in offset, letterpress, and typograhical printing, whereas rapidly evaporating solvents are employed in flexographic and gravure printing.

14.7.4. Extraction [14.239]–[14.241]

The distribution of dissolved substances in two solvent phases is employed on a large scale in the industrial separation of mixtures of substances. Examples are the removal of unsaturated constituents from vegetable oils with furfurol or methanol, the purification of animal and vegetable oils with liquid propane, and the removal of waxes from lubricants with liquid propane or ketones. Penicillin is similarly concentrated with methyl isobutyl ketone, and aqueous glycerol is purified with xylene. Preparative and analytical separations are also performed by liquid–liquid extraction. Inorganic salts can be extracted from aqueous solutions with suitable solvents, such as ethers, ketones, and esters. This method is particularly efficient for metal halides and nitrates, e.g., the separation of uranium compounds from aqueous solutions or the fractional extraction of rare earths.

14.7.5. Extractive Distillation [14.242], [14.243]

In extractive distillation a solvent is added to the mixture to be separated, its boiling point is higher than that of the components of the mixture. In the case of a binary mixture, the added solvent must interact more strongly with one of the components to lower its volatility. The other more volatile component can thus be distilled off, leaving the added solvent and the higher-boiling component at the bottom of the column. The added solvent (entrainer) must be miscible with the mixture at all temperatures, concentrations, and pressures.

A mixture of cyclohexane (bp 80.8 °C) and benzene (bp 80.1 °C) can, for example, be separated by distillation after adding aniline, since the interaction between benzene and aniline is greater than that between cyclohexane and aniline. Azeotropic mixtures can be separated similarly by extractive distillation (e.g., water–ethanol by adding glycerol). Hydrocarbons with similar boiling points can be separated by extractive distillation in the presence of polar liquids (nitrobenzene, phenol, furfurol).

14.7.6. Chromatography

Chromatography of a mixture on a stationary phase using a liquid eluent is a form of solid–liquid extraction (adsorption chromatography). The distribution of components between the surface of the carrier material and the solvent depends on their chemical structure and the nature of the carrier material and the migrating solvent [14.10], [14.244]–[14.246]. The solvents used in adsorption chromatography are arranged in an eluotropic series, the eluting action of the solvents increases with increasing polarity as follows: pentane, petroleum ether, hexane, heptane, cyclohexane, tetrachloromethane, trichloroethylene, benzene, dichloromethane, trichloromethane, diethyl ether, ethyl acetate, pyridine, acetone, propanol, ethanol, methanol, and water.

Liquid–liquid extraction is also possible in which distribution takes place between a liquid eluent and a liquid (generally water) that is adsorbed on the surface of the carrier material (distribution chromatography).

14.7.7. Solvents for Chemical Reactions

Solvents for chemical reactions must be chemically inert under the reaction conditions. The following solvents have proved suitable [14.240]:

1) *Hydrogenation*: alcohols, glacial acetic acid, hydrocarbons, dioxane
2) *Oxidation*: glacial acetic acid, pyridine, nitrobenzene
3) *Halogenation*: tetrachloromethane, tetrachloroethane, dichlorobenzene, trichlorobenzene, nitrobenzene, glacial acetic acid
4) *Esterification*: benzene, toluene, xylene, dibutyl ether
5) *Nitration*: glacial acetic acid, dichlorobenzene, nitrobenzene
6) *Diazotization*: ethanol, glacial acetic acid, benzene, dimethylformamide
7) *Coupling reactions of diazonium compounds*: methanol, ethanol, glacial acetic acid, pyridine
8) *Grignard reaction*: diethyl ether
9) *Friedel–Crafts reaction*: nitrobenzene, benzene, carbon disulfide, tetrachloromethane, tetrachloroethane, 1,2-dichloroethane
10) *Dehydration*: benzene, toluene, xylene
11) *Sulfonation*: nitrobenzene, dioxane
12) *Dehydrohalogenation*: quinoline
13) *Decarboxylation*: quinoline
14) *Acetal formation*: hexane, benzene
15) *Ketene condensation*: diethyl ether, acetone, benzene, xylene

The rate and course of the reaction often depend on the nature of the solvent [14.2], [14.247], [14.248]. Solvent effects have also been described in radical polymerizations [14.249].

14.7.8. Solvents for Recrystallization

Solvents for recrystallization should readily dissolve the substance to be purified as well as the impurities on heating; on cooling the solution the pure substance then crystallizes out, whereas the impurities remain in solution. If the temperature dependence of the solubility of the substance in the solvent is only slight, it can be precipitated by adding a non-solvent to the hot solution. The purification effect of recrystallization is greater the more slowly the solution is cooled and accordingly the more slowly the crystals form.

14.7.9. Solvents in Film Production

Diethyl ether–ethanol mixtures are used as solvents in the production of cellulose nitrate films. Acetone is most suitable for cellulose acetate or cellulose acetate butyrate films, and dichloromethane for cellulose triacetate films.

14.7.10. Solvents for Synthetic Fibers

Solvents used for the preparation of spinning solutions should be inexpensive, easily recoverable, volatile, have as low a viscosity as possible, and should readily dissolve the high molecular mass polymers to be spun. The following solvents have proved suitable: dichloromethane–methanol for cellulose triacetate, tetrahydrofuran or carbon disulfide–acetone for non-postchlorinated poly(vinyl chloride), acetone for postchlorinated poly(vinyl chloride), and dimethyl sulfoxide or propylene carbonate for polyacrylonitrile.

14.7.11. Solvents for Rubber, Plastics, and Resin Solutions

Aliphatic hydrocarbons, in some cases mixed with aromatic hydrocarbons, are suitable for rubber solutions.

Plastics and resin solutions usually contain highly or moderately volatile solvents. The solutions are used for example to coat articles with a plastic, as adhesive solutions [14.250], as nail polish, and for textile impregnation. Esters, ketones, alcohols, or cyclic ethers are suitable depending on the intended use.

14.7.12. Solvents for Degreasing

Ketones or chlorinated hydrocarbons, especially trichloroethylene, perchloroethylene, or dichloromethane, are used to degrease metals [14.251], [14.252]. The chlorinated hydrocarbons used to degrease light metals are stabilized to prevent the metal-catalyzed elimination of hydrogen chloride. The solvent mixture [14.253]–[14.256] for metal degreasing are being replaced by aqueous systems [14.257], [14.258] due to more stringent environmental protection requirements. See also Section 8.2.1.2.

14.7.13. Solvents for Dry Cleaning

Aliphatic hydrocarbons or perchloroethylene are used for the chemical cleaning of textiles. Resins are often added to the dipping baths to hydrophobize the textiles [14.259].

14.7.14. Solvents in Aerosol Cans and Dispensers

Solvents used in aerosol cans (spray cans for hair lacquer, cleansing agents, paints) must, of course, dissolve the substances that are to be sprayed. They must also be miscible with the propellants without causing the dissolved substance to precipitate. Previously, chlorofluorohydrocarbons were mainly used as propellants but they have a damaging effect on the stratospheric ozone layer and have now been largely replaced by alternatives (e.g., hydrocarbons such as butane, diethyl ether, fluorocarbons, carbon dioxide) [14.260].

14.8. Economic Aspects [14.261]

Solvents are mainly used for the following purposes:

1) For cleaning and degreasing: electronic components, metal surfaces, dry cleaning of textiles
2) As a raw material: for preparing other products; in the paint, adhesives, and plastics industries; for producing resin solutions
3) As a carrier or dispersion medium: pharmaceutical products, agrochemicals

Future solvent consumption is expected to decrease on account of worldwide environmental restrictions. There will be a shift from aromatic and aliphatic hydrocarbons to oxygen-containing solvents because the latter have a lower ozone formation potential and better dissolution properties.

Table 14. Development of consumption of solvents in Western Europe (as percentage of total)

	1980	1986	1990	2000
Oxygen-containing solvents	36.5	45	51	65
Aliphatics	28.5	22	20.5	18
Aromatics	20.5	20	19	15.5
Chlorinated hydrocarbons	14.5	13	9.5	1.5
Total consumption 10^6 t	5.1	4.75	4.7	3.3

14.9. Solvent Groups

Important information and data on individual solvents are listed in the following tables:

Table 15 Solubility parameters, dipole moments, and hydrogen bond parameters
Table 16 Miscibility with water
Table 17 Azeotropes of the most important solvents
Table 18 Relationships between solubility parameter and surface tension
Table 19 Gross calorific values
Table 20 Standards for solvents and solvent tests
Table 21 Abbreviations
Table 22 Physical properties
Table 23 Safety and physiological data

14.9.1. Aliphatic Hydrocarbons

Aliphatic hydrocarbons, naphthas, gasoline, or paraffin hydrocarbons are chemically inert and are thus very stable solvents [14.262], [14.263]. Aliphatic hydrocarbons exhibit a good solvency for mineral oils, fatty oils (with the exception of castor oil), waxes, and paraffin. They also dissolve rubber, polyisobutene, molten polyethylene, poly(butyl acrylate), poly(butyl methacrylate), and poly(vinyl ethers). However, most other polymers, polar resins, cellulose derivatives, and most paint binders are insoluble. Resins and binders with a low polarity dissolve less readily in aliphatic hydrocarbons than in aromatic hydrocarbons.

Aliphatic hydrocarbons are subdivided according to their boiling point ranges into special boiling point hydrocarbons (SBP, DIN 51 631), white spirit (DIN 51 632), and petroleum ether (DIN 51 630) (see Tables 20 and 22).

Special boiling point hydrocarbons are used for quick-drying paints, dipping solutions, and quick-drying adhesives. Their flash point is below 21 °C, which means that they must be used in explosion-proof areas. Petroleum ether is a special petroleum distillation fraction (bp 40–60 °C), and is used as a solvent in the chemical industry.

White spirit is used predominantly in the paint industry as a solvent or diluent for oil-based, alkyd resin, chlorinated rubber, and some vinyl chloride copolymer paints; light napththas have flash points above 21 °C. Varnish makers' and painters' (VMP) naphtha (bp 100–150 °C) is used mainly as diluent.

Table 15. Solubility parameters δ, dipole moments μ $(J/cm^3)^{1/2}$ and hydrogen bond parameters γ

Solvent	δ	δ_D	δ_P	δ_H	μ	γ
n-Hexane	14.9	14.7	0	0	0	0
n-Heptane	15.1	15.1	0	0	0	0
Cyclohexane	16.8	16.8	0	0	0	0
Toluene	18.2	18.0	1.4	2.0	0.4	4.5
Xylene	18.0	17.8	1.0	3.1	0.4	4.5
Ethyl benzene	18.0	17.8	0.6	1.4	0.6	1.5
Cumene	17.1	17.1	0.4	0		
Styrene	19.0	18.6	1.0	4.1	0	1.5
Dichloromethane	19.8	18.2	6.1	6.3	1.2	1.5
1,2-Dichloroethane	20.1	18.8	5.3	4.1	1.8	1.5
Methanol	29.7	15.1	11.3	22.9	1.7	18.7
Ethanol	26.0	15.1	8.0	20.1	1.7	18.7
1-Propanol	24.3	15.1	6.1	17.6	1.7	18.7
2-Propanol	23.5	15.3	6.1	17.2	1.7	18.7
1-Butanol	23.3	16.0	6.1	15.8	1.8	18.7
Isobutanol	21.9	15.3	5.7	15.8	1.8	17.9
2-Ethylhexanol	19.4	16.0	3.3	11.9	1.7	18.7
Cyclohexanol	23.3	17.4	4.1	13.5	1.7	18.7
Diacetone alcohol	18.8	15.8	8.2	10.8	3.2	13.0
Acetone	20.5	15.6	11.7	4.1	2.7	9.7
Methyl ethyl ketone	19.0	16.0	9.0	5.1	2.8	7.7
Methyl isobutyl ketone	17.2	15.3	6.1	4.1	2.7	7.7
Diisobutyl ketone	16.0	16.0	3.7	4.1	2.7	8.4
Cyclohexanone	19.8	17.8	7.0	7.0	3.0	11.7
Mesityl oxide	18.4	16.4	7.2	6.1	2.7	9.8
Isophorone	19.2	16.6	8.2	7.4		14.9
Methyl acetate	18.0	14.5	5.8	9.1	1.6	8.4
Ethyl acetate	18.6	15.1	5.3	9.2	1.7	8.4
Butyl acetate	17.4	15.8	3.7	6.3	1.8	8.8
Isobutyl acetate	17.0	15.1	3.7	7.6	1.9	8.8
Ethyl glycol acetate	17.8	16.0	4.7	10.6	2.3	9.4
Butyl glycol acetate	17.9	15.8	3.7	7.7		10.3
Ethyl diglycol acetate	19.0	16.0	5.4	8.7		
Butyl diglycol acetate	18.6	16.2	4.6	7.9		
Methoxypropyl acetate	18.1	15.3	4.6	8.5		
Ethoxypropyl acetate	17.9	15.4	4.1	8.1		
Ethylene carbonate	30.1	19.4	21.7	5.1	4.9	4.9
Propylene carbonate	27.2	20.1	18.0	4.1	5.0	4.9
Diethyl ether	15.6	14.3	5.1	2.0	1.1	13.0
Methyl tert-butyl ether	15.6	14.4	3.4	5.0		
Tetrahydrofuran	19.2	16.8	6.8	7.2		12.0
Dioxane	20.3	17.6	8.6	4.1	0.4	9.7
Ethyl glycol	20.3	16.2	9.2	14.3	2.1	13.0
Propyl glycol	22.4	16.2	5.7	14.4		
Butyl glycol	18.2	16.0	6.3	12.1	2.1	13.0
Ethyl diglycol	19.6	16.2	7.6	12.3	1.6	13.0
Butyl diglycol	18.2	16.0	7.0	10.6	1.6	13.0
1-Methoxy-2-propanol	19.8	15.6	8.2	14.3		

Table 15. (continued)

Solvent	δ	δ_D	δ_P	δ_H	μ	γ
1-Ethoxy-2-propanol	22.1	15.8	5.7	14.3		
1-Isopropoxy-2-propanol	21.1	15.6	5.0	13.4		
1-Isobutoxy-2-propanol	20.6	15.7	4.4	12.6		
Methyl dipropylene glycol	18.1	16.4	8.0	11.6		
Dimethylformamide	24.8	17.4	13.7	11.3	3.8	11.7
Dimethyl acetamide	22.1	16.8	11.5	10.2	2.0	12.3
Dimethyl sulfoxide	26.4	19.2	13.3	12.1	4.0	7.7
Carbon disulfide	20.5	20.5	0	0	0	0
1-Nitropropane	20.6	16.3	11.9	4.1	3.6	
2-Nitropropane	20.3	16.2	12.1	4.1	3.7	2.5
Nitrobenzene	20.5	17.6	12.3	4.1	4.3	2.8
N-Methyl-2-pyrrolidone	21.4	18.4	12.6	7.4		

Table 16. Miscibility of solvents with water (wt% at 20 °C)

Solvent	Solvent in water	Water in solvent
Hexane	0.53	0.1
Tetrahydronaphthalene		0.2
Dipentene		0.72
Toluene	0.035	0.05
p-Xylene	0.02	0.02
Ethylbenzene	0.02	0.02
Styrene		0.04
Methanol	∞	∞
Ethanol	∞	∞
Propanol	∞	∞
Isopropyl alcohol	∞	∞
Butanol	7.5	19.7
Isobutanol	8.4	16.2
sec-Butanol	12.5	44.1
tert-Butanol	∞	∞
Hexanol	0.58	7.2
Trimethylcyclohexanol	0.19	4.0
Cyclohexanol	3.6	3.6
Methylbenzyl alcohol	2.9	5.8
Ethylene glycol	∞	∞
Methyl glycol	∞	∞
Ethyl glycol	∞	∞
Propyl glycol	∞	∞
Butyl glycol	∞	∞
Ethyl diglycol	∞	∞
Methoxypropanol	∞	∞
Methyldipropylene glycol	∞	∞
Nitroethane	4.5	0.9

Table 16. (continued)

Solvent	Solvent in water	Water in solvent
1-Nitropropane	1.4	0.5
2-Nitropropane	1.7	0.6
Diethyl ether	6.9	1.2
Dibutyl ether	0.3	0.2
Methyl *tert*-butyl ether	4.8	1.3
Tetrahydrofuran	∞	∞
Dioxane	∞	∞
Methyl acetate	24.0	8.0
Ethyl acetate	6.1	3.3
Isopropyl acetate	2.9	1.9
Butyl acetate	0.83	0.62
Isobutyl acetate	0.67	1.65
Ethyl glycol acetate	23.5	6.5
Butyl glycol acetate	1.5	1.7
Cyclohexyl acetate	0.2	0.5
Butyl glycolate	7.5	25.0
Propylene carbonate	21.4	7.5
Acetone	∞	∞
Methyl ethyl ketone	26.0	12.0
Methyl isobutyl ketone	2.0	2.4
Diisobutyl ketone	0.04	0.42
Cyclohexanone	2.3	8.0
Isophorone	1.2	4.3
Trimethylcyclohexanone	0.3	1.4
Diacetone alcohol	∞	∞
Dichloromethane	2.0	0.16
1,1,1-Trichloroethane	0.44	0.05
Trichloroethylene	0.1	0.02
Tetrachloroethylene	0.02	0.01
Dimethylformamide	∞	∞
Dimethyl sulfoxide	∞	∞

Table 17. Azeotropes of the most important solvents

Solvent A	Solvent B	Azeotrope	
		bp (101.3 kPa), °C	Composition A/B, wt%
Hexane	water	61.6	94.4/5.6
	methanol	50.0	73.1/26.9
	ethanol	58.7	79/21
	isopropyl alcohol	61.0	78/22
	isobutanol	68.3	97.5/2.5
	acetone	49.8	41/59
	methyl ethyl ketone	64.2	63/37
	ethyl acetate	65.0	58/42

Table 17. (continued)

Solvent A	Solvent B	Azeotrope	
		bp (101.3 kPa), °C	Composition A/B, wt%
Cyclohexane	water	69.8	91.5/8.5
	methanol	54.2	63/37
	propanol	74.3	80/20
	isobutanol	78.1	86/14
m-Xylene	water	94.5	60/40
Methanol	methyl acetate	54.0	19/81
	ethyl acetate	62.3	44/56
	acetone	55.7	12/78
	methyl ethyl ketone	63.5	70/30
	dichloromethane	39.2	8/92
	trichloroethylene	60.2	36/64
Ethanol	water	78.2	95.6/4.4
	hexane	58.7	21/79
	toluene	76.7	68/32
	methyl ethyl ketone	74.8	34/66
	dichloromethane	41.0	3.5/96.5
	trichloroethylene	70.9	27/73
Butanol	water	93.0	55.5/44.5
	hexane	67.0	97/3
	cyclohexane	79.8	10/90
	toluene	105.6	27/73
	p-xylene	115.7	68/32
	isobutyl acetate	114.5	50/50
	methyl isobutyl ketone	114.35	30/70
Isobutanol	water	89.9	66.8/33.2
	hexane	68.3	97.5/2.5
	toluene	100.9	44.5/55.5
	p-xylene	107.5	83/17
	isobutyl acetate	107.6	95/5
tert-Butanol	water		88.2/11.8
2-Ethylhexanol	water	99.1	20/80
Cyclohexanol	water	97.8	20/80
	α-pinene	149.9	35.5/64.5
	m-xylene	143.0	14/86
	ethyl acetate	153.8	5/95
Ethyl glycol	water	99.4	28.8/71.2
	dibutyl ether	127.0	50/50
	butyl acetate	125.8	35.7/64.3
Butyl glycol	water	98.8	20.8/79.2
Ethyl acetate	water	70.4	91.8/8.2
	hexane	65.0	42/58

Table 17. (continued)

Solvent A	Solvent B	Azeotrope bp (101.3 kPa), °C	Composition A/B, wt%
	methanol	62.1	51.4/48.6
	ethanol	71.8	69/31
	isopropyl alcohol	74.8	77/23
Butyl acetate	water	90.7	73/27
	isopropyl alcohol	80.1	48/52
	butanol	117.6	32.8/67.2
	ethyl glycol	125.8	64.3/35.7
Methyl ethyl ketone	water	73.5	89/11
	hexane	64.2	37/63
	cyclohexane	72.0	40/60
	isopropyl alcohol	77.5	68/32
	ethyl acetate	76.7	22/78
Methyl isobutyl ketone	water	87.9	75.7/24.3
Isophorone	water	99.5	16.1/83.9
Tetrahydrofuran	water	64.0	94.7/5.3
Dichloromethane	water	38.1	98.5/1.5
Trichloroethylene	water	72.9	93.4/6.6
Tetrachloroethylene	water	87.1	84.1/15.9
1,1,1-Trichloroethane	water	65.0	95.7/4.3

Table 18. Relationships between solubility parameter and surface tension

Solvent	Surface tension, mN/m	Molecular volume, V	Solubility parameter, calculated, $(J/cm^3)^{1/2}$	Solubility parameter, actual, $(J/cm^3)^{1/2}$
Dichloromethane	29.0	64.0	22.4	19.8
Trichloroethylene	32.0	90.9	22.1	18.8
Tetrachloroethylene	31.7	102.2	21.5	19.0
Toluene	28.5	105.5	20.2	18.2
Xylene	28.3	121.5	19.6	18.0
Ethylbenzene	29.0	121.2	19.9	18.0
Methanol	22.6	40.5	21.2	29.7
Ethanol	22.3	58.4	19.6	26.0
Isopropyl alcohol	21.7	76.5	18.4	23.5
Butanol	24.6	91.4	19.6	23.3
Isobutanol	23.0	92.4	19.1	21.9
Acetone	23.7	73.4	19.3	20.5
Methyl ethyl ketone	24.6	89.6	19.2	19.0
Methyl isobutyl ketone	23.9	125.3	17.7	17.2

Table 18. (continued)

Solvent	Surface tension, mN/m	Molecular volume, V	Solubility parameter, calculated, $(J/cm^3)^{1/2}$	Solubility parameter, actual, $(J/cm^3)^{1/2}$
Cyclohexanone	34.5	103.9	22.5	19.8
Methyl acetate	24.8	79.5	19.7	19.6
Ethyl acetate	23.9	97.9	18.5	18.6
Butyl acetate	25.5	132.0	18.2	17.4
Isobutyl acetate	23.7	133.4	17.4	17.0
Ethyl glycol acetate	29.5	136.0	19.6	17.8
Ethyl glycol	28.2	96.9	20.4	20.3
Butyl glycol	29.0	131.2	19.6	18.2
Ethyl diglycol	31.8	135.5	20.5	19.6
Butyl diglycol	31.5	169.7	19.5	18.2
Diethyl ether	16.5	103.8	14.9	15.6
Dioxane	36.5	85.2	24.1	20.3

Table 19. Gross calorific values of solvents

Solvent	Gross calorific value, MJ/kg
Hexane	− 44.6
Heptane	− 47.7
Toluene	− 42.6
Xylene	− 43.1
Methanol	− 22.5
Ethanol	− 29.7
Butanol	− 36.0
Isobutanol	− 36.0
Methylbenzyl alcohol	− 35.9
Methyl acetate	− 22.1
Ethyl acetate	− 25.0
Isobutyl acetate	− 31.2
Butyl acetate	− 30.6
Ethyl glycol acetate	− 25.9
Propylene carbonate	− 14.3
Ethyl glycol	− 28.3
Butyl glycol	− 32.8
Ethyl diglycol	− 27.7
Butyl diglycol	− 31.7
Tetrahydrofuran	− 36.7
Methyl *tert*-butyl ether	− 38.2
Acetone	− 30.8
Methyl ethyl ketone	− 33.9
Methyl isobutyl ketone	− 37.2
Isophorone	− 38.1
Dichloromethane	− 7.1
Tetrachloroethylene	− 4.9

Table 20. Standards for solvents and solvent tests

Standard	Subject area	Standard	Subject area
DIN 51 405	gas chromatography	DIN 53 245	alcohols
DIN 51 413		ASTM D 1 152	methanol
DIN 55 682		ISO R 1 387	
DIN 55 683			
ASTM E 260		ISO R 1 388	ethanol
ASTM D 3545		ASTM D 770	isopropyl alcohol
		ISO R 756	
DIN 55 685	GC alcohols	ASTM D 304	butanol
DIN 55 686	GC acetic acid esters	ISO R 755	
ASTM D 3545		ASTM D 1 007	sec-butanol
		ISO 2496	
DIN 55 687	GC ketones	ASTM D 2 635	methyl isobutyl carbinol
DIN 55 688	GC ethylene glycol ether	ASTM D 2 627	diacetone alcohol
		ISO 2517	
DIN 55 689	GC propylene glycol ether	ISO 1843	higher alcohols
DIN 51 408	chlorine content, combustion	DIN 53 246	acetic acid esters
		ISO R 1386	
		ASTM D 302	ethyl acetate
		ASTM D 657	isopropyl acetate
		ASTM D 303	butyl acetate
		ASTM D 1 718	isobutyl acetate
		ASTM D 343	ethyl glycol acetate
DIN 51 423	refractive index	DIN 53 247	ketones
DIN 53 491		ASTM D 329	acetone
ASTM D 1 218		ISO R 757	
ASTM D 1 747		ASTM D 740	methyl ethyl ketone
		ISO 2497	
		ASTM D 1 153	methyl isobutyl ketone
		ISO 2499	ethyl isoamyl ketone
		ISO 2500	
DIN 51 630	petroleum ether	DIN 53 248	turpentine oil, pine oil
DIN 51 631	SBP spirits/aliphatic hydrocarbons	DIN 53 249	dipentene
ISO 1250			
DIN 51 632	light naphthas	DIN 53 401	saponification number
		ASTM D 464	
		ISO 2114	
		ISO 3681	
DIN 51 633	benzene and homologues	DIN 53 402	acid number
ASTM D 2600		ASTM D 1 980	
		ASTM D 1 613	
		ASTM D 664	
		ISO 3682	
DIN 51 635	mineral spirits		

Table 20. (continued)

Standard	Subject area	Standard	Subject area
DIN 51 755 ASTM D 56 ISO 1523	flash point AP (closed cup)	DIN 53 409 ASTM D 1 209 ISO 6271	color hazen unit
DIN 51 757 ASTM D 891 ASTM D 1 298 ASTM D 4052 ISO 1306	density	DIN 53 977	trichloroethylene
DIN 51 758 ASTM D 93 ISO 3679	flash point PM (closed cup)	DIN 53 978	tetrachloroethylene
DIN 51 777 ASTM D 1 364 ASTM D 1 744 ASTM E 203 ISO 760	water content	DIN 55 680 ISO 1516	danger class
DIN 51 794 ASTM D 2 155 ASTM E 659	ignition temperature	DIN 55 681	trichloroethylene, stability
DIN 51 848	precision		
DIN 52 900 ISO 11 014	safety data sheet		
DIN 53 015 DIN 53 177 DIN 53 214 ASTM D 1 200 ASTM D 445 ISO 1342	viscosity		
DIN 53 169	solvents, general		
DIN 53 170 DIN 53 249 ASTM D 3 539	evaporation number		
DIN 53 171	boiling curve/ boiling point range	DIN 55 690 DIN 55 999	ethylene glycol ether
DIN 51 751		DIN 55 998	propylene glycol ether
ASTM D 1 078		ASTM D 3 128	methyl glycol

Table 20. (continued)

Standard	Subject area	Standard	Subject area
ASTM E 133 ISO 4626		ASTM D 331 ASTM D 330	ethyl glycol butyl glycol
DIN 53 172 ASTM D 1353 ASTM D 2109 ISO 3251	dry residue	DIN 58 752	solvents for optics
DIN 53 173 ASTM D 2192 ASTM D 2119	carbonyl number	DIN ISO 6162 ISO 6162	iodine color number
DIN 53 174	solvent mixtures	DIN 53 241	iodine number
DIN 53 240 ASTM D 1957 ASTM E 222 ISO 4629	hydroxyl number	DIN 13 310 ASTM D 971	interfacial surface tension
DIN 53 242	raw materials	DIN 16 513	pure toluene for gravure inks
DIN 51 376 ASTM D 92	flash point (Cleveland), open cup	DIN 51 750 DIN EN 21 512 ASTM D 270 ISO 1512 ISO 842	sampling
DIN 55 651 (draft)	abbreviations for solvents		

Table 21. Abbreviations for solvents (DIN 55651 draft)

Abbreviation	Name	CAS-No.
BAC	butyl acetate	123-86-4
BDG	butyl diglycol	112-34-5
BDGA	butyl diglycol acetate	124-17-4
BG	butyl glycol	111-76-2
BGA	butyl glycol acetate	112-07-2
BLO	butyrolactone	96-48-0
BTG	butyl triglycol	143-22-6
DAA	diacetone alcohol	123-42-2
DBE	dibasic esters	
DCM	dichloromethane	74-95-3
DEE	diethyl ether	60-29-7
DEG	diethylene glycol	111-46-6
DEK	diethyl ketone	96-22-0
DIBK	diisobutyl ketone	108-83-8
DIP	diisopropylene glycol mono isopropylether	4039-63-8
DIPE	diisopropylether	108-20-3
DMAC	dimethylacetamide	127-19-5
DMF	dimethylformamide	68-12-2
DMK	acetone (dimethylketone)	67-64-1
DMSO	dimethylsulfoxide	67-68-5
DPB	dipropylene glycol butyl ether	35884-42-5
DPG	dipropylene glycol	25265-71-8
DPM	dipropylene glycol methyl ether	34590-94-8
DPMA	dipropylene glycol methyl ether acetate	
EAK	ethyl amyl ketone	106-68-3
EC	ethylene carbonate	96-49-1
EDG	ethyl diglycol	111-90-0
EDGA	ethyl diglycol acetate	112-15-2
EEP	2-ethoxy ethyl propionate	7737-40-8
EG	ethyl glycol	110-80-5
EGA	ethyl glycol acetate	111-15-9
2EH	2-ethyl hexanol	104-76-7
2EHA	2-ethyl hexyl acetate	103-09-3
EtAC	ethyl acetate	141-78-6
EtOH	ethanol	64-17-5
HG	hexyl glycol	122-25-4
IBA	isobutanol	78-83-1
IBAC	isobutyl acetate	110-19-0
IBIB	isobutyl isobutyrate	97-85-8
IP	isophorone	78-59-1
IPA	isopropanol	67-63-0
IPAC	isopropyl acetate	108-21-4
IPG	isopropyl glycol	109-59-1
IPP	isopropoxypropanol	3944-36-3
MAK	methyl amyl ketone	110-43-0
MBK	methyl butyl ketone	591-78-6

Table 21. (continued)

Abbreviation	Name	CAS-No.
MDG	methyl diglycol	111-77-3
MEG	monoethylene glycol	107-21-1
MEK	methyl ethyl ketone	78-93-3
MeAC	methyl acetate	79-20-9
MeOH	methanol	67-56-1
MG	methyl glycol	109-86-4
MGA	methyl glycol acetate	110-49-6
MIAK	methyl isoamyl ketone	110-12-3
MIBC	methyl isobutyl carbinol	108-11-2
MIBK	methyl isobutyl ketone	108-10-1
MIPK	methyl isopropyl ketone	563-80-4
MPG	monopropylene glycol	57-55-6
MPK	methyl propyl ketone	107-87-9
MTBE	methyl tert-butyl ether	1634-04-4
MTG	methyl triglycol	112-35-6
NBA	n-butanol	71-36-3
NMP	N-methyl-2-pyrrolidone	872-50-4
NPA	n-propanol	71-23-8
NPP	propoxypropanol	30136-13-1
PAC	propyl acetate	109-60-4
PB	butoxypropanol	29387-86-8
PBA	butoxy propyl acetate	57515-72-7
PC	propylene carbonate	108-32-7
PDG	propyl diglycol	6881-94-3
PER	perchloro ethylene	127-18-4
PG	propyl glycol	2807-30-9
PGDA	propylene glycol diacetate	632-84-7
PM	methoxypropanol	107-98-2
PMA	methoxypropyl acetate	108-65-5
PP	phenoxypropanol	770-35-4 (4169-04-4 [1])
PTB	tert-butoxypropanol	
SB	mineral spirits	
SBA	sec-butanol	78-92-2
SBAC	sec-butyl acetate	105-46-4
TB	white spirit	
TBA	tert-butanol	75-65-0
TEG	triethylene glycol	112-27-6
THF	tetrahydrofurane	109-99-9
TMC-on	trimethyl cyclohexanone	873-94-9
TPG	tripropylene glycol	1638-16-0
TPM	tripropylene glycol methyl ether	25498-49-1
TRI	trichloro ethene	79-01-6

[1] No. of isomers

Table 22. Physical properties of solvents

Solvent	M	bp range (101.3 kPa), °C	Solidification point, °C	Density d_4^{20}, g/mL	Refractive index, n_D^{20}	Viscosity (20 °C), mPa · s	Vapor pressure (20 °C), kPa	Flash point, °C
Aliphatic Hydrocarbons								
SBP spirit	~ 86	40–80	< −20	0.666	1.377	0.5	33.0	−30
SBP spirit	~ 100	64–138	< −20	0.708	1.399	0.6	12.5	−30
SBP spirit	~ 99	80–110	< −20	0.710	1.398	0.6	8.7	−20
SBP spirit	~ 112	100–140	< −20	0.740	1.418	0.8	4.0	−2
White spirit	~ 132	144–165	< −20	0.743	1.415	0.7	4.0	25
Petroleum ether	82.2	42–62	−150	0.653	1.370	0.46 (25 °C)	31.0	< 0
Pentane	72.2	35–38	−130	0.632	1.358	0.45 (25 °C)	65.0	< −20
Hexane	86.2	65–70	−95	0.664	1.372	0.46 (25 °C)	20.0	−22
Heptane	100.2	94–99	−91	0.684	1.387	0.62 (25 °C)	8.5	−4
Isododecane	170	164–188	−81	0.752	1.420	1.34	0.31	46
Cycloaliphatic Hydrocarbons								
Cyclohexane	84.2	80.5–81.5	6.5	0.778	1.426	0.94	10.4	−17
Methylcyclohexane	98.2	101–103	−126	0.769	1.423	0.68	5.1	−4
Ethylcyclohexane	112.2	132	−111.3	0.79				19
Tetrahydronaphthalene	132.2	200–209	−31	0.969	1.539	2.2	0.024	74
Decahydronaphthalene	138.2	185–195	−30.4	0.888	1.475	2.4	0.29	57
Terpenes and Terpenoids								
Wood turpentine oil	140	150–180	< −40	0.860	1.465		0.44	35
Root turpentine oil	140	155–170	< −30	0.862	1.465	1.48		34
Balsam turpentine oil	140	150–180	< −30	0.861	1.471		0.44	32
Pine oil	140	179–230		0.883–0.936	> 1.5			
α-Pinene	136.2	154–156	< −40	0.863	1.466		0.57	33
β-Pinene	136.2	164–166	−50	0.872	1.476	55	0.61	
Dipentene	136.2	174–181		0.864	1.474	1.7	0.22	50
D-Limonene	136.2	176	−96.9	0.843	1.473	3.5		49

Table 22. (continued)

Solvent	M	bp range (101.3 kPa), °C	Solidification point, °C	Density d_4^{20}, g/mL	Refractive index, n_D^{20}	Viscosity (20 °C), mPa·s	Vapor pressure (20 °C), kPa	Flash point, °C
Aromatic Hydrocarbons								
Benzene	78.1	79.8–80.8	5.5	0.879	1.501	0.65	10.1	−11
Toluene	92.1	110–111	−95	0.873	1.499	0.61	2.9	6
Xylene	106.2	137–142	−25	0.874	1.498	0.65	0.9	25
Ethylbenzene	106.2	135.5–136.5	−95	0.867	1.496	0.67	1.0	23
Cumene	120.2	152.9	−96	0.862	1.492	0.73	0.5	46
Mesitylene	120.2	164.7	−44.7	0.870	1.499	0.71	0.8 (38 °C)	46
Pseudocumene	120.2	169.3	−43.8	0.880	1.505		0.7 (38 °C)	52
Hemellitene	120.2	176.1	−25.3	0.900	1.513		0.5 (38 °C)	53
Cymol	134.2	174–177	−25	0.857	1.493	0.83		
Styrene	104.2	145	−31	0.907	1.547	0.76	0.71	31
Chlorinated Hydrocarbons								
Dichloromethane	84.9	40.2	−96	1.326	1.424	0.44	47.5	
Trichloromethane	119.4	61.2	−64	1.488	1.445	0.58	21.0	
Ethyl chloride	64.5	12.5	−138	0.924	1.379	0.96	133.0	−43
Isopropyl chloride	78.6	36.2	−117	0.859	1.378	0.29		−36
1,2-Dichloroethane	99.0	83–84	−36	1.254	1.444	0.74	8.2	13
1,1,1-Trichloroethane	133.4	74.0	−32	1.332	1.438	1.1	13.3	
Trichloroethylene	131.4	86.7	−86	1.446	1.477	0.58	7.3	
Perchloroethylene	165.9	121	−23	1.629	1.506	0.88	1.9	
1,2-Dichloropropane	113.0	97.0	−100	1.155	1.442	0.85	5.6	12
Alcohols								
Methanol	32.0	64.6	−98	0.791	1.329	0.61	12.8	10
Ethanol	46.1	78.3	−114	0.789	1.361	1.19	5.9	13
Propanol	60.1	97.2	−127	0.804	1.386	2.26	1.9	23
Isopropyl alcohol	60.1	82.4	−88	0.786	1.377	2.20	4.2	13

Table 22. (continued)

Solvent	M	bp range (101.3 kPa), °C	Solidification point, °C	Density d_4^{20}, g/mL	Refractive index, n_D^{20}	Viscosity (20 °C), mPa·s	Vapor pressure (20 °C), kPa	Flash point, °C
Butanol	74.1	117.7	−80	0.810	1.399	3.00	0.67	35
Isobutanol	74.1	107.7	−108	0.802	1.396	3.76	1.2	28
sec-Butanol	74.1	99.5	−115	0.807	1.397	4.21	1.6	21
tert-Butanol	74.1	82.5	24.3	0.774	1.382	3.4 (30 °C)	4.1	14
Amyl alcohol	88.2	137–139	−78.5	0.815	1.410	3.34	0.37	49
Isoamyl alcohol	88.2	130	−134	0.810	1.408	4.37	0.30	44
Hexanol	102.2	157.6	−57	0.820	1.418	4.3 (25 °C)	0.09	74
Methylisobutylcarbinol	102.2	131.4	−90	0.808	1.411	4.1	0.30	41
2-Ethylbutanol	102.2	149.5	−15	0.833	1.421	5.6	0.16	58
Isooctyl alcohol	130.2	184–190		0.832	1.431	10	0.04	75
2-Ethylhexanol	130.2	183–185	−76	0.833	1.432	11.9	0.02	82
Isononanol	144.3	200–207		0.844	1.440	19	0.04	93
Isodecanol	158.3	214–222		0.838	1.440	18		102
Diisobutylcarbinol	144.2	178	−65	0.812	1.423	2.8		74
Cyclohexanol	100.2	160.8	20	0.949	1.465	62.5	0.06	67
Methylcyclohexanol	114.2	160–195	−38	0.925	1.465	26.2		68
Trimethylcyclohexanol	142.2	190–196	cis 37.3, trans 57.3	0.861 (60 °C)	1.439 (60 °C)	5.8 (60 °C, cis)		> 76
Benzyl alcohol	108.1	205.2	−15	1.043	1.539	0.61	0.002	96
Methylbenzyl alcohol	122.2	200–205	21.5	1.008	1.525	11.2	0.003	90
Furfuryl alcohol	98.1	170.5		1.128	1.486	4.7		65
Tetrahydrofurfuryl alcohol	102.1	177	<−80	1.062	1.451	5.49 (25 °C)	0.03	80
Diacetone alcohol	116.2	168	−56	0.946	1.424	3.6	0.10	67

Ketones

Solvent	M	bp range (101.3 kPa), °C	Solidification point, °C	Density d_4^{20}, g/mL	Refractive index, n_D^{20}	Viscosity (20 °C), mPa·s	Vapor pressure (20 °C), kPa	Flash point, °C
Acetone	58.1	56.2	−95	0.791	1.359	0.33	24.1	−19
Methyl ethyl ketone	72.1	79.6	−86	0.805	1.379	0.40	0.1	−14
Methyl propyl ketone	86.1	102.3	−86	0.810	1.390	0.50	3.6	10
Methyl isopropyl ketone	86.1	92	−92	0.803	1.389	0.48	8.6	5
Methyl butyl ketone	100.2	127.2	−57	0.808	1.403	0.63	1.3	23

Table 22. (continued)

Solvent	M	bp range (101.3 kPa), °C	Solidification point, °C	Density d_4^{20}, g/mL	Refractive index, n_D^{20}	Viscosity (20 °C), mPa·s	Vapor pressure (20 °C), kPa	Flash point, °C
Methyl isobutyl ketone	100.2	114–117	−84	0.802	1.396	0.61	2.15	14
Methyl amyl ketone	114.2	147.0–153.5	−33	0.818	1.408	0.81	0.5	46
Methyl isoamyl ketone	114.2	141–148	−74	0.813	1.407	0.73	0.6	41
Methyl heptyl ketone	142.2	194	−9	0.820	1.422			60
Diethyl ketone	86.1	100–102	−41	0.814	1.391	0.47	1.3	4
Ethyl butyl ketone	114.2	147.6	−39	0.816	1.408	0.76	0.53	46
Ethyl amyl ketone	128.2	160.6	−58	0.822	1.412	0.80	0.27	44
Diisopropyl ketone	114.2	124.4		0.806	1.401	0.63	0.5	19
Diisobutyl ketone	142.2	169.0	−46	0.806	1.414	1.05	0.23	47
Cyclohexanone	98.1	153–156	−26	0.946	1.451	2.2	0.35	43
Methylcyclohexanone	112.2	168	< −60	0.919	1.442	1.78	0.07	48
Dimethylcyclohexanone	126.2	175–195		0.910				61
Trimethylcyclohexanone	140.2	188.8	−10	0.888	1.445	2.54	0.1	64
Mesityl oxide	98.2	128.0	−59	0.855	1.446	0.6	1.2	31
Isophorone	138.2	210–216	−8	0.920	1.477	2.6	0.04	85
Acetyl acetone	100.1	140.5	−23	0.973	1.451	0.6	1.0	35
Esters								
Methyl formate	60.1	31–32	−99	0.978	1.341			−18
Ethyl formate	74.1	54–55	−80.5	0.927	1.359	0.41		−19
Butyl formate	102.1	96–110	−92	0.889				
Isobutyl formate	102.1	97.5	−96	0.875	1.386	0.67	4.4	
Methyl acetate	74.1	55–57	−98	0.932	1.361	0.37	22.6	−13
Ethyl acetate	88.1	76–77	−84	0.900	1.372	0.45	10.3	−5
Propyl acetate	102.1	101–102	−92	0.888	1.385	0.59	3.1	10
Isopropyl acetate	102.1	88.4	−73	0.872	1.377	0.57	5.8	4
Butyl acetate	116.2	124–128	−77.9	0.880	1.394	0.74	1.11	23
Isobutyl acetate	116.2	114–118	−99	0.871	1.390	0.69	1.8	19
sec-Butyl acetate	116.2	112.2	−98.9	0.871	1.388	0.65	2.2	19

Table 22. (continued)

Solvent	M	bp range (101.3 kPa), °C	Solidification point, °C	Density d_4^{20}, g/mL	Refractive index, n_D^{20}	Viscosity (20 °C), mPa · s	Vapor pressure (20 °C), kPa	Flash point, °C
Amyl acetate	130.2	146–149	−100	0.876	1.403	0.92	0.6	40
Isoamyl acetate	130.2	142.1		0.874	1.401		0.6	23
Hexyl acetate	144.2	164–176		0.874	1.410	1.22 (25 °C)	0.16	59
Heptyl acetate	158.2	189–202		0.874	1.416	1.24 (25 °C)		66
2-Ethylhexyl acetate	172.3	192–205	−80	0.872	1.420	1.5	0.06	76
Cyclohexyl acetate	142.2	174.5	−72	0.970	1.440	2.27	0.9	58
Benzyl acetate	150.2	213–216	−52	1.055	1.506			93
Propylene glycol diacetate	160.1	190–192		1.057	1.414	3.0		87
Methyl propionate	88.1	75–82	−88	0.918	1.377			2
Ethyl propionate	102.1	95–99	−74	0.888	1.386 (15 °C)			8
Propyl propionate	116.2	121–123	−76	0.881	1.393	0.7		22
Butyl propionate	130.2	130–147	−90	0.870	1.404			38
Pentyl propionate	144.1	165–168		0.872	1.408	1.0		51
Ethyl butyrate	116.2	115–125	−98	0.880	1.393			22
Propyl butyrate	130.2	136–143	−95	0.872	1.398 (25 °C)			37
Butyl butyrate	144.2	159–162	−92	0.872	1.405			50
Isobutyl butyrate	144.2	142–156		0.860	1.403	0.97	0.5	46
Amyl butyrate	158.2	165–170	−73	0.864	1.414 (25 °C)			59
Methyl isobutyrate	102.1	99–101	−85	0.895				−1
Ethyl isobutyrate	116.2	109–111	−88	0.869				10
Isopropyl isobutyrate	130.2	119–121		0.844				15
Isobutyl isobutyrate	144.2	147.7–150.1	−80	0.855	1.400	1.02	0.4	49
Methyl lactate	104.1	145		1.085	1.414	2.9	0.30	55
Ethyl lactate	118.1	154.5	−25	1.036	1.412	2.44	0.2	48
Isopropyl lactate	132.2	157		0.98		2.9	0.10	55
Butyl lactate	146.2	188.0	−46	0.950	1.420	3.8	0.05	61
Butyl glycolate	132.2	182.0	−17	1.013	1.425	4.40	0.13	68
Methyl glycol acetate	118.1	142–148	−65	1.003	1.402	1.43	0.5	46
Ethyl glycol acetate	132.2	152–160	−62	0.975	1.405	1.35	0.3	52

Table 22. (continued)

Solvent	M	bp range (101.3 kPa), °C	Solidification point, °C	Density d_4^{20}, g/mL	Refractive index, n_D^{20}	Viscosity (20 °C), mPa·s	Vapor pressure (20 °C), kPa	Flash point, °C
Butyl glycol acetate	160.2	190–198	−65	0.945	1.415	1.8	0.04	75
Ethyl diglycol acetate	176.2	210–222	−25	1.011	1.421	2.8	0.013	107
Butyl diglycol acetate	204.3	235–250	−32	0.978	1.426	3.6	0.0013	108
Methoxypropyl acetate	132.2	143–150	−65	0.967	1.403	1.23	0.53	44
Ethoxypropyl acetate	146.2	158		0.941	1.405	1.4	0.23	54
3-Methoxybutyl acetate	146.2	166–172	−60	0.953	1.409	1.43	0.4	60
Ethyl 3-ethoxypropionate	146.2	165–172		0.947		1.0	0.23	58
Dibasic ester	160	196–225	−20	1.092	1.422	2.6	0.02	>75
Ethylene carbonate	88.1	246.7	36.4	1.322 (0 °C)	1.415 (50 °C)	1.68 (50 °C)	0.001	145
Propylene carbonate	102.1	243.4	−54	1.205	1.422	2.76	0.004	130
Butyrolactone	86.1	204	−42	1.124	1.436	2.0	0.09	98
Glycol Ethers								
Methyl glycol	76.1	124.5	−85	0.966	1.402	1.98	0.83	38
Ethyl glycol	90.1	134–137	<−80	0.930	1.408	2.08	0.5	43
Propyl glycol	104.2	149.5–153.5	−90	0.911	1.414	2.42	0.13	49
Isopropyl glycol	104.2	142.8	<−60	0.907	1.410	2.6 (23 °C)	0.35	46
Butyl glycol	118.2	168–172	<−70	0.901	1.419	3.26	0.1	67
Hexyl glycol	146.2	208	−50	0.887	1.429	4.4	<0.10	91
Phenyl glycol	138.2	244–250	−37	1.110	1.538	29.0		121
Methyl diglycol	120.2	194.2	−85	1.021	1.424	4.0	0.03	90
Ethyl diglycol	134.2	196–205	−90	0.990	1.429	4.38	0.02	94
Butyl diglycol	162.2	224–234	−68	0.956	1.431	5.85	0.01	105
Hexyl diglycol	190.3	240–265	−40	0.934	1.437	8.0		129
Methyl triglycol	164.2	245–255	−44	1.049	1.439	7.3	0.001	133
Ethyl triglycol	178.2	255.4	−19	1.04	1.440	7.8	<0.001	120
Butyl triglycol	206.3	255–295	−35.1	0.985	1.439	9.42	<0.001	130
Butyl tetraglycol	250.3	300–340		1.015	1.446		<0.001	170
1-Methoxy-2-propanol	90.1	119–122	−96	0.922	1.403	1.89	1.12	32

Table 22. (continued)

Solvent	M	bp range (101.3 kPa), °C	Solidification point, °C	Density d_4^{20}, g/mL	Refractive index, n_D^{20}	Viscosity (20 °C), mPa·s	Vapor pressure (20 °C), kPa	Flash point, °C
Ethoxypropanol	104.1	130–138	−90	0.904	1.404	2.1	1.3	42
Isopropoxypropanol	118.1	139–145		0.877	1.407	2.3		41
Butoxypropanol	132.2	170	−102	0.884	1.417	3.38	0.08	63
Isobutoxypropanol	132.2	172	−51	0.888			0.11	
tert-Butoxypropanol	132.2	151	−34	0.872			0.6	45
Phenoxypropanol	152.2	241–246	−29	1.064	1.411	3.3	< 0.001	120
Methyl dipropylene glycol	148.2	187–192	−83	0.957	1.524	34	0.025	80
Isopropyl dipropylene glycol	178.2	198–208		0.917	1.422	3.71		87
Butyl dipropylene glycol	192.2	229	−75	0.91	1.420	4.3	0.006	113
Methyl tripropylene glycol	206.3	235–251	−78	0.967		4.6	0.004	121
Butyl tripropylene glycol	248.3	274	< −75	0.92	1.428	6.16	< 0.001	136
Diglycol dimethyl ether	134.2	160		0.944		8.0	1.5	57
Dipropylene glycol dimethyl ether	162.2	175	< −71	0.904	1.407	2.0	0.7	65

Ethers

Solvent	M	bp range (101.3 kPa), °C	Solidification point, °C	Density d_4^{20}, g/mL	Refractive index, n_D^{20}	Viscosity (20 °C), mPa·s	Vapor pressure (20 °C), kPa	Flash point, °C
Diethyl ether	74.1	34.5	−116	0.714	1.353	0.24	58.8	−40
Diisopropyl ether	102.2	68.3	−85.5	0.724	1.368	0.32	21.0	−22
Dibutyl ether	130.2	142.2	−95	0.769	1.399	0.7	0.65	25
Methyl tert-butyl ether	88.2	55.3	−109	0.740	1.369	0.36	27.1	−28
Tetrahydrofuran	72.1	66	−108	0.888	1.407	0.61	17.3	−21.5
Dioxane	88.2	101.1	11.8	1.034	1.423	1.23	4.0	5
2,2-Dimethyl-4-hydroxymethyl-1,3-dioxolane	132.2	190		1.068	1.438			90
1,2-Propylene oxide	58.1	33.9		0.830	1.363	0.38	59.5	−37

Miscellaneous Solvents

Solvent	M	bp range (101.3 kPa), °C	Solidification point, °C	Density d_4^{20}, g/mL	Refractive index, n_D^{20}	Viscosity (20 °C), mPa·s	Vapor pressure (20 °C), kPa	Flash point, °C
Dimethyl acetal	90.1	56–65		0.839	1.430			−28
Dimethylformamide	73.1	153.0	−61	0.949	1.430	0.92	0.6	57
Dimethylacetamide	87.1	165.5	−20	0.959	1.435	0.92 (25 °C)	0.2	70

Table 22. (continued)

Solvent	M	bp range (101.3 kPa), °C	Solidification point, °C	Density d_4^{20}, g/mL	Refractive index, n_D^{20}	Viscosity (20 °C), mPa · s	Vapor pressure (20 °C), kPa	Flash point, °C
Dimethyl sulfoxide	78.1	189.0	18.5	1.100	1.478	1.99 (25 °C)	0.06	95
Tetramethylene sulfone	120.2	285	−36	1.173	1.483			177
Carbon disulfide	76.1	45.5−46.8	−111	1.263	1.625			−30
Furfurol	96.1	160−165	−36.5	1.155	1.526			57
Nitromethane	61.0	101.2	−29	1.139	1.381	0.65	0.15	33
Nitroethane	75.1	114.0	−90	1.052	1.391	0.68	3.7	29
1-Nitropropane	89.1	131		0.994	1.394		2.08	34
2-Nitropropane	89.1	120.3	−93	0.992	1.394	0.77	1.72	27
N-Methylpyrrolidone	99.1	202−205	−24.4	1.028	1.470	1.8	0.04	91
N-Ethylpyrrolidone	113.1	200	< −70	0.993	1.466	3.5		93
N-Cyclohexylpyrrolidone	167.3	284	12	1.026	1.495	11.5	< 0.005	> 100
N-(2-Hydroxyethyl)-pyrrolidone	129	295	20	1.139	1.495	53		> 100
1,3-Dimethyl-2-imidazolidinone	114.1	225	8.2	1.060	1.470	1.9		107
Hexamethylenephosphoric triamide	179.0	233	6					

Table 23. Safety and physiological data of solvents

Solvent	Evaporation no. (diethyl ether = 1)	Evaporation no. (butyl acetate = 1)	Ignition temperature, °C	MAK 1997 ppm	MAK 1997 mg/m³	TLV-TWA 1996 ppm	TLV-TWA 1996 mg/m³	Odor threshold, ppm	LD_{50}, rat, oral, mg/kg	LC_{50}, rat, inhalation, $\times 10^{-2}$ ppm
Pentane	<1	13	285	1000	3000	600	1770	900	>24 (mL/kg)	>90000/2 h Mouse
Hexane	1.4	8.4	240	200	720	50	176		24–49 (mL/kg)	54000–74000
Heptane	3.0	3.3	215	500	2100	400	1640	150	>25 (mL/kg)	14000–16000
Octane		1.23		500	2400	300	1400	200		
Cyclohexane	3.4	5.60		200	700	300	1030	2	1297	
Methylcyclohexane		2.99	285	500	2000	400	1610			
Turpentine oil	38	<0.005	252	100	^{f}560	100	556		2860	
Tetrahydronaphthalene	200		425							
Toluene	6.1	2.0	535	50	^{c}190	50, A4	188, A4	40	5000	40/4 h
Xylene	13.5	0.76	525	100	^{d}440	100, A4	434, A4	0.4–20	4300	63.5/4 h
Ethylbenzene	8.8	0.84	435	100	^{d,e}440	100	434	125	3500	40/4 h
Cumene		0.14	424	50	^{c,e}250	50	246	40 (ppb)	1400	80/4 h
Styrene	16	0.536	480	20	^{c}86	50	213	0.1	5000	240/2 h
Dichloromethane	1.8	25.0	662	100	^{d,h}350	^{g}50, A3	^{g}174, A3	550	2136	880/0.5 h
1,1,1-Trichloroethane	2.6	6.00	537	200	^{c}1100	350, A4	1910, A4	300	10300	10/4 h (LCLo)
Trichloroethylene	3.1	4.9	460		^{i}carcinogenic	50, A5	269, A5	200	7193	80/4 h (LCLo)
Perchloroethylene	6.0	2.59			^{e,h}—	25, A3	170, A3	320	8850	40/4 h (LCLo)
Methanol	6.3	4.1	463	200	^{c,e}270	200	262	6000	5628	640/4 h
Ethanol	8.3	2.4	419	1000	^{c}1900	1000, A4	1880, A4	6000	7060	380/10 h
Isopropyl alcohol	11	1.5	425	200	^{d}500	400	983	600	5045	160/8 h
Butanol	33	0.47	360	100	^{d}310	—	—	33	790	50/5 h
Isobutanol	25	0.85	390	100	^{c}310	50	152	80	2460	80/8 h (LCLo)
Isoamyl alcohol	62		365	100	^{c}370	100	361	15	1300	
Methylisobutylcarbinol	66	0.28	305	25	^{e}110	25	104		2600	20/4 h
2-Ethylhexanol	690	0.0018	330					0.2	800	
Cyclohexanol	150	0.08	300	50	210	50	206	50 (ppb)	2060	
Methylcyclohexanol	800		296	50	240	50	234	300		
Furfuryl alcohol		0.04	391	10	41	10	40	9	275	2.33/4 h

Table 23. (continued)

Solvent	Evaporation no. (diethyl ether = 1)	Evaporation no. (butyl acetate = 1)	Ignition temperature, °C	MAK 1997 ppm	MAK 1997 mg/m³	TLV-TWA 1996 ppm	TLV-TWA 1996 mg/m³	Odor threshold, ppm	LD_{50}, rat, oral, mg/kg	LC_{50}, rat, inhalation, $\times 10^{-2}$ ppm
Perchlorethylene				—	e,h —					
Trichlorethylene				i carcinogenic						
Diacetone alcohol	150	0.15	624	50	240	50	238	3	4000	7/8 h
Acetone	2.0	5.6	540	500	1200	750	1780	300	9750	
Methyl ethyl ketone	3.3	3.7	514	200	c,e 600	200	590	30	3400	20/4 h
Methyl propyl ketone	4.8	2.3	449	200	710	200	705	8	3730	20/4 h
Methyl butyl ketone	13	0.98	424	5	21	5	20	32	2590	
Methyl isobutyl ketone	7	1.4	460	20	c,e 83	50	205	8	2080	40/4 h (LCLo)
Diisobutyl ketone	48	0.2	345	50	300	25	145	50	5750	20/4 h (LCLo)
Cyclohexanone	41	0.25	420	—	f,h —	25	100	50	1620	20/4 h
Methyl cyclohexanone	53	0.18	595	50	e 230	50	229	0.5	2140	
Mesityl oxide	8.2	0.88	340	25	100	15	60	30	1120	10/4 h
Isophorone	330	0.023	462	2	c,h 11	—	—	0.7	2330	18.4/4 h (LCLo)
Methyl acetate	2.1	9.5	501	200	c 610	200	—	200	3700	
Ethyl acetate	3	4.2	460	400	c 1500	400, A4	1440, A4	50	6100	16/8 h
Propyl acetate	4.8	2.0	457	200	850	200	835	20	9800	80/4 h
Butyl acetate	12	1.0	370	100	480	150, A4	713, A4	15	14000	20/4 h
Isobutyl acetate	5.8	1.7	480	100	480	150	713	4	15000	80/4 h
Amyl acetate	15	0.8	380	50	270	100	532	20	7400	52
Methyl glycol acetate	21	0.35	400	5	b,e 25			0.9	3390	70/4 h
Ethyl glycol acetate	57	0.20	345	5	b,e 27			0.5	2900	15/8 h (LCLo)
Ethyl diglycol acetate	>1200	0.008	360						11 000	
Butyl glycol acetate	137	0.03	375	20	c,e 130				3200	
Butyl diglycol acetate	>3000	0.002	290						6500	
1-Methoxy-2-propyl acetate	34	0.32	354	50	c 270				>8500	>23.9/6 h
Ethoxypropyl acetate	70	0.19							>5000	
Ethyl-3-ethoxypropionate		0.12	377							
Butyrolactone	>1000	0.03							1580	
Ethyl lactate	80	0.22	294						>2000	>50

Table 23. (continued)

Solvent	Evaporation no. (diethyl ether = 1)	Evaporation no. (butyl acetate = 1)	Ignition temperature, °C	MAK 1997 ppm	MAK 1997 mg/m^3	TLV-TWA 1996 ppm	TLV-TWA 1996 mg/m^3	Odor threshold, ppm	LD$_{50}$, rat, oral, mg/kg	LC$_{50}$, rat, inhalation, ×10^{-2} ppm
Dibasic ester	4800	0.008	370						17000	10.7/1 h
Methyl glycol	34	0.53	290	5	^{b,e}16			90	2460	20/4 h
Cyclohexanone					^{f,h}—					
Ethyl glycol	43	0.38	235	5	^{b,e}19			50	3000	40/4 h (LCLo)
Propyl glycol	75	0.2	235	20	^{c,e}80				3089	> 21.3/6 h
Butyl glycol	119	0.082	240	20	^{c,e}98			0.5	1480	4.5/4 h
Ethyl diglycol	1200	0.017	220						5540	> 170/12 d
Butyl diglycol	3750	0.004	225						6560	> 30/7 h
1-Methoxy-2-propanol	22	0.75	290	100	^{c}370				5660	55/4 h
Ethoxypropanol	33	0.46	280							
Diethyl ether	1	28.0	180	400	^{d}1200	400	1210	0.3	1215	730/2.5 h
Diisopropyl ether	1.6	8.1	405	500	2100			0.06	8470	3000/2 h (LCLo)
Methyl tert-butyl ether	1.6	8.4	460			40, A3	144, A3	267	3866	850/4 h
Tetrahydrofuran	2.2	6.3	212	50	^{c}150	200	590	40	3000 (LDLo)	780/2 h
Dioxane	7.3	2.17	266	20	^{d,e,h}73	25	90	200	7120	
Dimethylformamide	120	0.20	445	10	^{b,e}30	10, A4	30, A4	90	2800	
Dimethylacetamide	172	0.138	400	10	^{c,e}36	10, A4	36, A4		4300	
Carbon disulfide	1.8	10.90	102	5	^{b,e}16	10	31			
Furfurol	75		391	—	^{e,h}—				127	1.53/4 h
Nitroethane	21	1.2		100	310	100	307		1100	
1-Nitropropane		0.9		25	92	25, A4	91, A4	300	800	
2-Nitropropane	10	1.5	428	^{g}carcinogenic		10, A3	36, A3	^{g}36	200	15.13/5 h
N-Methylpyrrolidone		0.04	287	19	^{c,e}80				3600	
Furfurol				—	^{e,h}—					

^{a,b,c,d} Pregnancy groups. ^{e} Danger of skin resorption. ^{f} Danger of sensitization. ^{g} III A2 Carcinogenic group. ^{h} III B Carcinogenic group. ^{i} III A1 Carcinogenic group.

14.9.2. Cycloaliphatic Hydrocarbons

The solvency of cycloaliphatic hydrocarbons is between that of aliphatic and aromatic hydrocarbons. They have a high solvency for fats, oils, oil-modified alkyd resins, styrene-modified oils and alkyd resins, bitumen, rubber, and other polymers. Polar resins (e.g., urea–, melamine–, and phenol–formaldehyde resins), as well as alcohol-soluble synthetic resins and cellulose esters are, however, insoluble.

Cycloaliphatic hydrocarbons are miscible with most other solvents, but are insoluble in water.

Cyclohexane [*110-82-7*] is a water-clear, colorless liquid with a gasoline-like smell; it is miscible with most organic solvents except methanol, dimethylformamide, and solvents of similar polarity.

Methylcyclohexane [*108-87-2*] is similar to cyclohexane but less volatile.

1,2,3,4-Tetrahydronaphthalene [*119-64-2*] (tetralin) is an aromatic-cycloaliphatic hydrocarbon. It is a colorless liquid with a naphthalene-like odor, insoluble in water, and miscible with all common organic solvents. It dissolves fats, oils, linoxyn, rubber, waxes, asphalt, bitumen, pitch, tar, phenol, naphthalene, iodine, sulfur, etc., and is used on a large scale in painting work, and in floor wax and shoe polish production. It also dissolves colophony, Congo copals, glyptal resins, coumarone resins, ketone–formaldehyde resins, and aminoplasts. It imparts good flow properties to paints and produces high-gloss, smooth film surfaces. It is autooxidative and thus acts as an oxygen carrier in drying oils.

Decahydronaphthalene [*91-17-8*] (decalin) is a colorless solvent with a pungent odor and fairly high volatility, its solvency is somewhat lower than that of tetrahydronaphthalene.

14.9.3. Terpene Hydrocarbons and Terpenoids

Turpentine oil [*8006-64-2*] (DIN 53248). Only pure ethereal oil obtained from the distillation of the resinous secretion of living pine trees, and from which no valuable constituents (e.g., pinene) have been extracted, may be used as balsam terpentine oil (RAL, Sheet 848C). All turpentine oils obtained in any other way must be specially labelled with details of their source. Oils of turpentine from different countries differ in composition. American and Greek oils contain predominantly D-pinene, whereas French, Spanish, and Portuguese oils contain L-pinene.

Balsam turpentine oil is a highly mobile liquid with a characteristic odor. On prolonged storage the boiling point falls, and resinification may subsequently occur. Turpentine oil has a good solvency for fats, oils, waxes, and hydrocarbon resins. It is miscible with ethanol and aliphatic and aromatic hydrocarbons. Its dissolution properties are better than those of white spirit. It is used in addition to white spirit in oil-based and alkyd resin paints. The higher boiling constituents of wood turpentine oil contain a large amount of terpineol and are treated as pine oil. Pine oil is used as a high-boiling solvent in stoving enamels.

Root turpentine oil is obtained by steam distillation of wood chips and shavings with a high resin content. Its properties are similar to those of sulfite cellulose oil of turpentine, which occurs in wood pulp production. Root turpentine oil is a colorless liquid with similar properties to balsams turpentine oil.

Wood oil is obtained by dry distillation of chips and shavings with a high resin content. It is a yellow, acidic liquid with an unpleasant odor. On exposure to air and also after purification it becomes discolored and the odor becomes more unpleasant. It is only suitable as a solvent for dark paints and lacquers.

Terpenoids are obtained by distillation of turpentine oils. α-Pinene [*80-56-8*] is used in the preparation of camphor. Higher-boiling constituents include pine oil [*8000-41-7*] (terpineol; trade names: Depanol, Solvenol, Terposol, Yarmor). Hydroterpine is prepared by hydrogenation of pine oil and is used as a substitute for turpentine oil. Terpenoids are used as high-boiling solvents in oil-based and alkyd resin paints. Dipentene [*138-86-3*] prevents skin formation in air-drying paints. D-Limonene [*5989-27-5*] see [14.264].

14.9.4. Aromatic Hydrocarbons

Compared to aliphatic hydrocarbons, aromatic hydrocarbons (DIN 51 633) have a higher solvency for oils, castor oil, oil-modified alkyd resins, styrene-modfied oils and alkyd resins, saturated polyester resins, polystyrene, poly(vinyl ethers), polyacrylate and polymethacrylate esters, poly(vinyl acetate), vinyl chloride and vinyl acetate copolymers, and many low-polarity resins.

Aromatic hydrocarbons are used as diluents in solutions of cellulose nitrate, cellulose esters, and ethers with true solvents such as esters and ketones. Rubber, polyisobutene, and molten polyethylene also dissolve in them. Poly(vinyl chloride), solid polyethylene, polyamides, and shellac are, however, insoluble or only swell.

Toluene [*108-88-3*] is mainly used in cellulose nitrate lacquers; in heat-curing paints based on urea–, melamine–, or phenol–formaldehyde resins; in alkyd resin paints; and in paints based on chlorinated rubber, polystyrene, polyacrylates, or poly(vinyl acetate). Mixtures of toluene with esters are used to dissolve vinyl chloride copolymers and postchlorinated poly(vinyl chloride) [14.265].

Xylene (mixture of *o*-, *m*-, and *p*-isomers, [*56004-61-6*], [*108-38-3*], [*41051-88-1*]) generally contains very small amounts of toluene and larger amounts of ethylbenzene. Xylene is the most important aromatic solvent in the paint industry. It dissolves poly(vinyl acetate) only in combination with alcohols or glycol ethers, but is otherwise similar to toluene as regards dissolution properties.

Ethylbenzene [*100-41-4*] (phenylethane) has a water-clear appearance and characteristic odor. It is miscible with practically all organic solvents, but is insoluble in water. Ethylbenzene is of limited importance as a paints solvent. Its main use is as an industrial starting material for the production of styrene by catalytic dehydrogenation. Catalytic oxidation of ethylbenzene with air in the presence of heavy-metal oxides yields acetophenone and phenylmethylcarbinol. It also improves the antiknock properties of fuels for Otto engines.

Cumene [*98-82-8*] is the principal constituent of heavy naphtha which is the feedstock for phenol and acetone synthesis by the Hock process. It is also a byproduct in the production of sulfite pulp.

Styrene [*100-42-5*] is a colorless liquid that acts as a solvent for unsaturated polyester resins. Styrene is stabilized by the manufacturer to avoid polymerization on storage. Feedstock for chemical syntheses (polymers and copolymers, styrene-modified alkyd resins and oils).

14.9.5. Chlorinated Hydrocarbons

Chlorinated hydrocarbons have a better solvency than the corresponding nonchlorinated compounds for most resins, polymers, rubber, waxes, asphalt, and bitumen. Chlorinated hydrocarbons are miscible with other organic solvents, but are insoluble in water. They have a sweetish odor. Increasing the number of chlorine substituents reduces the combustibility and improves the solvency, but also increases the toxicity.

All chlorinated hydrocarbons may decompose under the action of light, air, heat, and water. Decomposition can be reduced but not completely prevented by adding stabilizers. On account of their health hazard, some chlorinated hydrocarbons may no longer be used as conventional solvents, e.g., tetrachloromethane, tetrachloroethane, and pentachloroethane. Dichloromethane, trichloroethylene, perchloroethylene, and 1,1,1-trichloroethane are increasingly being replaced for reasons of industrial hygiene and environmental protection, particularly of water.

Dichloromethane [*75-09-2*] (methylene chloride) is a colorless, highly volatile, neutral liquid with a characteristic odor. It is insoluble in water but miscible with organic solvents. It has a very good solvency for many organic substances, such as fats, oils, waxes, and resins. Bitumen, rubber, chlorinated rubber, polystyrene, postchlorinated poly(vinyl chloride), vinyl chloride copolymers, polyacrylates, and cellulose esters are also soluble. The solubility spectrum can be expanded by adding other solvents. A mixture of methanol or ethanol and dichloromethane is a good solvent for cellulose ethers and acetyl cellulose. Cellulose nitrate is, however, insoluble.

Dichloromethane is used as a noncombustible solvent and extractant for oils, fats, waxes, fish oil, etc. from industrial and animal products, as well as caffeine, hops, castor oil, cocoa butter, and ethereal oils from substances of plant origin. It is used in the deparaffination of petroleum and the azeotropic dewatering of solvents. Dichloromethane is a constituent of paint-removal pastes and baths, but is increasingly being replaced by aqueous systems. It is used as a solvent in the production of cellulose-acetate-based or cellulose-acetobutyrate-based films and is employed industrially in the leather, metal, rubber, adhesives, and plastics industries.

Trichloromethane [*67-66-3*] (chloroform) is a solvent having a powerful narcotic effect and is not widely used.

Tetrachloromethane [*56-23-5*] (carbon tetrachloride, "Tetra"), is a colorless, neutral, nonflammable liquid. It is toxic and forms phosgene, hydrogen chloride, and chlorine when heated. It dissolves fats, oils, resins and waxes, but not poly(vinyl acetate). Tetrachloromethane is used in the production of rubber-adhesive solutions, for diluting sulfur monochloride in rubber vulcanization, and in the leather and shoe industries.

1,2-Dichloroethane [*107-06-2*] (ethylene chloride) is one of the most stable chlorinated hydrocarbons, with a good solvency for fats, oils, resins, rubber, bitumen, asphalt, and tar. Highly toxic, the odor threshold is higher than the MAK value. 1,2-Dichloroethane is used in the production of building protection agents, roofing felts, and cold asphalt. It is used as an extracting agent for alkaloids and ethereal oils.

1,1,2,2-Tetrachloroethane [*79-34-5*] is a good solvent for resins, rubber, and cellulose acetate. It is used only to a limited extent on account of its potential health hazard.

1,1,1-Trichloroethane [*71-55-6*] (methylchloroform) is insoluble in water, but miscible with organic solvents. It dissolves fats, oils, resins, waxes, bitumen, and asphalt. It is used to degrease metals; as a solvent in the paint, adhesives, and plastics industries; as well as an extracting agent for fats and ethereal oils. It is being increasingly replaced for environmental protection reasons by other systems.

1,2-Dichloroethylene [*75-35-4*] (vinylidene chloride) is a highly volatile liquid with a low flammability although the hot vapors can ignite.

Trichloroethylene [*79-01-6*] ("Tri") is insoluble in water, but is miscible with organic solvents. It dissolves fats, oil, waxes, rubber, and many resins; it is used as a solvent and extracting agent. In paints, trichloroethylene raises the flash point, and is used as a solvent in paint removal pastes. It is increasingly being replaced by other systems for environmental protection reasons.

Perchloroethylene [*127-18-4*] (tetrachloroethylene, "Per") is a colorless, water-insoluble, nonflammable solvent having a very good solvency for greases, fats, waxes, oils, bitumen, tar, and many natural and synthetic resins. It is used in chemical cleaning systems, to degrease light and heavy metals, to degrease pelts and leather (tanning), extraction of animal and vegetable fats and oils, and textile dyeing (solvent for dye baths). It is also used in the preparation of secondary chemical products (fluorinated hydrocarbons, trichloroacetic acid, etc.). Perchloroethylene is increasingly being replaced by other systems for environmental protection reasons.

1,2-Dichloropropane [*78-87-5*] is a solvent for bitumen, asphalt, and tar in the production of road-surfacing materials, building protective agents, and roofing felts.

Chlorobenzene [*108-90-7*] is a colorless, neutral liquid with a weak, benzene-like odor. It is insoluble in water and miscible with organic solvents. Chlorobenzene has a good solvency for fats, oils, resins, polymers, binders, rubber, and chlorinated rubber. Cellulose ethers dissolve in the presence of small amounts of alcohols; cellulose nitrate is insoluble. Chlorobenzene is a solvent in the production of bitumen and asphalt coatings for building protection.

14.9.6. Alcohols

On account of their hydroxyl groups alcohols (DIN 53245) differ from aliphatic, aromatic, and chlorinated hydrocarbons by having a higher polarity and a stronger tendency to form hydrogen bonds. The relationship between the nonpolar hydrocarbon chain and the hydroxyl group is decisive for the solvency of alcohols [14.266]. Lower alcohols accordingly exhibit a pronounced solvency for strongly polar resins such as shellac, copal, glyptal resins, urea–, melamine–, and phenol–formaldehyde resins, cellulose nitrate, slightly ethylated ethyl cellulose, and poly(vinyl acetate). Lower alcohols are, however, poor solvents for nonpolar substances such as fats, oils, oil-modified alkyd resins, styrene-modified oils, and hydrocarbon resins.

The solvency for polar substances decreases with increasing hydrocarbon chain length, propanol and higher alcohols are therefore no longer particularly good solvents for cellulose nitrate and poly(vinyl acetate). Higher alcohols are mainly used as diluents, their solvency is substantially improved by combination with the corresponding acetate esters. On account of their gentle dissolution properties, higher alcohols are highly suited for topcoats that are applied to a primer; the primer is not softened and lifted by these alcohol solvents. In the painting of plastics, alcohols produce only weak superficial swelling of the plastic, resulting in good adhesion without softening of the plastic. In the roller coating of paper alcohols prevent the rubber rollers from swelling, and bleeding of the coating material into the paper, high-gloss prints and paper coatings are therefore obtained.

Methanol [67-56-1] (methyl alcohol) is a clear, colorless liquid with a characteristic odor. It is hygroscopic and miscible in all proportions with water as well as with many organic solvents. It is less soluble in fats and oils, and is only partially miscible with aliphatic hydrocarbons. Numerous inorganic substances (e.g., many salts) are soluble in methanol. Methanol has a good solvency for polar resins, cellulose nitrate, and ethyl cellulose. However, oil-modified alkyd resins and polymers, with the exception of poly(vinyl acetate), poly(vinyl ethers), polyvinylpyrolidone and polymethacrylamide, are insoluble.

Methanol is used for methylation, and is the alcohol most easily esterified. It is used in the preparation of formaldehyde and methyl esters (e.g., dimethyl terephthalate, methyl methacrylate, methyl formate). It is also employed as a solvent for cellulose nitrate, colophony, shellac, and urea resins in the explosives and paint industries. Furthermore, it is used as an antifreeze, fuel, and extracting agent. Methanol is toxic although cases of poisoning are extremely rare if it is correctly used.

Ethanol [64-17-5] (ethyl alcohol) is available as an ethanol–water azeotrope and in anhydrous form. Both forms are supplied completely or partially denatured. Complete denaturation is effected by adding methyl ethyl ketone. Toluene, petroleum ether, and special gasolines are usually used for partial denaturation.

Ethanol is a colorless, clear liquid with a characteristic, pleasant odor. It is miscible in all proportions with water and readily miscible with many organic solvents (e.g., ethers, hydrocarbons, acids, esters, ketones, carbon disulfide, glycols, and other alcohols). Ethanol dissolves castor oil, cellulose nitrate with a low nitrate content, polar resins, and polymers. Ethanol in combination with aromatic compounds dissolves cellulose acetate. Mixtures of ethanol, aromatic hydrocarbons, and water are good solvents for some polyamides. Ethanol is extensively used in the chemical and pharmaceutical industries. It is employed as a raw material for many chemical syntheses (e.g., esterification, as an ethylating agent, and reaction medium). Ethanol is an excellent solvent, diluent, and extracting agent for fats, oils, paints, and natural substances, such as aroma substances, fragrances, perfumes, and pharmaceutical preparations. On account of its pleasant odor and weak substrate solvency it is employed in the formulation of paper and plastics coatings and paints, as well as for flexographic inks. It is also used in the form of methylated spirit for cooking or heating.

Propanol [71-23-8] (1-propanol) is a colorless liquid that is soluble in water and miscible with organic solvents. It has better dissolution properties than ethanol for fats and oils, and dissolves polar resins in the same way as ethanol. Cellulose nitrate and poly(vinyl acetate) are, however, almost insoluble. For economic reasons propanol is of only limited use as a solvent, and is a starting material for esters.

Isopropyl alcohol [67-63-0] (2-propanol, isopropanol) is a colorless liquid that is miscible in all proportions with water and with commonly used organic solvents. It forms binary and ternary azeotropes with water and many organic solvents.

Isopropyl alcohol is extensively used in cosmetics, particularly for hair and skin lotions, and in pharmacy for preparations intended for external use. A further large area of use is the paint and printing ink industry. Isopropyl alcohol is added to fuels to prevent the icing up of the carburetor and increase the octane number. It is an important feedstock for the chemical industry, for example, in the production of acetone (particularly in the United States), esters, plasticizers, and ethers. Isopropyl alcohol is also important as a solvent (e.g., for recrystallization and extraction) and as a moistening agent for cellulose nitrate. It is furthermore used in the aerosol sector.

Butanol [71-36-3] (1-butanol) is a colorless neutral liquid that has a limited miscibility with water and a characteristic odor. It is miscible with organic solvents. Butanol has a high solvency for most known natural and synthetic resins, fats, oils, linseed oil, saturated polyesters, and poly(vinyl acetate). It considerably increases the dilutability of cellulose nitrate solutions with non-solvents. Cellulose esters, cellulose ethers, chlorinated rubber, poly(vinyl chloride), vinyl chloride copolymers, and polystyrene are not dissolved by butanol.

Butanol is one of the most important additives in the production of cellulose nitrate lacquers. It ensures haze-free drying and prevents blushing of drying lacquer films. Butanol improves not only the flow of cellulose nitrate lacquers but also acts as a solubilizer between the individual lacquer constituents. It is used to prevent stringing in the production of spirit varnishes. Although most alkyd resins are only sparingly soluble in butanol, even small additions of butanol to alkyd resin or oil-based paints considerably decrease viscosity and improve flow and coatability. Butanol is used in electrostatically sprayed paints to regulate the conductivity. It can be employed to a limited extent as diluent tender in vinyl chloride copolymers. Butanol serves as a viscosity-reducing auxiliary solvent in waterborne paints and prevents foaming of colorants (e.g., in the paper industry). It is suitable for moistening cellulose nitrate wool and is used to a large extent as a feedstock in the production of plasticizers and urea– and melamine–formaldehyde resins.

Isobutanol [78-83-1] (2-methyl-1-propanol) is a colorless, neutral liquid with a characteristic odor. It exhibits limited miscibility with water. Most organic solvents are miscible in practically all proportions with isobutanol. Isobutanol readily dissolves most natural and synthetic resins. Waxes dissolve satisfactorily only on heating. Cellulose esters and ethers, natural rubber, neoprene, chlorinated rubber, and polymers such as polystyrene and poly(vinyl chloride) are insoluble in isobutanol.

Isobutanol prevents blushing in drying paint films, and also improves flow and gloss. When added in amounts of 5 to 10% it reduces the viscosity of oil-based paints, alkyd paints, and cellulose nitrate lacquers. Isobutanol can replace butanol as a moistening agent for cellulose nitrate. It is also used in the production of spirit varnishes and printing inks, and as a raw material for the production of plasticizers.

sec-Butanol [*78-92-2*] (2-butanol) has a peppermint-like odor and a considerably higher water solubility than butanol. It is used to a limited extent as a solvent, e.g., in paint-removal pastes, spirit varnishes, and printing inks. On account of its good water absorption it is also suitable as a cosolvent in waterborne paints.

tert-Butanol [*75-65-0*] (2-methyl-2-propanol) is a colorless liquid that solidifies at room temperature. It is soluble in most organic solvents such as alcohols, ethers, ketones, esters, and hydrocarbons. It is miscible in all proportions with water. The azeotrope with water is a liquid at room temperature.

tert-Butanol is widely used as a solvent and solubilizer because of its extremely high resistance to oxidizing agents and halogens in comparison with other alcohols. On account of its good solvency for waxes it is used as a wax remover and as an additive for the deparaffination of oils. *tert*-Butanol is widely used as an alkylating agent for aromatic compounds and phenols, isomerization of the *tert*-butyl group does not occur. *tert*-Butanol is employed in the preparation of *tert*-butyl hydroperoxides, which are used as polymerization catalysts. It also acts as a stabilizer in chlorinated hydrocarbons and as a denaturing agent for ethanol. *tert*-Butanol is used in the production of aromas (including *tert*-butyl esters and *tert*-butyl aromatics) and antioxidants.

Amyl Alcohol. Of the eight isomers of amyl alcohol, three have asymmetrically substituted carbon atoms. The amyl alcohol obtained from fusel oil consists of 3-methylbutanol [*123-51-3*] (isoamyl alcohol) and 2-methylbutanol [*137-32-6*] (active amyl alcohol); the mixture is sold as fermentation amyl alcohol or technical grade isoamyl alcohol. Isoamyl alcohol has similar dissolution properties to butanol. It is used as a solvent, and as a raw material for acetate esters, plasticizers, and lubricant components. Amyl alcohol containing *n*-pentanol (1-Pentanol [*71-41-0*]) as the principal constituent is used on a small scale as a solvent, like butanol. It has a better oil solubility and is therefore used as an additive in lubricating and hydraulic oils. It is used as an extracting agent for uranium salts and bituminous constituents of coal tar. It also serves as a raw material in the production of plasticizers and amyl xanthogenates (flotation auxiliaries for sulfidic ores). Pentasols are commercially available mixtures of primary and secondary pentanols.

Hexanol. The most important isomers are hexanol [*111-27-3*], 2-ethyl-1-butanol [*97-95-0*] and methylisobutylcarbinol [*108-11-2*] (4-methyl-2-pentanol). Hexanol is a high boiler and is used to improve the flow and surface properties of paints; it is also used as a solvent for fats, waxes, and dyes.

2-Ethylhexanol [*104-76-7*] is a colorless, waterwhite liquid with a characteristic odor. It is practically insoluble in water, but is miscible with all usual organic solvents. It is a good solvent for many vegetable oils and fats, for some dyes, and for many synthetic and natural paint raw materials.

2-Ethylhexanol is esterified to produce plasticizers for various uses. The esters of 2-ethylhexanol with phthalic, adipic, azelaic, and phosphoric acid are particularly important. Maleate and acrylate esters are used in copolymerization for the internal plastification of copolymers. Di-2-ethylhexyl esters of adipic, sebacic, and decanedicarboxylic acids are base oils used for the production of synthetic, thermally-stable lubricants, in particular for jet engines and gas turbines. 2-Ethylhexyl sulfate and diethylhexyl sulfosuccinate are used as wetting and leveling agents in the textile industry. Ethoxylates of this alcohol are used as raw materials for emulsifiers, detergents, and cleansing agents. Esterification with hydrochloric acid yields 2-chloro-2-ethylhexane, which is used in the production of tin-containing stabilizers and as an intermediate for introducing the 2-ethylhexyl group into other compounds. 2-Ethylhexanol is a defoaming agent for aqueous systems in the paper, ceramics, and textile industries. It also serves as a grinding auxiliary for pigments, acts as a surface impregnation agent, and facilitates the dispersion of pigments in nonaqueous solvents. As a high-boiling solvent, small amounts of 2-ethylhexanol are added to stoving enamels to improve flow and gloss formation.

Benzyl alcohol [100-51-6] is miscible with organic solvents apart from aliphatic hydrocarbons. It dissolves cellulose esters and ethers, fats, oils, alkyd resins, natural and synthetic resins, and colorants. Polymers—with the exception of lower poly(vinyl ethers) and poly(vinyl acetates)—are not dissolved. Small amounts of benzyl alcohol improve the flow and gloss paints, delay the evaporation of other solvent components, and have a plasticizing effect in physically drying paints. It is used in ballpoint pen inks. Benzyl alcohol reduces the viscosity in two-pack epoxy resin systems.

Methylbenzyl alcohol [98-85-1] (1-phenyl-ethanol, 1-phenylethyl alcohol, phenylmethylcarbinol) is an almost colorless, neutral liquid that has limited miscibility with water and a weak, bitter, almond-like odor. It has a high solvency for alcohol-soluble cellulose nitrate, cellulose acetate, and cellulose acetobutyrate; for many natural and synthetic resins; and for fats and oils. In contrast to benzyl alcohol, it is miscible with white spirit.

Methylbenzyl alcohol can be used like benzyl alcohol, and can be employed advantageously in stoving enamels. In cellulose nitrate and acetyl cellulose lacquers methylbenzyl alcohol helps to improve flow and film formation and prevents blushing at relatively high atmospheric humidity levels. It has also proved to be a useful additive in paint-removal agents on account of its dissolution properties and long evaporation time. The solvency of methylbenzyl alcohol for colorants is similar to that of benzyl alcohol.

Cyclohexanol [108-93-0] has a camphor-like odor with a water solubility of 2%. It is miscible with other solvents and dissolves fats, oils, waxes, and bitumen, but not cellulose derivatives.

Cyclohexanol is used in cellulose nitrate combination lacquers and oil-based paints. It delays drying, prevents blushing, and improves flow and gloss. In topcoats and clear varnishes, cyclohexanol can prevent undesirable dissolution of the primer coats. Cyclohexanol is used to remove paraffin from mineral oil; as a solvent in waxing, cleansing, and polishing agents; and as a wetting agent for spray liquids.

Methylcyclohexanol [583-59-5] is marketed as a mixture of isomeric methylcyclohexanols with a camphor-like odor that is insoluble in water but miscible with all organic solvents. Its dissolution properties are similar to those of cyclohexanol. On account of its fat-dissolving properties, methylcyclohexanol improves the adhesion of paints to articles that cannot be completely degreased before being painted.

3,3,5-Trimethylcyclohexanol is a cycloaliphatic alcohol that is available as a mixture of *cis–trans* isomers [*933-48-2*], [*767-54-4*] or in its *cis* form. Trimethylcyclohexanol is insoluble in water, but miscible in all proportions with most organic solvents. It has an odor reminiscent of menthol and readily dissolves waxes, resins, fats, and oils. Trimethylcyclohexanol is used in the production of special plasticizers. Bis(trimethylcyclohexyl) phthalate is for example a "hard" plasticizer for poly(vinyl chloride) with favorable elasticity and extracting properties. Trimethylcyclohexanol esters of salicylic acid act as UV absorbers and are used in sunscreen oils. The esters of mandelic acid are spasmolytics, the ester of *cis*-trimethylcyclohexanol being said to have a lower toxicity. The esters formed from trimethylcyclohexanol and polychlorophenoxyacetic acids are intermediates in the preparation of synthetic hormones. Trimethylcyclohexanol is used under the name homomenthol as a substitute for camphor and menthol. Trimethylcyclohexanol is used in the rubber industry on account of its ability to coagulate latex. It also acts as an antifoaming agent (e.g., in hydraulic fluids, textile soaps, and emulsion paints). Oxidation of trimethylcyclohexanol with nitric acid yields trimethyladipic acid, a starting product in the production of plasticizers, lubricants, dinitriles, diamines, glycols, and diisocyanates.

Furfuryl alcohol [*98-00-0*] is a colorless liquid that dissolves cellulose nitrate and many other paint binders.

Tetrahydrofurfuryl alcohol [*97-99-4*] is a colorless liquid with a weak odor that is miscible with water and organic solvents except aliphatic hydrocarbons. It dissolves cellulose nitrate and acetate, chlorinated rubber, shellac, resin esters, and many other binders.

Diacetone alcohol [*123-42-2*] (4-hydroxy-4-methyl-2-pentanone) is an almost odorless ketone alcohol that is weakly acidic as a result of rearrangement to the enol form. It is miscible with water and organic solvents except aliphatic hydrocarbons. It acts as a good solvent for cellulose esters and ethers, alcohol-soluble resins, castor oil, and plasticizers. Poly(vinyl acetate) and chlorinated rubber are partially dissolved or swollen. Polystyrene, poly(vinyl chloride), vinyl chloride copolymers, damar resins, resin esters, rubber, bitumen, mineral oils, ketone resins, and maleate resins are insoluble. Diacetone alcohol is used as a high boiler in stoving enamels to improve flow and gloss.

14.9.7. Ketones

The ketones used as solvents are water clear, highly mobile liquids having a characteristic odor (DIN 53 247). They are highly volatile and chemically very stable. On account of the carbonyl group ketones are hydrogen acceptors and have an outstanding solvency. The lower ketones dissolve polar resins, fats, oils, and less polar substances. The higher ketones have a more pronounced hydrocarbon character and are particularly good solvents for nonpolar resins, polymers, and copolymers.

Acetone [*67-64-1*] (2-propanone, dimethyl ketone) is miscible with water and organic solvents. It has excellent solvency for cellulose esters and ethers, cellulose ether esters, ketone resins, ketone–formaldehyde resins, vinyl chloride copolymers, polyacrylates, polystyrene, chlorinated rubber, fats, natural and synthetic oils, and resins. Poly(vinyl chloride) and poly(methyl methacrylate) are swollen. Polyacrylonitrile, polyamides, and rubber are insoluble. Acetone is used as a highly volatile solvent in fast-drying cellulose lacquers, in adhesives, as an extracting agent and detergent, as a cellulose nitrate swelling agent in the production of smokeless gun powder, as a solvent in the production of acetate silk and cellulose acetate films. It is a solvent for acetylene in steel cylinders (dissolved acetylene).

Methyl ethyl ketone [*78-93-3*] (MEK, 2-butanone) has a similar solvency to acetone. Cellulose acetate is only slightly soluble, natural and synthetic waxes are insoluble. Methyl ethyl ketone is often used instead of ethyl acetate in paints (especially in wood paints), although its more pungent odor mitigates against its widespread use.

Methyl propyl ketone [*107-87-9*] (MPK, 2-pentanone) and *methyl isopropyl ketone* [*563-80-4*] (MIPK, 3-methyl-2-butanone) are comparable to methyl ethyl ketone, but have a somewhat lower solvency. They are more expensive than methyl ethyl ketone and are therefore of only minor importance as solvents.

Methyl butyl ketone [*591-78-6*] (MBK, 2-hexanone) is slightly soluble in water, miscible with organic solvents. Being a medium boiler it dissolves cellulose nitrate, vinyl polymers and copolymers, and natural and synthetic resins. It increases the dilutability with non-solvents and diluents. As a paint solvent, methyl butyl ketone is important only in hot-spraying and coil-coating paints. It has a very low MAK value and is recommended as a solvent because it is photochemically inactive and does not, therefore, contribute to the formation of "photosmog."

Methyl isobutyl ketone [*108-10-1*] (MIBK, 4-methyl-2-pentanone) is a colorless liquid with a sweet odor. It is partially miscible with water, but is completely miscible with organic solvents.

Methyl isobutyl ketone is a solvent for many natural and synthetic resins such as cellulose nitrate, poly(vinyl acetate), vinyl chloride copolymers, epoxy resins, most acrylic, alkyd, ketone, coumarone and indene resins, aminoplasts, phenoplasts, rubber and chlorinated rubber, colophony, colophony esters, elemi, manila and damara (dewaxed) resins, gum sandarac and copal esters, fats, and oils. Methyl isobutyl ketone has a poor solvency, however, for most poly(vinyl acetate) resins, poly(vinyl chloride), cellulose esters and ethers, shellac, and natural waxes. Methyl isobutyl ketone is used in the paint industry as a universal medium boiler. It imparts good flow and gloss properties to cellulose nitrate lacquers, improves blushing resistance, and permits the production of highly concentrated solutions containing high proportions of inexpensive diluents. In combination with alcohols and aromatics, methyl isobutyl ketone is an important constituent of practically all epoxy resin paint formulations. Being a good solvent for low molecular mass PVC and vinyl chloride copolymers, it is used to prepare low-viscosity solutions with a high dilutability for aromatic hydrocarbons. Methyl isobutyl ketone is used as a medium-boiling component in, for example, the production of stamping varnishes for steel, tinplate sheet metal, or aluminum, as well as for printing PVC films. Methyl isobutyl ketone reduces the viscosity of alkyd resin paints and is used in all types of acrylic paints.

It has proved very useful as an anhydrous and hydroxyl group-free solvent in polyurethane paints.

Methyl isobutyl ketone is advantageously used for the preparation of pesticide concentrates, including natural (pyrethrin) and synthetic insecticides and herbicides. It is a versatile extracting agent and is used to dewax mineral oils, and in the refining of tall oil, and the purification of stearic acid. Methyl isobutyl ketone is a particularly important extracting agent for metals and metal chlorides as well as for separating niobium and tantalum ores. It plays an important role in the purification and extraction of penicillin and other antibiotics. Methyl isobutyl ketone is used for the concentration of aqueous formaldehyde solutions by azeotropic distillation under reduced pressure. Isopropyl alcohol can be freed from ethanol or butanol by extractive distillation with methyl isobutyl ketone. Finally, a mixture of equal parts of methyl isobutyl ketone, ethyl acetate, and gasoline (*bp* 74.5–79.5 °C) is used as a denaturing agent for ethanol.

Methyl isobutyl ketone has become increasingly important as a chemical feedstock. The oxime is used as an antiskinning agent in oxidatively drying paints, and the peroxide is used to cure unsaturated polyester resins. Methyl isobutyl ketone is an intermediate in the preparation of imidazolones and acetal esters. It is also used to block amines (ketimines) that act as moisture-initiated latent epoxy resin hardeners. Finally, alkaline autocondensation at elevated temperature yields higher molecular mass products that are employed as fragrances and perfumes.

Methyl amyl ketone [*110-43-0*] (MAK, 2-heptanone) and *methyl isoamyl ketone* [*110-12-3*] (MIAK, 5-methyl-2-hexanone) are high boilers with good solvency. They are comparable to methyl isobutyl ketone as regards dissolution properties.

Ethyl amyl ketone [*541-85-5*] (EAK, 5-methyl-3-heptanone) is insoluble in water, but miscible with organic solvents. It is a high boiler with good solvency and improves the flow properties of paints.

Dipropyl ketone [*110-43-0*] (DPK, 4-heptanone) is a high boiler with a sweet odor.

Diisopropyl ketone [*565-80-0*] (DIPK, 2,4-dimethyl-3-pentanone) is a high-boiling solvent used in the production of cellulose nitrate emulsion lacquers for coating leather, chlorinated rubber paints, and diluents for poly(vinyl chloride) organosols [14.267].

Diisobutyl ketone (DIBK) is a colorless, low-viscosity liquid, consisting of a mixture of two isomeric dimethyl heptanones (2,6-dimethyl-4-heptanone [*108-83-8*] and 2,4-dimethyl-6-heptanone [*19549-80-5*]). It is practically immiscible with water, but is miscible in all proportions with all common organic solvents. It is a high boiler with good solvency properties for cellulose nitrate, vinyl resins, waxes, and many natural and synthetic resins.

Diisobutyl ketone is used as a high boiler in cellulose nitrate lacquers. It acts as a leveling agent and has outstanding properties as a solvent for cellulose nitrate emulsion lacquers used to coat leather. Diisobutyl ketone is used as a solvent for vinyl chloride copolymers and as a dispersant for poly(vinyl chloride) organosols. It also serves as an extracting agent and feedstock for chemical syntheses.

Mesityl oxide [*141-79-7*] (4-methyl-4-penten-2-one) has a strong, peppermint-like odor. It is insoluble in water, but miscible with organic solvents. It has a good solvency for cellulose nitrate, vinyl chloride and vinyl ether copolymers, polyacrylates, and many resins.

Cyclohexanone [*108-94-1*] (Hexanone) is insoluble in water, miscible with organic solvents. It is a high boiler with very good solvency for cellulose nitrate, cellulose ethers and esters, colophony, shellac, alkyd resins, natural and synthetic resins, chlorinated rubber, rubber, vinyl polymers and copolymers, polystyrene, ketone and ketone–formaldehyde resins, fats, oils, waxes, blown oils, and bitumen. [14.268].

Methylcyclohexanone [*1331-22-2*] is an industrial isomer mixture. It is comparable to cyclohexanone as regards solvency and miscibility, but does not dissolve cellulose acetate.

Dimethylcyclohexanone is an industrial, high-boiling mixture of *cis* and *trans* isomers; [*54396-54-2*] and [*54396-55-3*], respectively. It is comparable to methylcyclohexanone as regards solvency and miscibility.

Trimethylcyclohexanone [*873-94-9*] (TMC-one, 3,3,5-trimethylcyclohexanone) is a saturated cyclic ketone. It is a colorless high boiler with an aromatic odor reminiscent of menthol. Trimethylcyclohexanone is only moderately miscible with water, but is miscible in all proportions with all organic solvents. It is chemically closely related to isophorone. Trimethylcyclohexanone dissolves cellulose nitrate, low molecular mass PVC grades, poly(vinyl acetate), vinyl chloride–vinylacetate copolymers, chlorinated rubber, alkyd resins, unsaturated polyester resins, epoxy resins, acrylic resins, etc.

In paints it is used as a leveling agent for air- and oven-drying systems. It counteracts bubble and crater formation and improves flow and gloss. Vinyl paints based on low molecular mass poly(vinyl chloride) or vinyl chloride copolymers that are used to coat metals or plastics exhibit good storage stability without gelling when trimethylcyclohexanone is added, even with high contents of diluents. Trimethylcyclohexanone is used in combination with suitable diluents as a temporary plasticizer in the processing of poly(vinyl chloride). In pastes consisting of poly(vinyl chloride) and plasticizers it serves as a thinner with a low gelling tendency. Trimethylcyclohexanone is used as a solvent in cellulose nitrate emulsion lacquers for coating leather and as a cosolvent for pesticide formulations. It is a starting material for various syntheses. Oxidation yields peroxides which are used as catalysts for the polymerization of unsaturated compounds with a long pot life. Trimethylcyclohexanone oxime prevents skin formation in air-drying paints. It can be used in the synthesis of polymers such as polyamides, polyesters, or polyurethanes. A substituted cyclic monoamine (3,3,5-trimethylcyclohexylamine) is formed by aminating hydrogenation, which is used as an anticorrosive agent, in cosmetic articles, or for special chemical syntheses.

Isophorone [14.268], [14.269] is an unsaturated cyclic ketone. It consists of α-isophorone [*78-59-1*] (3,5,5-trimethyl-2-cyclohexen-1-one), which contains about 1–3% of the isomer β-isophorone [*471-01-2*] (3,5,5-trimethyl-3-cyclohexen-1-one). Isophorone is a stable, water-white liquid with a mild odor that is miscible in all proportions with organic solvents. It dissolves many natural and synthetic resins and polymers, such as poly(vinyl chloride) and vinyl chloride copolymers, poly(vinyl acetate), polyacrylates, polymethacrylates, polystyrene, chlorinated rubber, alkyd resins, saturated and unsaturated polyesters, epoxy resins, cellulose nitrate, cellulose ethers and esters, damar resin (dewaxed), kauri, waxes, fats, oils, phenol–, melamine–, and urea–formaldehyde resins, as well as plant protection agents. However, isophorone does not dissolve polyethylene, polypropylene, polyamides, polyurethanes, polyureas, polyacrylonitrile, polycarbonate, cellulose triacetate, or sandarac.

Isophorone is used in the paint industry as a high-boiling solvent for physically- and oven-drying paints. It improves flow, gloss, adhesion, and the wetting of the substrate and pigments.

Isophorone is used as a high-boiling solvent in cellulose nitrate emulsion lacquers for coating leather. In printing inks for plastic substrates that are based on vinyl resins isophorone reduces the viscosity at high solids contents, it also improves the adhesion due to its swelling ability for poly(vinyl chloride) and other plastics. In the formulation of plant protection agents isophorone is used as a solvent having a high dilutability, good emulsifiability, and good emulsion stability. It is employed in adhesives on account of its universal solvency for many thermoplastic polymers. Isophorone is used in admixture with low-boiling compounds particularly for bonding articles made of poly(vinyl chloride) or polystyrene. The special chemical structure of isophorone with its keto group and an olefinic double bond means that isophorone can be used for the synthesis of a large number of compounds and products including solvents, plasticizers, glycols, amines, diamines, diisocyanates, and dicarboxylic acids.

14.9.8. Esters

Esters are clear liquids that often have a pleasant, fruity odor. They are neutral and very stable, but can be hydrolyzed on heating in the presence of strong acids or bases. They are less polar than the corresponding alcohols. Esters have an extremely good solvency for polar substances; with increasing carbon chain length in the alcohol and acid groups the solvency for polar products decreases, but increases for less polar products. Lower esters are partially water-soluble. The acetates are the most important ester solvents for the paint industry (DIN 53 246). Formates are used less because they have a more powerful physiology and are more readily hydrolyzed. Propionates, butyrates, and isobutyrates are of minor importance on account of their intense fruity odor. Certain esters of glycolic and lactic acids, as well as dimethyl esters of dicarboxylic acid mixtures are also important.

Methyl formate [*107-31-3*] is a colorless, water-soluble liquid. It dissolves cellulose nitrate, cellulose acetate, oils, and fatty acids.

Ethyl formate [*109-94-4*] is partially miscible with water. It dissolves cellulose nitrate, cellulose acetate, and many resins.

Butyl formate [*592-84-7*] is slightly soluble in water. It dissolves cellulose nitrate, fats, oils, many polymers, and chlorinated rubber. It does not, however, dissolve cellulose acetate.

Isobutyl formate [*542-55-2*] has dissolution properties similar to those of butyl formate. It is a constituent of commercially available mixed solvents for paints.

Methyl acetate [*79-20-9*] is partially miscible with water and readily miscible with most organic solvents. It has a good solvency for cellulose esters and ethers, colophony, urea–, melamine–, and phenol–formaldehyde resins, poly(vinyl acetate), alkyd resins, ketone resins, and other resins. It does not dissolve shellac, damar resin, copal, or poly(vinyl chloride). Methyl acetate is used as a highly volatile solvent by itself or mixed with alcohols or other esters to reduce the viscosity of paints, as a substitute for acetone for adhesives, and as a softening agent for stiff shoe caps in the shoe industry.

Ethyl acetate [*79-20-9*] is a colorless, neutral liquid that is partially miscible with water and has a pleasant, fruity odor. It has a good solvency for cellulose nitrate, cellulose ethers, chlorinated rubber, poly(vinyl acetate), vinyl chloride copolymers, polyacrylates, polystyrene, fats, oils, and many natural and synthetic resins (alkyd resins, saturated polyesters, ketone resins). Cellulose acetate is, however, dissolved only in the presence of small amounts of ethanol. Poly(vinyl chloride) is insoluble.

Ethyl acetate is one of the most important solvents for quick-drying paints (e.g., wood paints based on cellulose nitrate). It is also commonly employed in isocyanate paints. It increases the dilutability with non-solvents and diluents. Ethyl acetate is also used in adhesives, leather lacquers, leather impregnating agents, as an extracting and cleansing agent, as a fragrance in the perfume industry, in polishes, in the production of glazed paper and transparent paper, and in organic syntheses.

Propyl acetate [*109-60-4*] and *isopropyl acetate* [*108-21-4*] [14.270] are only slightly soluble in water. They are less volatile than ethyl acetate and have an improved solubility for less polar resins. Cellulose acetate is insoluble.

Butyl acetate [*123-86-4*] is a colorless, neutral, water-immiscible liquid with a pleasant, fruity odor. It has a good solvency for cellulose nitrate, cellulose ethers, chlorinated rubber, postchlorinated poly(vinyl chloride), poly(vinyl acetate), polyacrylates, polymethacrylates, vinyl chloride copolymers, polystyrene, natural and synthetic resins, alkyd resins, fats, and oils. Cellulose acetate is insoluble.

Butyl acetate is the most important moderately volatile solvent for the paint industry. Its volatility is sufficiently high for it to evaporate rapidly from the paint film, but is low enough to prevent cratering, blushing, and flow disorders. The solvency of butyl acetate is increased considerably by adding butanol. Butyl acetate is generally used together with aromatic hydrocarbons in paint formulations. The product has a low intrinsic viscosity and is therefore suitable as an auxiliary solvent in high-solids paints; it is also the most widely used solvent in polyurethane paints.

Isobutyl acetate [*110-19-0*] is a colorless, neutral liquid with a pleasant, fruity odor. It is miscible with organic solvents, but immiscible with water. Isobutyl acetate has a high solvency for cellulose nitrate, colophony, damar resin, ketone and ketone–formaldehyde resins, maleate resins, urea and melamine resins, and phenolic and alkyd resins. Polymers such as polystyrene, poly(vinyl ethers), poly(vinyl acetate), polyacrylates, chlorinated rubber and vinyl chloride copolymers, as well as fats, greases, and oils are readily dissolved. Postchlorinated poly(vinyl chloride) is less soluble. Polyisobutene, cellulose ethers, polymethacrylates, poly(vinyl butyrals), natural rubber, and manila copal are swollen. Shellac, cellulose acetate, cellulose acetobutyrate, poly(vinyl chloride), and poly(vinyl formal) are insoluble.

Isobutyl acetate has a higher volatility and lower flash point than butyl acetate, and is therefore used in the production of quick-drying cellulose nitrate lacquers. On account of its low water content it can be used as a solvent and diluent for polyurethane paints. Isobutyl acetate is used as a viscosity-reducing auxiliary solvent in low-solvent paints.

sec-Butyl acetate [*105-46-4*] is comparable to isobutyl acetate, but is highly volatile and less thinnable. It is of only minor importance as a solvent.

Amyl acetate [*628-63-7*] and *isoamyl acetate* [*123-92-2*] are comparable to butyl acetate as regards solvency and dilutability, but are less volatile. They have a rather intense odor and are used in nail polishes.

2-Ethylhexyl acetate [*103-09-3*] is a high-boiling solvent with good solvency properties. It is used in stoving enamels.

Octyl acetate is a high-boiling mixture of isomers that is used in stoving enamels.

Nonyl acetate is a high-boiling mixture of isomers that is used in stoving enamels.

Hexyl acetate [*142-92-7*] has a good solvency for resins, polymers, cellulose derivatives, fats, and oils. Cellulose acetate is insoluble. Small amounts of hexyl acetate are used as a leveling agent in paints.

Cyclohexyl acetate [*622-45-7*] is very slightly miscible with water, but completely miscible with common organic solvents. Its solvency properties are comparable to those of amyl acetate. Cyclohexyl acetate dissolves oils, fats, resins, waxes, cellulose nitrate, cellulose tripropionate and acetobutyrate, alkyd resins, unsaturated and saturated polyester resins, phenolic resins and aminoplasts, poly(vinyl chloride), vinyl chloride copolymers, poly(vinyl acetate), poly(vinyl ethers), epoxy resins, and acrylic resins, basic dyes, blown oils, crude rubber, metallic soaps, shellac, and bitumen.

In metal paints cyclohexyl acetate improves wetting and flow properties and prevents blistering. On account of its low water miscibility and high surface tension, cyclohexyl acetate promotes the formation of stable emulsions when used as a cosolvent (e.g., in pest control agents and cellulose nitrate emulsion lacquers for coating leather). It is used for the production of pigment pastes on account of its good solvency for basic dyes. It is also employed as a fragrance component in the perfume industry.

Benzyl acetate [*140-11-4*] is an aromatic smelling, water-insoluble liquid. It is used as a high boiler in printing inks.

Methyl glycol acetate [*110-49-6*] (2-methoxyethyl acetate) is a neutral liquid with a mild odor. It is miscible with water and organic solvents and has a good solvency for many natural and synthetic resins. It does not dissolve poly(vinyl chloride), rubber, polyisobutene, poly(vinyl butyrate), shellac, and damar resin. Methyl glycol acetate imparts good flow properties and high dilutability to paints and prevents blistering and blushing. On account of its teratogenic properties it is replaced by other solvents in paints.

Ethyl glycol acetate [*111-15-9*] (2-ethoxyethyl acetate) is a colorless, neutral liquid that is partially miscible with water and readily miscible with organic solvents. It has a slight odor. Ethyl glycol acetate dissolves many natural and synthetic resins. Cellulose acetate is swollen. Poly(vinyl chloride), rubber, and polyisobutene are insoluble. It is no longer used in paints due to its teratogenic properties.

Butyl glycol acetate [*112-07-2*] (2-butoxyethyl acetate) is insoluble in water, but miscible with organic solvents. It is a high boiler that is used to improve flow properties in stoving enamels. It has a good solvency for slightly polar resins and polymers, fats, and oils.

Ethyl diglycol acetate [*111-90-0*] [2-(2-ethoxyethoxy)ethyl acetate] is used as a high boiler to improve the flow properties and prevent blistering in industrial paints, automotive finishes, and coil coating of metals.

Butyl diglycol acetate [*124-17-4*] [2-(2-butoxyethoxy)ethyl acetate] is a high boiler that is used as a temporary plasticizer, as a leveling agent in stoving enamels and in printing inks, as a film-forming auxiliary in acrylate copolymer dispersions, and to reduce the viscosity of poly(vinyl chloride) plastisols.

1-Methoxypropyl acetate [*108-65-6*] and *2-methoxypropyl acetate* are pleasantly smelling liquids that are partially soluble in water, and miscible with organic solvents. Their dissolution properties and evaporation behavior are comparable but not identical to those of ethyl glycol acetate. 1-Methoxypropyl acetate is less toxic than methyl and ethyl glycol acetates; 2-methoxypropyl acetate is teratogenic.

Ethoxypropyl acetate [*54839-24-6*] is an isomeric mixture containing 1-ethoxypropyl acetate as the principal component. It is less toxic than ethyl glycol acetate, with which it shares many physical properties. Ethoxypropyl acetate differs from ethyl glycol acetate however as regards dissolution properties.

3-Methoxybutyl acetate [*4435-53-4*] (butoxyl) is a weakly smelling liquid that is partially soluble in water and miscible with organic solvents. Dissolution properties resemble those of ethyl glycol acetate, but volatility is lower.

Propionate esters are comparable to acetate esters having the same number of carbon atoms as regards dissolution, mixing, and evaporation properties. Propionates have a stronger odor than acetates and are therefore of only limited importance as solvents.

Ethyl 3-ethoxypropionate [*763-69-9*] is a high-boiling solvent that is a less toxic alternative to ethyl glycol acetate, with a comparable property spectrum.

Butyrate esters are comparable to acetate esters with the same number of carbon atoms as regards dissolution, mixing, and evaporation properties, but have a more intense odor, in some cases reminiscent of pineapple or banana. Butyl butyrate [*109-21-7*] is important as a high-boiling solvent.

Isobutyrate esters are comparable to butyrate esters, but have a slightly lower boiling point and higher volatility. Isobutyl isobutyrate [*97-85-8*] has some importance as a high-boiling solvent.

Lactate esters. Lactic acid esters are produced from sugar fermentation. In comparison to petroleum-derived solvents they are lower in toxicity and more biodegradable, but also higher in costs. Due to a possible improvement of the pigment wetting they can reduce viscosity in paints, which make them suitable for High-Solids-systems. Ethyl- and butyllactate are additional especially useful for enhancing penetration in wood caotings based on nitrocellulose [14.270a].

Ethyl lactate [*97-64-3*] is miscible with water and most organic solvents. It is a slowly evaporating solvent with a mild odor. It dissolves cellulose nitrate, cellulose acetobutyrate, strongly polar resins, and polar polymers, [e.g., poly(vinyl acetate) and polyacrylates]. Ethyl lactate has limited solvency for slightly polar resins, polystyrene, poly(vinyl chloride), and vinyl chloride copolymers. It is used as a powerful high-boiling solvent that can be diluted to a high degree with non-solvents.

Butyl lactate [*138-22-7*] has a good solvency for less polar resins and polymers and good dilutability with aliphatic hydrocarbons. It does not dissolve cellulose acetate. Butyl lactate is a high-boiling solvent that is used in small amounts as a leveling agent.

Butyl glycolate [*7397-62-8*] (butyl hydroxyacetate) is comparable to butyl acetate as regards dissolution properties and evaporation behavior. It has a good solvency for fats, oils, alkyd resins, cellulose nitrate, and many resins. It has a high dilutability with alcohols and aromatic hydrocarbons.

Butyl glycolate reduces the viscosity of alkyd resin paints. It improves the flow properties and gloss in air-drying and oven-drying paints, especially in coil coating and leather lacquers. It is also used to lower the film-forming temperature in poly(vinyl acetate) dispersions.

Dimethyl adipate [627-93-0], *glutarate* [1119-40-0], and *succinate* [106-65-0] are used as a high-boiling mixture with good dissolution properties in stoving enamels, coil-coatings, and in the automobile industry.

Ethylene carbonate [96-49-1] is soluble in water and organic solvents and solid at room temperature. It is a high boiler that is insoluble in gasoline and turpentine oil. It has a high solvency for polyacrylonitrile, polyamides, glycol terephthalates, and poly(vinyl chloride). Cellulose nitrate and cellulose acetobutyrate only dissolve in the presence of alcohol or esters. Ethylene carbonate is used for polyacrylonitrile spinning solutions.

Propylene carbonate [108-32-7] is liquid at room temperature. It is a high boiler with a solvency resembling that of ethylene carbonate. It dissolves polyacrylonitrile only in combination with ethylene carbonate. Propylene carbonate is used in paints as a high-boiling solvent and film-forming auxiliary, especially in poly(vinyl fluoride) and poly(vinylidene fluoride) systems. It is also employed as an auxiliary in the pigment and dye industry.

Butyrolactone [96-48-0] has a mild odor with good dissolution properties. It is a high-boiling solvent that is used to improve properties.

14.9.9. Ethers

Straight-chain aliphatic ethers are water-clear, highly mobile liquids with a high volatility and characteristic, anesthetizing odor. On account of their high volatility they are of only minor importance in paint technology. Ethers are usually used as extracting agents in chemical syntheses and for isolating natural substances. They are highly flammable and form explosive peroxides in the presence of oxygen, especially on exposure to light.

Cycloaliphatic ethers are better solvents for binders and are therefore more important in paint technology.

Ethylene and propylene glycol ethers are extremely good solvents (DIN 55999). For toxicity reasons monoethylene glycol ethers are being replaced in the paint, dye, and printing ink industries by other solvents, including the corresponding monopropylene glycol ethers.

Diethyl ether [60-29-7] (DEE, ethoxyethane, ether) is a clear, colorless, highly mobile liquid with a low boiling point and a characteristic pleasant odor. It is miscible in all proportions with common organic solvents such as alcohols, ketones, hydrocarbons, esters, and ethereal oils. Diethyl ether has an evaporation number of 1.

Diethyl ether is widely used in the chemical industry as a solvent, extracting agent, and reaction medium on account of its chemical resistance in acid and alkaline media, its low boiling point, and favorable dissolution properties (e.g., for organometallic compounds). Diethyl ether mixed with ethanol is an excellent solvent for cellulose nitrate. It is used in large amounts in the gelatinization of cellulose nitrate and preparation of collodion solution. Ether is employed in medicine as an anesthetic, but this use is declining.

Diisopropyl ether [*108-20-3*] is miscible with most organic solvents, but immiscible with ethylene glycol and glycerol. It is only slightly water soluble. Diisopropyl ether is an excellent solvent for natural oils, mineral oils, and waxes. It is used as an extracting agent and reaction medium in chemical and pharmaceutical syntheses.

Dibutyl ether [*142-96-1*] and *di-sec-butyl ether* [*6863-58-7*] are very slightly soluble in water but miscible with organic solvents except ethylene glycol and glycerol. They are used as extracting agents and reaction media in the chemical and pharmaceutical industries.

Methyl tert-butyl ether [*1634-04-4*] (MTB) is a combustible, clear, colorless liquid with a low viscosity and characteristic, turpentine-like odor. It is miscible with organic solvents, and is only slightly soluble in water. It forms azeotropic mixtures with some solvents such as methanol. MTB does not undergo autoxidation and, in contrast to other ethers, it does not form peroxides with atmospheric oxygen. It improves the antiknock properties when added to motor gasoline.

Tetrahydrofuran [*109-99-9*] (oxolane, tetramethylene oxide, oxacyclopentane, THF) [14.271] is a colorless, highly volatile liquid having an outstanding solvency. Under prolonged exposure to air it tends to form peroxides. Tetrahydrofuran is completely miscible with most organic solvents and water. It forms azeotropic mixtures with a number of solvents. Tetrahydrofuran dissolves practically all plastics apart from certain polyamides (e.g., polyamide 6 and 12) and polytetrafluoroethylene.

Tetrahydrofuran is used alone or mixed with other solvents to prepare poly(vinyl chloride) and polyurethane solutions for coating films, sheets, and textiles (leather imitates). The special surface properties of video, computer, and recording tapes are obtained by using appropriate coatings that are prepared from tetrahydrofuran-containing solutions. Offset printing plates are produced by coating them with light-sensitive compounds that are applied from tetrahydrofuran-containing solutions. Tetrahydrofuran is also highly suitable as a solvent for printing inks based on poly(vinyl chloride). Its outstanding solvency for many plastics makes tetrahydrofuran a particularly suitable solvent for a number of special finishes that are used to coat plastics. Tetrahydrofuran has proved to be especially useful in cold cleansers for eliminating impurities and contaminants that can otherwise only be removed with difficulty.

Tetrahydrofuran is extremely suitable for extracting active agents from natural or synthetic raw materials. Substances can be separated on the basis of the specific solvency properties of tetrahydrofuran [e.g., insoluble polyamides and soluble poly(vinyl chloride)]. On account of its good dissolution properties for poly(vinyl chloride) tetrahydrofuran can be used alone or with other solvents to bond PVC pipes and for solution welding of PVC sealant films. Solutions of PVC in tetrahydrofuran are employed for gap-filling bonding of PVC moldings. Tetrahydrofuran is used as a stripper to remove coats of paint. It also serves as a reaction medium in the preparation of chemical intermediates with organometallic compounds.

1,4-Dioxane [*123-91-1*] has a slight odor reminiscent of butanol, and is miscible with water and organic solvents. It is a good solvent for cellulose derivatives, polymers, chlorinated rubber, and resins [14.272].

Metadioxane [*505-22-6*] (2,2-dimethyl-4-hydroxymethyl-1,3-dioxolane) is a high-boiling solvent with an excellent solvency for resins and polymers. It is used in printing inks, adhesives, and in the textile sector.

14.9.10. Glycol Ethers

Methyl glycol [*109-86-4*] (2-methoxyethanol, ethylene glycol monomethyl ether) has a slight odor, and is miscible with water and organic solvents except aliphatic hydrocarbons. It is a very good solvent for many natural and synthetic resins. It does not dissolve fats, oils (except castor oil), damar resin, rubber, bitumen, hydrocarbon resins, polystyrene, poly(vinyl chloride), and vinyl chloride copolymers. On account of its teratogenic properties it is replaced as a solvent in the paint and colorants sector by other solvents or solvent mixtures.

Ethyl glycol [*110-80-5*] (2-ethoxyethanol, ethylene glycol monoethyl ether) is a neutral, colorless liquid with a slight odor, and is miscible in all proportions with water and organic solvents. Ethyl glycol has a high solvency for many natural and synthetic resins and binders. Damar resin, rubber, bitumen, cellulose acetate, polystyrene, and poly(vinyl chloride) are insoluble. It is no longer used as a solvent in the paint and colorants industry due to its teratogenic properties.

Propyl glycol [*2807-30-9*] (2-propoxyethanol, ethylene glycol monopropyl ether) and *isopropyl glycol* [*109-59-1*] (2-isopropoxyethanol, ethylene glycol monoisopropyl ether) correspond to ethyl glycol as regards dissolution and miscibility properties, but evaporate more slowly and have a better solvency for slightly polar resins. They are less toxic than ethyl glycol and are therefore increasingly used as alternatives.

Butyl glycol [*111-76-2*] (2-butoxyethanol, ethylene glycol monobutyl ether) is a neutral, colorless liquid with a very weak, pleasant odor. It is miscible with water at room temperature, but has a miscibility gap with water at higher temperatures. It is miscible with organic solvents, and a very good solvent for cellulose nitrate, cellulose ethers, colophony, shellac, chlorinated rubber, polyacrylates, polymethacrylates, alkyd resins, phenol–, urea–, and melamine–formaldehyde resins, oils, fats, and synthetic resins. Damar resin, rubber, bitumen, polystyrene, and poly(vinyl chloride) are insoluble. Poly(vinyl acetate), poly(vinyl butyral), postchlorinated poly(vinyl chloride), and cellulose acetate are sparingly soluble or swellable. Butyl glycol is widely used in solvent-containing paint systems to improve the flow and surface quality of coatings. It is increasingly used in water-thinnable paints on account of its solubilizing action.

Methyl diglycol [*111-77-3*] [2-(2-methoxyethoxy)ethanol, diethylene glycol monomethyl ether] is miscible with water and most organic solvents. It is a good solvent for cellulose nitrate, colophony, phenol–, urea–, and melamine–formaldehyde resins, alkyd resins, alcohol-soluble maleate resins, ketone–formaldehyde resins, poly(vinyl butyral), poly(vinyl acetate), poly(vinyl ethers), epoxy resins, castor oil, linseed oil, and turpentine oil. However, it does not dissolve damar resin, rubber, bitumen, polyisobutylene, polystyrene, waxes, mineral oil, polyacrylates, and polymethacrylates. Small amounts are used as a low-volatility solvent in paints to improve the gloss and flow, in the printing ink industry, and for ballpoint pen pastes.

Ethyl diglycol [*111-90-0*] [2-(2-ethoxyethoxy)ethanol, diethylene glycol monoethyl ether] is a neutral, colorless, practically odorless liquid. It is miscible in all proportions with alcohols, esters, ethers, ketones, and water. It is partially miscible with aromatic and aliphatic hydrocarbons. Ethyl diglycol dissolves many paint resins, such as cellulose nitrate, chlorinated rubber, shellac, colophony, phenolic resins, urea– and melamine–formaldehyde resins, polyacrylates, esters, poly(vinyl acetate), as well as alkyd, ketone, and maleate resins. However, it does not dissolve cellulose acetate, poly(vinyl chloride), vinyl chloride copolymers, polystyrene, polymethacrylates, asphalt, and rubber. On account of its low solidification point and low viscosity at low temperatures, ethyl diglycol is a constituent of many brake fluids. It is used in the paint industry to improve gloss and flow properties, and thereby increase the dilutability. Ethyl diglycol can be used to predissolve alcohol-soluble dyes and is a constituent of many wood stains. This solvent is also used in printing inks and as a cleaning agent in offset printing. It is a constituent of Indian inks and ballpoint pen pastes. Ethyl diglycol phthalate is an excellent plasticizer for vinyl resins. Ethyl diglycol is used as a solvent in the printing and dyeing of textile fibers and fabrics. It also acts as a solubilizer for detergent raw materials in aqueous solutions and in drilling and cutting fluids. In the determination of the saponification value of waxes, resins, and fats, ethyl diglycol has the advantage that it can substantially accelerate saponification because it is a high-boiling solvent.

Butyl diglycol [*112-34-5*] [2-(2-butoxyethoxy)ethanol, diethylene glycol monobutyl ether] is a clear, colorless, neutral liquid with a pleasantly mild odor. It is miscible with water and organic solvents, including aliphatic compounds. Butyl diglycol has a high solvency for cellulose nitrate, cellulose ethers, chlorinated rubber, poly(vinyl acetate), polyacrylates, and some oils, as well as for many synthetic resins, natural resins, and dyes. Polystyrene, poly(vinyl chloride), fats, and most oils are not dissolved.

Butyl diglycol is used as a high-boiling solvent to improve gloss and flow properties. On account of its high evaporation number even additions of < 5% considerably improve paint properties, without noticeably increasing the stoving time. In cellulose nitrate and cellulose ether lacquers even smaller amounts are effective. In emulsion paints and cold-hardening paints butyl diglycol improves the coatability and enhances the surface gloss. On account of its good solvency for resins and dyes, butyl diglycol is used with other low-volatility solvents for coatings and printing inks. It is used as a solubilizer in mineral oil products.

Ethyl triglycol [*112-50-5*] [2-(2-ethoxyethoxy)ethoxy]ethanol, triethylene glycol monoethyl ether] is an almost colorless, neutral, mild-smelling liquid with a low hygroscopicity. It is soluble in water and most organic solvents, but is only partially miscible with aromatic and aliphatic hydrocarbons. Ethyl triglycol dissolves cellulose nitrate, shellac, colophony, ketone resins, maleate resins, chlorinated rubber, alkyd resins, and many other paint resins. It does not dissolve cellulose acetate, poly(vinyl chloride), vinyl chloride copolymers, fats, oils, and rubber.

The uses of ethyl triglycol are similar to those of ethyl diglycol. It serves as a solubilizer for incompatible liquids, in the production of insecticides, and in the detergent industry in handwash pastes. It is employed in printing inks and, in small amounts, in wood paints to prevent the wood fibers from standing perpendicular of the surface during painting. Ethyl triglycol is used in hydraulic and brake fluids because its viscosity remains largely constant under temperature fluctuations. It is also used as an intermediate in the production of plasticizers.

Butyl triglycol [*143-22-6*] [2-[2-(butoxyethoxy)ethoxy]ethanol, triethylene glycol monobutyl ether] is an almost colorless, neutral, weakly smelling liquid. It is soluble in water and most organic solvents, but is only partially miscible with aromatic and aliphatic solvents. Its dissolution properties are comparable to those of butyl diglycol. Butyl triglycol dissolves cellulose nitrate, cellulose ethers, chlorinated rubber, poly(vinyl acetate), colophony, many other natural and synthetic resins, as well as drying greases and oils. Bitumen, nondrying fats and oils, cellulose acetate, poly(vinyl chloride), and rubber are, however, insoluble. Butyl triglycol serves as a solubilizer for mutually incompatible liquids, and is used in the production of household and metal cleansing agents as well as for wood preservation. It is employed in cutting and hydraulic oils. Butyl triglycol is used in the production of Indian inks, writing inks, and ballpoint pen pastes. In the paint industry it is suitable as a high-boiling solvent in stoving enamels, as a leveling agent, and as an additional solvent in wood paints, in order to prevent wood fibers standing proud from the surface. Butyl triglycol is a useful solvent for the formulation of pharmaceutical and cosmetic preparations and in the leather auxiliaries industry.

Butyl tetraglycol [2-[2-(butoxyethoxyethoxy)ethoxy]ethanol, tetraethylene glycol monobutyl ether] is a neutral, practically odorless liquid that is soluble in water and most organic solvents, but only partially miscible with aromatic and aliphatic solvents. Its properties are comparable to those of butyl diglycol. Butyl tetraglycol dissolves cellulose nitrate, cellulose ethers, chlorinated rubber, poly(vinyl acetate), colophony, and many other natural and synthetic resins. However, it does not dissolve nondrying fats and oils, cellulose acetate, poly(vinyl chloride), and rubber. Butyl tetraglycol is used as a solubilizer on account of its good solvency for organic solvents, water, and natural and synthetic resins. It is also a constituent of hydraulic and brake fluids.

Diethylene glycol dimethyl ether [*111-96-6*] is a hygroscopic, weakly smelling liquid that is miscible with water and organic solvents. It dissolves vinyl chloride copolymers, postchlorinated poly(vinyl chloride), polymethacrylate, polystyrene, polychloroprene, and cellulose acetate. It does not dissolve rubber and polyethylene, and swells poly(vinyl chloride). Diglycol dimethyl ether also dissolves sodium borohydride, other covalent inorganic metal compounds, and sodium. It is used as a solvent in chemical reactions involving metals and organometallic compounds.

Methoxypropanol [*107-98-2*] (1-methoxy-2-propanol). Commercially available methoxypropanol also contains small amounts of 2-methoxy-1-propanol. It is a colorless, neutral liquid with a weak pleasant odor that is miscible in all proportions with water and organic solvents. The properties of methoxypropanol are largely comparable to those of ethyl glycol [14.273]. It has a somewhat higher volatility and, as a solvent component in paints and printing inks, improves the wetting of some pigments and colorants. It has a good solvency for cellulose nitrate, cellulose ethers, chlorinated rubber, poly(vinyl acetate), poly(vinyl butyral), ketone and ketone–formaldehyde resins, shellac, colophony, phenol–, melamine–, and urea–formaldehyde resins, alkyd resins, polyacrylates, polymethacrylates, castor oil, linseed oil, and some vinyl chloride copolymers. However, it does not dissolve copal resins, rubber, bitumen, cellulose acetate, polystyrene, poly(vinyl alcohol), and poly(vinyl chloride). On account of its mild, pleasant odor, methoxypropanol is extremely suitable for cellulose nitrate lacquers used to coat wood, paper, and metals. Being a moderately volatile solvent, it improves paint penetration, flow properties, and the gloss of paint coats; it also prevents blushing and formation of fish eyes and blisters. Addition of methoxypropanol does not delay the drying of paint systems. It increases the dilutability with inexpensive non-solvents, reduces and stabilizes the paint viscosity, and improves the coatability of paints. Methoxypropanol only slowly swells dry paint coats, so that it facilitates painting over previous coats. It is used as a solvent in printing inks (flexographic, gravure packaging printing) to improve their gloss and flow. On account of its good dissolution properties for dyes and ability to wet pigments, it intensifies the color of the inks.

Ethoxypropanol is a mixture of 1-ethoxy-2-propanol [*1569-02-4*] and 2-ethoxy-1-propanol. Its physical properties are very similar to those of ethyl glycol, but is a less toxic alternative.

Isopropoxypropanol consists of a mixture of 1-isopropoxy-2-propanol [*3944-36-3*] and 2-isopropoxy-1-propanol [*3944-37-4*]. It is a weakly smelling liquid with a relatively high boiling that is miscible with water and organic solvents. Its dissolution properties, evaporation properties, and uses are similar to those of butyl glycol. It is an effective solubilizer and used as an auxiliary solvent in high-solids and waterborne paints.

Isobutoxypropanol is a mixture of 1-isobutoxy-2-propanol [*26447-43-8*] and 2-isobutoxy-1-propanol. It is a weakly smelling liquid with a high boiling point that is miscible with water and organic solvents. Its dissolution properties, evaporation properties, and uses are similar to those of butyl glycol. Isobutoxypropanol is a highly effective solubilizer and is used as an auxiliary solvent in high-solids and waterborne paints. Propylene glycol *tert*-butyl ether is similar to isobutoxypropanol [14.274].

Methyl dipropylene glycol [*34590-94-8*] [2-(2-methoxypropoxy)propanol, dipropylene glycol methyl ether] (isomer mixture) is a high-boiling solvent. It is waterclear and miscible in all proportions with water, and has only a slight intrinsic odor. It readily dissolves cellulose nitrate, cellulose esters and ethers, as well as many natural and synthetic resins. On account of its slow evaporation rate small amounts of methyl dipropylene glycol are added to paints to control their flow properties, evaporation rate, and dilutability with non-solvents. It improves the coatability of cellulose nitrate lacquers, chlorinated rubber paints, and alkyd resin paints. Methyl dipropylene glycol is also suitable as a film-forming auxiliary in dispersions and as an auxiliary solvent in waterborne paints. It is highly effective as a solubilizer in cleansing agents and as a solvent in writing inks, pastes, and printing inks.

Methoxybutanol [*30677-36-2*] (3-methoxy-1-butanol) is a mild-smelling liquid that is miscible with water and organic solvents. It has a good solvency for cellulose nitrate, cellulose esters, poly(vinyl butyral), ketone resins, phenol–, urea–, and melamine–formaldehyde resins, alkyd and maleate resins, plasticizers, fats, and drying oils. However, it does not dissolve mineral oils, waxes, rubber, chlorinated rubber, cellulose acetate, polyisobutene, polystyrene, poly(vinyl chloride), and vinyl chloride copolymers. It is used like butanol as a slow-evaporating solvent.

14.9.11. Miscellaneous Solvents [14.275]

Methylal [*109-87-5*] (formaldehyde dimethyl acetal, dimethoxymethane) is a pleasantly smelling, extremely volatile solvent. It dissolves polystyrene, poly(vinyl acetate), vinyl chloride copolymers, acrylates, methacrylates, and synthetic and natural resins. It is used in paints and lacquers, adhesives, and aerosols. Important extracting agent for natural substances, essences, and oils. It serves as a reaction medium in the chemical industry and as an intermediate for chemical syntheses.

Dimethylacetal [*534-15-6*] (acetaldehyde dimethyl acetal, 1,1-dimethoxyethane) is a neutral liquid that is miscible with water and organic solvents. It dissolves cellulose nitrate, cellulose ethers, poly(vinyl acetate), polyacrylates, polymethacrylates, poly(vinyl ethers), some vinyl chloride copolymers, and synthetic and natural resins. It does not dissolve poly(vinyl chloride), polystyrene, chlorinated rubber, and cellulose acetate. It is used in the production of paints, adhesives, and shoe-cap stiffeners.

N,N-Dimethylformamide [*68-12-2*] (DMF) [14.276] is miscible with water and organic solvents except aliphatic hydrocarbons. It is a good high-boiling solvent for cellulose esters, cellulose ethers, poly(vinyl chloride), vinyl chloride copolymers, poly(vinyl acetate), polyacrylonitrile, polystyrene, chlorinated rubber, polyacrylates, ketone resins, and phenolic resins. Alkyd resins and resin esters are partially soluble. Dimethylformamide does not dissolve polyethylene, polypropylene, urea–formaldehyde resins, rubber, and polyamides. It is used as a solvent in printing inks, for polyacrylonitrile spinning solutions [14.277], and as a solvent in the synthesis of acetylene.

N,N-Dimethylacetamide [*127-19-5*] (DMA) is miscible with water and organic solvents. It has a very good solvency for many resins and polymers. It is used in the production of acrylic fibers, films, sheets, and coatings, and as a reaction medium and intermediate in organic syntheses.

Dimethyl sulfoxide [*67-68-5*] (DMSO) is a clear, colorless liquid that is miscible with water and organic solvents except aliphatic hydrocarbons. It is a good high-boiling solvent for cellulose esters, cellulose ethers, poly(vinyl acetate), polyacrylates, vinyl chloride copolymers, polyacrylonitrile, chlorinated rubber, and many resins. It is used in polyacrylonitrile spinning solutions, in paint-removal agents, as a film-forming auxiliary in dispersions, as an extracting agent, and as a reaction medium in organic syntheses [14.278], [14.279].

Tetramethylene sulfone [*126-33-0*] (sulfolane) is a waterclear liquid that is miscible with water and most organic solvents except aliphatic hydrocarbons. It dissolves cellulose acetate, cellulose nitrate, cellulose ethers, poly(vinyl chloride), vinyl chloride copolymers, poly(vinyl acetate), polystyrene, and many resins. Its solvency is lower than that of dimethyl sulfoxide. Tetramethylene sulfone is of only minor importance as a solvent.

Carbon disulfide [75-15-0] is a clear, highly mobile liquid that is highly refractive and very flammable. It is insoluble in water, but generally miscible with organic solvents. It has a good solvency for fats, oils, waxes, rubber, and many resins. Cellulose esters are insoluble, vinyl polymers dissolve only in mixtures with ketones or esters. Carbon disulfide is used in the production of silk, rubber solutions, and poly(vinyl chloride) spinning solutions. It is of no importance for paints, plastics solutions, or as a fat-extracting agent.

2-Nitropropane [79-46-9] is a colorless, nonhygroscopic liquid with a mild odor. It dissolves cellulose nitrate, cellulose ethers, alkyd resins, chlorinated rubber, poly(vinyl acetate), vinyl chloride copolymers. Poly(vinyl chloride), colophony, polyacrylonitrile, waxes, rubber, and shellac are insoluble. It is used as a cosolvent in paints to improve pigment wetting, flow properties, and electrostatic processing; it also reduces the paint drying time. 2-Nitropropane is classified as carcinogenic.

N-Methylpyrrolidone [872-50-4] (NMP) has a fairly mild, amine-like odor, and is miscible with water and most organic solvents. It has a good solvency for cellulose ethers, butadiene–acrylonitrile copolymers, polyamides, polyacrylonitrile, waxes, polyacrylates, vinyl chloride copolymers, and epoxy resins. It is used in paint removers and stripping paints to reduce paint viscosity, and to improve the wettability of paint systems. *N*-Methylpyrrolidone is also employed for the extraction of hydrocarbons, and as a solvent in the synthesis of acetylene.

Hexamethylphosphoric triamide [680-31-9] [14.280] is a basic, highly polar, nonflammable solvent with a very good solvency. Its dissolution properties are comparable to those of dimethyl sulfoxide and dimethylformamide. It is used for selective dissolution of organic compounds, as a reaction medium in chemical syntheses, as an antifreeze, and as an antistatic agent.

1,3-Dimethyl-2-imidazolidinone [80-73-9] (DMI) is a colorless, high-boiling, highly polar aprotic solvent of low toxicity with good chemical and thermal stability. It is miscible with water and most organic solvents. It is used as a reaction medium in preparative organic chemistry, in the production of polyamides, as a solvent in the production of dyes, agrochemicals, and pharmaceuticals and in the electronics industry. It is also a constituent of nail-polish, ballpoint pen pastes, and coating compositions.

15. References

References for Chapter 1

General References

[1.1] S. LeSoto et al.: *Paint/Coatings Dictionary*, Fed. of Soc. for Coatings Technology, Philadelphia, Pennsylvania 1978.
[1.2] H. Clausen: *Wörterbuch der Lacktechnologie*, Edition Lack und Chemie, Elvira Moeller GmbH, Filderstadt, 1980.
[1.3] D. H. Solomon: *The Chemistry of Organic Film Formers*, J. Wiley and Sons, New York 1967.
[1.4] H. Kittel: *Lehrbuch der Lacke und Beschichtungen*, vol. I–VIII, Verlag W. A. Colomb, Stuttgart – Berlin 1971–1980.
[1.5] S. T. Harris: *The Technology of Powder Coatings*, Portcullis Press Ltd., London 1976.
[1.6] A. V. Patsis (ed.): *Advances in Organic Coatings Science and Technology – Proceedings of the Intern. Conf. in Org. Coatings Sci. and Techn.*, vols. 1–10, Technomic Publishing Co., Westport, Connecticut 1979 – Lancaster, Pennsylvania 1988.
[1.7] G. Odian: *Principles of Polymerization*, J. Wiley and Sons, New York 1981.
[1.8] R. A. Dickie, F. L. Floyd: *Polymeric Materials for Corrosion Control*, Am. Chem Soc., Washington, DC, 1986.
[1.9] D. R. Randell: *Radiation Curing of Polymers*, The Royal Society of Chemistry, London 1987.
[1.10] E. V. Schmid: *Exterior Durability of Organic Coatings*, FMJ International Publications Ltd., Redhill 1988.

References for Chapter 2

[2.1] J. Bentley in R. Lambourne (ed.): *Paint and Surface Coatings: Theory and Practice*, Ellis Horwood, Chichester 1988, pp. 46–52.
A. E. Rheineck: "Drying Oil – Modification and Use," in R. R. Myers, J. S. Long (eds.): *Treatise on Coatings*, vol. 1, part II, Marcel Dekker, New York 1969.
[2.2] Wolff Walsrode, EP 0076443, 1986 (E. Lühmann, L. Hoppe, K. Szablikowski); EP 0154241, 1989 (E. Lühmann, L. Hoppe, K. Szablikowski, K.-F. Lampert).
[2.3] DIN 53179.
[2.4] A. Kraus: *Handbuch der Nitrocelluloselacke*, vol. 2, Westliche Berliner Verlagsgesellschaft Hunemann KG, Berlin 1963, p. 248ff.
[2.5] L. G. Curtis, J. D. Crowley: *Applied Polymer Science*, American Chemical Society, Washington, DC, 2nd ed., p. 1053.
[2.6] C. J. Malm, G. H. Hiatt: *Cellulose and Cellulose Derivatives*, 2nd ed., Wiley Interscience, New York 1955; vol. 5, part 2, p. 798.
[2.7] M. Salo, *Official Digest*, 31 Sept. 1959.
[2.8] Bayer AG: Bayer Produkte für die Lackindustrie, 1967.
[2.9] K. Hoehne, Lt. Nordt: "Neues über chlorierte Polymere" *Farbe + Lack* **76** (1970) no. 2, 123–126.

[2.10] B. Dodson, I. C. McNeill: "Thermal Degradation of Chlorinated Rubber, Evidence for an Alternative Cyclic Structure for Chlorinated Rubber," *J. Polym. Sci. Polym. Chem. Ed.* **12** (1974) 2305–2315.

[2.11] R. Lapasin, G. Torriano, S. Volpe: *Ind. Vernice* **29** (1975) no. 4, 2–19. (Thick-film paints based on chlorinated rubber and inert pigments – special aspects of anticorrosive properties and the film-forming process)

[2.12] G. C. Reid: "Chlorinated Rubber, the Solution to Many Corrosion Control Problems," *Anti-Corros. Methods Mater.* **22** (1975) no. 12, 8–13.

[2.13] K. Hoehne: "Zusammensetzung und Eigenschaften von Chlorkautschuk-Dickschichtanstrichen," *Farbe + Lack* **77** (1971) no. 9, 866–869.

[2.14] M. Yamabe, H. Higaki, G. Kojima in G. D. Parfitt, A. V. Patsis (eds.): *Organic Coatings Science and Technology*, vol. 7, Marcel Dekker, New York – Basel 1984, p. 25.

[2.15] V. Handforth, *J. Oil. Colour Chem. Assoc.* **73** (1990) no. 4, 145–148.

[2.16] Pennwalt Chemical Co., Technical Data, M. Kevin: *Building Design & Construction*, March 1983, p. 2.

[2.17] L. Stonberg, *Conf. Proc. Alum. Finish* **87** (1986) 26.1–26.17.

[2.18] J. E. McCann, *Surf. Coat. Aust.* **27** (1990) no. 1–2, 8–12.

[2.19] K. V. Summer, *Conf. Proc. Alum. Finish* **87** (1986) 36.1–36.19.

[2.20] Nippon Oil and Fats, JP-Kokai 86 238 863 A2, 1986 (A. Hikita et al.).

[2.21] Pennwalt Chemical Co., Technical Data, Philadelphia, June 1984.

[2.22] Du Pont, US 4 011 361, 1975 (E. Vassiliou et al.).

[2.23] G. Kojima et al. in A. V. Patsis (ed.): *11th International Conference in Organic Science and Technology*, vol. 9, Technomic Publishing Company, Lancaster, Pa., 1987, pp. 120–128.

[2.24] Du Pont, GB 1 064 840, 1967 (J. C. S. Fang).

[2.24a] CYTOP Technical Bulletin (1990), Asahi Glass. W. H. Buck and P. R. Resnick, TEFLON AF Technical Information, DuPont (1993).

[2.25] Du Pont, US 3 318 850, 1967 (F. B. Stilmar).

[2.26] L. Harold et al., *157th ACS Meeting, Minneapolis* **29** (1969) April, 221.

[2.27] Du Pont, US 2 468 664, 1949 (E. William et al.).

[2.28] Du Pont, US 3 651 003, 1972 (M. F. Bechtold et al.).

[2.29] S. Munekata et al., *Report of Research Laboratory, Asahi Glass Co.* **34** (1984) no. 2, 205.

[2.30] N. Miyazaki, T. Takayanagi, *Report of Research Laboratory, Asahi Glass Co* **36** (1986) no. 1, 155.

[2.31] S. Munekata, *Prog. Org. Coat.* **16** (1988) 113–134.

[2.32] Dainippon Ink and Chemicals, JP-Kokai 84 102 962, 1984 (M. Ooka et al.).

[2.33] Central Glass Co., JP-Kokai 86 57 609, 1986 (T. Koishi et al.).

[2.34] Mitsui Petrochemical Industries, JP-Kokai 86 141 713, 1986 (S. Honma et al.).

[2.35] Daikin Industries, JP-Kokai 84 189 108, 1984 (A. Oomori et al.).

[2.35a] Toagosei Chemical Industry, US 5064920.

[2.36] N. Miyazaki, M. Kamba, *First Pacific Polymer Conference*, Hawaii, Dec. 1989, p. 557.

[2.36a] M. Yamauchi, T. Hirono, S. Kodama and M. Matsuo, Surface Coating International, 79, 312–318 (1996).

[2.37] J. R. Griffith, *CHEMTECH* **12** (1982) 290.

[2.38] S. J. Shaw et al., *Polym. Mater. Sci. Eng.* **56** (1987) 209.

[2.39] J. R. Griffith et al., *Ind. Eng. Chem. Prod. Res. Dev.* **25** (1986) no. 4, 572.

[2.40] W. S. Zimmt, *CHEMTECH* **11** (1981) 681.

[2.41] M. K. Yousuf, *Mod. Paint Coat.* **79** (1989) 48.

[2.42] J. L. Gerlock, D. R. Bauer, L. M. Briggs, *Prog. Org. Coat.* **15** (1987) 197.

[2.43] A. J. Tortorello, M. A. Kinsella, *J. Coat. Technol.* **55** (1983) 99.

[2.44] W. Brushwell, *Farbe + Lack* **86** (1980) 706.

[2.45] R. Zimmermann, *Farbe + Lack* **82** (1976) 383.

[2.46] A. Mercurio, S. N. Lewis, *J. Paint Technol.* **47** (1975) 37.
[2.47] C. K. Schoff, *Prog. Org. Coat.* **4** (1976) 189.
[2.48] M. Takahashi, *Polym. Plast. Technol. Eng.* **15** (1980) 1.
[2.49] K. O'Hara, *J. Oil. Colour Chem. Assoc.* **71** (1988) 413.
[2.50] H. Kittel, *Adhäsion* **21** (1977) 162.
[2.51] D. W. Aldinger: *Congr. FATIPEC I*, 1984, p. 69.
[2.52] J. P. H. Juffermans, *Prog. Org. Coat.* **17** (1989) 15.
[2.53] D. R. Bauer, R. A. Dickie, *J. Coat. Technol.* **58** (1986) 41.
[2.54] S. Günther, *Ind. Lackierbetr.* **57** (1989) 167.
[2.55] J. R. Erickson, *J. Coat. Technol.* **48** (1976) 58.
[2.56] R. Dowbenko, D. P. Hart, *Ind. Eng. Chem. Prod. Res. Dev.* **12** (1973) 14.
[2.57] G. Y. Tilak, *Prog. Org. Coat.* **13** (1985) 333.
[2.58] L. W. Hill, A. Kaul, K. Kozlowski, J. O. Sauter, *Polym. Mater. Sci. Eng.* **59** (1988) 283.
[2.59] T. Nakamichi, M. Ishidoya, *J. Coat. Technol.* **60** (1988) 33.
[2.60] H. Sander, R. Kroker, *Farbe + Lack* **82** (1976) 1105.
[2.61] R. D. Athey, *Farbe + Lack* **95** (1989) 475.
[2.61a] A. Noomen, *Prog. Org. Coat.* **17** (1989) 27.
[2.62] E. Karsten, O. Lückert: *Lackrohstoff-Tabellen*, Curt R. Vincentz Verlag, Hannover 1987.
Resin Index, British Resin Manufacturer's Association, Redhill, Surrey 1986.
Raw Materials Index-Resins, National Paint and Coatings Association, Washington, DC, 1988.
[2.63] Lewis Berger & Sons, EP 573 809, 1942; 573 835, 1943; 580 912, 1942; DE 975 352, 1949.
[2.64] Röhm und Haas, DE 1 022 381, 1953.
[2.65] Röhm und Haas, DE 912 752, 1951; F. Schlenker, *Farbe + Lack* **58** (1952) 174.
[2.66] W. Krauß, *Farbe + Lack* **64** (1958) 39; **64** (1958) 209; **70** (1964) 876.
[2.67] W. Götze, *Fette Seifen Anstrichm.* **65** (1963) 493.
[2.68] W. B. Winkler, US 2 663 649, 1953.
[2.69] I. G. Farben, DE 738 254, 1940.
[2.70] Shell International Research, EP 942 465, 1963 (Dutch prior. 1959 and 1960).
[2.71] Shell International Research, EP 1 269 628, 1960.
[2.72] K. Hamann, *Farbe + Lack* **81** (1975) 907.
[2.73] R. Küchenmeister, *Farbe + Lack* **78** (1972) 550; E. Knappe, *Fette Seifen Anstrichm.* **77** (1975) 149; J. Schoeps, *J. Oberflächentechnik* **11** (1975) 32.
[2.74] J. E. Glass (ed.): *Water Soluble Polymers – Beauty with Performance*, American Chemical Society, Washington, D.C., 1986.
[2.75] Akzo, DE-OS 3 803 141 A 1, 1989.
[2.76] Coates Brothers, EP-A 0 253 474 A 2, 1988 (Brit. prior. 1986, GB 8 612 193).
[2.77] K. Holmberg: *High Solids Alkyd Resins*, Marcel Dekker, New York, NY.
[2.78] S. J. Storfer, J. T. Di Piazza, R. E. Moran, *J. Coat. Technol.* **60** (1988) no. 761, 37.
[2.79] J. Spauwen, *J. Oil Colour Chem. Assoc.* **71** (1988) no. 2, 47.
[2.80] S. J. Bellettiere, D. M. Mahoney, *J. Coat. Technol.* **59** (1987) no. 752, 101.
[2.81] *Ullmann*, 4th ed., **19**, 61.
[2.82] R. Seidler, H. G. Stolzenbach, *Farbe + Lack* **81** (1975) 281.
[2.83] V. Mirgel, K. Nachtkamp, *Farbe + Lack* **89** (1983) 928.
[2.84] W. Marquardt, H. Gempeler, *Farbe + Lack* **84** (1978) 301.
[2.85] J. Dörffel, *Farbe + Lack* **81** (1975) 10.
[2.86] E. Häring, O. Lückert, in: *Emissionsarm lackieren*, Curt. R. Vincentz Verlag, Hannover 1987, p. 51.
[2.87] K. Brüning, K. G. Sturm, *Adhäsion* **3** (1973) 83.
[2.88] Dynamit Nobel AG, DE 2 126 048, 1971 (G. Schade, H. Hermsdorf, H. Vltavsky).
[2.89] Hüls AG, DE 2 521 791, 1975 (K. Burzin, K.-H. Haneklaus, D. Stoye).

[2.90] K. H. Hornung, U. Biethan, *Farbe + Lack* **76** (1970) 461.
[2.91] U. Biethan, J. Dörffel, D. Stoye, *Farbe + Lack* **77** (1971) 988.
[2.92] Hüls AG, DE 2346818, 1973 (K. Schmitt, J. Disteldorf, F. Schmitt); DE 2938855, 1979 (R. Gras, E. Wolf).
[2.93] M. Schmitthenner: *XIV. Internat. Conference in Org. Coatings Sc. and Techn.*, Athens, 1988.
[2.94] Code of Federal Regulations (annual publication), § 175.300, Chap. (b) (3) (vii), "Polyester resins".
[2.95] D. Stoye, W. Andrejewski, J. Dörffel: *Congr. FATIPEC XIII*, 1976, p. 605.
[2.96] Hüls AG, EP 45994, 1981 (J. Disteldorf, R. Gras, H. Schnurbusch).
[2.97] S. Tochihara, *Prog. Org. Coat.* **10** (1982) 195.
[2.98] DE Bayer: 2804216 (1978).
[2.99] Stoye/Freitag: Resins for Coatings, Carl Hanser Verlag, Munich (1996), pp. 84–85, 92.
[2.100] I. G. Farben, DRP 540101 (1930), 544326 (1930); 571665 (1930); 598732 (1932).
[2.101] E. Karsten, O. Lückert: Lackrohstoff-Tabellen, 9th edition, C. R. Vincentz Verlag, Hannover (1992), pp. 522–549 "Resin Index" of the British Resin Manufacturers Association "Raw Materials Index – Resins", National Paint and Coatings Association, USA.
[2.102] BASF, DBP 948816 (1951) Scott-Bader, EP 713312 (1951).
[2.103] W. Gebhardt, W. Hermann and K. Hamann: farbe + lack 64 (1958), p. 303.
[2.104] Chem. Werke Hüls, EP 842958 (1960).
[2.105] ICI, EP 578867 (1994).
[2.106] Bayer, DAS 1024654 (1955), Bayer, DT 1494437 (1956) H. J. Traenckner, H. U. Pohl: Angew. Makromolek. Chemie 108 (1982), pp. 61–78.
[2.107] BASF, DE 1195491 (1963).
[2.108] Bayer, DE – OS 1927320 (1971).
[2.109] H. G. Heine, H.-J. Traenckner: Prog. Org. Coat. 3, pp. 115–139 (1975) J. Ohngemach, K. H. Neisius, J. Eichler, C. P. Herz: Kontakte 3, pp. 15–20 (1980) M. Jakobi, A. Henne, A. Böttcher: Adhäsion, Issue 4, pp. 6–10 (1986).
[2.110] K. Demmler, E. Müller, M. Schwarz: Defazet Dtsch. Farben Z. 31, pp. 115–128 (1977).
[2.111] Hallack: 5. Möbellackierforum Detmold 9/96.
[2.112] W. Kremer: farbe + lack 94 (1988), pp. 205–208 Stoye/Freitag: Resins for Coatings, Carl Hanser Verlag, Munich (1996), pp. 94–95. W. Fischer: I-Lack 2/93, pp. 43–58.
[2.113] K. H. Hauck, F. Hecker-Over: DAS 1025302 (1954).
[2.114] H. Kittel: Lehrbuch der Lacke und Beschichtungen, Volume 5, Verlag W. A. Colomb, pp. 311–315 (1977).
[2.115] G. Oertel: *Polyurethane Handbook*, Hanser Publishers, New York 1985.
[2.116] M. Bock, R. Halpaap, *Farbe + Lack* **93** (1987) 264.
[2.117] W. Wieczorrek, *Polym. Paint Colour J.* **178** (1988) no. 4225, 827.
[2.118] B. Taub, G. H. Patschke, *Modern Paints Coat.* **79** (1989) 41.
[2.119] H. Lee, K. Neville: *Handbook of Epoxy Resins*, McGraw Hill, New York 1982.
[2.120] A. M. Paquin: *Epoxidverbindungen und Epoxidharze*, Springer Verlag, Berlin-Göttingen-Heidelberg 1958.
[2.121] H. Wagner, H. F. Sarx: *Lackkunstharze*, Hanser Verlag, München 1971.
[2.122] H. Batzer, F. Lohse, *Kunststoffe* **66** (1976) no. 10, 637.
[2.123] W. Marquardt, H. Gempeler, *Farbe + Lack* **84** (1978) 301–306.
[2.124] W. Marquardt, H. Gempeler, *Farbe + Lack* **86** (1980) 696–698.
[2.125] W. Schneider, *Farbe + Lack* **85** (1979) 925–929.
[2.126] F. Lohse, H. Zweifel, *Adv. Polym. Sci.* **78** (1986) 61–81.
[2.127] M. Gaschke, B. Dreher, *J. Coat. Technol.* **48** (1976) no. 617, 46–51.
[2.128] K. Brugger, R. Schmid: *Congr. FATIPEC XI*, 1972, p. 307.
[2.129] J. Bersier, P. M. Bersier, B. Dreher, A. Heer: *Congr. FATIPEC X, 1970*, p. 255.
[2.130] K. Buser, *Plaste Kautsch.* **3** (1963) 189.

[2.131] A. Manz, *Beckacite-Nachr.* **25** (1966) no. 3, 70.
[2.132] W. J. van Westrenen et al.: *Congr. FATIPEC VIII, 1966,* p. 126.
W. Weger: *Congr. FATIPEC XVII,* 1986, p. 43.
[2.133] P. Kordemenos, J. D. Nordstrem, *J. Coat. Technol.* **54** (1982) no. 686.
[2.134] P. Robinson, *J. Coat. Technol.* **53** (1987) no. 674.
[2.135] F. Lohse, H. Zweifel, *Adv. Polym. Sci* **78** (1986). J. V. Crivello in N. S. Allen (ed.): *Developments in Polymer Photochemistry 2,* Applied Science, London 1981.
[2.136] W. A. Finzel, E. P. Plueddemann: "The History of Silicone Coatings," *ACS Meeting,* Miami Beach, Fla., September 13, 1989.
L. H. Brown in R. R. Myers, J. S. Long (eds.): *Treatise on Coatings,* vol. 1, part III, chap. 3, "Silicones in Protective Coatings," Marcel Dekker, New York, 1972.
S. Paul: *Surface Coatings,* chap. 2.3., J. Wiley & Sons, Chichester 1985.
[2.137] T. C. Kendrick, B. Parbhoo, J. W. White in S. Patai, Z. Rappoport (eds.): *The Chemistry of Organic Silicon Compounds,* part 2, chap. 21, "Siloxane Polymers and Copolymers," J. Wiley & Sons, Chichester 1989.
[2.138] N. Grassie, I. G. Macfarlane, *Eur. Polym. J.* **14** (1978) no. 11, 875.
[2.139] W. A. Finzel: "Properties of High Temperature Silicone Coatings," *J. Protective Coatings Linings* **4** (1987) no. 8, 38–43.
[2.140] Dow Corning Corp.: *Silicones for Protective Coatings,* 48686-0994, Midland, Mich., 1985.
[2.141] W. A. Finzel: "Silicone Coatings: Popular Choice for the Tough Jobs," *Am. Paint Coat. J.* **65** (1981) 18.
[2.142] Dow Corning, EP 279 513, 1988 (A. G. Short).
[2.143] R. W. Cope, M. A. Glaser: *Silicone Resins for Organic Coatings,* Federation of Society for Paint Technology, Philadelphia 1970.
[2.144] L. H. Brown: *Handbook of Coating Additives,* Marcel Dekker, New York 1987.
[2.145] Dow Corning Europe: *Silicone Additives: That Little Bit Extra,* 48686-0994, 1988.
[2.146] G. C. Sawicki, J. W. White: "Speciality Silicones for Foam Control," *Chemspec Europe 89 BACS Symposium,* Manchester 1989.
[2.147] N. J. Albrecht, W. J. Blank: "The Use of Triazine Resins in High Solids Coatings," *Adv. Org. Coat. Sci. Technol. Ser.* **4** (1982).
[2.148] W. J. Blank: "Amino Resins in High Solids Coatings," *J. Coat. Technol.* **54** (1982) no. 687, 26–41.
[2.149] W. J. Blank: "Reaction Mechanism of Melamine Resins," *J. Coat. Technol.* **51** (1979) no. 656, 61–70.
[2.150] A. Berge, S. Gudmunden, J. Ugelstad, *Eur. Polym. J.* **5** (1969) 171.
[2.151] A. Berge, B. Kvaeven, J. Ugelstad, *Eur. Polym. J.* **6** (1970) 981.
[2.152] L. J. Calbo: "Effect of Catalyst Structure on the Properties of Coatings Crosslinked with Hexa(methoxymethyl)melamine," *J. Coat. Technol.* **52** (1980) no. 660, 75–83.
[2.153] W. Schedlbauer: "Melamin-Formaldehyd-Lackharze," *Farbe + Lack* **79** (1973) no. 9, 846–852.
[2.154] W. F. Herbes, S. J. O'Brien, R. G. Weyker: "Melamine-Formaldehyde," *Chemical Aftertreatment of Textiles,* Wiley Interscience, New York 1971.
[2.155] G. G. Parekh: "Chemistry of Glycoluril-Formaldehyde Resins and their Performance in Coatings," *J. Coat. Technol.* **51** (1979) no. 658, 101–110.
[2.156] G. Christensen: "Analysis of Functional Groups in Amino Resins," *Prog. Org. Coat.* **5** (1977) 255–276.
[2.157] H. Schindlbauer, J. Anderer: "Eine einfache Charakterisierung von Melamin-Formaldehyd-Kondensaten mit Hilfe von NMR-Messungen," *Fresenius Z. Anal. Chem.* **301** (1980) 210–214.
[2.158] M. Chiavarini, N. Del Fanti, R. Bigatto: "Compositive Characterization of Urea-Formaldehyde Adhesives by NMR Spectroscopy," *Angew. Makromol. Chem.* **46** (1975) no. 695, 151–162.

[2.159] M. Chiavarini, N. Del Fanti, R. Bigatto: "Compositive Characterization of Melamine-Formaldehyde Condensates by NMR Spectroscopy," *Angew. Makromol. Chem.* **56** (1976) no. 818, 15–25.
[2.160] B. Tomita, H. Ono: "Melamine-Formaldehyde Resins: Constitutional Characterization by Fourier Transform 13C-NMR," *J. Polym. Sci. Polym. Chem. Ed.* **17** (1979) 3205–3215.
[2.161] A. J. J. de Breet, W. Dankelman, W. G. B. Huysmans, J. De Wit: "^{13}C-NMR Analysis of Formaldehyde Resins," *Angew. Makromol. Chem.* **62** (1977) no. 877, 7–21.
[2.162] Technical literature of Monsanto Chemical Company.
[2.163] Estimates of market size were obtained from private industry sources. Published information on amino resin usage bundles other end uses and does not permit any reasonable estimates of market size.
[2.164] S. E. Feinman: *Formaldehyde Sensitivity and Toxicity,* CRC Press, Boca Raton, Fla., 1988.
[2.165] K. Hultzsch: *Chemie der Phenolharze,* Springer Verlag, Berlin 1950.
[2.166] Robert W. Martin: *The Chemistry of Phenolic Resins,* J. Wiley & Sons, New York 1956.
[2.167] P. Nylén, E. Sunderland: *Modern Surface Coatings,* J. Wiley & Sons, London 1965.
[2.168] H. Abraham: *Asphalts and Allied Substances,* D. van Nostrand Comp., New Jersey 1960.
[2.169] F. M. Depke: "Lacktechnische Schriften," in vol. 8: *Bitumen- und Teerlacke,* Verlag W. A. Colomb, Stuttgart–Berlin 1970.
[2.170] H. Walther: *Bituminöse Stoffe im Bauwesen, Straßenbau,* Chemie und Technik Verlag, Heidelberg 1962.
[2.171] H.-G. Franck, U. G. Collin: *Steinkohlenteer, Chemie, Technologie und Verwendung,* Springer Verlag, Berlin–Heidelberg–New York 1968.
[2.172] L. Bernhard: *Das Wasserglas, seine Darstellung und Anwendung,* H. Bechold, Frankfurt 1893.
[2.173] H. Trillich: *Die Wasserglas-Anstrich- und Mal-Verfahren,* B. Heller, München 1929.
[2.174] A. W. Keim, DE 4315, 1878.
[2.175] R. Wenda, W. Lukas, *Farbe + Lack* **93** (1987) 186–189.
[2.176] H. Wulf: *Große Farbwarenkunde,* R. Müller, Köln 1967.
[2.177] E. Bagda, *Farbe + Lack* **90** (1984) 443.
[2.178] Tiedemann (ed.): *Grundlagen zur Formulierung von Dispersions-Silikat-Systemen,* Viernheim 1988.
[2.179] Technical Service Report on Trasol: *Production of Trasol Paints, Plasters and Primers,* Henkel KGaA, Düsseldorf 1986.
[2.180] W. Haller, *Adv. Org. Coat. Sci. Technol. Ser.* **12** (1990) 55–57.
[2.181] J. W. Lyons: *The Chemistry and Uses of Fire Retardants,* Wiley-Interscience, New York 1970.
[2.182] A. Williams: *Flame Retardant Coatings and Building Materials,* Chemical Technology Review 25, Noyes Data, Park Ridge 1974.
[2.183] T. Ginsberg, *J. Coat. Technol.* **53** (1981) no. 677, 23–32.
[2.184] T. Ginsberg, L. G. Kaufmann, *Mod. Paint Coat.* (1981) October, 138–143.
[2.185] W. C. La Course et al., *J. Can. Ceram. Soc.* **52** (1983) 18–23.
[2.186] T. Ginsberg, C. N. Merriam, L. M. Robeson, *J. Oil Colour Chem. Assoc.* **59** (1976) 315–321.
[2.187] J. Loomans, K. van Leberghe, *Chim. Peint.* **12** (1949) 184–190.
[2.188] H. Anders, *DEFAZET Dtsch. Farben Z.* **8** (1954) 452–453.
[2.189] Carboline, US 3 056 684, 1959 (S. L. Lopata, W. R. Keithler).
[2.190] Zinc Lock Corp., US 3 392 130, 1962 (R. A. Rucker, J. B. Heymes).
[2.191] Hempel's Marine Paints, US 3 442 824, 3 546 155, 1967 (J. W. Chandler).
[2.192] Stauffer Chem. Corp., US 3 392 036, 1965 (G. D. McLeod).
[2.193] Dynamit Nobel, DE 2 000 199, 1970 (W. Dittrich).
[2.194] Stauffer Chem. Corp., US 3 730 743, 1970 (G. D. McLeod).
[2.195] The Anderson Development Co., DE-OS 2 147 865, 1971 (A. R. Anderson).
[2.196] D. H. Gelfer, P. Vandorsten, *Mater Perform.* **20** (1981) no. 5, 23–28.

[2.197] B. V. Gregorovich and I. Hazan, "Environmental etch performance, and scratch and mar resistance of automotive clearcoats", Progress in Organic Coatings, Vol. 24, 1994, pp. 131–146.
[2.198] U. Roeckrath, K. Brockkötter, Th. Frey, U. Poth, G. Wigger. Int. Conf. Org. Coatings. Proc. 22. Pg 273–289. (1996).
[2.199] D. L. Singer, S. Swarup, M. A. Mayo, Appl. WO 9410213 (1994).

References for Chapter 3

[3.1] *Ullmann*, 4th ed., **15**, 652.
[3.2] Verband der deutschen Lackindustrie (VdL), information 9/90, 17.04.1990.
[3.3] R. B. Seymour, *J. Coat. Technol.* **61** (1989) no. 776, 73.
[3.4] J. Hambrecht, *Coat. Mag.* **12** (1990) 44.
[3.5] ASTM D 3960-87, Determining Volatile Organic Content (VOC) of Paints and Related Coatings.
[3.6] K. Tachi, C. Okuda, S. Suzuki, *J. Coat. Technol.* **62** (1990) no. 782, 43.
[3.7] H. J. Drexler, B. Biallas, *I-Lack* **58** (1990) 208.
[3.8] J. Ramsbotham, *Prog. Org. Coat.* **8** (1980) 113.
[3.9] R. C. Nelson, V. T. Figurelli, J. G. Walsham, G. D. Edwards, *J. Paint. Technol.* **42** (1970) 644.
[3.10] C. M. Hansen, R. W. Tess (eds.): "Solvents Theory and Practice," *Adv. Chem. Ser.* **124** (1973) 1 ff.
[3.11] D. W. Richards, *Am. Paint Coat. J.* **73** (1988) 126.
[3.12] J. Ramsbotham, *J. Oil Colour Chem. Assoc.* **26** (1979) 359.
[3.13] R. A. Waggoner, F. D. Blum, *J. Coat. Technol.* **61** (1989) no. 768, 51.
[3.14] M. Shimbo, M. Ochi, K. Arai, *J. Coat. Technol.* **57** (1985) no. 728, 93.
[3.15] R. Henning, *Proc. Polyurethane World Congr.* **1987**, 626.
[3.16] B. Henshaw, *Surf. Coat. Aust.* **25** (1988) 12.
[3.17] M. Kashimoto, *JSAE Rev.* **11** (1990) 88.
[3.18] V. Handforth, *J. Oil Colour Chem. Assoc.* **73** (1990) 145.
[3.19] J. A. Simms, H. J. Spinelli, *J. Coat. Technol.* **59**, (1987) no. 752, 125.
[3.20] A. Noomen, *Proc. Int. Conf. Org. Coat. Sci. Technol. 13th* **1989**, 251.
[3.21] H. R. Lucas, *J. Coat. Technol.* **57** (1985) no. 731, 49.
[3.22] PPG, US 4650718, 1987 (D. Simpson et al.).
[3.23] ICI, EP 0134691, 1984 (S. M. Horley et al.).
[3.24] U. Zorll, *Farbe + Lack* **95** (1989) 801.
[3.25] M. Schmitthenner, W. Funke, *Proc. 2nd Solvents Symp. "Solvents: The Neglected Parameter,"* University of Manchester 1977, 101.
[3.26] M. J. Diana, *SAE Tech. Pap. Ser.* (1991) 910089.
[3.27] H. Scheene, T. Heere, *Chem. Ing. Tech.* **61** (1989) 330.
[3.28] K. Holmberg: *High Solids Alkyd Resins,* Marcel Dekker, New York 1987.
[3.29] Du Pont, US 4224202, 1980 (P. Heiberger).
[3.30] Rohm and Haas, US 4261872, 1981 (W. D. Emmons).
[3.31] Bayer, EP 057834, 1982 (E. Eimers, R. Dhein, K. Fuhr).
[3.32] Bayer, EP 032657, 1981 (E. Eimers, R. Dhein).
[3.33] D. J. Love, *Paint and Resin* **59** (1989) 19–22.
[3.34] J. P. H. Juffermans, *Prog. Org. Coat.* **17** (1989) 15–26.
[3.35] Akzo N.V., EP-A 234641, 1987 (D. J. F. Geerdes, P. J. C. Nelen, K. H. Zabel).
[3.36] T. A. Potter, *Polym. Mater. Sci. Eng.* **55** (1986) 425–433.
[3.37] W. J. Blank, *J. Coat. Techn.* **60** (1988) no. 764, 43–50.

[3.38] W. Heitz, *Angew. Chem.* **145/146** (1986) 37–68.
[3.39] K. Hönel, *Farbe + Lack* **59** (1953) 174.
[3.40] K. Weigel: *Elektrophorese-Lacke*, Wissenschaftl. Verlagsgesellschaft, Stuttgart 1967.
[3.41] H. J. Luthardt, *Congr. FATIPEC XIV*, 1978, p. 413.
[3.42] W. Burckhardt, H. J. Luthardt, *J. Oil Colour Chem. Assoc.* **62** (1979) 375.
[3.43] EP-B 0015035, 1979 (G. Günter, H. Haeufler).
[3.44] EP-B 0038127, 1980 (A. J. Backhouse).
[3.45] EP-A 0238108, 1986; EP-A 0273530, 1986; EP-A 0287144, 1987 (R. Buter).
[3.46] B. D. Meyer in: *Umweltfreundliche Lackiersysteme für die industrielle Lackierung*, Expert-Verlag, Kontakt und Studium, vol. 271, Ehningen 1989, pp. 146–244.
[3.47] E. Engler: *Pulverlackmarkt*, Pulverlacktagung Hamburg, Oct. 7, 8, 1991.
[3.48] *VDI-Handbuch Kunststofftechnik*, VDI 2538, Sept. 1979, VDI Düsseldorf.
[3.49] *VDI-Handbuch Reinhaltung der Luft*, vol. 3, Entwurf VDI 2588, May, 1986, VDI Düsseldorf.
[3.50] B. D. Meyer, *Farbe + Lack* **89** (1983) no. 10, 763–771.
[3.51] The Powder Coating Institute; *Powder Coating 88*, Alexandria, Virginia 1988.
[3.52] H. Dörr F. Holzinger: *Kronos Titandioxid in Dispersionsfarben*, Kronos Titan, Leverkusen 1989.
[3.53] E. Nägele: *Dispersionsbaustoffe*, Rudolf Müller Verlag, 1989.
[3.54] K. E. J. Barrett (ed.): *Dispersion Polymerization in Organic Media*, J. Wiley & Sons, London 1975.
[3.55] Chap. 2 in [3.54].
[3.56] R. H. Ottewill, T. Walker, *Kolloid Z. Z. Polym.* **227** (1968) 108.
[3.57] A. Doroszkowski, R. Lambourne, *J. Poly. Sci. Part C* **34** (1971) 253.
[3.58] F. Th. Hesselink, *J. Phys. Chem.* **73** (1969) 3488.
[3.59] D. H. Napper, *J. Colloid Interface Sci.* **29** (1969), 168.
[3.60] Chap. 3 in [3.54].
[3.61] M. W. Thompson in R. Buscall, T. Corner and J. Stagemen (eds.): *Polymer Colloids*, J. F. Elsevier, London 1985.
[3.62] ICI, GB 893429, 1957; 990144, 1960; 1017931, 1966 (D. W. J. Osmond).
[3.63] ICI, GB 1052241, 1962 (D. W. J. Osmond).
[3.64] F. Bueche, *J. Colloid. Interface Sci.* **41** (1972) 374.
[3.65] F. A. Waite, *J. Oil Colour Chem. Assoc.* **54** (1971) 342.
[3.66] ICI, GB 1096912, 1963 (M. W. Thompson, F. A. Waite).
[3.67] ICI, GB 1174391, 1966 (D. W. J. Osmond, F. A. Waite, D. J. Walbridge).
[3.68] J. V. Dawkins, S. A. Shakir, T. G. Croucher, *Eur. Polym. J.* **23** (1987) no. 2, 173.
[3.69] J. V. Dawkins, G. Taylor in R. M. Fitch (ed.): *Polymer Colloids II*, Plenum Press, New York 1980, p. 447.
[3.70] ICI, GB 1373531, 1971 (P. F. Nicks, P. G. Osborne).
ICI, EP-A 466310, 1992 (E. Nield, R. Ahmed, R. A. Choudery).
ICI, EP-A 465001, 1992 (E. Nield, R. A. Choudery).
ICI, EP-A 464998, 1992 (E. Nield, R. Ahmed, R. A. Choudery).
[3.71] BALM Paints, GB 1231614, 1967.
[3.72] S. J. Barsted, L. J. Novakowska, I. Wagstaff, D. J Walbridge, *Trans. Faraday Soc.* **67** (1971) 3598.
[3.73] J. V. Dawkins, G. Taylor, *J. Chem. Soc. Faraday Trans. 1* **76** (1986) 1263.
[3.74] D. J. Cebula et al., *Colloid Polym. Sci.* **261** (1983) no. 7, 555.
[3.75] R. J. R. Cairns, R. H. Ottewill, D. W. J. Osmond, I. Wagstaff, *J. Colloid Interface Sci.* **54** (1976) 45.
[3.76] D. H. Napper, *Trans. Faraday Soc.* **64** (1968) 1701.
[3.77] B. Vincent, J. Clarke, K. G. Barnett, *Colloids Surf.* **17** (1986) no. 1, 51.

[3.78] G. J. Fleer, J. H. M. H. Scheutjens, B. Vincent, *ACS Symp. Ser.* **240** (1984) 245.
[3.79] J. V. Dawkins et al., *ACS Symp. Ser.* **165** (1981) 189.
[3.80] J. S. Higgins et al., *Polym. Commun.* **29** (1988) no. 5, 122.
[3.81] K. E. J. Barrett, H. R. Thomas, *J. Polym. Sci. Part A* **7** (1969) 2621.
[3.82] K. E. J. Barrett, H. R. Thomas, *Br. Polym. J.* **2** (1970) 45.
[3.83] K. E. J. Barrett, H. R. Thomas, *Kinet. Mech. Polyreactions Int. Symp. Macromol. Chem. Prep.* **3** (1969) 375.
[3.84] G. Chin, M. A. Winnik, M. D. Croucher, *Colloid Polym. Sci.* **264** (1986) 25.
[3.85] M. D. Croucher, L. S. Egan, M. A. Winnik, *Polym. Eng. Sci.* **26** (1986) 15.
[3.86] T. A. Strivens, *Colloids Surf.* **18** (1986) no. 2–4, 395.
[3.87] ICI, GB 1 157 630, 1965 (M. W. Thompson, D. J. Walbridge).
[3.88] R. Dowbenko, D. P. Hart, *Ind. Eng. Chem. Prod. Res. Dev.* **12** (1973) 14.
[3.89] Akzo, EP-A 337 522, 1989 (R. Butler).
Kansal Paint, EP-A 292 262. 1988 (A. Yabuta, Y. Hiramatsu, A. Kasari)
[3.90] A. J. Backhouse, *J. Coat. Technol.* **54** (693) (1982) 83.
[3.91] ICI, US 4 180 489, 1979; US 4 242 384, 1980 (M. S. Andrew, A. J. Backhouse).
Hitachi Chem., JP 02 053 803, 1990 (M. Hasebe, I. Moribe).
[3.92] ICI, GB 1 242 054, 1967 (N. D. P. Smith, M. W. Thompson, E. J. West).
[3.93] C. B. Arends, D. K. Hoffman, G. C. Kolb, *Polym. Prepr. Am. Chem. Soc. Div. Polym. Chem.* **26** (1985) no. 1, 232.
S. W. Rees, *Adv. Org. Coat. Sci. Technol. Ser.* **12** (1990) 27–33.
[3.94] Rohm & Haas, GB 956 454, 1959.
[394a] C. W. A. Bromley, J. Coat. Technol. **61** (1989) 39–43.
[3.95] Rohm & Haas, GB 993 794, 1961.
[3.96] Union Carbide, GB 1 199 651, 1966.
[3.97] Du Pont, GB 1 272 890, 1969.
[3.98] ICI, DE 2 139 315, 1970.
[3.99] BASF, GB 1 203 427, 1966.
[3.100] Hercules Inc., BE 877 607, 1980.
[3.101] Nippon Oils and Fats, GB 2 071 116, 1981.
[3.102] Xerox Corp., US 4 617 249, 1985.
[3.103] C. K. Ober, K. P. Lok, *Macromolecules* **20** (1987) no. 2, 268.
[3.104] Mobil Oil Corp., EP-A 290 195 A1, 1988.
[3.105] Allied Colloids Ltd., EP-A 150 933 A2, 1985.
[3.106] G. C. Burnside, G. E. P. Brewer, *J. Paint Technol.* **38** (1960) 96.
[3.107] L. R. LeBras, R. M. Christensen, *J. Paint Technol.* **44** (1972) 63.
[3.108] Verband der Lackindustrie e. V., Mitteilung 9/90, April 17th 1990.
[3.109] A. D. Wilson, J. W. Nicholson, H. J. Posser (eds.): *Surface Coatings – 2*, Elsevier Applied Science Publishers, Barling 1988, p. 39.
[3.110] M. S. ElAasser et al., *J. Coat. Technol.* **56** (1984) 37.
[3.111] V. I. Eliseeva, V. N. Cherny, *Prog. Org. Coat.* **17** (1989) 251.
[3.112] F. Beck, *Prog. Org. Coat.* **4** (1976) 1.
[3.113] H. Spoor, H. U. Schenck, *Farbe + Lack* **88** (1982) 94.
[3.114] P. E. Pierce, J. Kovacs, C. Higgenbothan, *Ind. Eng. Chem.* **17** (1978) 317.
[3.115] P. E. Pierce, *Ind. Coat. Techn.* **53** (1981) 52.
[3.116] W. Rausch, *Ind. Lackierbetrieb* **49** (1981) 413.
[3.117] W. Woebcken: *Kunststoffhandbuch, -Duroplaste,* Hanser Verlag, München–Wien 1988, p. 1023.
[3.118] ICI, GB 1 109 979, 1964 (B. A. Cooke).
[3.119] PPG Ind., FR 1 394 007, 1962 (D. P. Hart, R. H. Christensen).
[3.120] H. Verdin, *J. Oil Colour Chem. Assoc.* **59** (1986) 81.

[3.121] P. I. Kordomenos, J. D. Nordstrom, *J. Coat. Technol.* **54** (1982) 33.
[3.122] BASF, EP 125438, 1983 (E. Schupp et al.).
[3.123] Z. W. Wicks, *Prog. Org. Coat.* **3** (1985) 73.
[3.124] PPG, US 4423166, 1982 (T. Moriarity, W. Geiger).
[3.125] H. Kraus, *Metalloberfläche* **36** (1982) 531.
[3.126] H. J. Streitberger, K. Arlt, F. J. Pötter, *Fahrzeug und Karosserie* **7** (1984) 34.
[3.127] B. A. Baker, *Light Metal Age* **41** (1983) 20.
[3.128] Th. Brücken, *Canmaker* **2** (1989) 31.
[3.129] Mutsutseshi Denhi, Kansai Paint, DE 3628340 (M. Hoshino et al.).

References for Chapter 4

[4.1] W. Büchner, R. Schiebs, G. Winter, K. H. Büchel: *Industrielle anorganische Chemie*, Verlag Chemie, Weinheim 1984.
[4.2] G. Buxbaum et al.: "Anorganische Pigmente," *Winnacker-Küchler, Chemische Technologie*, 4th. ed., vol. 3, Carl Hanser Verlag, München–Wien 1983.
[4.3] P. A. Lewis: *Pigment Handbook*, J. Wiley & Sons, New York 1988.
[4.4] G. Benzing et al.: *Pigmente für Anstrichmittel*, expert verlag, Grafenau 1988.
[4.5] Bayer AG: "Anorganische Pigmente für Anstrich- und Beschichtungsstoffe," company bulletin, Leverkusen 1989.
[4.6] H. Kittel: *Lehrbuch der Lacke und Beschichtungen*, vol. 2, Verlag W. A. Colomb, Berlin 1974.
[4.7] A. Goldschmidt et al: "Lacke und Farben," *Glasurit-Handbuch*, C. R. Vincentz Verlag, Hannover 1984.
[4.8] G. D. Parfitt: *Dispersions of Powders in Liquids*, Applied Science Publishers, London 1981.
[4.9] T. C. Patton: *Paint Flow and Pigment Dispersion*, J. Wiley & Sons, New York 1978.
[4.10] *TRGS, UVV, MAK-Werte (TRGS 900)*, C. Heymanns Verlag KG, Köln 1989.
[4.11] Kunststoff LMBG XL: Gesundheitliche Beurteilung von Kunststoffen im Rahmen des Lebensmittel- und Bedarfsgegenständegesetzes: XL Lacke und Anstrichstoffe für Lebensmittelbehälter und -verpackungen, Bundesgesundheitsblatt, no. 13, 1968, pp. 192–194.
[4.12] O. Lückert: *Pigment+Füllstoff-Tabellen*, 4th ed., Laatzen 1989.
[4.13] H. Haagen, D. Martinovic, *Farbe+Lack* **95** (1989) 892–895.
[4.14] H. S. Katz, J. V. Milewski: *Handbook of Fillers and Reinforcements for Plastics*, Van Nostrand Reinhold Co., New York 1978.
[4.15] H. Kittel: *Lehrbuch der Lacke und Beschichtungen*, vol. II, Verlag W. A. Colomb in der H. Heenemann GmbH, Berlin–Oberschwandorf 1974, pp. 457.

References for Chapter 5

[5.1] L. J. Calbo (ed.): *Handbook of Coatings Additives*, Marcel Dekker, New York, vol. 1 1987 vol. 2 1992.
[5.2] D. R. Karsa (ed.): *Additives for Waterbased Coatings*, The Royal Society of Chemistry, Cambridge 1991.
[5.3] R. Lambourne (ed.): *Paint and Surface Coatings: Theory and Practice*, Ellis Horwood, Chichester 1987.
[5.4] T. C. Patton: *Paint Flow and Pigment Dispersion*, J. Wiley & Sons, New York 1979.
[5.5] H. Nagorny: *Lack-Additiv-Tabellen 1993–1994* Hilden 1993.
[5.6] E. Karsten, O. Lückert: *Lackrohstoff-Tabellen*, Curt R. Vincentz Verlag, Hannover 1992.
[5.7] National Paint and Coatings Association (ed.): *Raw Materials Index*, NPCA, Washington 1989.

[5.8] G. Berner, M. Rembold, *Farbe+Lack* **89** (1983) 840.
[5.9] A. Jurgetz, H. Rothbächer, *Farbe+Lack* **91** (1985) 921.
[5.10] G. Berner, G. Deslex, K. Schoohf: *Congr. FATIPEC*, vol. II, Amsterdam 1980, p. 61.
[5.11] H. J. Heller, H. R. Blattmann, *Pure Appl. Chem.* **30** (1972) 145; **36** (1973) 141.
[5.12] A. Valet, F. Sitek, G. Berner, *Farbe+Lack* **94** (1988) 734.
[5.13] J. Otterstedt, *J. Chem. Phys.* **58** (1973) 5716.
[5.14] W. Klöppfer, *J. Polym. Sci. Polym. Symp.* **57** (1978) 205.
[5.15] F. Werner, H. E. A. Kramer, *Eur. Polym. J.* **13** (1977) 501.
[5.16] M. Rembold, G. Berner, *Org. Coat. Appl. Polym. Sci. Proc.* **47** (1982) 624.
[5.17] A. Valet, Light Stabilizers for Paints, Vinzenz, Hannover 1997.
[5.18] D. M. Wiles, D. J. Carlson: *New Trends in the Photochemistry of Polymers*, Elsevier, London–New York 1985.
[5.19] J. L. Gerlock, D. R. Bauer, D. F. Mielewski, *11th Int. Conf. on Advances in the Stabilization and Controlled Degradation of Polymers*, Luzern 1989, p. 25.

References for Chapter 6

[6.1] Kirk-Othmer, 3rd. ed., **16**, pp. 762–768.
[6.2] V. Wagenseil, *Oberfläche±JOT* **12** (1988) 26–28.
[6.3] W. Antony, *Oberfläche±JOT* **8** (1988) 16–18.
[6.4] G. P. Wahl, *Das deutsche Malerblatt* **7** (1990) 41–44.

References for Chapter 7

General References

[7.1] C. L. Prasher: *Crushing and Grinding Process Handbook*, J. Wiley & Sons, New York 1987.
[7.2] G. D. Parfitt: *Dispersion of Powders in Liquids*, Elsevier, Amsterdam–London–New York 1969.
[7.3] H. Rumpf: *Mechanische Verfahrenstechnik*, C. Hanser Verlag, München–Wien 1975.
[7.4] W. Jänicke, *Comp. Chem. Eng.* **3** (1987) 303–309.
[7.5] ISO 9000–9004.
[7.6] N. Stehr, J. Schwedes, *Aufbereit. Tech.* **10** (1983) 597–604.
[7.7] N. Stehr, *Keram. Z.* **42** (1990) no. 3, 162–167.
[7.8] H. Weit, J. Schwedes, *Chem.-Ing.-Tech.* **58** (1986) no. 10, 818–819.
[7.9] J. Winkler, L. Dulog, *Farbe+Lack* **90** (1984) no. 4, 244–250 (1), no. 5, 355–359 (2).
[7.10] O. Schmitz, J. Luo: *Congr. FATIPEC XVIII*, vol. I b, Venice 1986, pp. 623–655.

Specific References

[7.11] *Galasurit-Handbuch "Lacke+Farben"*, BASF Lacke+Farben AG, C. R. Vincentz Verlag, Hannover 1984.
[7.12] M. Bruhm in W. Masing (ed.): *Handbuch der Qualitätssicherung*, C. Hanser, München–Wien 1988.
[7.13] M. H. Pahl, *Chem.-Ing.-Tech.* **57** (1985) no. 5, 421–430 (part 1), *Chem.-Ing.-Tech.* **57** (1985) no. 6, 506–510 (part 2).
[7.14] J. McErlane, *Pigm. Resin Technol.* **8** (1987) 4–7.
[7.15] O. Wälty, *Farbe+Lack* **92** (1986) no. 8, 734–773.
[7.16] H.-P. Wilke et al.: *Rührtechnik*, Hüthig-Verlag, Heidelberg 1988.
[7.17] R. Pörtner, U. Werner, *Chem.-Ing.-Tech.* **61** (1989) no. 3, 250–251.
[7.18] R. Geisler, A. Mersmann, H. Voit, *Chem.-Ing.-Tech.* **60** (1988) no. 12, 947–955.

[7.19] J. Raasch, K. Sommer, *Chem.-Ing.-Tech.* **62** (1990) no. 1, 17–22.
[7.20] L. Pedrocchi, *Swiss Chem.* **11** (1989) no. 5, 23–32.
[7.21] L. Pedrocchi, F. Widmer, *Chem.-Ing.-Tech.* **61** (1989) no. 1, 82–83.
[7.22] W. Müller, *Plaste Kautsch.* **33** (1986) no. 1, 32–34.
[7.23] G. Nöltner, *Chem.-Ing.-Tech.* **60** (1988) no. 12, 986–994.
[7.24] H. Reichert, *Chem.-Ing.-Tech.* **47** (1975) no. 7, 313.
[7.25] H. Reichert, *Chem.-Ing.-Tech.* **45** (1973) no. 6, 391–395.
[7.26] W. Herbst, *Farbe + Lack* **77** (1971) no. 11, 1072–1080 (I); no. 12, 1197–1203 (II).
[7.27] B. Heinrich, L. Kreitner, *Aufbereit. Tech.* **10** (1981) 556–624.
[7.28] J. Johnson, A. Szegvari, M. Li, *Polym. Paint Colour J.* (1982) no. 7, 459–463.
[7.29] H. Mölls, R. Hörnle: *Wirkungsmechanismus der Naßzerkleinerung in der Rührwerkskugelmühle*, Dechema Monographie 69, 1972.
[7.30] K. Engels, *Farbe + Lack* **87** (1981) no. 1, 18–25.
[7.31] K. Engels, *Farbe + Lack* **94** (1988) no. 5, 350–364.
[7.32] I. Berg, *Polym. Paint Colour J. Suppl.* **23** (1983) 1–28.
[7.33] R. Stadler, R. Polke, J. Schwedes, F. Vock, *Chem.-Ing.-Tech.* **62** (1990) no. 11, 907–915.
[7.34] F. Vock, K. Warnke, K. Sollik, *Farbe + Lack* **11/12** (1987) 905–910, 995–1000.
[7.35] N. Stehr, *Chem.-Ing.-Tech.* **61** (1989) no. 5, 422–423.

References for Chapter 8

[8.1] A. Goldschmidt, B. Hantschke, E. Knappe, G.-F. Vock: *Glasurit-Handbuch*, 11th ed., Curt R. Vincentz Verlag, Hannover 1984.
[8.2] D. Ondratschek, K. Ortlieb: *Taschenbuch für Lackierbetriebe*, 47th ed., Curt R. Vincentz Verlag, Hannover 1989 (published annually).
[8.3] K. W. Thomer, D. Ondratschek: "Lackieren," in: *Handbuch der Fertigungstechnik*, vol. 4/1, Carl Hanser Verlag, München 1987.
[8.4] D. Oppen: "Die Vorbehandlung des metallischen Anstrichuntergrundes," Seminar: Lackierungen in der Metallindustrie, Technische Akademie Esslingen, Esslingen 1983.
[8.5] W. Rausch: *Die Phosphatierung von Metallen*, 2nd ed., Eugen G. Leuze Verlag, Saulgau 1988.
[8.6] U. Hoffmann: "Grundsätzliche verfahrenstechnische Gesichtspunkte bei der Kunststofflackierung," Seminar: Lackieren von Kunststoffoberflächen, Württembergischer Ingenieurverein Stuttgart, Stuttgart 1984.
[8.7] K. Wittel, *Ind. Lackierbetr.* **54** (1986) no. 3, 121–127.
[8.8] R. Hartmann: "Physikalische Vorbehandlungsmethoden von Polymeren," Seminar: 4th Deutsche Kunststoff-Finish, Bad Nauheim 1987, Praxis-Forum, Berlin 1987.
[8.9] W. Burckhardt, *VDI-Ber.* **449** (1982) 89–98.
[8.10] W. Burckhardt, *Ind. Lackierbetr.* **51** (1983) no. 5, 162–164.
[8.11] B. D. Meyer in: *Umweltfreundliche Lackiersysteme für die industrielle Lackierung*, Expert-Verlag, Kontakt und Studium, vol. 271, Ehningen 1989, pp. 146–244.
[8.12] *Fachtagung Pulverlack*, Curt R. Vincentz Verlag, Hannover 1985.

References for Chapter 9

[9.1] F. Wilborn et al.: *Physikalische und technologische Prüfverfahren für Lacke und ihre Rohstoffe*, Berliner Union, Stuttgart 1953.
[9.2] H. A. Gardner, G. S. Sward: *Physical and Chemical Examination of Paints, Varnishes, Lacquers and Colors*, Bethesda, Md., 1962.
[9.3] G. G. Sward (ed.): *Paint Testing Manual*, 13th ed., ASTM STP 500, ASTM, Philadelphia, Pa., 1972.

[9.4] H. Knittel et al.: "Untersuchung und Prüfung," *Lehrbuch der Lacke und Beschichtungen*, vol. VII, part 1, Verlag W. A. Colomb, Berlin–Oberschwandorf 1980.
[9.5] O. Lückert (ed.): *Prüftechnik bei Lackherstellung und Lackverarbeitung*, Curt R. Vincentz Verlag, Hannover 1992.
[9.6] H. Schmidbauer: *Congr. FATIPEC XX.*, Nice 1990.
[9.7] M. Klarskov, A. Saarnack, J. Jakobsen: *Congr. FATIPEC XX*, Nice 1990.
[9.8] K. Armbruster, M. Breucker, *Farbe + Lack* **95** (1989) 896–899.
[9.9] D. Baumgarten, *Farbe + Lack* **87** (1981) 375–381.
[9.10] H. Plog: *Schichtdicken-Messung, Verfahren und Geräte*, 2nd ed., Lenze Verlag, Saulgau 1968.
[9.11] U. Zorll, *Dtsch. Farben Z.* **30** (1976) 334–339.
[9.12] M. Cremer, *Progr. Org. Coat.* **9** (1981) 241–279.
[9.13] H. Schene: *Congr. FATIPEC XX*, Nice 1990.
[9.14] F. X. D. O'Donnell, F. W. Billmeyer Jr: "Psychrometric Scaling of Gloss," in I. J. Rennilson, W. N. Hale (eds.): *Review and Evaluation of Appearance*, ASTM, Philadelphia, Pa., 1986.
[9.15] R. S. Hunter, D. B. Judd, *ASTM Bulletin* **97** (1939) 11–14.
[9.16] H. K. Hammond III, J. Nimeroff, *J. Res. Nat. Bur. Stand. (U.S.)* **44** (1950) 585–598.
[9.17] U. Zorll, *Prog. Org. Coat.* **1** (1972) 113–155.
[9.18] U. Zorll, *Farbe + Lack* **79** (1973) 191–201.
[9.19] R. S. Hunter, R. W. Harold: *The Measurement of Appearance*, Wiley-Interscience, New York–Chichester–Brisbane–Toronto–Singapore 1987.
[9.20] H. K. Hammond III, *ASTM Standardization News* **1987**, 36–40.
[9.21] W. Czepluch, *Farbe + Lack* **93** (1987) 119–123.
[9.22] W. Czepluch: *Vergleiche von Tastschnitten, optischen und rechnerischen Verfahren zur Kennzeichnung von Oberflächenprofilen*, VDI-Verlag, Düsseldorf 1987.
[9.23] F. Finus: *Congr. FATIPEC XI*, Garmisch-Partenkirchen 1974.
[9.24] F. Sadowski, *Farbe + Lack* **91** (1985) 702–705.
[9.25] W. N. Hale Jr., *ASTM Standardization News* **1987** 46–49.
[9.26] W. Schultze: *Farbenlehre und Farbenmessung*, 3. ed., Springer Verlag, Berlin–Heidelberg–New York 1975.
[9.27] H. G. Völz: *Industrielle Farbprüfung*, VCH Verlagsges., Weinheim 1990.
[9.28] H. Schmelzer, *J. Coat. Technol.* **58** (1968) 53–59.
[9.29] H. Cloppenburg, D. Schmittmann, *Farbe + Lack* **95** (1989) 631–636.
[9.30] J. Sickfeld, *Industrie-Lackierbetrieb* **58** (1990) 53–60.
[9.31] H. Kraus, J. Sickfeld, *Industrie-Lackierbetrieb* **58** (1990) 12–20.
[9.32] W. König, *Farbe + Lack* **59** (1953) 435–443.
[9.33] W. Weiler, H. Fischer, *Materialprüfung* **29** (1987) 308–311.
[9.34] A. Zosel, *Farbe + Lack* **94** (1988) 809–815.
[9.35] M. Osterbrock, O. T. de Vries, *J. Oil Colour Chem. Assoc.* **73** (1990) 398–402.
[9.36] G. W. Becker, K. Frankenfeld, *Farbe + Lack* **66** (1960) 433–448.
[9.37] U. Zorll, *Materialprüfung* **31** (1989) 381–384.
[9.38] E. Ladstädter, *Farbe + Lack* **90** (1984) 646–653.
[9.39] R. Harlfinger, *Farbe + Lack* **94** (1988) 179–183.
[9.40] E. Bagda, *Farbe + Lack* **94** (1988) 606–609; **95** (1989) 176–178.
[9.41] R. R. Wiggle, A. G. Smith, J. V. Petrocelli, *J. Paint Technol.* **40** (1968) 40–42.
[9.42] W. Göring, *Industrie-Lackierbetrieb* **57** (1989) 397–402.
[9.43] J. G. Calvert, J. N. Pitts: *Photochemistry*, J. Wiley & Sons, New York 1967.
[9.44] P. A. Mullen, N. D. Searle, *J. Appl. Polym. Sci.* **14** (1970) 765–776.
[9.45] B. Ranby, F. J. Rabek: *Photodegradation, Photooxidation and Photostabilisation of Polymers*, J. Wiley & Sons, London 1975.
[9.46] N. D. Searle in A. V. Patsis (ed.): *Advances in the Stabilization and Controlled Degradation of Polymers*, vol. I, Technomic Publishing Co., Lancaster 1989, pp. 62–74.

[9.47] P. Trubiroha in A. V. Patsis (ed.): *Advances in the Stabilization and Controlled Degradation of Polymers*, vol. I, Technomic Publishing Co., Lancaster 1989, pp. 236–241.
[9.48] G. Kämpf, K. Sommer, E. Zirngiebl, *Farbe + Lack* **95** (1989) 883–886.
[9.49] W. Papenroth, *Farbe + Lack* **76** (1970) 467–471.
[9.50] E. Preininger, *Farbe + Lack* **87** (1981) 100–104.
[9.51] J. A. Simms, *J. Coat. Technol.* **59** (1987) 45–53.
[9.52] Association of Automobile Industries, *J. Coat. Technol.* **58** (1986) 57–65.
[9.53] D. R. Bauer, M. C. Paputa Peck, R. O. Carter III, *J. Coat. Technol.* **59** (1987) 103–109.
[9.54] P. Trubiroha, *Industrie-Lackierbetrieb* **58** (1990) 180–184.

References for Chapter 10

[10.1] J. Boxall, *Pigm. Resin. Technol.* **14** (1985) no. 9, 4–10.
[10.2] J. B. Lear, *J. Coat. Technol.* **53**, (1981) 51–57.
[10.3] Oil and Colour Chemist's Association, Australia: *Surface Coatings*, vol. 2, Paints and their Applications, Tafe Educational Books, Randwick, Australia, 1984, pp. 759–790.
[10.4] A. Goldschmidt, B. Hantschke, E. Knappe, G.-F. Vock: *Glasurit-Handbuch*, Curt R. Vincentz Verlag, Hannover 1984, p. 323.
[10.5] S. Hvilsted, *Prog. Org. Coat.* **13** (1985) 253–271.
[10.6] Standard Recommended Practice for Qualitative Identification of Polymers in Emulsion Paints, ASTM D 3168-73 (reapproved 1979).
[10.7] H. Wagner, H. F. Sarx: *Lackkunstharze*, Carl Hanser Verlag, München 1971.
[10.8] G. Ellis, M. Clayborn, S. E. Richards, *Spectrochimica Acta* **46 A** (1990) 227–241.
[10.9] D. O. Hummel, F. Scholl: *Atlas of Polymer and Plastics Analysis*, vol. 2, Carl Hanser Verlag, München 1984.
[10.10] Chicago Society for Coatings Technology: *An Infrared Spectroscopy Atlas for the Coatings Industry*, Federation of Societies of Coatings Technology, Philadelphia, Pa., 1980.
[10.11] R. G. Messerschmidt, M. A. Harthcock: *Infrared Microspectroscopy*, Marcel Dekker, New York 1988.
[10.12] P. R. Griffiths, J. A. de Haseth: *Fourier Transform Infrared Spectrometry*, J. Wiley & Sons, New York 1986.
[10.13] L. J. Mathias, *Mod. Paint Coat.* **75** (1985) no. 11, 38–42, 44.
[10.14] L. J. Mathias, *Proc. Water-Borne Higher-Solids Coat. Symp.* **12th** (1985) 25–43.
[10.15] I. L. Davis, *J. Oil Colour Chem. Assoc.* **68** (1985) no. 5, 109–115.
[10.16] D. M. Doddrell, D. T. Pegg, M. B. Bendall, *J. Magn. Reson.* **48** (1982) 323–327.
[10.17] S. Hvilsted in O. Kramer (ed.): *Biological Synthetic Polymer Networks*, Elsevier, London 1988, pp. 243–254.
[10.18] G. L. Marshall, J. A. Lander, *Eur. Polym. J.* **21** (1985) 949–958.
[10.19] K. J. Ivin, S. Pitchumani, C. R. Reddy, S. Rajadurai, *Eur. Polym. J.* **17** (1981) 341–346.
[10.20] K. H. Matthews, A. McLennaghan, R. A. Pethrick, *Br. Polym. J.* **19** (1987) 165–179.
[10.21] J. T. K. Woo, J. E. Evans, *J. Coat. Technol.* **49** (1977) 42–50.
[10.22] D. Wendisch, H. Reiff, D. Dietrich, *Angew. Makromol. Chem.* **141** (1986) 173–183.
[10.23] C. F. Poranski, Jr., W. B. Moniz, *J. Coat. Technol.* **49** (1977) 57–61.
[10.24] R. D. Smith, B. W. Wright, C. R. Yonker, *Anal. Chem.* **60** (1988) 1323 A–1336 A.
[10.25] C. Y. Kuo, T. Provder, *ACS Symp. Ser.* **352** (1987) 2–28.
[10.26] J. W. Hellgeth, L. T. Taylor, *J. Chromatogr. Sci* **24** (1986) 519–527.
[10.27] U. Schernau, *Eur. Coat. J.* **4** (1989) no. 6, 510–520.
[10.28] C. Schenk, A. H. M. Kayen: *Congr. FATIPEC XV*, vol. III, Amsterdam 1980, pp. 345–360.
[10.29] J. H. Charlesworth, *Anal. Chem.* **50** (1978) 1414–1420.

[10.30] U. Schernau, M. Freitag: *Congr. FATIPEC XIX*, vol. I, Aachen 1988, pp. 419–430.
[10.31] C. P. A. Kappelmeier: *Chemical Analysis of Resin-Based Coating Materials*, Interscience Publishers Inc., New York 1959.
[10.32] IUPAC, Recommended Methods for the Analysis of Alkyd Resins, *Pure Appl. Chem.* **33** (1973) 413–435.
[10.33] J. K. Haken, *Prog. Org. Coat.* **14** (1986) 247–295.
[10.34] W. Butte, *J. Chromatogr.* **261** (1983) 142–145.
[10.35] R. W. Scott, *J. Paint Technol.* **41** (1969) 422–430.
[10.36] P. Kamarchik, Jr., *J. Coat. Technol.* **52** (1980) 79–82.
[10.37] W. Kugler, Landeskriminalamt Baden-Württemberg, Stuttgart, personal communication.
[10.38] R. Jenkins, M. Holomany, *Powder Diffr.* **2** (1987) 215–219.
[10.39] E. Baier, *Farbe + Lack* **96** (1990) 173–179.
[10.40] M. Saltzmann, R. Kumar, F. W. Billmeyer, *Color. Res. Appl.* **7** (1982) 327–337.
[10.41] A. Whitaker, *J. Soc. Dyers Colour.* **102** (1986) 66–76.
[10.42] R. Goebel, W. Stoecklein, *Scanning Microsc.* **1** (1987) 1007–1015.
[10.43] K. J. Hyver, P. Sandra: *High Resolution Gas Chromatography*, 3rd ed., Hewlett-Packard Company, Avondale 1989.
[10.44] R. Moulder et al., *HRC CC, J. High Resolut. Chromatogr. Chromatogr. Commun.* **12** (1989) 688–691.
[10.45] S. B. Hawthorne, *Anal. Chem.* **62** (1990) 633A–642A.
[10.46] J. B. Lear, *J. Coat. Technol.* **53** (1981) 63–65.
[10.47] W. Stoecklein, M. Gloger, *Nicolet FT-IR*, *Spectral Lines* **9** (1988) no. 1, 2–6.
[10.48] R. Blaschke, N. Dingerdissen, R. Göcke, K. Franke: *Congr. FATIPEC XVII*, vol. II, Lugano 1984, pp. 363–387.
[10.49] A. Benninghoven, *J. Vac. Sci. Technol.* **A3** (1985) 451–460.
[10.50] N. Basener, *Farbe + Lack* **94** (1988) 821–825.
[10.51] J. A. Gardella, Jr., J. J. Pireaux, *Anal. Chem.* **62** (1990) 645A–661A.

References for Chapter 11

[11.1] *Glasurit Handbuch*, 11th ed., Curt R. Vincentz Verlag, Hannover 1984.
[11.2] K. A. van Oeteren: *Korrosionsschutz durch Beschichtungsstoffe*, vols. 1 and 2, Carl Hanser Verlag, München 1980.
[11.3] W. Göldner, "Wasserverdünnbare Lacke für die Autoserienlackierung," *Ind. Lackierbetr.* **58** (1990) no. 6, 205–207.
[11.4] K. Wittel: "Vorbehandlungsverfahren beim Einsatz verschiedener Metalle," *Ind. Lackierbetr.* **57** (1989) no. 3, 97 ff.
[11.5] R. D. Athey: "Electrocoating Review," *Met. Finish.* **87** (1989) no. 4, 23–27.
[11.6] G. Lovell: "Electrocoat Basics," *Prod. Finish.* **54** (1990) no. 7, 58 ff.
[11.7] J. Goebbels: "Bedingungen für die Anwendung von Wasserbasislack," *Oberfläche ± JOT* **29** (1989) no. 10, 48 ff.
[11.8] R. W. Blackford: "Corrosion Protective Coatings in Aeorospace," *Schriftenreihe Praxisforum*, vol. 1, Berlin, 18–21 March 1991, pp. 242–249.
[11.9] L. W. Jedermann: "Chromatfreie Primer für die Luftfahrt–Luftfahrtspezifischen Anforderungen an das Korrosionsschutzvermögen und an die Mediumbeständigkeit," *Schriftenreihe Praxisforum*, vol. 1, Berlin, 18–21 March 1991, pp. 251 ff.
[11.10] J. Hirst, C. R. Hegedus: "Synergistic Soiling, Cleaning and Weathering Effects on Aircraft Polyurethane Topcoats," *Met. Finish.* **87** (1989) no. 1.
[11.11] D. M. James: "Marine Paints," in R. R. Myers, J. S. Long (eds.): *Treatise on Coatings*, vol. 4, part 1, Marcel Dekker, New York 1975, pp. 415–475.

[11.12] *Marine Fouling and Its Prevention*, Woods Hole Oceanographic Institute & United States Naval Academy, Annapolis, Md., 1952.
[11.13] R. Lambourne: "The Painting of Ships," chapter 13 in R. Lambourne (ed.): *Paint and Surface Coatings*, Ellis Horwood, Chichester 1987, pp. 528–546.
[11.14] A. V. Robinson: "Tank Coatings," *Proc. MariChem.* **80** (1980) 169.
[11.15] D. Banks: "The Effects of Low Molecular Weight Cargoes on Tank Coating Systems," *Proc. MariChem.* **85** (1985) 169–172.
[11.16] C. E. M. van der Kolk: "OCL Epoxy Tanklining," *Proc. MariChem.* **89** (1989) session 8, paper 8.6.
[11.17] O. B. Sorensen: "The Factors Influencing Ventilation and Drying of Epoxy Coated Cargo Tanks," *Proc. MariChem.* **89** (1989) session 8, paper 8.7.
[11.18] R. L. Townsin, D. Byrne, T. Svensen, A. Milne: "Speed, Power and Roughness; The Economics of Outer Bottom Maintenance," *Trans RINA* **122** (1980) 459–474.
[11.19] J. D. Castlow, R. C. Tipper: *Marine Biodeterioration, and Interdisciplinary Study*, Naval Institute Press, Annapolis 1984.
[11.20] J. A. Montemarano, J. Dyckman: "Performance of Organo-Metallic Polymers," *J. Paint Technol.* **47** (1975) 59.
[11.21] A. Deeks, D. Hudson, D. James, B. W. Sparrow: "Tributyltin Methacrylate Copolymers in Antifouling Paints," *2nd International Congress of Marine Corrosion and Fouling*, Athens 1968.
[11.22] D. James, GB 1 124 297, 1968.
[11.23] A. Milne, G. Hails, GB 1 457 590, 1976.
[11.24] A. Milne: "Self-Polishing Coatings in Marine Antifouling Paints," Royal Inst. Chem. Annual Chemical Congress, Durham, April 1980.
[11.25] A. Milne, M. E. Callow, R. Pitchers: "Control of Fouling by Non-Biocidal Systems," in Evans & Hoagland (eds.): *Algal Biofouling*, Elsevier, Amsterdam 1986, pp. 145–158.
[11.26] A. Milne, GB 1 470 465, 1989.
[11.27] R. R. Brooks, EP 329 375, 1989.
[11.28] A. Milne, I. S. Millichamp, GB 2 218 708 A, 1989.
[11.29] B. Meuthen et al.: "25 Years of Wide Strip Coil Coating," *ECCA Congress Brussels*, Nov. 1985.
[11.30] S. Fabriz: "Neue Lacksysteme im systematischen Vergleich der Einsatzchancen unter heutiger Gesetzgebung," *Ind. Lackierbetr.* **57** (1989) no. 4, 177 ff.
[11.31] W. Baulmann: "Durchbruch für die Elektronenstrahlhärtung von Möbeloberflächen," *Ind. Lackierbetr.* **57** (1989) no. 6, 208 ff.
[11.32] B. Riberi: "Polyurethanlacke für die Möbelindustrie,", *Ind. Lackierbetr.* **55** (1987) no. 10, 341 ff.
[11.33] H. Kittel: *Lehrbuch der Lacke und Beschichtungen*, vol. 1, Verlag W. A. Colomb, Berlin 1977.
[11.34] A. Brasholz: *Handbuch der Anstrich- und Beschichtungstechnik*, Bauverlag, Wiesbaden 1978.
[11.35] Institut für Fenstertechnik, Report 1983.
[11.36] Bundesausschuß Farbe und Sachwertschutz, " Beschichtungen auf nichtmaßhaltigen Bauteilen aus Holz," Data sheet no. 3, 1991; "Beschichtungen auf Fenstern und Außentüren," Data sheet no. 18, Frankfurt/Main 1989.
[11.37] H. Kittel: *Lehrbuch der Lacke und Beschichtungen*, vol. 5, Verlag W. A. Colomb, Berlin 1977.
[11.38] Bundesausschuß Farbe und Sachwertschutz, *"Beschichtungen auf Zink und verzinktem Stahl,"* Data sheet no. 5, Frankfurt/Main 1982.
[11.39] Bundesausschuß Farbe und Sachwertschutz, "Beschichtungen auf Kunststoff im Hochbau," Data sheet no. 22, Frankfurt/Main 1984.

References for Chapter 12

[12.1] G. Thöresz, *Oberfläche + JOT* **33** (1993) no. 3, 16–22.
[12.1a] Oberfläche + *JOT*, 1997, **9**, 16–28.
[12.2] "Sanitär-Armaturen farbig pulverbeschichten," *Ind. Lackierbetr.* **57** (1989) 455–458.
[12.3] G. Senn, *Oberfläche ± JOT* **29** (1989) no. 9, 38–42.
[12.4] N. Thißen, *Oberfläche + JOT* **32** (1992) no. 10, 44–49.
[12.5] F.-J. Schanze: "Betriebserfahrungen mit dem Thermoreaktor," *VDI Ber.* **730** (1989) 61–77.
[12.6] W. Koch, M. Angrick: "Neue Entwicklungen im Bereich der Biofiltertechnik," *VDI Ber.* **735**, 349–355.
A. Windsperger, R. Bucher, K. Stefan, *Staub Reinh. d. Luft* **50** (1990) 465–470.
[12.7] I. Hanhoff-Stemping: *Die neue TA-Luft – Aktuelle immisionsschutzrechtliche Anforderungen an den Anlagenbetreiber*, vol. 1, part 5, WEKA-Verlag, 1992.
[12.8] D. Schneider, *Oberfläche ± JOT* **29** (1989) no. 9, 61–64.
[12.9] T. R. May, *Farbe + Lack* **95** (1989) 816–822.
[12.10] H. Sutter, *Oberfläche + JOT* **32** (1992) no. 10, 32–34 and **33** (1993) no. 3, 26–32.
[12.11] A. Askergren et al., *Arb. Hälsa* **4** (1988) 1–64.
[12.12] "Some Organic Solvents, Resin Monomers and Related Compounds, Pigments and Occupational Exposures in Paint Manufacture and Painting," *IARC Monogr. Carcinog. Risks Chem. Humans* **47** (1989) 385–425.
[12.13] K. Ringen: "Health Hazards among Painters," in A. Englund, K. Ringen, M. A. Mehlman (eds.): *Occupational Health Hazard of Solvents, Advances in Modern Environmental Toxicology*, vol. II, Princeton Sci. Publ., Princeton 1982, pp. 111–131.

References for Chapter 13

[13.1] The Surface Coating Industry world wide 1997–2002, Materials Technology Publications, Watford, UK.
[13.2] *Paint & Ink International* **2** (1989), no. 3, 14–46.

References for Chapter 14

General References

T. H. Durrans, O. Merz: *Lösungsmittel und Weichmachungsmittel*, W. Knapp Verlag, Halle 1933.
T. H. Durrans: *Solvents*, Chapman and Hall, London 1971.
H. Gnamm, W. Sommer: *Die Lösungs- und Weichmachungsmittel*, Wissenschaftliche Verlagsgesell., Stuttgart 1958.
O. Fuchs: *Gnamm-Fuchs, Lösungsmittel und Weichmachungsmittel*, 8th ed., vol. 1, 2, Wissenschaftliche Verlagsgesell., Stuttgart 1980.
L. Scheflan, M. B. Jacobs: *The Handbook of Solvents*, van Nostrand, Toronto 1953.
I. Mellan: *Handbook of Solvents*, Reinhold Publ. Co., New York 1957.
I. Mellan: *Source Book of Industrial Solvents*, Reinhold Publ. Co., vol. 2, 3, New York 1957, 1959.
I. Mellan: *Industrial Solvents Handbook*, 2nd ed., Noyes Data Corp., Park Ridge, N.Y. 1977.
E. W. Flick: *Industrial Solvents Handbook*, 3rd ed., Noyes Data Corp., Park Ridge/USA 1985.
E. W. Flick: *Industrial Solvents Handbook*, 4th ed., Noyes Data Corp., Park Ridge/USA 1991.
O. Jordan: *Chemische Technologie der Lösemittel*, Springer Verlag, Berlin 1932.
A. Krause, *Handbuch der Nitrocelluloselacke*, Part 1, "Lösungsmittel", Wilhelm Pansegrau Verlag, Berlin 1955.
H. Guinot: *Solvants et Plastifiants*, Dunod, Paris 1948.

15. References

I.G. Farbenindustrie AG: *Lösungs- und Weichmachungsmittel*, Frankfurt 1937.

E. Schwarz: *Wie setzen wir die Verluste an flüchtigen Lösungsmitteln herab?*, Allgemeiner Industrieverlag Knorre & Co., Berlin 1943.

F. Fritz: *Die wichtigsten Lösungs- und Weichmachungsmittel*, VEB Verlag Technik, Berlin 1957.

J. A. Riddick, W. B. Bunger: *Techniques of Chemistry*, vol. 2, Organic Solvents, Wiley-Interscience, London 1970.

M. R. J. Dack: *Techniques of Chemistry*, vol. 8, "Solutions and Solubilities", J. Wiley & Sons, London 1976.

E. Krell: *Einführung in die Trennverfahren*, VEB Deutscher Verlag für Grundstoffindustrie, Leipzig 1975.

K. Thinius: *Anleitung zur Analyse der Lösungsmittel*, 2nd ed., J.A. Barth-Verlag, Leipzig 1957.

H. Haase: *Statische Elektrizität als Gefahr*, Verlag Chemie, Weinheim 1968.

R. W. Tess: *Solvents Theory and Practice*, American Chemical Society, Washington 1973.

H. Kittel: *Lehrbuch der Lacke und Beschichtungen*, vol. 3, Verlag W.A. Colomb, Berlin-Oberschwandorf 1976, pp. 17–129.

A. F. M. Barton: *Handbook of Solubility Parameters and Other Cohesion Parameters*, CRC Press, Boca Raton, Fla. 1983.

J. Roire: *Les Solvants*, Erec, Puteaux 1989.

R. Lambourne: *Paint and Surface Coatings, Theory and Practice*, Ellis Horwood, Chichester 1987, pp. 195–211.

M. Ash, I. Ash: *Dispersants, Solvents and Solubilizers*, vol. 2, Edward Arnold, London 1988.

K. Dören, W. Freitag, D. Stoye: *Wasserlacke—Umweltschonende Alternative für Beschichtungen*, Verlag TÜV Rheinland, Köln 1992, pp. 54–56, 107–117.

Specific References

[14.1] J. Laakmann, G. Hasselkuß, *Ind. Lackierbetr.* **58** (1990) no. 5, 185–186.
[14.2] C. Reichardt: *Solvents and Solvent Effects in Organic Chemistry*, 2nd ed., VCH Verlagsgesellschaft, Weinheim 1988.
[14.3] O. Fuchs, *Farbe Lack* **71** (1965) no. 2, 104–112.
[14.4] O. Fuchs, *CLB Chem. Labor Betr.* **36** (1985) no. 8, 378–380.
[14.5] I. Vavruch, *CLB Chem. Labor Betr.* **35** (1984) no. 11, 536–543.
[14.6] W. H. Keesom, *Phys. Z.* **22** (1921) 129, 643; **23** (1922) 225.
[14.7] P. Debye, *Phys. Z.* **21** (1920) 178; **22** (1921) 302.
[14.8] F. London, *Z. Phys. Chem. Abt. B* **11** (1931) 222; *Z. Phys.* **63** (1930) 245.
[14.9] G. C. Pimentel, A. L. McClellan: *The Hydrogen Bond*, W. H. Freeman, London 1960.
[14.10] W. A. P. Luck, *Naturwissenschaften* **54** (1967) 601.
[14.11] A. A. Orr, *J. Paint Technol.* **47** (1975) no. 607, 45–49.
[14.12] P. L. Huyskens, M. C. Haulait-Prison, L. D. Brandts Buys, X. M. van der Borght, *J. Coat. Technol.* **57** (n 724) (1985) 57–67.
[14.13] P. L. Huyskens, M. C. Haulait-Pirson, X. van der Borght, *Farbe Lack* **92** (1986) no. 12, 1145–1149.
[14.14] B. Ziolkowsky, *Seifen Öle Fette Wachse* **96** (1970) no. 22, 801–806.
[14.15] M. L. Huggins, *J. Paint Technol.* **41** (1969) 509.
[14.16] J. H. Hildebrand, R. L. Scott: *The Solubility of Non-Electrolytes*, 3rd ed., Reinhold Publ. Co., New York 1950.
[14.17] H. Hildebrand, *J. Am. Chem. Soc.* **51** (1929) 66; *Nature London* **168** (1951) 868.
[14.18] E. A. Guggenheim, *Proc. R. Soc. London, Ser. A* **183** (1944) 203, 213.
[14.19] H. Mauser, G. Kortüm, *Z. Naturforsch. A* **10**A (1955) 317.
[14.20] K. L. Hoy, *J. Paint. Technol.* **42** (1970) no. 541, 76–118.
[14.21] G. Scatchard, *Chem. Rev.* **8** (1931) 321.
[14.22] J. H. Hildebrand, *J. Am. Chem. Soc.* **57** (1935) 866.

[14.23] A. E. Rheineck, K. F. Lin, *J. Paint Technol.* **40** (1968) no. 527, 613–616.
[14.24] H. Ahmad, M. Yaseen, *J. Oil Colour Chem. Assoc.* **60** (1977) 99–103.
[14.25] H. Ahmad, *J. Oil Colour Chem. Assoc.* **63** (1980) 2.
[14.26] C. M. Hansen, *J. Paint Technol.* **39** (1967) 104.
[14.27] Shell Chemical: Technical Bulletin, "Solvent System Design," Rotterdam 1978, "Solubility Parameters," Rotterdam 1978, "Solvent Power," Rotterdam 1979.
[14.28] E. B. Bagley, T. P. Nelson, J. M. Scigliano, *J. Paint Technol.* **43** (1971) no. 555, 35–42.
[14.29] A. Jayasri, M. Yaseen, *J. Coat. Technol.* **52** (n 667) (1980) 41–45; *J. Oil Colour Chem. Assoc.* **63** (1980) 61–69.
[14.30] C. M. Hansen, *Ind. Eng. Chem. Prod. Res. Dev.* **8** (1969) 2.
[14.31] C. M. Hansen, *J. Paint Technol.* **42** (1970) 660.
[14.32] J. H. Meyer zu Bexten, *Farbe Lack* **78** (1972) no. 9, 813–822.
[14.33] B. Marti, *Schweiz. Arch. Angew. Wiss. Tech.* **33** (1967) no. 10, 297–304.
[14.34] R. C. Nelson, V. F. Figurelli, J. G. Walsham, G. D. Edwards, *J. Paint Technol.* **42** (1970) no. 550, 644–652.
[14.35] J. P. Teas, *J. Paint Technol.* **41** (1969) no. 516, 19–25.
[14.36] ASTM D 3132–72, Standard Test Method for Solubility Range of Resins and Polymers, reapproved 1976.
[14.37] D. W. van Krevelen: *Properties of Polymers, Cohesive Properties and Solubility*, Elsevier, Amsterdam 1976, pp. 129–159.
[14.38] C. M. Kok, A. Rudin, *J. Coat. Technol.* **55** (n 704) (1983) 57–62.
[14.39] M. V. Ram Mohan Rao, M. Yaseen, *Farbe Lack* **91** (1985) no. 9, 810–812.
[14.40] G. Narender, M. Yaseen, *J. Coat. Technol.* **61** (n 773) (1989) 41–51.
[14.41] J. K. Rigler: "Maßgeschneiderte Lösemittel für Polymere," Chemietag des Vereins österreichischer Chemiker, Vortrag, Graz 1977.
[14.42] O. Lückert, *Farbe Lack* **75** (1969) no. 2, 128–138.
[14.43] E. F. Meyer, R. E. Wagner, *J. Phys. Chem.* **70** (1966) 3162.
[14.44] J. L. Gardon, *J. Paint Technol.* **38** (1966) no. 492, 43–57.
[14.45] H. A. Papazian, *J. Am. Chem. Soc.* **93** (1971) 5634.
[14.46] C. F. Holmes, *J. Am. Chem. Soc.* **95** (1973) 1014.
[14.47] R. C. Nelson, R. W. Hemwall, G. D. Edwards, *J. Paint Technol.* **42** (1970) no. 550, 636–643.
[14.48] H. Burrell, *Interchem. Rev.* **14** (1955) 31.
[14.49] H. Burrell, *Off. Dig. Fed. Paint Varn. Prod. Clubs* **27** (1955) 726.
[14.50] H. Burrell, *J. Paint Technol.* **40** (1968) no. 520, 197–208.
[14.51] E. P. Liebermann, *Off. Dig. Fed. Soc. Paint Technol.* **34** (1962) 30.
[14.52] M. Dyck, P. Hoyer, *Farbe Lack* **70** (1964) no. 7, 522–532.
[14.53] E. B. Bagley, S. A. Chen, *J. Paint Technol.* **41** (1969) 494.
[14.54] T. B. Epley, R. S. Drago, *J. Paint Technol.* **41** (1969) 500.
[14.55] C. M. Hansen, K. Skaarup, *J. Paint Technol.* **39** (1967) 505.
[14.56] *Kirk-Othmer*, 2nd ed., Suppl. vol., 889.
[14.57] J. D. Crowley, G. S. Teague, J. W. Lowe, *J. Paint Technol.* **38** (1966) no. 496, 269–280; **39** (1967) 19; *Farbe Lack* **73** (1967) no. 2, 120–131.
[14.58] I. Brandrup, E. H. Immergut: *Polymer Handbook*, 2nd ed., Interscience, New York 1974.
[14.59] P. A. Small, *J. Appl. Chem.* **3** (1953) 71.
[14.60] O. Fuchs, *Dtsch. Farben Z.* **23** (1969) 114.
[14.61] O. Fuchs, *Kunststoffe* **43** (1953) 409.
[14.62] U. Biethan, K. H. Hornung: *X. FATIPEC Congress Book*, Montreux 1970, pp. 277–281.
[14.63] E. Bagda, *Colloid Polym. Sci.* **255** (1977) 384–392.
[14.64] I. Prigogine, A. Bellemans, V. Mathot: *The Molecular Theory of Solutions*, North Holland Publ., Amsterdam 1957.

[14.65] B. A. Wolf, J. W. Breitenbach, J. K. Rigler, *Angew. Makromol. Chem.* **34** (1973) 177–182.
[14.66] L. H. Horsley: *Azeotropic Data, Advances in Chemistry Series*, Am. Chem. Soc., Washington, D.C. 1952.
[14.67] L. H. Horsley: *Azeotropic Data, II, Advances in Chemistry Series*, Am. Chem. Soc., Washington, D.C. 1962; III, Washington, D.C. 1973.
[14.68] C. H. Fisher, *J. Coat. Technol.* **63** (n 799) (1991) 79–83.
[14.69] T. E. Gilbert, *J. Paint Technol.* **43** (1971) no. 562, 93–97.
[14.70] G. M. Sletmoe, *J. Paint. Technol.* **42** (1970) no. 543, 246–249.
[14.71] G. Marwedel, O. Hauser, *Farbe Lack* **54** (1948) no. 6, 115–118; no. 8, 175–181.
[14.72] G. Marwedel, *Farbe Lack* **56** (1950) no. 11, 485–493; **59** (1953) 276–282, 317–320; **63** (1957) no. 10, 481–487, no. 11, 542–548.
[14.73] Shell Chemie: Firmenschriften, "Voraussage der Verdunstungseigenschaften von Lösungsmittelgemischen," Frankfurt 1970, "Verdunstung von organischen Lösemitteln aus Oberflächenbeschichtungen," Frankfurt 1979.
[14.74] A. L. Rocklin, *J. Coat. Technol.* **48** (n 622) (1976) 45–47.
[14.75] H. L. Jackson, *J. Coat. Technol.* **58** (n 741) (1986) 87–89.
[14.75a] Z. W. Wicks, F. N. Jones, S. P. Pappas, *Organic Coatings*, 224.
[14.76] A. A. Sarnotsky, *J. Paint. Technol.* **41** (1969) no. 539, 692–701.
[14.77] J. Ramsbotham, *J. Oil Colour Chem. Assoc.* **62** (1979) 359–366.
[14.78] G. Rosen, L. M. Andersson, *J. Oil Colour Chem. Assoc.* **73** (1990) no. 5, 202–211.
[14.79] J. F. Rankl, M. Huber, G. Grienberger, G. Bonn, *J. Oil Colour Chem. Assoc.* **73** (1990) no. 8, 334–336.
[14.80] C. F. Böttcher: *The Theory of Electric Polarisation*, Elsevier, Amsterdam 1952.
[14.81] L.-H. Lee, *J. Paint Technol.* **42** (1970) no. 545, 365–370.
[14.82] D. C. Busby et al., *J. Oil Colour Chem. Assoc.* **74** (1922) no. 10, 362–368.
[14.83] W. Ormandy, E. Craven, *J. Inst. Pet. Technol.* **8** (1922) 145.
[14.84] A. Blom, *Farben Ztg.* **36** (1931) 873.
[14.85] J. Rybicky, J. R. Stevens, *J. Coat. Technol.* **53** (n 676) (1981) 40–42.
[14.86] H. A. Wray, *J. Coat. Technol.* **56** (n 717) (1984) 37–43.
[14.87] K. G. Probst, *J. Paint. Technol.* **41** (1969) no. 539, 670–679.
[14.88] W. H. Ellis, *J. Coat. Technol.* **48** (n 614) (1976) 45–57.
[14.89] H. Magdanz, H. Wacha, *Plaste Kautsch.* **25** (1978) no. 12, 707–712.
[14.90] J. L. McGovern, *J. Coat. Technol.* **64** (n 810) (1992) 33–38, 39–44.
[14.91] R. Smith, *J. Paint Technol.* **44** (1972) no. 565, 38–42.
[14.92] Kansas City Society for Paint Technol., *J. Paint Technol.* **44** (1972) 565.
[14.93] E. Oehley, *Chem. Ing. Tech.* **26** (1954) 97.
[14.94] Le Chatelier, *Compt. Rend.* **126** (1898) 1344.
[14.95] G. Hommel: *Handbuch der gefährlichen Güter*, Springer Verlag, Berlin 1992.
[14.96] Kühn-Birett, Merkblätter Gefährliche Arbeitsstoffe, Ecomed, Landsberg 1992.
[14.97] Sorbe, Sicherheitstechnische Kenndaten chemischer Stoffe, Ecomed, Landsberg 1992.
[14.98] K. Nabert, G. Schön: *Sicherheitstechnische Kennzahlen brennbarer Gase und Dämpfe*, 2nd ed., Deutscher Eichverlag, Braunschweig 1970.
[14.99] P. H. Noll, *Ind. Lackierbetr.* **40** (1972) 354.
[14.100] Anonymus, *Ind. Lackierbetr.* **44** (1976) 320.
[14.101] J. J. Engel, T. J. Byerley, *J. Coat. Technol.* **57** (n 723) (1985) 29–33.
[14.102] S. D. Seneker, T. A. Potter, *J. Coat. Technol.* **63** (n 793) (1991) 19–23.
[14.103] S. F. Thames, P. C. Boyer, *J. Coat. Technol.* **62** (n 784) (1990) 51–63.
[14.103a] H. F. Bender, Sicherer Umgang mit Gefahrstoffen, VCH Weinheim, 1995.
[14.104] H. Greim, W. Dessau: *Kombinationswirkung organischer Lösungsmittel. Toxische Wirkung auf Leber und Nervensystem*, Schriftenreihe der Bundesanstalt für Arbeitsschutz, Dortmund 1985.

[14.105] L. Browning: *Toxicity and Metabolism of Industrial Solvents,* Elsevier, New York 1965; *Registry of Toxic Effects of Chemical Substances,* National Institute for Occupational Safety and Health, Rockville/USA 1976.
[14.106] D. Henschler: *Gesundheitsschädliche Arbeitsstoffe, Toxikologisch-arbeitsmedizinische Begründung von MAK-Werten,* Deutsche Forschungsgemeinschaft, VCH Verlagsgesellschaft, Weinheim 1991.
[14.107] H. Altenkirch, *Adhäsion* (1984) no. 10, 9–11.
[14.108] K. Kampmann, R. D. Henkler: "Arbeitsmedizinische Untersuchungen von lösemittelexponierten Chemiearbeitern in einer Lackfabrik," *Lack im Gespräch,* vol. 9, Informationsdienst Deutsches Lackinstitut, Frankfurt/M 1986.
[14.109] P. Grasso, M. Sharrat, D. M. Davies, D. Irvine, *Fd. Chem. Toxicol.* **22** (1984) no. 10, 819–852.
[14.110] R. J. Lewis: *Sax's Dangerous Properties of Industrial Materials,* 8th ed., vol. 1–3, Van Nostrand Reinhold, New York 1992.
[14.111] Maximale Arbeitsplatzkonzentrationen und Biologische Arbeitsstofftoleranzwerte, Senatskommission zur Prüfung gesundheitsschädlicher Arbeitsstoffe, Deutsche Forschungsgemeinschaft, Bonn 1992.
[14.112] K.-H. Cohr, J. Stokholm, *Dan. Kemi* **58** (1977) no. 10, 316–322.
[14.113] VDI-Guideline VDI 3881 Olfactometry, Odour Threshold Determination, Part 1: "Fundamentals," Part 2: "Sampling," Part 3: "Olfactometers with Gas Jet Dilution," Part 4: "Instructions for application and performance characteristics."
[14.114] A. H. Walter, *Farbe Lack* **77** (1971) no. 10, 1009–1011; **80** (1974) 23.
[14.115] A. H. Walter, *Adhäsion* (1975) 178.
[14.116] A. H. Walter, *Dtsch. Farben Z.* **22** (1968) 235.
[14.117] F. Mutter, *Eurocoat* (1990) no. 11, 690–695.
[14.118] H. Haagen, *Farbe Lack* **91** (1985) no. 9, 795–801.
[14.119] W. Lohrer, G. Pahlke, *Ind. Lackierbetr.* **53** (1985) no. 3, 75–78.
[14.120] Globus, *Chem. Rundsch.* **43** (1990) no. 16, 2.
[14.121] G. W. Fiero, *J. Paint Technol.* **40** (1968) no. 520, 222–228.
[14.122] H. H. G. Jellinek, F. Flajsman, F. J. Kryman, *J. Appl. Polym. Sci.* **13** (1969) 107–116.
[14.123] National Paint, Varnish and Lacquers Assoc., Inc., Washington D.C.: "Paint Industry Smog Chamber," *Final Technical Report on the Role of Solvents in Photochemical Smog Formation,* Washington, D.C. 1970.
[14.124] K.-H. Weinert, *Ind. Lackierbetr.* **47** (1979) no. 6, 216.
[14.125] *Patty,* 4th ed., **1**, Part B, 195–258.
[14.126] W. S. Zimmt: Der Einfluß von Lösungsmittel-Emissionsregelungen auf die amerikanische Farbenindustrie, Vortragsmanuskript, Ivry sur Seine 1970.
[14.127] Eastman Chemical Co.: Formulator's Notes No. M-6.1, Kingsport, TN 1977.
[14.128] R. Kreisler, W. Zöllner: Merkblatt zur TA Luft, 6th ed., Verband der Lackindustrie, Frankfurt/M 1990.
[14.129] W. Lohrer, G. Pahlke, *Ind. Lackierbetr.* **49** (1981) no. 4, 135–140.
[14.130] R. Schaaf, *Ind. Lackierbetr.* **53** (1985) no. 4, 111–112.
[14.131] L. M. Keinert, *Sicherheitsingenieur* (1975) no. 10, 1–8.
[14.132] K. R. Smith, *J. Coat. Technol.* **62** (n 788) (1990) 71–81.
[14.133] H. Aichinger: Die österreichische Lösungsmittelverordnung, Bundesministerium für Umwelt, Jugend und Familie, Wien 1991. 492. Verordnung des Bundesministers für die Umwelt, Jugend und Familie über Verbote und Beschränkungen von organischen Lösungsmitteln (Lösungsmittelverordnung), Bundesgesetzblatt für die Republik Österreich, 1991, p. 179.
[14.134] Proposal for a Council Directive on The Limitation of the Emission of Organic Solvents, Council of the European Union, Brussel, 1996.

[14.135] U. Zorll, *Farbe Lack* **97** (1991) no. 4, 382–383.
[14.136] U. Sayler, *Ind. Lackierbetr.* **42** (1974) no. 11, 417–420.
[14.137] H. Schene, *Z. Ind. Fertigung* **64** (1974) 282–287.
[14.138] W. Lohrer, *Oberfläche JOT* **32** (1992) no. 4, 108–119.
[14.139] K. Verschueren: *Handbook of Environmental Data on Organic Chemicals*, 2nd ed., Van Nostrand Reinhold, New York 1983.
[14.140] Katalog wassergefährdender Flüssigkeiten der Länderverordnungen über das Lagern wassergefährdender Flüssigkeiten (VLwFl).
[14.141] H.-J. Erberich, *Farbe Lack* **90** (1984) no. 2, 146.
[14.142] OECD Guidelines for Testing Chemicals, Paris 1981/1987: no. 201–209 Effects on Biotic Systems, no. 301 Ready Biodegradability, no. 302 Inherent Biodegradability, no. 303 Simultation Test, no. 304 Biodegradability in Soil, no. 305 Bioaccumulation, no. 401–482 Health Effects.
[14.143] DIN 38405, Deutsches Einheitsverfahren (DEV) zur Wasser-, Abwasser- und Schlammuntersuchung.
[14.144] E. Montorsi, *Pitture Vernici Eur.* **68** (1992) no. 1, 29–31.
[14.145] Richtlinie des Rates der Europäischen Gemeinschaft zur Angleichung der Rechts- und Verwaltungsvorschriften für die Einstufung, Verpackung und Kennzeichnung gefährlicher Stoffe, Grundrichtlinie 67/548 EWG, Amtsblatt der Europäischen Gemeinschaften no. 196, pp. 1 ff (1967).
[14.146] L. H. Keith, D. B. Walters: *Compendium of Safety Data Sheets for Research and Industrial Chemicals*, Part 1–2, VCH Publishers, New York 1985/1989.
[14.147] Gebrüder Sulzer AG, Lösungsmittel aus der Abluft, Winterthur, CH 1991.
[14.148] H. E. Koenen, *Chem. Rundsch.* **43** (1990) no. 45, 19.
[14.149] Anonymus, *Oberfläche JOT* **32** (1992) no. 2, 54.
[14.150] P. Toepke, *Farbe Lack* **97** (1991) no. 8, 700–701.
[14.151] Buss-SMS GmbH, Butzbach, *Chem. Rundsch.* **43** (1990) no. 45, 4.
[14.152] The Weir Group Ltd., Glasgow, *Ind. Lackierbetr.* **45** (1977) no. 11, 439.
[14.153] B. Hamill, *Paint Resin* **39** (1985) no. 5, 34–37.
[14.154] A. Schmidt, M. Ulrich, *Chem. Ing. Tech.* **62** (1990) no. 1, 43–46.
[14.155] D. Rasenack, ILV-Mitteilungen des Fraunhofer-Instituts für Lebensmitteltechnologie und Verpackung **8–9**, (1984) 116–133.
[14.156] OFRU-Recycling, *Oberfläche JOT* **23** (1983) no. 10, 8.
[14.157] W. Schnabel, *Oberfläche JOT* **32** (1992) no. 6, 40–42.
[14.158] H. Erler, K. Hermanns, M. Karthaus, *Coating* (1992) no. 4, 120–122.
[14.159] R. W. Baker, Yoskioka, J. M. Mohr, A. J. Khan, *J. Membr. Sci.* **31** (1987) 259.
[14.160] H. Strathmann, C.-M. Bell, K. Kimmerle, *Pure Appl. Chem.* **58** (1986) no. 12, 1663–1668.
[14.161] VDI-Richtlinie 3674, Abgasreinigung durch Adsorption, Oberflächenreaktion und heterogene Katalyse. VDI-Richtlinie 3675, Abgasreinigung durch Adsorption. VDI-Richtlinie 2442, Abgasreinigung durch thermische Verbrennung. VDI-Richtlinie 3476, Abgasreinigung durch katalytische Verfahren. VDI-Richtlinie 3477/3478, Biologische Abluftreinigung. VDI-Richtlinie 2588, Emissionsminderung: Beschichtung von metallischen Oberflächen mit organischen Stoffen. VDI-Richtlinie 2280, Auswurfbegrenzung: Organische Verbindungen, insbesondere Lösemittel. VDI-Richtlinien 2460, 3481, 2457, Emissionsmeßtechnik Organische Stoffe.
[14.162] S. P. P. Ottengraf et al., *Bioprocess Eng.* **1** (1986) 61–69.
[14.163] D. A. Carlson, C. P. Leiser, *J. Water Pollut. Control Fed.* **38** (1966) 829.
[14.164] U. Bäuerle, K. Fischer, D. Bardtke, *Staub-Reinhalt. Luft* **46** (1986) 233–235; CJB Developments Ltd., *J. Oil Colour Chem. Assoc.* **75** (1992) no. 11, 439–441.
[14.165] G. Heck, G. Müller, M. Ulrich, *Chem. Ing. Tech.* **60** (1988) no. 4, 286–297.
[14.166] H. Buri et al.: *Lösemitelemission*, Verein zur Förderung der Wasser- und Lufthygiene, Zürich 1990.

[14.167] P. Murr, *Oberfläche JOT* **32** (1992) no. 3, 34–39.
[14.168] J. Halbartschlager, *Oberfläche JOT* **32** (1992) no. 3, 28–33.
[14.169] R. F. Piepho, H. Schuster, *Ind. Lackierbetr.* **58** (1990) no. 5, 172–174.
[14.170] *Patty*, 4th ed., 1, Part B, 259–287.
[14.171] E. Mink, *Oberfläche JOT* **31** (1991) no. 3, 22–26.
[14.172] J. Diebold, *Oberfläche JOT* **30** (1990) no. 11, 857–859.
[14.173] H. Deuster, *Ind. Lackierbetr.* **58** (1990) no. 9, 325–326.
[14.174] K. Panzer, *Ind. Lackierbetr.* **58** (1990) no. 9, 323–324.
[14.175] M. B. Hundert, *J. Paint Technol.* **40** (1968) no. 516, 33–48.
[14.176] D. G. Anderson, *J. Paint. Technol.* **40** (1968) no. 527, 549–557.
[14.177] G. G. Esposito, *J. Paint. Technol.* **40** (1968) no. 520, 214.
[14.178] L. Rohrschneider, *Anal. Chem.* **45** (1973) 1241–1247.
[14.179] D. Groß, H.-J. Kretzschmar, H. Weigel, *Farbe Lack* **82** (1976) no. 8, 690–692.
[14.180] K. L. Olson, C. A. Wong, L. L. Fleck, D. F. Lazar, *J. Coat. Technol.* **60** (n 762) (1988) 45–50.
[14.181] K.-H. Pannwitz: *Ein neuer Probenehmer für organische Lösemitteldämpfe*, Dräger-Publikationen, Lübeck 1990.
[14.182] DIN-Arbeitskreis "Lösemittel," *Farbe Lack* **97** (1991) no. 2, 127–129; *J. Oil Colour Chem. Assoc.* **75** (1992) no. 4, 154–156.
[14.183] L. R. Snyder, *J. Chromatogr.* **16** (1964) 55–88.
[14.184] M. Gruenfeld, *J. Paint Technol.* **42** (1970) no. 543, 237–242.
[14.185] R. Prakash, I. C. Shukla, *J. Oil Colour Chem. Assoc.* **63** (1980) 370–372.
[14.186] D. Downing, *J. Oil Colour Chem. Assoc.* **74** (1991) no. 8, 284–285.
[14.187] Shell Chemicals: Thinner Formulation for the Paint Industry, Technical Bulletin, Rotterdam 1981.
[14.188] New York Society for Paint Technol., *J. Paint Technol.* **41** (1969) 692.
[14.189] J. K. Rigler, B. A. Wolf, J. W. Breitenbach, *Angew. Makromol. Chem.* **57** (1977) no. 854, 15–27.
[14.190] W. H. Ellis, Z. Saary, D. G. Lesnini, *J. Paint Technol.* **41** (1969) no. 531, 249–258.
[14.191] G. Marwedel, *Dtsch. Farben Z.* **26** (1972), no. 7, 8, 365–418.
[14.192] E. Bagda, *Dtsch. Farben Z.* **32** (1978) no. 10, 372–378.
[14.193] K. M. A. Shareef, M. Yaseen, *J. Coat. Technol.* **55** (n 701) (1983) 43–52.
[14.194] A. Toussaint, I. Szigetvari, *J. Coat. Technol.* **59** (n 750) (1987) 49–59.
[14.195] R. A. Waggoner, F. D. Blum, *J. Coat. Technol.* **61** (n 768) (1989) 51–56.
[14.196] A. Vinther, P. Sörensen, *Farbe Lack* **77** (1971) no. 4, 317–325.
[14.197] M. Camina, D. M. Howell, *J. Oil Colour Chem. Assoc.* **55** (1972) 929.
[14.198] P. Fink-Jensen, *Farbe Lack* **68** (1962) 155.
[14.199] H. A. Dierstead, J. Turkevich, *J. Chem. Phys.* **12** (1944) 24.
[14.200] A. R. H. Tawn, *Farbe Lack* **75** (1969) no. 4, 311–318.
[14.201] H. P. Schreiber, *J. Paint Technol.* **46** (1974) no. 598, 35–39.
[14.202] G. Holtmann, *Schweizer Maschinenmarkt* **43** (1978) 52–55.
[14.203] W. Andrejewski, J. Dörffel, D. Stoye, Hüls AG, Der Lichtbogen, company brochure, no. 175, Marl 1974.
[14.204] G. Walz, Kunstharz Nachr. **33** (1974) no. 6, 30–39.
[14.205] G. Sprinkle, *J. Coat. Technol.* **53** (n 680) (1981) 67–72.
[14.206] G. P. Sprinkle, *Mod. Paint Coat.* (1983) no. 4, 44–48.
[14.207] W. H. Ellis, *J. Coat. Technol.* **55** (n 696) (1983) 63–72.
[14.208] J. R. Erickson, *J. Coat. Technol.* **48** (n 620) (1976) 58.
[14.209] A. Goldschmidt, *Farbe Lack* **84** (1978) no. 9, 675–680.
[14.210] J. J. Stratta, P. W. Dillon, R. H. Semp, *J. Coat. Technol.* **50** (n 647) (1978) 39–47.
[14.211] D. Stoye, *Coating* (1987) no. 6, 223.

[14.212] J. Spauwen, *J. Oil Colour Chem. Assoc.* **71** (1988) no. 2, 47–49.
[14.213] W. Appelt, *Farbe Lack* **96** (1990) no. 3, 200–201; *J. Oil Colour Chem. Assoc.* **73** (1990) no. 5, 200–201.
[14.214] M. R. C. Gerstenberger, D. K. Kruse, *Farbe Lack* **90** (1984) no. 7, 563–568.
[14.215] F. Müller, *Ind. Lackierbetr.* **58** (1990) no. 2, 48–52.
[14.216] A. L. Rocklin, *J. Coat. Technol.* **50** (n 646) (1978) 47–55.
[14.217] A. L. Rocklin, D. C. Bonner, *J. Coat. Technol.* **52** (n 670) (1980) 27–36.
[14.218] H. Magdanz, K. Berger, G. Schumann, *Farbe Lack* **75** (1969) no. 3, 221–235.
[14.219] A. L. Rocklin, G. D. Edwards, *J. Coat. Technol.* **48** (n 620) (1976) 68–74.
[14.220] P. M. Grant, *J. Coat. Technol.* **53** (n 677) (1981) 33–38.
[14.221] L. O. Kornum, *J. Oil Colour Chem. Assoc.* **63** (1980) 103–123.
[14.222] T. Imai, K. Tsubouchi, *J. Coat. Technol.* **52** (n 666) (1980) 71–78.
[14.223] R. F. Eaton, F. G. Willeboordse, *J. Coat. Technol.* **52** (n 660) (1980) 63–70.
[14.224] A. Jones, L. Campey, *J. Coat. Technol.* **56** (n 713) (1984) 69–72.
[14.225] W. J. Blank, *J. Coat. Technol.* **61** (n 777) (1989) 119–128.
[14.226] J. Dörffel, *Farbe Lack* **81** (1975) 10–15.
[14.227] P. Sörensen, *Dtsch. Farben Z.* **24** (1970) 473.
[14.228] H. Jebsen-Marwedel, G. Marwedel, *Farbe Lack* **66** (1960) no. 6, 314–330.
[14.229] G. Marwedel, *Farbe Lack* **66** (1960) no. 7, 379–385.
[14.230] G. Marwedel, H. Jebsen-Marwedel, *Dtsch. Farben Z.* **24** (1970) no. 3, 103–116.
[14.231] R. Haug, *Dtsch. Farben Z.* **23** (1969) no. 10, 467–472.
[14.232] J. E. Weigel, E. G. Sabino, *J. Paint Technol.* **41** (1969) no. 529, 81–88.
[14.233] W. Z. Borer, R. L. Bartosiewicz, P. J. Secrest, *J. Paint Technol.* **43** (1971) no. 559, 61–64.
[14.234] W. Götze, *IX. Fatipec Congress Book* no. 2, 79–83, Brussels 1968.
[14.235] D. Y. Perera, D. van den Eynde, *J. Coat. Technol.* **55** (n 699) (1983) 37–43.
[14.236] D. Stoye, K.-H. Haneklaus, *Fette Seifen Anstrichstoffe* **4** (1972) 217.
[14.237] L. M. Keinert, *Oberfläche* (1975/1976), no. 12/1.
[14.238] S. Knödler, *Papier (Darmstadt)* **35** (1981) no. 10 A, 74–79.
[14.239] O. Jordan: *Chemische Technologie der Lösungsmittel*, Springer Verlag, Berlin 1932, pp. 91–95.
[14.240] A. K. Doolittle: *The Technology of Solvents and Plasticizers*, Wiley-Interscience, New York 1954.
[14.241] L. Alders: *Liquid-Liquid Extraction*, Elsevier, Amsterdam 1955.
[14.242] A. Weissberger: *Technique of Organic Chemistry*, vol. 4, Interscience, New York 1965. E. S. Perry, A. Weissberger: *Distillation*, Interscience, New York 1965.
[14.243] Winnacker-Küchler, vol. **4**, **5**, Hanser Verlag, München 1972/1981.
[14.244] K. Randerath: *Dünnschichtchromatographie*, Verlag Chemie, Weinheim 1965.
[14.245] E. S. Perry, A. Weissberger: *Technique of Organic Chemistry*, vol. 12; J. G. Kirchner: *Thin-Layer Chromatography*, Interscience, London 1967.
[14.246] F. Cramer: *Papierchromatographie*, Verlag Chemie, Weinheim 1962.
[14.247] E. S. Amis: *Solvents Effect on Reaction Rates and Mechanism*, Academic Press, New York 1966.
[14.248] C. A. Bunton: *Nucleophilic Substitution at a Saturated Carbon Atom*, Elsevier, Amsterdam 1963, p. 111.
[14.249] L. Küchler: *Polymerisationskinetik*, Springer Verlag, Berlin 1951.
[14.250] W. C. Aten, Technische Akademie Esslingen: Einsatzmöglichkeiten von Computern bei der Formulierung von Lösemittelgemischen für Klebstoffe, Vortrag, Esslingen 1984.
[14.251] P. Goerlich, *Ind. Lackierbetr.* **43** (1975) no. 11, 383–386.
[14.252] G. Ohlendorf, *Ind. Lackierbetr.* **51** (1983) no. 2, 49–54.
[14.253] H. Gölz, *Oberfläche JOT* **32** (1992) no. 2, 56–58.
[14.254] K.-H. Adams, *Ind. Lackierbetr.* **58** (1990) no. 1, 5–7.

[14.255] J. Kitlar, *Oberfläche JOT* **32** (1992) no. 1, 50–51.
[14.256] G. Ohlendorf, *Ind. Lackierbetr.* **51** (1983) no. 2, 49–54.
[14.257] E. Schmidt, *Oberfläche JOT* **30** (1990) no. 3, 28–31.
[14.258] W. Hater, *Ind. Lackierbetr.* **59** (1991) no. 5, 156–158.
[14.259] B. Rohland, W. Majewski, H. Krüssmann, *WRP* (1989) no. 5, 46–47.
[14.260] P. L. Bartlett, *Seifen Fette Öle Wachse* **113** (1987) no. 13, 437–442.
[14.261] Frost & Sullivan, *Polym. Paint Colour J.* **178** (1988) no. 4207, 57.
[14.262] M. R. C. Gerstenberger, D. K. Kruse, *Farbe Lack* **90** (1984) no. 1, 30–34.
[14.263] W. H. Ellis, C. D. McLaughlin, *J. Coat. Technol.* **52** (n 662) (1980) 61–72.
[14.264] E. Montorsi, *Pitture Vernici Eur.* **68** (1992) no. 11, 85–90.
[14.265] J. R. Kelsey, D. Lyons, S. R. Smithey, *J. Oil Colour Chem. Assoc.* **74** (1991) no. 8, 278–287.
[14.266] J. Falbe, B. Cornils, *Fortschr. Chem. Forsch.* **11** (1968) no. 1, 101–134.
[14.267] C.-D. Mengler, *Chem. Z.* **95** (1971) 309–316.
[14.268] C. Parducci, *Riv. Colore Verniciatura Ind.* **10** (1977) no. 114, 329–332.
[14.269] G. S. Salvapati, M. Janardanarao, *J. Sci. Ind. Res.* **42** (1983) 261–267.
[14.270] M. Peignier, P. Lasnet de Lanty, R. Zimmermann, *Farbe Lack* **86** (1980) no. 11, 973–975.
[14.270a] M. Carstensen, Coatings World May/June 1997, 36–37.
[14.271] P. I. Smith, "The Use of Tetrahydrofuran in Coating Formulations", *Ind. Finish. and Surf. Coat.* (1971).
[14.272] W. Stumpf: *Chemie und Anwendung des 1,4-Dioxans*, Verlag Chemie, Weinheim 1956.
[14.273] J. E. Hamlin, *Paint Resin* **37** (1983) no. 5. 42–47.
[14.274] R. A. Heckman, *Mod. Paint Coat.* (1986) no. 6, 36–42.
[14.275] H. Kittel, *Adhäsion* (1987) no. 5, 33–34.
[14.276] H. W. Lang, *Chem. Ztg.* **84** (1960) no. 8, 239–247.
[14.277] S. S. Pizey: *Synthetic Reagents*, vol. 1, Wiley-Interscience, New York 1974.
[14.278] D. Martin, H. G. Hauthal: *Dimethylsulfoxid*, Akademie Verlag, Berlin 1971.
[14.279] Gaylord Chemical Corp., Technical Bulletin Dimethylsulfoxid, Slidell/USA.
[14.280] H. Normant, *Angew. Chem.* **79** (1967) 1029; *Angew. Chem. Int. Ed. Engl.* **6** (1967) 1046.

Index

abrasion 156
    resistance 230
accelerators 59
acetone 359
acid-catalyzed coatings 80
acid paint removers 174
acrylated acrylics 136
acrylated epoxy system 136
acrylate–methacrylate copolymers 126
acrylate-modified polyester 55
acrylate resins
    cross-linking 38
    properties 37
acrylates
    radiation-curable systems 136
acrylic coatings 37
    cross-linking 39
acrylic paints 40
acrylic resins 108
    uses 41
acute toxicity 305
additives 5, 159 ff
    analysis 240
    classification 159
    electrodeposition paints 140
    radiation-curable systems 136
    testing 171
    uses 171
adhesion 229
adhesives
    radiation-curable 138
aerosol cans 326
aircraft coatings 252
air draught sensitivity 164
air-drying paints
    solids contents 46
air-drying UP resins 58
airless atomization 204
airmix atomization 204
airmix process 205
Air Pollution Control Act 311
Air Quality Act 312

alcohols 353 ff
alcohol-soluble nitrocellulose 13
aliphatic epoxy compounds 70
aliphatic hydrocarbons 327
alkaline paint strippers 175
alkali phosphating 199
alkyd coatings 41
alkyd paints 42
alkyd resins 41, 50
    binders 42
    classification 42
    combination with other binders 43
    emulsions 46
    health protection 5
    high-solids paints 46, 107
    modification 43 f
    testing 47
    types 42
    uses 42
alkylated amino resins
    production 82
    properties 81
alkyl silicates 96
    composition 96
    production 98
    properties 97
amino resins 54, 72, 80
    formaldehyde release 86
    market 85
    specifications 83
    storage 83
    use 84
amyl alcohol 356
analysis
    additives 240
    binders 236
    coatings 235, 241
    extenders 239
    pigments 239
    solvents 240, 318
anhydrides 72
anodic electrodeposition 140, 208

anticorrosion pigments   170
anticorrosive primers   246
antifloating agents   6
antiflooding agents   6
antifoaming agents   6
antifouling   257
antisetting agents   5, 48
antiskinning additives   165
antiskinning agents   6
    in alkyd resins   48
application
    methods   104, 203
    of paints   7
    pretreatment   195
    technology   104
    with brush   210
aqueous nitrocellulose lacquers   15
aqueous systems   149
aqueous UV-curing   61
architectural coatings   275
architectural paints   262
aromatic hydrocarbons   351 f
asphalt coatings   91
asphalt combination coatings   91
atomization   203
    methods   204 ff
automotive coatings   18
automotive fillers   41
automotive finishes   55
automotive paints   245 ff
auxiliaries
    emulsion paints   127
azeotropes   292, 330 ff
bag filter   192
ball mills   49
Ballotini method   226
barrier agents   59
basecoats   18
    metallic   247
benzoguanamine   80
benzoguanamine resins   54
benzyl acetate   364
benzyl alcohol   357
binders
    analysis   236 ff
    coating powders   118
    definition   3
    effect of molecular mass   4
    electrodeposition paints   140
    emulsified   112
    low solvent paints   106

    styrene content   38
    types   4, 11
    water soluble   110 f
bismuth pigments   145
bisphenol A resins   69
bisphenol epoxy resins   73
bisphenol F resins   69
bitumen   92
    coatings   91 f
    combination coatings   93
    emulsions   93
bituminous coatings   23
bituminous substances   23
blasting   175
block and graft copolymer stabilizers   130
blocked polyisocyanates   64, 66
blue iron pigments   146
blushing   321
boiling point   294
branched carboxylic acids   45
Bricks   263
brightness   156
brushing   211
buldings
    coatings   262
buses
    coatings   251
butanol   355
    physical properties   110
butyl acetate   363
butyl diglycol   369
butyl diglycol acetate   365
butyl formate   362
butyl glycol   368
butyl glycol acetate   364
butyl glycolate   365
butyl lactate   365
butyl tetraglycol   370
butyl triglycol   370
butyrate esters   365
butyrolactone   366
CAB   17 ff
CAB-modified urethane   18
cadmium pigments   145
calorific value   303
can coatings   55, 77, 260
car body paints   245 f
carbon disulfide   373
carboxy-functional polyesters   72
carcinogenicity   306
car repair paints   248

cartridge filter   193
catalysts   6, 165
catalytically curing compounds   72
catalytic driers   11
cathodic electrodeposition   140f, 209
cellulose acetate   16
cellulose acetate butyrates   16, 18
   hydroxyl content   17
   properties   17
   uses   17f
cellulose acetate propionates   16, 19
   properties   17
cellulose-based coatings   12
cellulose butyrate   16
cellulose nitrate
   dilution ratios   288
   lacquers   12
cement   263
centrifugation   211f
chemical cleaning   196
chemical curing   216
chemical drying   8
   solventborne paints   103
chemical paint removal   173
chemical plants   244
chemical properties
   coatings   231
chemical resistance   304
   pigments   148
chemical workup   238
chlorinated hydrocarbons   352f
chlorinated polyethylene and polypropylene   25
chlorinated poly(vinyl chloride)   26
chlorinated rubber (CR)   19
   acrylic resin combinations   22
   alkyd resin combinations   22
   viscosity grade   20
chlorinated rubber coatings   19
chlorinated rubber paints
   application   21
   combination paints   22
   composition   20
   production   21
   uses   22
chlorobenzene   353
chlorosulfonated polyethylene   25
chromate pigments   146
chromating   200
chromatography   237, 323
chromium oxide pigments   145

chronic toxicity   305
CIELAB color space   228
Clean Air Act   311
clean air measures   267
cleaning   196
clearcoat   18
clear-on-base finishes   18
coating materials   216
coating powders   115, 118
   binders   118
   colorants   120
   constituents   117
   environmental aspects   122
   preparation   184
   storage   122
   testing   121
   transportation   122
   types   115
coatings
   analysis   235, 241
   definition   1
   economic aspects   10, 275
   for cans   87
   for plastic   18, 216
   for wood   18, 216
   producers   275
   properties   219, 222ff
   requirements   1
   types   11
   uses   243ff
cohesive energy density   282
coil coatings   55, 150, 212, 258
cold paint removal   174
colloidal solutions   278
color   227
   values   228
colorants
   coating powders   120
comb gauge   225
combustible liquids   315
combustion energy   303
complex inorganic colour pigments   145
composition of paints   3
compressed air   203
compressed air atomization   204
concrete   263
continuous mixer   186
conventional dipping   208f
conversion layers   198
copolyesters   51
copolymerization   132

copolymers of ethylene with maleic acid 25
corona discharge 201
corrosion
    inhibitors 6, 170
    protection 243ff
    resistance 231
Coulomb forces 278
critical pigment volume concentration CPVC 150
cross-linking 54
    acrylic coatings 39f
cubic expansion coefficient 301
cumene 352
curable fluoropolymers 29
curing 62
    agents 70ff
    methods 217
    with heat carriers 217
    with radiation 217
curtain coatings 62
cyclic perfluoropolymers 29
cycloaliphatic epoxy resins 70, 77
cycloaliphatic hydrocarbons 350f
cyclohexane 350
cyclohexanol 357
cyclohexanone 361
cyclohexyl acetate 364
damping test 229
dangerous substances
    classification 314
deaeration agents 59
deepfreezers 259
defoamers 127, 160
degreasing 198, 325
density 297
    paints 221f
diacetone alcohol 358
dibutyl ether 367
1,2-dichloroethane 353
1,2-dichloroethylene 353
dichloromethane 352
1,2-dichloropropane 353
dielectric constant 300f
diethylene glycol dimethyl ether 370
diethyl ether 366
diisobutyl ketone 360
diisopropyl ether 367
diisopropyl ketone 360
dilutability 288f
dilution processes 281

dilution ratio 288
    temperature dependence 289
dimethylacetal 372
dimethyl adipate 366
dimethylcyclohexanone 361
1,3-dimethyl-2-imidazolidinone 373
dimethyl sulfoxide 372
1,4-dioxane 368
dipole–dipole forces 279
dipole moment 285, 328f
dipping 207
    methods 208
dipropyl ketone 360
dishwashers 259
dispensers 326
dispersants 5, 127
dispersibility 156
dispersing additives 161
dispersion 32, 35, 38, 112, 126, 157
    forces 280
    medium 130
    methods 189f
    polymerization 130f
    preparation 183
    steric stabilization 130
dissolution 291
dissolver disks 187
dissolvers 186f
domestic appliances
    coatings 259
driers 11, 165
    in alkyd resins 48
drum application 211
drum coating 212
dry cleaning 326
dry film thickness 224
drying
    chemical 8
    electrical methods 218
    physical 8
    polyaddition 9
    polycondensation 9
    polymerization 9
drying equipment 226
drying oils 11, 12
drying time 226
EB curing 137
economic aspects 275
electrical conductivity 301
electrical insulation paints 89
electrodeposition coatings 208, 246

electrodeposition paints
    applications 140
    properties 139
electron-beam accelerators 137
electron beams 218
electrostatic atomization 205, 207
electrostatic coating process 206
electrostatic powder coating 115
electrostatic spraying 104, 214
emulsified binders 112
emulsifiers 127
emulsion paints 125, 127
    production 128
    properties 128
    thickeners 167
    types 129
emulsion polymers 112
emulsions
    defoamers 160
engines 249
environmental protection 182
epoxy coatings 69
epoxy esters
    solvent containing and air drying 75
epoxy hardeners 54
epoxy novolacs 69
epoxy paints 108
epoxy resin–curing agent combinations 71
epoxy resin esters 73
epoxy resins 88
    chemically modified 73
    solvent-containing coatings 74
    solvent-free coatings 73
    types 69
    waterborne coatings 75
esters 362
    of epoxy resin and methacrylic acid 73
    of epoxy resins and acrylic acid 73
ester-soluble nitrocelluloses 13
ETFE 28
ethanol 354
ethers 366
ethoxypropanol 371
ethoxypropyl acetate 365
ethyl acetate 363
ethyl amyl ketone 360
ethylbenzene 351
ethyl diglycol 369
ethyl diglycol acetate 364
ethylene carbonate 366
ethylene copolymers 24

ethylene-vinyl acetate copolymers 24
ethyl 3-ethoxypropionate 365
ethyl formate 362
ethyl glycol 368
ethyl glycol acetate 364
2-ethylhexanol 356
2-ethylhexyl acetate 364
ethyl lactate 365
ethyl silicate 98
ethyl silicate binders 100
ethyl silicate zinc dust coatings 100
ethyl triglycol 369
evaporation 293, 295 f
    numbers 295
    rate 295
extenders 4 f, 60, 143, 150, 153, 155
    activity 156
    analysis 239
    classification 151
    composition 152
    density 153
    economic aspects 151
    influence 127
    modification 157
    occupational health 151
    properties 152, 154 ff
    refractive indices 150
exterior-use coatings 263
exterior-use paints 129
extruder process 117
fatty acids 42
FDA 55
fillers 60, 247
film
    formation 1, 8
    production 325
    properties 103
    testing 222
film-formation promoters 5
film-forming auxiliaries 128
film thickness
    measurement 224
filters 192
fineness of grind 222
finishing coats
    buildings 263
fire hazard 315
fire retardant coatings 266
flash point 221, 301 f
flexibility 230
flexible packaging 57

floating 162
flocculation 163
flooding 162
floor coatings 211 f, 265
flow-coating method 213
flow promoters 59
flow properties 321
fluidized-bed coating 215
fluidized-bed paint removal 174
fluidized-bed process 115
fluoroolefin–vinyl ether terpolymers 31
fluoroolefin–vinyl terpolymers 30
fluoropolymer coatings 27
Food and Drug Administrations 87
food packaging 55
fouling 257
freight cars 250
freight containers 251
FTIR spectroscopy 238
fully alkylated amino resin 82
furniture coatings 261
gas chromatography 238
gel permeation chromatography 237
German Chemikaliengesetz 314
glass transition temperature 37, 52
gloss 226, 321
glycidyl esters 70
glycol ethers 368
glyptal resins 50
gold lacquers 89
grinding 157
grinding mills 189
HALS 169
hardeners 58
hardness 229
heating oil storage premises 265
heats of combustion 303
heats of mixing 281
heavy-duty coatings 243 ff
heavy-metal pigments 244
heterocyclic epoxy compounds 70
hexakis(methoxymethyl)melamine HMMM 81
hexamethylphosphoric triamide 373
hexanol 356
hexyl acetate 364
hiding power 225
high boilers 293
high-build 61
high-frequency curing 218

high molecular mass copolyesters
   properties 51
high-pressure removal with water 175
high-solids paints 102, 105, 320
   alkyd resins 46
   production 107
high-solvent paints 101
high-viscosity nitrocellulose 13
HMMM 54, 86
hot alkaline paint removal 173
hot-curing epoxy resin esters 76
hot spraying 204 f
HPLC 238
hybrid systems 112
hydraulic atomization 204 f
hydrogen abstraction grafting 131
hydrogen bond parameters 286, 328 f
hydrogen bonds 280
hydroquinone 59
hygroscopicity 296 f
ignition limit 303
ignition temperature 302
immersed body method 222
in-can preservatives 166
indentation test 229
indoor furniture 261
induction forces 279
inductive curing 218
industrial paints 56
in-film preservation 166
infrared (IR) radiation 217
infrared spectroscopy 236
inks
   radiation-curable 138
inorganic pigments 143 ff
integrated pollution prevention and control 312
interference pigments 146
interior-use
   coatings 265
   paints 129
intermediate coats 246
intermolecular forces 278 f
ionic forces 278
iron oxide pigments 144
isobutanol 355
isobutoxypropanol 371
isobutyl acetate 363
isobutyl formate 362
isobutyl vinyl ether 26
isobutyrate esters 365

isocyanates   72
isophorone   361
isophorone diisocyanate   44
isopropoxypropanol   371
isopropyl alcohol   355
ketones   358
kneader mixers   186
kneaders   186f
knife coating   210, 213
knifing fillers   62
lactate esters   365
laminated panels   60
laser beams   218
latent solvents   287
laundry dryers   259
leveling   221
    agents   5
    properties   164
lightfastness   148
light stabilization
    methods   168
light stabilizers   167
    testing   170
linear polyester resins   54
London-Van der Waals forces   280
long oil alkyd resins   47
long oil drying alkyd resins   42
low boilers   293
low-solids paints   102
low-solvent binders   106
    coatings   3, 267
    emissions   267f
    paints   38, 105
low-temperature paint removal   175
low-viscosity nitrocellulose   13
low-zinc phosphating   199
Lumiflon   30
MAK
    Committee   306
    list   307
    values   309
maleate esters   26
marble   62
marine coatings   252ff
    antifouling   257
    fouling   257
marine tank coatings   256
mass spectrometry   238
material flow   178f, 181
matting agents   6
mechanical cleaning   196

mechanical paint removal   175
mechanical properties
    coatings   229
media mills   188
medium- and high-solids systems   40
medium boilers   293
medium oil alkyd resins   42, 47
medium-viscosity nitrocelluloses   13
melamine resins   80, 84
mesityl oxide   360
metadioxane   368
metal compounds   107
metallic basecoats   41, 247
metallic effect pigments   147
metallic substrates
    pretreatment   196
metals
    coatings   264
    paint removal   173
methanol   354
methoxybutanol   371
3-methoxybutyl acetate   365
methoxypropanol   370
1-methoxypropyl acetate   365
methyl acetate   362
methylal   372
methyl amyl ketone   360
methylbenzyl alcohol   357
methyl butyl ketone   359
methylcyclohexanol   357
methylcyclohexanone   361
methyl diglycol   368
methyl dipropylene glycol   371
methyl ethyl ketone   359
methyl formate   362
methyl glycol   368
methyl glycol acetate   364
methyl isobutyl ketone   359
methyl propyl ketone   359
methyl *tert*-butyl ether   367
microencapsulated polyisocyanate   66
microwaves   217
migration   148
mills   188f
mineral substrates coatings   263
minimum film-forming temperature MFT   128
miscibility
    behavior   283
    polymer–solvent pairs   285
    solvents   284

miscibility gap   281, 291
mixers   185
mixing enthalpy   281
mixtures
    physical properties   292
modified phenolic resins   90
moisture   66
molybdate pigments   146
monomers
    radiation-curable systems   136
multicoat paints   322
multicoat systems   9
mutagenicity   307
nacreous pigments   146
National Environmental Policy Act
   312
natural asphalts   91
natural resin   2
neutralizing agents   6
neutral paint strippers   174
newtonian liquids   166
nitrocellulose
    nitrogen content   13
    viscosity   13
nitrocellulose lacquers
    applications   15
    components   12
    formation   14
    preparation   15
    raw materials   13
    uses   15
2-nitropropane   373
N-methylpyrrolidone   373
N,N-dimethylacetamide   372
N,N-dimethylformamide   372
nonaqueous dispersion paints   129
    stabilization   130
nonaqueous dispersions (NAD)   38
    characterization   133
    properties   134
nonelectrostatic atomization   204
non-solvents   288
nonvolatile matter   221
nonyl acetate   364
novolacs   89 f
nuclear magnetic resonance spectroscopy
  (NMR)   237
octyl acetate   364
odor threshold   309
oil absorption   157
oil-based coatings   11

oil paints   11
    binders   11
    film formation   11
    properties   12
    thickness   11
one component mixture   64
one-pack mixture   64 ff
optical properties   226 f
orange peel   164
organic cellulose ester   16
organic fluoropolymers   27
organic pigments
    properties   148
organic solvents   2
overspray   271
packaging   84, 260
paint additives   159
    definition   5
    types   5
paint application   7, 195
paint curing   216
paint factory   180
paint-making processes   182 f
paint markets   275
paint pigments   148
paint removal   173 ff
    mineral substrates   175
    types   173
    wood   175
paint removers   322
paint residues   271
paints
    application   2, 104
    definition   1
    density   221 f
    mechanical properties   321
    preparation   183
    properties   219
    rheological properties   220
    solvent retention   321
    systems   101
    testing   219
    types   11
    viscosity   219, 319
paint-solvent mixtures   318
paint systems   101
paint transfer efficiency   104
paraffin-type UP resins   58
particle shape   153
passenger cars
    coatings   249

pearlescent pigments 146
perchloroethylene 353
Pergut 20
peroxide 58, 60
personal safety precautions 309
phenolic aldehyde resins 87
phenolic resins 72, 86
phosphating processes 199
photochemical ozone creation potential 312
photoinitiators 59, 137
pH value 156
physical curing 216
physical drying 8
physically drying solventborne paints 103
physiological data 347 ff
pigment dispersions
    stabilization 161
pigment distribution 162
pigment pastes
    homogenization 191
pigments 4, 60, 127, 143
    analysis 239
    anticorrosion 170
    effect on viscosity 149
    types 5
    uses in paints 149
pigment wetting 321
pinholing 114
pitch coatings 91, 94
plasma arc 218
plasticizers 13, 33
    definition 4
    for nitrocellulose 14
    for resins 14
plastics
    coatings 216, 264
    pretreatment 201
pneumatic atomization 204
polarity 285
polarizability 285
polyaddition 9
polyalcohols 51
polyamide 125
polyamines 71, 74
polybutadienes 12
polycarboxylic acids 50 f
polycondensation 9, 53
polyepoxides 39
polyester
    acrylates 136
    solubility 53

polyester–amine systems 52
polyester–amino resin 54
polyester coatings
    saturated 50
polyester resins 108
    cross-linking 54
    production 53
polyester–styrene resins 135
polyethylene 24, 125
polyhydroxyl 67
polyisobutenes 24
polyisocyanate resins 54
polyisocyanates 39, 64, 67 f
polymer dispersions 32, 112
polymerization 9
polymers
    waterborne dispersion paints 126
polyolefins 24
polyols 64
polyphenols 72
polysiloxanes 78
polystyrene 35
polytetrafluoroethylene 28
polyurea 66
polyurethane coatings 63
    properties 68
    raw materials 64
    starting products 63
    uses 68
polyurethane paints 39, 63
polyurethanes 108
    aqueous systems 67
    physically drying 67
polyurethane systems 65
poly(vinyl acetals) 34
poly(vinyl acetate) 31, 33
poly(vinyl acetate) dispersions 32
poly(vinyl alcohol) 33
poly(vinyl butyrals) 34 f, 88
poly(vinyl chloride) 25, 125
poly(vinyl ester) dispersions 32
poly(vinyl esters) 31, 33
poly(vinyl ethers) 35
poly(vinyl formals) 34
poly(vinyl halides) 25
poly(vinylidene fluoride) 28
pot life 221
pouring 211
pouring method 210, 212
powder binders 118

powder coatings   56, 75f, 150, 214
   characteristics   215
   production   185
precipitated extenders   157
prepolymers
   radiation-curable systems   136
preservatives   6, 127, 165
pretreatment   195
   car bodies   246
   metallic substrates   196
   wood   202
pretreatment methods   197
   plastics   201f
primers
   buildings   263
priming   253
   marine coating   253
priming coats   89
print-free test   226
printing   211
printing inks   322
product finishes   275
   of paints   115
   strategy   177
   technology   177
product types   179
promoters   59
propanol   355
propionate esters   365
propyl acetate   363
propylene carbonate   366
propyl glycol   368
protective colloids   127
proton acceptors   286
proton donors   286
pseudoplastic flow   166
PTFE   28
p-toluenesulfonic acid   82
pure acrylates   126
putties   60
PVC   25f
PVC plastisols   109
PVDF   28
pyrolysis   238
quality assurance   181
radiation-curable polyester coatings   56
radiation-curable resins
   classification   135
radiation-curable systems
   based on acrylates   136

radiation curing   55, 77, 137
   applications   138
radiation-curing coating   3
radiation-curing systems   18, 135
radiator coatings   265
radical scavengers   169
railroad rolling stock
   coatings   249
railroad vehicles
   paint application   250
random copolymers   130
rate of dissolution   291
reactive diluents   106
real solutions   278
recrystallization   324
red lead   21
refractive index   297ff
refrigerators   259
reproduction toxicity   307
resins   15
   analysis   236
   chemical workup   238
   definition   3
   for low-solvent paints   38
   for nitrocellulose combination lacquers   15
resin solutions   325
resistance to liquids   232
resol resins   90
resols   87
   plasticizing co-resins   88
   uses   87
rheological properties
   paints   220
rheology additives   166
rinsing   200
road transport vehicles
   coatings   251
roller application   211
roller coatings   61, 210, 212
roller mills   191
rolling   211
root turpentine   351
rotating viscometer   220
sagging   221
salt bath paint removal   175
sample conditioning   224
saturated polyester coatings   50
   uses   55
saturated polyester resins
   classification   52
   properties   51

scratch test   229
sec-butanol   356
sec-butyl acetate   364
self-cross-linking acrylates   41
shear-thinning   166
ship paints   255
short oil alkyd resins   42, 47
siccatives   165
sieves   192
silanols   97
silicate coatings   94
silicate paints
    applications   95
    composition   95
silicone coatings   78
silicone-modified alkyd resins   44
silicone organic blends   79
silicone paint additives   80
silicone resins   78
size distribution   153
slip properties   164
solar radiation   232
solid bisphenol A resins   74
solid epoxy resins   78
solid residues   271
solid resins   31
solids content   106
solubility
    behaviour   283
    influence of molecular mass   290
    parameter   282 ff, 328 f
    polymers   284
    properties   285
soluble polystyrene   36
soluble styrene copolymers   36
solution enthalpy   281
solution properties   291
solutions
    theory   278
solvation   287
solvency   289
solventborne coatings
    defoamers   160
solventborne paints   101, 103
solventborne systems   149
solvent-containing coatings   74
solvent-containing paints   76
solvent-containing paint strippers   176
solvent content reduction   105
solvent densities
    temperature dependence   297

solvent emission   104
solvent-free coatings   73
solvent-free paints   105
solvent-free powder coatings   3
solvent groups   327
solvent miscibility   281 ff
solvents   2, 15, 59, 277 ff, 287
    abbreviations   337
    analysis   240, 318
    calorific values   333
    chemical properties   304
    chemical reactions   324
    chromatography   323
    classification   293
    definition   277
    dilutability   289
    dilution   281 f
    dilution ratio   289
    dissolution   291
    EC classification   307
    economic aspects   326
    electrical properties   300 f
    environmental protection   311
    extraction   323
    extractive distillation   323
    fire hazard   315
    hydrogen bonds   286
    in alkyd resins   47
    in coatings   7
    in paints   102, 318
    MAK classification   307
    miscibility with water   329 f
    mixtures   291
    occupational health   309
    physical properties   328 f, 339
    physiological data   347 ff
    polarity   285
    properties   277, 293 ff.
    purification   318
    recrystallization   324
    recycling   316
    safety regulations   315
    solubility   282 ff, 290
    solution properties   291
    thermal properties   300 f.
    toxicology   305
    uses   318
solvent tests   334 ff
special-purpose coatings   275
specific heat   300
specific surface area   156

spot fillers   62
spray coatings   60
spray fillers   62
spraying   62, 203
stabilizers   59
    precursor   132
stackability   226
standards for solvents   334 ff
steel constructions
    galvanized   244
    protection   243
sterically hindered amines   169
steric stabilization
    dispersions   130
stone   263
stoving paints   46
strip coil coating   211
styrene   352
styrene–acrylate copolymers   126
styrene–butadiene dispersions   36
styrene–butadiene polymers   126
styrene copolymer dispersions   35
styrene copolymers   35
styrene-modified alkyd resins   43
substrates   195
surface additives   163 ff
surface flow control additives   164
surface preparation
    marine coating   253
    steel constructions   244
surface tension   164, 298 f, 332 f
surface-treated extenders   157
synthetic binders
    phenolic resins   86
    properties   4
synthetic fibers   325
synthetic resin-bound plasters   129
T_g 52
TA Luft   313
tamped density   153
tamped volume   153
tank coatings   256
Technische Regeln für Gefahrstoffe TRGS   314
terpene hydrocarbons   350
terpenoids   350 f
tert-butanol   356
testing
    coating powders   121
test panels   223
test samples
    preparations   222

1,1,2,2-tetrachloroethane   353
tetrachloromethane   352
tetrafluoroethylene copolymers   28
tetrahydrofuran   367
tetrahydrofurfuryl alcohol   358
tetramethylene sulfone   372
TGIC   76, 118
thermal conductivity   300 f
thermal paint removal   174
thermodynamic principles   280
thermoplastic binders   118
thermoplastic coating powders   116
    uses   125
thermoplastic fluoropolymers   28
thermoplastic polyesters   125
thermoplastic powders   120
thermosetting binders   118
thermosetting coating powders   115 f
    extruder process   117
    uses   123 f
thermosetting powders   119
thick-coat paints   322
thickeners   128, 167
thickening agents   6
thin-coat   60
thiols   71
thixotropic alkyd resins   44
thixotropic materials   166
thixotropy   167
tin catalysts
    effect on polyisocyanate resins   54
tinting paints   129
titanium dioxide   143 f
TLV values   310
toluene   351
toluene diisocyanate   44
topcoats   247
toxicology   272
    solvents   305
toxic properties
    classification   305
Toxic Substances Control Act TSCA   314
trailers
    coatings   251
transparent pigments   147
transport vehicles
    coatings   249
1,1,1-trichloroethane   353
trichloromethane   352
3,3,5-trimethylcyclohexanol   358

trimethylcyclohexanone 361
troweling 211 f
trucks
    coatings 251
turpentine oil 350
two-coat metallic coatings 167
two component mixture 64
two-pack coating 205
two-pack mixture 64 ff
two-pack polyurethane paints 56
ultramarine pigments 146
undercoating systems 245
underwater coatings 22
unsaturated polyester binders 57
unsaturated polyester coatings 57
UP resins 57, 60
    applications 60 ff
    for fillers 58
    formulation 60
    raw materials 58 ff
    storage 63
    transport 63
    types 58
urea 80
urethane alkyd resins 44 f
urethane elastomers 18
urethane oils 66
UV absorbers 168
UV-curable acrylate 136
UV-curable systems 137
UV-curing 137
UV-curing roller primers 61
UV-curing systems 59
UV radiation 217
vapor density 300
vaporization 293
vapor pressure 294
varnishes
    preparation 182
VDA 230
vegetable oils 42
vibration method 222
vinyl acetate copolymers 32, 126
vinyl chloride copolymers 25 f
vinyl chloride terpolymers 27
vinyl coatings 23
vinyl compounds 43
vinyl halide copolymers 25
vinylidene chloride copolymers 27
vinyl propionate copolymers 126
viscometers 220

viscosity 298
    paints 219
viscosity grade 20
wall coatings 265
washing machines 259
waste 271, 316
waste air treatment 269
waste disposal 316
waste-gas purification 316
wastewater treatment 270
water
    physical properties 110
waterborne acrylic paints 40
waterborne alkyd resins 45
    solvents 48
waterborne bitumens 92
waterborne coatings 3, 75 f
waterborne dispersion paints 125
waterborne paints 109, 320
    applications 111, 113
    binders 113
    defoamers 160
    dispensions 112
    environmental aspects 114
    hybrid systems 112
    organic co-solvents 111
    polyester resins 53
    production 113
    properties 110
    solids content 111
    storage 114
waterborne paint systems 160
water glass binder 95
water glass coatings 94
water resistance 231
water-soluble binders 110 f
water vapor transmission 231
weathering conditions 170
weathering tests 232
weather resistance 148
weld seams 125
wet coating methods 211
wet film thickness 224
wet paint coating methods 210
wettability 156
wetting additives 161
wetting agents 5, 127
    in alkyd resins 49
white pigments 143
    refractive indices 150

white spirit   327
wiping   211
wood   60
    coatings   216, 263
    pretreatment   202
wood oil   351
world paint markets   275

xylene   351
    physical properties   110
yellowness   156
yield point   166
zinc dust paints   97f
    binders   99
zinc phosphating   199